Experimental Methods in the Physical Sciences
Volume 44
Neutron Scattering – Fundamentals

Experimental Methods in the Physical Sciences

Thomas Lucatorto, Albert C. Parr and Kenneth Baldwin
Editors in Chief

Founding Editors
L. MARTON
C. MARTON

Experimental Methods in the Physical Sciences
Volume 44

Neutron Scattering – Fundamentals

Edited by

Felix Fernandez-Alonso
ISIS Facility, Rutherford Appleton Laboratory,
Chilton, Didcot, Oxfordshire, and
Department of Physics and Astronomy,
University College London, London, United Kingdom

David L. Price
Synchrotron Radiation and Neutron Research Group,
CEMHTI, Centre National de la Recherche Scientifique,
Orléans, France

AMSTERDAM • BOSTON • HEIDELBERG • LONDON
NEW YORK • OXFORD • PARIS • SAN DIEGO
SAN FRANCISCO • SINGAPORE • SYDNEY • TOKYO
Academic Press is an imprint of Elsevier

Academic Press is an imprint of Elsevier
525 B Street, Suite 1800, San Diego, CA 92101-4495, USA
225 Wyman Street, Waltham, MA 02451, USA
The Boulevard, Langford Lane, Kidlington, Oxford, OX5 1GB, UK
32 Jamestown Road, London NW1 7BY, UK
Radarweg 29, PO Box 211, 1000 AE Amsterdam, The Netherlands

Copyright © 2013 Elsevier Inc. All rights reserved

No part of this publication may be reproduced, stored in a retrieval system
or transmitted in any form or by any means electronic, mechanical, photocopying,
recording or otherwise without the prior written permission of the publisher

Permissions may be sought directly from Elsevier's Science & Technology Rights
Department in Oxford, UK: phone (+44) (0) 1865 843830; fax (+44) (0) 1865 853333;
email: permissions@elsevier.com. Alternatively you can submit your request online by
visiting the Elsevier web site at http://elsevier.com/locate/permissions, and selecting
Obtaining permission to use Elsevier material

Notice
No responsibility is assumed by the publisher for any injury and/or damage to persons
or property as a matter of products liability, negligence or otherwise, or from any use
or operation of any methods, products, instructions or ideas contained in the material
herein. Because of rapid advances in the medical sciences, in particular, independent
verification of diagnoses and drug dosages should be made

Library of Congress Cataloging-in-Publication Data
Application submitted

British Library Cataloguing in Publication Data
A catalogue record for this book is available from the British Library

ISBN: 978-0-12-398374-9
ISSN: 1079-4042

For information on all Academic Press publications
visit our website at store.elsevier.com

Printed and bound in United States of America
13 14 15 16   10 9 8 7 6 5 4 3 2 1

Contents

Contributors	xi
Volumes in Series	xiii
Preface	xvii
Symbols	xxi

1. An Introduction to Neutron Scattering — 1
David L. Price and Felix Fernandez-Alonso

1.1 **Fundamentals**	2
1.1.1 Why Are Neutrons so Unique?	2
1.1.2 Thermal Neutrons for Condensed Matter Research	5
1.1.3 Conservation Laws	11
1.1.4 The Structure of Materials	11
1.1.5 Adding Motion: Dynamics and Spectroscopy	16
1.2 **Scattering Foundations**	22
1.2.1 The Master Formula and Fermi's Golden Rule	22
1.2.2 Nuclear Scattering	25
1.2.3 The Double Differential Cross Section in the Time Domain	26
1.2.4 Farewell to Nuclear Physics	27
1.2.5 Coherent and Incoherent Scattering	28
1.2.6 Scattering Functions	30
1.3 **Canonical Solids**	31
1.3.1 Normal Modes of Vibration	31
1.3.2 Scattering Under the Harmonic Approximation	32
1.3.3 Purely Elastic Events	33
1.3.4 Inelastic (One-Phonon) Scattering	35
1.3.5 Multiphonon Scattering	39
1.3.6 Beyond Harmonic Vibrations	41
1.4 **Beyond Canonical Solids**	42
1.4.1 Space–Time (Van Hove) Correlation Functions	42
1.4.2 Pair Distribution Functions	43
1.4.3 Properties of the Dynamic Structure Factor	47
1.4.4 From Order to Disorder: Diffuse Scattering	58
1.4.5 Stochastic Diffusion	61
1.4.6 Beyond Atoms and Molecules: Large-Scale Structures	75
1.5 **Magnetic Structure and Polarized Neutrons**	82
1.5.1 Basic Principles	82
1.5.2 Polarized Neutrons	88
1.5.3 Magnetic Bragg Scattering	95
1.5.4 Diffuse Scattering from Magnetic Disorder	97
1.5.5 Large-Scale Magnetic Structures	101

1.6 Spin Dynamics	106
1.6.1 Generalized Susceptibility	106
1.6.2 Spin Waves	108
1.6.3 Crystal Fields and Magnetic Clusters	111
1.6.4 Spin Fluctuations	114
1.6.5 Interband Transitions	115
1.6.6 Critical Scattering	117
1.7 Nuclear Spin: Order and Disorder	120
1.7.1 A Closer Look at Nuclear Spins	121
1.7.2 Scattering Cross Sections	121
1.7.3 Uncorrelated and Correlated Spin Ensembles	123
1.8 Outlook	125
References	127
2. Neutron Sources	**137**
Francisco J. Bermejo and Fernando Sordo	
2.1 Scope	138
2.2 Useful Neutron Production Reactions	140
2.2.1 Fission	141
2.2.2 Direct and Stripping Reactions	142
2.2.3 Bremsstrahlung	144
2.2.4 Spallation Reactions	146
2.3 Neutron Slowing Down and Moderators	153
2.3.1 Moderators	158
2.4 Basic Building Blocks of Accelerators to Drive Neutron Sources	166
2.4.1 Beam Injectors	167
2.4.2 Targets	178
2.5 Accelerator-Driven Sources: Some Predecessors	197
2.6 State-of-the-Art Accelerator Drivers for Neutron Sources	199
2.6.1 Last-Generation Megawatt-Range Sources	199
2.6.2 Medium-Power (100 kW) Sources	205
2.6.3 Compact, Accelerator-Driven Sources	209
2.7 Research Reactors	212
2.7.1 Core Designs	213
2.7.2 Reactor Vessel	217
2.8 Future Prospects	219
2.8.1 Accelerators	219
2.8.2 Hybrid Systems	225
2.8.3 Reactors	226
2.9 Nonneutron-Scattering Uses of Neutron Sources	227
2.9.1 Isotope Production, In-Vessel Irradiation, γ-Radiation, and Neutron Activation Analysis	228
2.9.2 Nuclear Physics and Engineering: Astroparticle Physics, Nuclear Structure and Reactions, and Transmutation of Nuclear Waste	228
2.9.3 Hadron Physics: Neutrino-Related Phenomena	229

Contents

    2.9.4 Fundamental Physics: Foundations of Quantum Mechanics, Effects of Gravity on Isolated Particles, Search for Dark Matter Using Ultracold Neutrons, Tests, and Validations of the Standard Model of Particle Physics   230
    2.9.5 Use of Muon Beams for Condensed Matter and Fusion Research   231
Acknowledgments   231
Appendix A Some Basic Relationships   232
Appendix B The Transport Equation for Neutrons   234
References   237

3. Experimental Techniques   245
Masatoshi Arai

3.1 Introduction   246
3.2 **Scattering Measurements**   247
    3.2.1 Cross Section   247
    3.2.2 Integrated Intensity and the Lorentz Factor   249
3.3 **Useful Neutrons for Condensed Matter Science**   254
    3.3.1 Neutron Flux from Moderators   254
    3.3.2 Pulse Peak Structure   255
    3.3.3 Pulse Peak Width   257
    3.3.4 Choice of Parameters in Spallation Sources   257
    3.3.5 High-Energy Neutron Background   259
3.4 **Diffraction Techniques**   261
    3.4.1 Powder Diffraction   262
    3.4.2 Single-Crystal Diffractometers   271
3.5 **Inelastic Scattering Techniques**   275
    3.5.1 Triple-Axis Spectrometer   275
    3.5.2 Chopper Instruments   277
    3.5.3 Inverted-Geometry Instrument   288
3.6 **Instruments for Semi-Macroscopic Structures**   293
    3.6.1 Small-Angle Neutron Scattering Instruments   293
    3.6.2 Neutron Spin-Echo Spectrometers   296
3.7 **Neutron Detectors**   300
    3.7.1 ^{3}He-Gas Detectors   301
    3.7.2 Scintillation Detectors   301
3.8 **Beam Transport and Tailoring**   305
    3.8.1 Neutron Optics   305
    3.8.2 Choppers   313
References   318

4. Structure of Complex Materials   321
Silvia C. Capelli

4.1 Introduction   321
4.2 **Useful Properties of Neutrons**   324

	4.2.1 Neutron Scattering Length	324
	4.2.2 A Particle with a Mass	324
4.3	What can be Learnt from Neutron Diffraction Experiments?	326
	4.3.1 Hydrogen Bonding	326
	4.3.2 Proton Migration	329
	4.3.3 Transition Metal Hydrides	332
	4.3.4 Porous Materials	333
	4.3.5 Diffuse Scattering	335
4.4	Outlook	339
	4.4.1 Neutron Sources	339
	4.4.2 Neutron Optics	340
	4.4.3 Detectors	341
	4.4.4 Samples and Sample Environment	342
	4.4.5 Software	347
4.5	Conclusions	348
	References	349

5. Large-Scale Structures — 353

Jeffrey Penfold and Ian M. Tucker

5.1	Introduction	353
5.2	Experimental Details	356
	5.2.1 Fundamentals of Neutron Reflectivity	356
	5.2.2 Fundamentals of Small-Angle Neutron Scattering	358
	5.2.3 Experimental Details for Neutron Reflection	361
	5.2.4 Experimental Details for SANS	363
5.3	Thin Films, Interfaces, and Solutions	365
	5.3.1 Adsorption at the Air–Solution Interface	365
	5.3.2 Adsorption at the Liquid–Solid Interface	374
	5.3.3 Structure of Biological Membranes	378
	5.3.4 Micelles	382
	5.3.5 Lamellar Phases and Vesicles	385
	5.3.6 Colloidal Particles	392
	5.3.7 Polymers in Solution, Melt, and Thin Films	394
	5.3.8 Proteins and Biomacromolecules in Solution and at Interfaces	404
5.4	Summary and Future Prospects	409
	References	409

6. Dynamics of Atoms and Molecules — 415

Mark R. Johnson and Gordon J. Kearley

6.1	Introduction	416
6.2	Brief Review of Theoretical Concepts	418
6.3	Modeling	419
	6.3.1 Mapping Potential Energy Surfaces	419
	6.3.2 Molecular Dynamics Simulation	420
	6.3.3 Empirical and *Ab Initio* Energy Calculation	420

6.4	Instrumentation	422
	6.4.1 Three-Axis Spectrometers	422
	6.4.2 Time of Flight	422
	6.4.3 Neutron Compton Scattering Spectrometers	423
	6.4.4 Molecular Spectrometers	424
	6.4.5 Backscattering Spectrometers	425
	6.4.6 Neutron Spin-echo Instruments	426
	6.4.7 The Measured Neutron-Scattering Signal	427
6.5	Oscillatory Motion, Incoherent Scattering	427
	6.5.1 Molecular Vibrations of Benzene	429
	6.5.2 Hydrogen-Bonded Systems	430
	6.5.3 Complex Hydrides	431
	6.5.4 Polymers	433
6.6	Oscillatory Motion, Coherent Scattering	434
	6.6.1 Classic Phonons and Soft Modes in $SrTiO_3$	435
	6.6.2 Negative Thermal Expansion	435
	6.6.3 Nanostructured Materials	438
	6.6.4 Oxygen-Ion Conductors—Brownmillerites	439
	6.6.5 Thermoelectrics—Skutterudites	441
	6.6.6 Pnictides	442
	6.6.7 Strontium Gallium Oxides	443
	6.6.8 Deoxyribonucleic acid	445
6.7	Tunneling	446
	6.7.1 Rotational Tunneling	446
	6.7.2 Translational Tunneling	453
6.8	Stochastic Relaxation/Dynamics	453
	6.8.1 Complex Diffusion	455
	6.8.2 Ligand Water Rotation	456
	6.8.3 Coherent QENS, Rotation	456
	6.8.4 Dynamical Transitions from Elastic Scans	457
	6.8.5 Diffusion of Coherent Scatterer CO_2	459
	6.8.6 Water and Complex Diffusion	460
	6.8.7 Ionic Liquids	463
6.9	Conclusion and Perspectives	464
	References	466

Appendix: Neutron Scattering Lengths and Cross Sections   471

*Javier Dawidowski, José R. Granada, Javier R. Santisteban,
Florencia Cantargi, and Luis A. Rodríguez Palomino*

A.1	Introduction	471
A.2	Theoretical Background	472
	A.2.1 Scattering Length	472
	A.2.2 Spin-Dependent Scattering Lengths	476
	A.2.3 Neutron–Atom Interactions	477
A.3	Methods of Measurement of Scattering Lengths	482
	A.3.1 Transmission	482
	A.3.2 Bragg Diffraction	484

A.3.3 Dynamical Diffraction ... 486
A.3.4 Prism Refraction ... 488
A.3.5 Christiansen Filter ... 489
A.3.6 Neutron Gravity Refractometer ... 490
A.3.7 Neutron Interferometry ... 491
A.3.8 Small-Angle Scattering ... 492
A.3.9 Total Reflection ... 493
A.3.10 Pseudomagnetic Method ... 493
A.3.11 High-Energy Experiments ... 494
A.4 Tables of Neutron Scattering Lengths and Cross Sections ... 495
References ... 527

Index ... 529

Contributors

Numbers in Parentheses indicate the pages on which the author's contributions begin.

Masatoshi Arai (245), Materials and Life Science Division, J-PARC Center, Japan Atomic Energy Agency, Tokai, Ibaraki, 319-1195, Japan

Francisco J. Bermejo (137), Consejo Superior de Investigaciones Científicas, C.S.I.C., Instituto de Estructura de la Materia, Madrid, and UAI Física Aplicable CSIC—UPV/EHU, Facultad de Ciencia y Tecnología, Campus de Leioa-Erandio, Bilbao, Spain

Silvia C. Capelli (321), Institut Laue Langevin, Grenoble, France

Felix Fernandez-Alonso (xvii and 1), ISIS Facility, Rutherford Appleton Laboratory, Chilton, Didcot, Oxfordshire, and Department of Physics and Astronomy, University College London, London, United Kingdom

Mark R. Johnson (415), Institute Laue Langevin, Grenoble, France

Gordon J. Kearley (415), The Bragg Institute, Australian Nuclear Science and Technology Institute, Kirrawee DC, New South Wales, Australia

Jeffrey Penfold (353), ISIS Facility, Rutherford Appleton Laboratory, Chilton, Didcot, Oxon, and Physical and Theoretical Chemistry Laboratory, Oxford University, Oxford, United Kingdom

David L. Price (xvii and 1), Synchrotron Radiation and Neutron Research Group, CEMHTI, Centre National de la Recherche Scientifique, Orléans, France

Fernando Sordo (137), Consorcio ESS-Bilbao, Leioa, and Instituto de Fusión Nuclear, José Gutierrez Abascal 2, Madrid, Spain

Ian M. Tucker (353), Unilever Research and Development Laboratory, Port Sunlight, Bebington, Wirral, United Kingdom

Volumes in Series

Experimental Methods in the Physical Sciences
(Formerly Methods of Experimental Physics)

Volume 1. Classical Methods
Edited by Immanuel Estermann

Volume 2. Electronic Methods, Second Edition (in two parts)
Edited by E. Bleuler and R. O. Haxby

Volume 3. Molecular Physics, Second Edition (in two parts)
Edited by Dudley Williams

Volume 4. Atomic and Electron Physics - Part A: Atomic Sources and Detectors; Part B: Free Atoms
Edited by Vernon W. Hughes and Howard L. Schultz

Volume 5. Nuclear Physics (in two parts)
Edited by Luke C. L. Yuan and Chien-Shiung Wu

Volume 6. Solid State Physics - Part A: Preparation, Structure, Mechanical and Thermal Properties; Part B: Electrical, Magnetic and Optical Properties
Edited by K. Lark-Horovitz and Vivian A. Johnson

Volume 7. Atomic and Electron Physics - Atomic Interactions (in two parts)
Edited by Benjamin Bederson and Wade L. Fite

Volume 8. Problems and Solutions for Students
Edited by L. Marton and W. F. Hornyak

Volume 9. Plasma Physics (in two parts)
Edited by Hans R. Griem and Ralph H. Lovberg

Volume 10. Physical Principles of Far-Infrared Radiation
Edited by L. C. Robinson

Volume 11. Solid State Physics
Edited by R. V. Coleman

Volume 12. Astrophysics - Part A: Optical and Infrared Astronomy
Edited by N. Carleton
Part B: Radio Telescopes; Part C: Radio Observations
Edited by M. L. Meeks

Volume 13. Spectroscopy (in two parts)
Edited by Dudley Williams

Volume 14. Vacuum Physics and Technology
Edited by G. L. Weissler and R. W. Carlson

Volume 15. Quantum Electronics (in two parts)
Edited by C. L. Tang

Volume 16. Polymers - Part A: Molecular Structure and Dynamics; Part B: Crystal Structure and Morphology; Part C: Physical Properties
Edited by R. A. Fava

Volume 17. Accelerators in Atomic Physics
Edited by P. Richard

Volume 18. Fluid Dynamics (in two parts)
Edited by R. J. Emrich

Volume 19. Ultrasonics
Edited by Peter D. Edmonds

Volume 20. Biophysics
Edited by Gerald Ehrenstein and Harold Lecar

Volume 21. Solid State Physics: Nuclear Methods
Edited by J. N. Mundy, S. J. Rothman, M. J. Fluss, and L. C. Smedskjaer

Volume 22. Solid State Physics: Surfaces
Edited by Robert L. Park and Max G. Lagally

Volume 23. Neutron Scattering (in three parts)
Edited by K. Skold and D. L. Price

Volume 24. Geophysics - Part A: Laboratory Measurements; Part B: Field Measurements
Edited by C. G. Sammis and T. L. Henyey

Volume 25. Geometrical and Instrumental Optics
Edited by Daniel Malacara

Volume 26. Physical Optics and Light Measurements
Edited by Daniel Malacara

Volume 27. Scanning Tunneling Microscopy
Edited by Joseph Stroscio and William Kaiser

Volume 28. Statistical Methods for Physical Science
Edited by John L. Stanford and Stephen B. Vardaman

Volume 29. Atomic, Molecular, and Optical Physics - Part A: Charged Particles; Part B: Atoms and Molecules;
Part C: Electromagnetic Radiation
Edited by F. B. Dunning and Randall G. Hulet

Volume 30. Laser Ablation and Desorption
Edited by John C. Miller and Richard F. Haglund, Jr.

Volume 31. Vacuum Ultraviolet Spectroscopy I
Edited by J. A. R. Samson and D. L. Ederer

Volume 32. Vacuum Ultraviolet Spectroscopy II
Edited by J. A. R. Samson and D. L. Ederer

Volume 33. Cumulative Author Index and Tables of Contents, Volumes 1-32

Volume 34. Cumulative Subject Index

Volume 35. Methods in the Physics of Porous Media
Edited by Po-zen Wong

Volume 36. Magnetic Imaging and its Applications to Materials
Edited by Marc De Graef and Yimei Zhu

Volume 37. Characterization of Amorphous and Crystalline Rough Surface: Principles and Applications
Edited by Yi Ping Zhao, Gwo-Ching Wang, and Toh-Ming Lu

Volume 38. Advances in Surface Science
Edited by Hari Singh Nalwa

Volume 39. Modern Acoustical Techniques for the Measurement of Mechanical Properties
Edited by Moises Levy, Henry E. Bass, and Richard Stern

Volume 40. Cavity-Enhanced Spectroscopies
Edited by Roger D. van Zee and J. Patrick Looney

Volume 41. Optical Radiometry
Edited by A. C. Parr, R. U. Datla, and J. L. Gardner

Volume 42. Radiometric Temperature Measurements. I. Fundamentals
Edited by Z. M. Zhang, B. K. Tsai, and G. Machin

Volume 43. Radiometric Temperature Measurements. II. Applications
Edited by Z. M. Zhang, B. K. Tsai, and G. Machin

Volume 44. Neutron Scattering – Fundamentals
Edited by Felix Fernandez-Alonso, and David L. Price

Preface

Just over 80 years ago, a brief letter from James Chadwick to *Nature* [1,2] presented conclusive experimental evidence unveiling the existence of a neutral particle (nearly) isobaric with the proton. The discovery of the henceforth-to-be-known-as "neutron" had profound consequences for both scientific research and the destiny of humankind, as it led to the unleashing of the might of nuclear power in less than a decade [3].

The first use of these "neutral protons" to probe the microscopic underpinnings of the materials world around us also dates back to those early years, with pioneering neutron-diffraction experiments at Oak Ridge National Laboratory (USA) in the mid 1940s, and the subsequent development of neutron spectroscopy at Chalk River (Canada) in the 1950s. Since then, neutron-scattering techniques have matured into a robust and increasingly versatile toolkit for physicists, chemists, biologists, materials scientists, engineers, or technologists. At the turn of the last century, the 1994 Nobel Prize in Physics awarded to C.G. Shull and B.N. Brockhouse recognized their ground-breaking efforts toward the development and consolidation of neutron science as a discipline in its own right [4]. This milestone also served to define neutron scattering as the technique *par excellence* to investigate *where atoms are* (structure) and *what atoms do* (dynamics), a popular motto across generations of neutron-scattering practitioners.

Sustained and continued developments in experimental methods over the past few decades have greatly increased the sensitivity and range of applications of neutron scattering. While early measurements probed distances on the order of interatomic spacings (fractions of a nm) and characteristic times associated with lattice vibrations (ps), contemporary neutron-scattering experiments can cover length scales from less than 0.01 to 1000s of nanometers, and timescales from the attosecond to the microsecond. These advances have been made possible via a significant expansion of the range of neutron energies available to the experimenter, from microelectron-volts (particularly at cold sources in research reactors) to hundreds of electron-volts (at pulsed spallation sources), as well as by unabated progress in the implementation of a variety of novel and ingenious ideas such as position- and polarization-sensitive detection or backscattering and spin-labeling methods. As a result, neutron science has grown beyond traditional research areas, from the conventional determination of crystal structures and lattice dynamics of half-a-century ago (not to forget their magnetic analogs), to high-resolution structural studies of disordered thin films, liquid interfaces, biological structures, macromolecular and supramolecular architectures and devices, or the unraveling of the dynamics and energy-level structure of complex molecular solids, nanostructured materials and surfaces, or magnetic clusters and novel superconductors. Along with these scientific and technical developments, the community of neutron scientists has also expanded and diversified beyond recognition. Whereas the early stages of neutron scattering had its roots in condensed-matter physics and crystallography, present-day users of central neutron-scattering facilities include chemists, biologists, ceramicists, and metallurgists, to name a few, as well as physicists with an increasingly diverse range of transdisciplinary interests, from the foundations of quantum mechanics to soft matter, food science, biology, geology, or archeometry.

The present and subsequent volumes in this series seek to cover in some detail the production and use of neutrons across the aforementioned disciplines, with a particular emphasis on technical and scientific developments over the past two decades. As such, it necessarily builds upon an earlier and very successful three-volume set edited by K. Sköld and D.L. Price, published in the 1980s by *Academic Press* as part of *Methods of Experimental Physics* (currently *Experimental Methods in the Physical Sciences*). Furthermore, with the third-generation spallation sources recently constructed in the United States and Japan, or in the advanced construction or planning stage in China and Europe, there has been an increasing interest in time-of-flight and broadband neutron-scattering techniques. Correspondingly, the improved performance of cold moderators at both reactors and spallation sources has extended long-wavelength capabilities to such an extent that a sharp distinction between fission- and accelerator-driven neutron sources may no longer be of relevance to the future of the discipline.

On a more practical front, the chapters that follow are meant to enable you to identify aspects of your work in which neutron-scattering techniques might contribute, conceive the important experiments to be done, assess what is required to carry them out, write a successful proposal to a user facility, and perform these experiments under the guidance and support of the appropriate facility-based scientist. The presentation is aimed at professionals at all levels, from early-career researchers to mature scientists who may be insufficiently aware or up to date with the breadth of opportunities provided by neutron techniques in their area of specialty. In this spirit, it does not aim to present a systematic and detailed development of the underlying theory, which may be found in superbly written texts such as those of Lovesey [5] or Squires [6]. Likewise, it is not a detailed hands-on manual of experimental methods, which in our opinion is best obtained directly from experienced practitioners or, alternatively, by attending practical training courses at the neutron facilities. As an intermediate (and highly advisable) step, we also note the existence of neutron-focused thematic schools, particularly those at Grenoble [7] and Oxford [8], both of which have been running on a regular basis since the 1990s. With these primary objectives in mind, each chapter focuses on well-defined areas of neutron science and has been written by a leading practitioner or practitioners of the application of neutron methods in that particular field.

In this first volume, we start out in Chapter 1 with a self-contained survey of the theoretical concepts and formalism of the technique, and also take the opportunity to establish the notation that will be used throughout. Chapters 2 and 3 review neutron production and instrumentation, respectively, areas which have profited enormously from recent developments in accelerator physics, materials research and engineering, or computing, to name a few. The remaining chapters of this volume treat several basic applications of neutron scattering, including the structure of complex materials (Chapter 4), large-scale structures (Chapter 5), and dynamics of atoms and molecules (Chapter 6). The Appendix goes back to some requisite fundamentals linked to neutron–matter interactions, along with a detailed compilation of neutron-scattering lengths and cross sections across the periodic table.

In closing this preface, we wish to thank all authors for taking time out of their busy schedules to be part of this venture, Drs. Thomas Lucatorto and Albert C. Parr for inviting us to undertake this work, and the staff of *Academic Press* for their encouragement, diligence, and forbearance along the way.

Felix Fernandez-Alonso
David L. Price

REFERENCES

[1] Chadwick J. Nature 1932;129:312.
[2] www.nobelprize.org/nobel_prizes/physics/laureates/1935/ [last accessed on 01/07/2013].
[3] MacKay A. The making of the atomic age. Oxford: Oxford University Press; 1984.
[4] www.nobelprize.org/nobel_prizes/physics/laureates/1994/ [last accessed on 01/07/2013].
[5] Lovesey SW. Theory of neutron scattering from condensed matter, volumes I and II. Oxford: Oxford University Press; 1986.
[6] Squires GL. Introduction to the theory of thermal neutron scattering. 3rd ed. Cambridge: Cambridge University Press; 2012.
[7] http://hercules-school.eu/ [last accessed on 02/07/2013].
[8] www.oxfordneutronschool.org/ [last accessed on 02/07/2013].

Symbols

LIST OF COMMONLY USED SYMBOLS IN ALPHABETICAL ORDER[1]

a	scattering length in center-of-mass frame
b	bound scattering length
A	atomic-mass (or nucleon) number
$\bar{b}$	coherent scattering length
b^+	scattering length for $I+\frac{1}{2}$ state
b^-	scattering length for $I-\frac{1}{2}$ state
b_i	incoherent scattering length
b_N	spin-dependent scattering length
c	speed of light in vacuum $= 299{,}792{,}458$ m s^{-1} [1]
$\mathbf{D}$	dynamical matrix
$\mathbf{D}_\perp(\mathbf{Q})$	magnetic interaction operator
d	mass density
$\mathbf{d}$	equilibrium position of atom in unit cell
$d\sigma/d\Omega$	differential scattering cross section
$d^2\sigma/d\Omega dE_f$	double differential scattering cross section
E	neutron energy transfer $(E_i - E_f)$
E_n	neutron energy
E_i, E_f	initial, final (scattered) neutron energy
e	elementary charge $= 1.602176565 \times 10^{-19}$ C
$e^k(\mathbf{q})$	polarization vector of normal mode k ($e_d^k(\mathbf{q})$ for a non-Bravais crystal)
$F(\boldsymbol{\tau})$	unit-cell structure factor
f	force constant
$f(\mathbf{Q})$	form factor
$G(\mathbf{r},t)$	space–time (van Hove) correlation function $[G_d(\mathbf{r},t)+G_s(\mathbf{r},t)]$
$G_d(\mathbf{r},t)$	"distinct" space–time correlation function
$G_s(\mathbf{r},t)$	"self" space–time correlation function
$g(\mathbf{r})$	pair distribution function
g_e	electron g factor $= -2.00231930436153$
g_n	neutron g factor $= -3.82608545$
h	Planck constant $= 6.62606957 \times 10^{-34}$ J s

---

1. Note: other symbols may be used that are unique to the chapter where they occur.

Symbol	Description
$\hbar$	Planck constant over 2π $(h/2\pi) = 1.054571726 \times 10^{-34}$ J s
$\mathbf{I}$	angular-momentum operator for nucleus
$I(\mathbf{Q},t)$	intermediate scattering function
$I_s(\mathbf{Q},t)$	self-intermediate scattering function
k_B	Boltzmann constant $= 1.3806488 \times 10^{-23}$ J K^{-1}
k_i, k_f	initial and final (scattered) neutron wave vectors
k_n	neutron wave vector $(k_n = 2\pi/\lambda_n)$
L_i, L_f	initial (primary) and final (secondary) flight path
$\mathbf{l}$	unit-cell position vector (lowercase "L")
M	atom mass
$\mathbf{M}$	magnetization operator
m_e	electron mass $= 9.10938291 \times 10^{-31}$ kg
m_n	neutron mass $= 1.674927351 \times 10^{-27}$ kg
m_p	proton mass $= 1.672621777 \times 10^{-27}$ kg
N	number of atoms in sample
N_A	Avogadro constant $= 6.02214129 \times 10^{23}$ mol^{-1}
N_c	number of unit cells in crystal
$\mathbf{R}_j$	position vector for particle j
$\mathbf{Q}$	scattering vector $(\mathbf{k}_i - \mathbf{k}_f)$
$\mathbf{q}$	reduced wave vector $(\mathbf{Q} - \boldsymbol{\tau})$
r_0	classical electron radius $(e^2/4\pi\varepsilon_0 m_e c^2) = 2.8179403267 \times 10^{-15}$ m
$\mathbf{S}$	atomic-spin operator
$S(\mathbf{Q})$	static structure factor $I(\mathbf{Q},0)$
$S(\mathbf{Q},E)$	dynamic structure factor or scattering function
$S_c(\mathbf{Q},E)$	coherent dynamic structure factor or scattering function
$S_i(\mathbf{Q},E)$	incoherent dynamic structure factor or scattering function
$\mathbf{s}$	electron-spin operator
T	temperature
u	unified atomic-mass unit $= 1.660538921 \times 10^{-27}$ kg
$\mathbf{u}^l$	vibrational amplitude ($\mathbf{u}_d^l$ for non-Bravais crystals)
V	volume of sample
v_i, v_f	initial, final (scattered) neutron velocity
v_n	neutron velocity
v_0	unit-cell volume
$2W$	exponential argument in the Debye–Waller factor
Z	atomic (or proton) number
$Z(E)$	phonon density of states
Γ	spectral linewidth
γ	neutron magnetic moment to nuclear magneton ratio $(\mu_n/\mu_N) = -1.91304272$
γ_n	neutron gyromagnetic ratio $= 1.83247179 \times 10^8$ s^{-1} T^{-1}
Θ	Debye temperature
θ	Bragg angle
λ_i, λ_f	initial and final (scattered) neutron wavelength

Symbols

$\boldsymbol{\mu}$	atomic magnetic moment
μ_N	nuclear magneton $(e\hbar/2m_p) = 5.05078353 \times 10^{-27}\,\mathrm{J\,T^{-1}}$
μ_B	Bohr magneton $(e\hbar/2m_e) = 927.400968 \times 10^{-26}\,\mathrm{J\,T^{-1}}$
μ_n	neutron magnetic moment $= -0.96623647 \times 10^{-26}\,\mathrm{J\,T^{-1}}$
ν	frequency $(\omega/2\pi)$
$\rho(\mathbf{r},t)$	particle-density operator
ρ_0	average number density
Σ	macroscopic cross section associated with a given cross section σ (see below)
σ_a	absorption cross section
σ_t	bound total cross section $(\sigma_s + \sigma_a)$
σ_c	bound coherent scattering cross section
σ_i	bound incoherent scattering cross section
σ_s	bound scattering cross section $(\sigma_c + \sigma_i)$
$\tfrac{1}{2}\boldsymbol{\sigma}$	neutron-spin (Pauli) operator
$\boldsymbol{\tau}$	reciprocal lattice vector $\{2\pi[(h/a),(k/b),(l/c)]\}$
Φ	neutron flux (typically defined as neutrons crossing per unit area per unit time)
ϕ	scattering angle $(=2\theta)$
χ	susceptibility
$\chi(\mathbf{Q},E)$	generalized susceptibility
Ω	solid angle
ω	radian frequency associated with neutron energy transfer $E = \hbar\omega$
$\omega_k(\mathbf{q})$	frequency of normal mode k

REFERENCE

[1] All numerical values have been taken from the CODATA Recommended Values of the Fundamental Physical Constants, as detailed in Mohr PJ, Taylor BN, Newell DB. Rev Mod Phys 2010;84:1527–605. Updated (2010) values can be found at, physics.nist.gov/cuu/Constants/ [last accessed on July 14, 2013].

Chapter 1

An Introduction to Neutron Scattering

David L. Price* and Felix Fernandez-Alonso[†,‡]
*Synchrotron Radiation and Neutron Research Group, CEMHTI, Centre National de la Recherche Scientifique, Orléans, France
[†]ISIS Facility, Rutherford Appleton Laboratory, Chilton, Didcot, Oxfordshire, United Kingdom
[‡]Department of Physics and Astronomy, University College London, London, United Kingdom

Chapter Outline

1.1. Fundamentals 2
  1.1.1. Why Are Neutrons so Unique? 2
  1.1.2. Thermal Neutrons for Condensed Matter Research 5
  1.1.3. Conservation Laws 11
  1.1.4. The Structure of Materials 11
  1.1.5. Adding Motion: Dynamics and Spectroscopy 16
1.2. Scattering Foundations 22
  1.2.1. The Master Formula and Fermi's Golden Rule 22
  1.2.2. Nuclear Scattering 25
  1.2.3. The Double Differential Cross Section in the Time Domain 26
  1.2.4. Farewell to Nuclear Physics 27
  1.2.5. Coherent and Incoherent Scattering 28
  1.2.6. Scattering Functions 30
1.3. Canonical Solids 31
  1.3.1. Normal Modes of Vibration 31
  1.3.2. Scattering Under the Harmonic Approximation 32
  1.3.3. Purely Elastic Events 33
  1.3.4. Inelastic (One-Phonon) Scattering 35
  1.3.5. Multiphonon Scattering 39
  1.3.6. Beyond Harmonic Vibrations 41
1.4. Beyond Canonical Solids 42
  1.4.1. Space–Time (Van Hove) Correlation Functions 42
  1.4.2. Pair Distribution Functions 43
  1.4.3. Properties of the Dynamic Structure Factor 47

1.4.4. From Order to Disorder: Diffuse Scattering	58	1.6.1. Generalized Susceptibility	106	
1.4.5. Stochastic Diffusion	61	1.6.2. Spin Waves	108	
1.4.6. Beyond Atoms and Molecules: Large-Scale Structures	75	1.6.3. Crystal Fields and Magnetic Clusters	111	
		1.6.4. Spin Fluctuations	114	
		1.6.5. Interband Transitions	115	
1.5. Magnetic Structure and Polarized Neutrons	**82**	1.6.6. Critical Scattering	117	
1.5.1. Basic Principles	82	**1.7. Nuclear Spin: Order and Disorder**	**120**	
1.5.2. Polarized Neutrons	88	1.7.1. A Closer Look at Nuclear Spins	121	
1.5.3. Magnetic Bragg Scattering	95	1.7.2. Scattering Cross Sections	121	
1.5.4. Diffuse Scattering from Magnetic Disorder	97	1.7.3. Uncorrelated and Correlated Spin Ensembles	123	
1.5.5. Large-Scale Magnetic Structures	101			
1.6. Spin Dynamics	**106**	**1.8. Outlook**	**125**	
		References	**127**	

1.1 FUNDAMENTALS

1.1.1 Why Are Neutrons so Unique?

The neutron is a stable subatomic particle (a hadron) when bound in atomic nuclei. Free neutrons are unstable and undergo β decay with a mean (exponential) lifetime of ca. 15 min [1], a limitation which is barely of significance for most research applications. Neutrons have zero charge, mass $m_n = 1.0087$ atomic-mass units [2], one-half spin (i.e., a fermion with $\sigma = 1/2$), and magnetic moment $\mu_n = \gamma \mu_N$, where $\mu_N = e\hbar/2m_p$ is the nuclear magneton and γ is a constant equal to -1.913. Notwithstanding over 80 years of neutron research following its discovery in the early 1930s [3–5], this particle still remains at the heart of cosmology and particle physics [6,7].

The four fundamental physical properties presented above also make the neutron a highly effective probe of condensed matter. The zero charge and nonzero spin imply that neutrons can interact with materials via nuclear (short-ranged, $fm = 10^{-15}$ m) and magnetic interactions, respectively. Both nuclear and magnetic interaction probabilities (or *cross sections*) are small, so the neutron can usually penetrate well into the bulk of the specimen under investigation. In the language of scattering theory, neutron-matter interactions (both nuclear and magnetic) can then be described within the first Born approximation and thus given in explicit form by relatively simple expressions. We shall return to this particular point in Section 1.2.

Neutrons for condensed-matter research are usually obtained by slowing down or "moderating" high-energy (MeV) neutrons produced by nuclear reactions. The slow neutrons thus produced have kinetic energies on the order of $k_B T$, where T is the temperature of the moderating medium (or "moderator") and k_B is the Boltzmann's constant. Given the wave nature of the neutron, there is an associated *de Broglie* wavelength λ_n for a given kinetic energy E_n obeying the well-known relation

$$E_n = \frac{h^2}{2m_n \lambda_n^2} = \frac{\hbar^2 k_n^2}{2m_n} = k_B T, \qquad (1.1)$$

where h is the Planck's constant and k_n is the neutron wave vector. The value of m_n is such that for $T \sim 300$ K, $\lambda_n \sim 2$ Å ($= 2 \times 10^{-10}$ m), a distance comparable to interatomic spacings in solids and dense fluids. Such neutrons are therefore ideally suited to studies of the *atomic structure* of condensed matter via diffraction (elastic scattering) studies, a fact that was recognized soon after the discovery of the neutron by Chadwick in 1932. Furthermore, the associated kinetic energy of such neutrons, $k_B T$, is on the order of 25 meV, a typical energy for collective vibrational modes (or phonons) in condensed-matter systems. Thus, both thermal-neutron wavelengths and energies are ideally suited to studies of the *atomic dynamics* via spectroscopy (inelastic scattering) experiments (cf. Sections 1.3 and 1.4).

The magnetic moment μ_n of the neutron also makes it a unique probe of *magnetism* on atomic and molecular scales, as neutrons may be scattered from the magnetic moments associated with unpaired electron density. Again, the wavelength and the energy of typical thermal neutrons are such that both magnetic structure and the associated spin dynamics can be studied via neutron-scattering experiments (cf. Sections 1.5 and 1.6). The spin of the neutron has further important consequences. When a neutron is scattered by a nucleus with nonzero spin, the strength of the interaction depends on the relative orientation of neutron and nuclear spins. This makes the neutron a unique probe of *nuclear-spin correlations* (both static and dynamic) at low temperatures (cf. Section 1.7).

Figure 1.1 shows neutron-wavelength spectra for the two most common types of high-intensity source, namely, a fission reactor and an accelerator-driven spallation source. It is customary to classify neutrons by the moderator temperature, that is, cold, thermal ($T = 300$ K), and hot. Neutrons that have not slowed down to characteristic velocities for a given moderator temperature are called epithermal and are readily available at spallation sources. Cold(hot) moderators give rise to spectra centered at longer(shorter) neutron wavelengths λ_n, as shown in Figure 1.1. A detailed comparison between the performance of steady and pulsed neutron sources ultimately depends

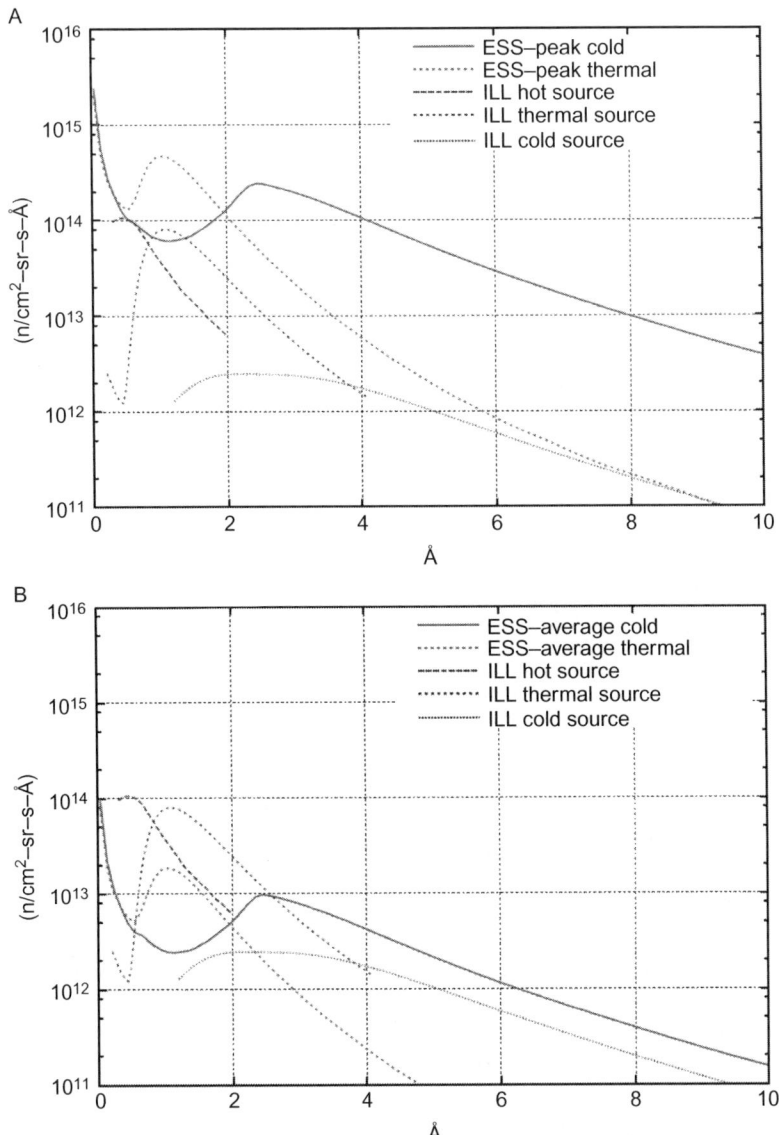

FIGURE 1.1 Neutron flux per unit solid angle per unit wavelength as a function of wavelength for cold, thermal, and hot moderators at the Institut Laue-Langevin (ILL) compared to those expected for the European Spallation Source (ESS), a next-generation pulsed spallation facility. The ILL is a steady reactor source whereas ESS is envisaged to deliver 2.86 ms pulses at a repetition rate of 14 Hz. (a) and (b) A comparison between these two neutron sources in terms of peak and time-averaged fluxes, respectively. *Reprinted with permission from Ref. [8].* (For color version of this figure, the reader is referred to the online version of this chapter.)

TABLE 1.1 Properties of the Neutron at Selected Kinetic Energies

Quantity	Unit	Definition	Ultracold	Cold	Thermal	Epithermal
Energy E_n	meV [a]		2.5×10^{-4}	1	25	1000
Temperature T	K	E_n/k_B	2.9×10^{-3}	12	290	12,000
Wavelength λ_n [b]	Å	$h/(2m_n E_n)^{1/2}$	570	9.0	1.8	0.29
Wave vector k_n [c]	Å^{-1}	$(2m_n E_n)^{1/2}/\hbar$	0.011	0.7	3.5	22
Velocity v_n [d]	m s^{-1}	$(2E_n/m_n)^{1/2}$	6.9	440	2,200	14,000

For definitions of the various symbols and associated numerical values, see the *List of Commonly Used Symbols* and references therein.
[a] $1\ meV = 1.6022 \times 10^{-22}$ J, the amount of energy necessary to move an electron against a potential difference of one millivolt (1 mV).
[b] λ_n (Å) $= 9.0446\ [E_n\ (meV)]^{-1/2}$.
[c] k_n (Å^{-1}) $= 0.69469\ [E_n\ (meV)]^{1/2}$.
[d] v_n (m s^{-1}) $= 437.39\ [E_n\ (meV)]^{1/2}$.

on application and, therefore, it requires defining the *useful neutron flux* at point of use, as presented in Chapter 3. To a first approximation, the peak neutron flux shown in Figure 1.1a can be used as a rough figure of merit provided that it is possible to exploit the full time structure of the pulsed neutron beam [9]. In this case, next-generation accelerator-based pulsed sources offer the prospect of a significant increase in useful flux relative to current capabilities. Neutron production and moderation are discussed in detail in Chapters 2 and 3.

For reference purposes, Table 1.1 gives the energy, temperature, wavelength, wave vector, and velocity for typical neutrons in each category, along with useful relations to convert between these quantities using commonly used physical units. Likewise, the energies associated with atomic and molecular excitations in condensed matter are expressed in a multitude of ways across the scientific literature, and these can always be related to the amount of energy E transferred to/from the neutron during a scattering experiment. To aid the reader, Table 1.2 provides a conversion table between commonly used units.

1.1.2 Thermal Neutrons for Condensed Matter Research

We consider a simple scattering experiment as schematically shown in Figure 1.2. Suppose that a beam of neutrons characterized by wave vector $\mathbf{k}_i$ falls on the sample. As pointed out in the previous section, the interaction probability is rather small and in a typical experimental situation most

TABLE 1.2 Common Quantities Used to Denote Energy or Energy Transfer

Quantity	Definition	Value at $E = 1$ meV
Radian frequency ω	$E/\hbar$	1.5193×10^{12} rad s^{-1}
Frequency ν	E/h	0.24180×10^{12} Hz $= 0.24180$ THz
Spectroscopic wave number (or kayser) $\tilde{\nu}$	E/hc	8.0655 cm^{-1}
Temperature T	E/k_B	11.605 K

For definitions of the various symbols and associated numerical values, see the *List of Commonly Used Symbols* and references therein.

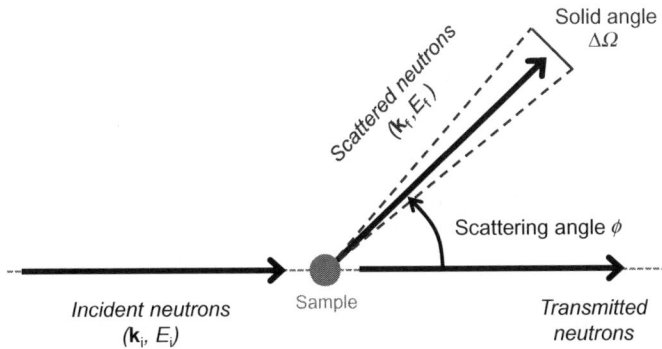

FIGURE 1.2 Schematic diagram of a generic arrangement for neutron-scattering experiments. (For color version of this figure, the reader is referred to the online version of this chapter.)

neutrons are transmitted without any interaction. Some, however, will be scattered and can be measured with a neutron detector placed, let us say, along direction $\hat{\mathbf{k}}_f$.

If the incident beam is characterized by a uniform flux Φ (neutrons crossing per unit area and per unit time), the sample has N identical atoms in the beam, and a detector of efficiency η subtends a solid angle $\Delta\Omega$, the count rate C on the detector is proportional (if $\Delta\Omega$ is small enough) to all these quantities. The proportionality constant is called the *differential scattering cross section* (denoted as $d\sigma/d\Omega$). Mathematically, we write

$$C = \eta \Phi N \left(\frac{d\sigma}{d\Omega}\right) \Delta\Omega. \tag{1.2}$$

The differential cross section is a function of both $\mathbf{k}_i$ and $\mathbf{k}_f$ (technically speaking, it describes the correlation between these two vector quantities).

As such, it is a property of the sample being measured and it may also depend on the spin state of incident and detected neutrons. The most important types of interaction for condensed-matter studies are the nuclear interaction and the magnetic-dipole interaction. Other (much weaker) interactions are discussed in detail in the Appendix.

1.1.2.1 Atomic Nuclei

The interaction between a slow neutron and an atom through the nuclear force can be expressed in a very simple form, provided that the neutron wavelength is much longer than the range of the nuclear force. This condition is easily fulfilled by thermal neutrons (for a more detailed discussion, see the Appendix). To illustrate this feature, we consider the case where the atoms in the sample are identical and noninteracting. In this case, the differential cross section is a constant:

$$\frac{d\sigma}{d\Omega} = b^2, \qquad (1.3)$$

where the scattering length b is a property only of the nucleus (characterized by atomic and atomic-mass numbers Z and weight A, respectively) and, in general, its spin state relative to that of the neutron. Typical values of b are on the order of 1–10 fm and the resulting cross sections are proportional to the square of the scattering length. Scattering cross sections are measured in barn (b), where $1\,\text{b} = 100\,\text{fm}^2 = 10^{-28}\,\text{m}^2$. Physically speaking, Equation (1.3) tells us that neutron deflection following a collision is equally probable in all directions, that is, isotropic scattering. It is noteworthy that scattering lengths for thermal neutron scattering are for the most part (e.g., away from resonances) independent of collision energy. Therefore, scattering observables such as total and differential cross sections can be linked quite effortlessly to the properties of the material under investigation via the dependence of scattering probabilities on energy and momentum transfers and not their absolute values (see discussion in Section 1.1.3). This invariance of thermal neutron-scattering lengths with energy has also important consequences in the design of neutron-scattering instruments, that is, the equivalence between *so-called* direct- and indirect-geometry instrumentation, covered in Chapter 3.

The scattering length is a quantity that depends on the details of the interaction between the neutron and the constituents of the nucleus. For this reason, both the magnitude and sign of b change in a rather irregular fashion as a function of Z or A, with a relatively mild increase in absolute magnitude as one moves across the periodic table of the elements (see Figure 1.3). In contrast, the atomic scattering length for X-ray scattering is a monotonically increasing function of Z, as X-ray scattering from an isolated electron is characterized by a (bound) scattering length equal to the classical electron radius $r_0 \approx 2.8\,\text{fm}$. These features of neutron scattering relative to X-rays have some

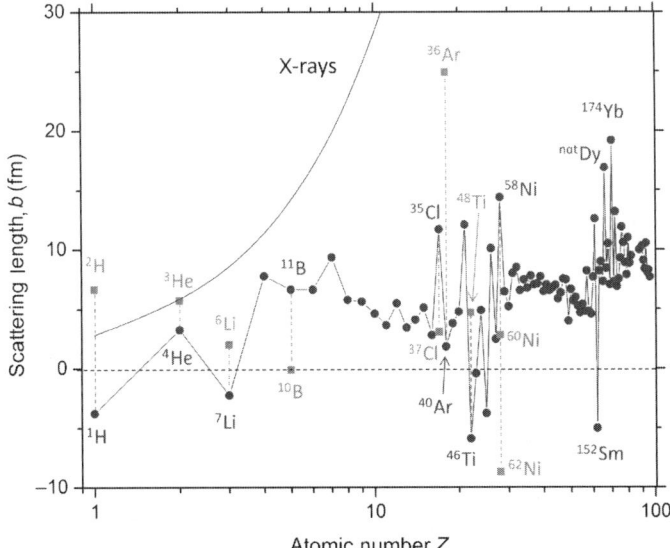

FIGURE 1.3 Neutron (blue circles and red squares) and X-ray (black) scattering lengths across the periodic table of the elements. For neutrons, the blue circles correspond to the most abundant isotope for a given element or its average value across naturally occurring isotopes where these data are not available. Red squares show the scattering lengths for specific isotopes of a given element, so as to illustrate their variation with A for a given Z. The X-ray data assume electrically neutral species (i.e., Z = number of electrons). Neutron-scattering lengths have been taken from the Appendix. (For interpretation of the references to color in this figure legend, the reader is referred to the online version of this chapter.)

powerful consequences, making neutrons sensitive to the presence of light atoms, most notably hydrogen, and to the difference between atoms with similar atomic number, for example, adjacent transition metals. The variation in scattering length between different isotopes of the same element can be large, as in the case of hydrogen, lithium, chlorine, argon, titanium, or nickel, and can be exploited in so-called *isotopic-substitution* experiments. The most widely used case corresponds to hydrogen versus deuterium, discussed in Chapter 4 and the Appendix. Notwithstanding the merits of this approach for the study of hydrogenous matter (organic materials, polymers, biomolecules, etc.), physicochemical changes associated with deuteration must always be kept in check, as recently reviewed in Ref. [10]. An in-depth discussion of isotopic substitution beyond the case of hydrogen can be found in Ref. [11] (see, in particular, Table 1.1 of this work).

In many cases, the scattering lengths for the two spin states ($I \pm 1/2$) resulting from the interaction of a nucleus of spin I with the neutron are also quite different, which, in general, leads to "incoherent" scattering as discussed in Section 1.2. If the nuclei are polarized, the spin dependence of the scattering length can be exploited to measure nuclear-spin correlations and structure (Section 1.7). In some cases, the scattering length can also be a complex quantity, where the imaginary part describes the absorption of neutrons by a particular nuclide. This situation arises when there is a low-lying resonance of the neutron–nucleus system, leading to scattering and absorption cross sections that depend on wavelength. More details can be found in the Appendix.

Also, we point out that the outcome of the scattering experiment discussed above and leading to isotropic scattering, cf. Equation (1.3), will depend on whether the nuclei are fixed or free to recoil during the scattering process. The scattering length measured in the laboratory-fixed frame is smaller by a factor $(A/A+1)$ if the nucleus is free to recoil [12]. For most condensed-matter neutron-scattering experiments, the fixed atom is a more appropriate limiting case, and it is conventional to quote the corresponding "bound-atom" values for scattering lengths and cross sections. We will return to the free-atom case when discussing scattering from the ideal gas in Section 1.4.

A list of scattering lengths and corresponding cross sections is given in the Appendix, which also includes a theoretical discussion and a description of the various methods to measure these quantities.

1.1.2.2 Electron and Nuclear Spins

With a magnetic moment $\mu_n = -1.913\mu_N$, neutrons can interact with unpaired electrons in magnetic materials. For the simple case of a paramagnet with electrons localized on specific ions with random spin orientations and no external magnetic field, the differential cross section is given by

$$\frac{d\sigma}{d\Omega} = (\gamma r_0)^2 [f(\mathbf{k}_i - \mathbf{k}_f)]^2 S(S+1), \tag{1.4}$$

where r_0 is the classical electron radius defined above and $|\gamma r_0| \approx 5.4$ fm, S is the spin quantum number for the ions, and $f(\mathbf{k}_i - \mathbf{k}_f)$, the so-called *magnetic form factor*, is given by the Fourier transform of the spatial distribution of unpaired-electron density about the ion centre. The form factor appears because this spatial distribution is of comparable dimensions to interatomic distances and, thus, to neutron wavelengths. This situation is in stark contrast to the short-range nuclear force, characterized by a delta-function-like spatial distribution relative to thermal-neutron wavelengths (i.e., a constant). Unlike nuclear scattering lengths, however, magnetic form factors can often be calculated quite accurately from the electronic wave functions of either atoms or molecules. A classic example from the pioneering work of Shull and Yamada is shown in Figure 1.4. Conversely, measurements of the form factor via the use of Equation (1.4) yield direct information about the wave functions associated with magnetic electrons. This particular case is discussed in more detail in Section 1.5.3.1.

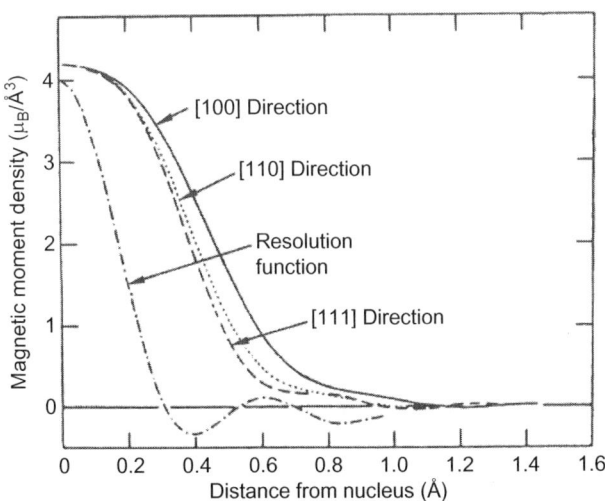

FIGURE 1.4 Distance dependence of the magnetic moment density for bulk iron along different crystallographic directions. These data have been obtained from magnetic-form-factor measurements using neutron scattering. *Reprinted with permission from Ref. [13].*

In addition to scattering from intrinsic electron spins S, the magnetic moment of the neutron can also interact with the current associated with a moving electron, giving rise to an orbital contribution to the magnetic interaction. In view of the more complicated form of the expressions for magnetic scattering, we develop the formalism of the following sections in terms of the nuclear interaction, returning to magnetism in Sections 1.5 and 1.6.

1.1.2.3 Nuclear Absorption

At the beginning of this section, we discussed two possible outcomes for a neutron incident on a sample: transmission and scattering. A third possibility is neutron absorption by a nucleus, either through a direct nuclear reaction or as a result of compound-nucleus formation. In most scattering experiments, these phenomena can be treated as a simple attenuation of the transmitted and scattered beams via the addition of an extra (multiplicative) factor in Equation (1.2), along with other "nonideal" contributions such as beam removal due to scattering and multiple-scattering events involving the same neutron. Strong absorption due to resonant capture is a special case that leads to a wavelength-dependent (complex) scattering length. Resonance scattering by epithermal neutrons has been proposed as a means to access higher-order correlations than those accessible via thermal neutron scattering [14], yet no experimental implementation of this approach has been demonstrated to date. In general, there is also a dependence on the spin state of the neutron–nucleus system. Absorption cross sections are listed along with scattering cross sections in the Appendix.

1.1.3 Conservation Laws

Returning to our schematic experimental arrangement shown in Figure 1.2, we see that the incident neutron wave vector $\mathbf{k}_i$ undergoes a change in direction during scattering. This deflection implies a change in the neutron linear momentum, which, as we know from classical mechanics, must be exchanged with the sample. The momentum transferred to the sample is conventionally described in terms of the corresponding scattering vector $\mathbf{Q}$, and the law of conservation of linear momentum is written as

$$\mathbf{k}_i - \mathbf{k}_f = \mathbf{Q}. \tag{1.5}$$

The momentum $\hbar \mathbf{Q}$ may be taken up by the scattering atom and subsequently shared with the rest of the sample, or in certain cases as discussed below, the sample as a whole can recoil; either outcome depends on the specific circumstances of the scattering event. In general, both the magnitude and direction of the linear momentum $\mathbf{k}$ will change, and so energy is also exchanged with the sample. Energy conservation implies

$$\frac{\hbar^2 k_i^2}{2m_n} - \frac{\hbar^2 k_f^2}{2m_n} = E_i - E_f = E, \tag{1.6}$$

where the neutron energy transfer E is defined to be positive when the sample takes energy from the neutron (neutron energy loss). The two terms on the left-hand side represent the initial (E_i) and final (E_f) neutron energies. A process whereby the neutron is scattered from $\mathbf{k}_i$ to $\mathbf{k}_f$ is therefore associated with a set of values of $\mathbf{Q}$ and E. The probability of scattering as a function of the variables ($\mathbf{Q}$, E) is a property of the particular sample and its environment (temperature, pressure, magnetic field, etc.) in the scattering experiment. Most of this book is devoted to a description of the form and properties of this function for different materials and environments.

1.1.4 The Structure of Materials

Let us consider a very simple case where the sample consists of individual noninteracting scattering centers of mass M. Then, the relation between E and $\mathbf{Q}$ is of the form

$$E = \frac{\hbar^2 Q^2}{2M}. \tag{1.7}$$

We note that Equations (1.5) and (1.6) represent the familiar two-body collision problem [15], with the solution

$$E = \frac{2E_i m_n}{M}(1 - \cos \phi) + O\left(\frac{m_n}{M}\right)^2, \tag{1.8}$$

where ϕ is the angle between $\mathbf{k}_i$ and $\mathbf{k}_f$ and E_i is the incident neutron energy. Thus, E tends to zero as the mass M of the scattering unit becomes large. If the sample is a solid, a significant amount of the scattering is "elastic," that is, $E=0$ exactly, and the sample recoils as a rigid unit. This phenomenon is analogous to recoil-less emission in Mossbauer spectroscopy where the scattering unit becomes the entire sample (or at least a macroscopic fraction of it). Hereafter, we shall also assume that $u=m_n$, where u is the unified atomic-mass unit defined in the *List of Commonly Used Symbols*.

From Equation (1.6), we must have $|\mathbf{k}_i|=|\mathbf{k}_f|$ in the case of purely elastic scattering, and thus from Equation (1.5)

$$2k_i\sin(\phi/2) = Q. \tag{1.9}$$

For crystalline materials, elastic scattering is strong when $\mathbf{Q}$ is equal to a reciprocal lattice vector of the crystal:

$$\mathbf{Q} = 2\pi\left(\frac{h}{a},\frac{k}{b},\frac{l}{c}\right) \text{ or } Q = \frac{2\pi}{d}, \tag{1.10}$$

where d is the spacing of the (h, k, l) set of crystal planes. Putting $k_i = 2\pi/\lambda_i$, we derive the familiar *Bragg condition* for diffraction:

$$\lambda_i = \lambda_f = 2d\sin(\phi/2) = 2d\sin\theta. \tag{1.11}$$

where θ is the Bragg angle. This expression describes the situation when waves scattered from successive crystallographic planes interfere constructively to give an intensity maximum in the diffraction pattern. In the case of a single crystal, there is a further condition that the crystal orientation be such that $\mathbf{Q}$ is parallel to the reciprocal lattice vector. The study of crystalline materials with thermal neutrons, leading to the identification of plane spacings from the peaks in the diffraction pattern, and the inference of crystal structures from these data represent a major research activity at neutron sources around the world. This particular application of neutron scattering is discussed in more detail in Chapter 4.

Equation (1.11) shows that there are two possible methods of scanning through the different plane spacings in the material: varying ϕ at constant λ_i or varying λ_i at constant ϕ. Both methods are used extensively. In both cases, the value of λ_i must be either defined by the experimental configuration or measured during the experiment for each scattering event. The most commonly used methods are based on a determination of either neutron wavelength using the Bragg condition for a single-crystal monochromator or analyzer (cf. Equation 1.11), or of velocity by measuring the time a neutron takes to travel a known distance. These methods are discussed in detail in Chapter 3.

Strictly speaking, to measure truly elastic scattering one should also determine the wavelength of both incident and scattered beams to ensure that

$|\mathbf{k}_i| = |\mathbf{k}_f|$. In practice, however, structural measurements in crystalline materials are usually made under conditions such that elastic scattering is the dominant contribution and the *total scattering* corresponding to a certain λ_i and ϕ is measured regardless of energy transfer. The inelastic ($E \neq 0$) contribution in this case is either neglected or somehow accounted for during data analysis. For noncrystalline samples such as glasses or liquids, the total scattering is usually the quantity of interest, although we shall see that, in practice, inelasticity corrections still have to be made to account for $E \neq 0$ scattering. Figure 1.5 shows typical time-dependent diffraction data for a polycrystalline (powder) sample of a hydrogen-storage material during *in situ* gas desorption. These data have been measured with the variable-λ_i, constant-ϕ method at a pulsed spallation source. The development of large-area detector arrays over the past couple of decades has made these detailed kinetic studies on complex multicomponent systems quite routine at high-intensity neutron sources.

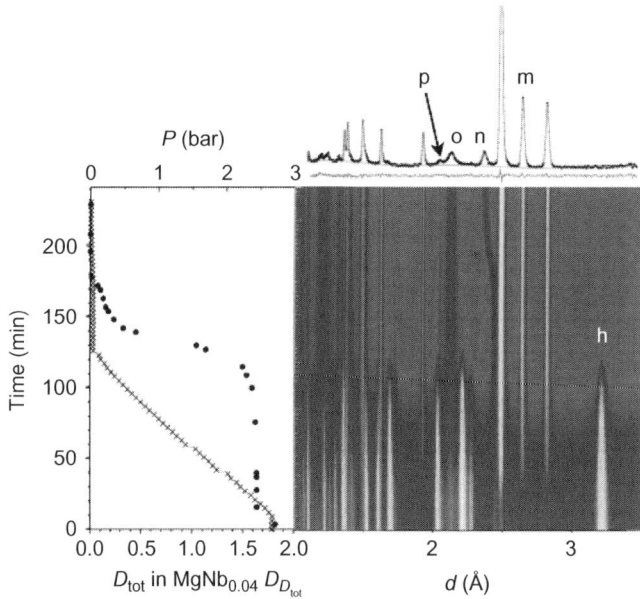

FIGURE 1.5 Time-resolved neutron diffraction data during the *in situ* desorption of D_2 from niobium-doped nanostructured magnesium at 593 K. The graph on the left shows the total deuterium content in the sample as inferred from these neutron measurements (crosses), as well as the corresponding gas pressures (black circles). The representative diffraction pattern displayed at the top shows characteristic reflections from the five different phases present during gas desorption: Mg (m), MgD_2 (h), MgO (o), Nb (n), and a perovskite phase (p). The red line in this diffraction pattern corresponds to a fit to the experimental data using Rietveld refinement methods. *Reprinted with permission from Ref. [16].* (For interpretation of the references to color in this figure legend, the reader is referred to the online version of this chapter.)

1.1.4.1 Comparison with Other Probes

It is instructive at this juncture to provide rough estimates of the total count rate in a typical neutron diffraction experiment. Combining Equations (1.2) and (1.3), we can write

$$C = \eta \Phi N \left(\frac{d\sigma}{d\Omega}\right) \Delta\Omega \approx \eta \Phi(\lambda_i) \Delta\lambda_i N b^2 \Delta\Omega, \quad (1.12)$$

where $\Phi(\lambda_i)$ is the incident neutron flux per unit wavelength and $\Delta\lambda_i$ is the wavelength bandwidth used in the measurement. If we require a 1% resolution, for a typical diffractometer we have $\Phi(\lambda_i)\Delta\lambda_i \approx 10^9$ n/cm^2 s, $\Delta\lambda_i/\lambda_i \approx 10^{-2}$, $b^2 \approx 2 \times 10^{-25}$ cm^2, $\Delta\Omega \approx 10^{-4}$, and so $C \approx 10^{-22}$ N. For an average count rate of 1 count per second we therefore require $\sim 10^{22}$ atoms or ~ 1 g of sample. Of course these numbers can vary considerably, and the efficiency of the measurement can be improved with devices such as multiplexed and position-sensitive detection systems to lower these requirements by orders of magnitude, as detailed in Chapter 3. It remains true, however, that the amount of sample required for neutron diffraction is generally in the milligram-to-gram range. The corresponding sample dimensions lie in the range of millimeter to centimeter, which are not a limitation as long as the material is available, since neutrons easily penetrate such distances in most materials.

The above situation is in marked contrast to the use of X-ray or electron scattering techniques where much smaller samples can be used and where the penetration depth is generally considerably less [17,18]. Experimentally, the much-higher scattering and absorption probabilities of X-rays and electrons must be taken into account when designing sample containers and other sample-environment equipment, a requirement that can turn into a real limitation when using laboratory sources or typical synchrotrons. For X-rays with energies on the order of 100 keV (available at third-generation synchrotron sources) absorption can be less of an issue.

When assessing the merits and strengths of (widely available) X-ray diffraction techniques relative to neutron scattering, the following considerations might be useful:

- Unlike neutron scattering (governed by the scattering length b and delta-like nuclear interactions), the analogous X-ray form factors $f(Q)$ display an intrinsic dependence on momentum transfer Q, falling off rapidly with increasing Q. Therefore, measurements at high-Q values (>5 Å^{-1}) become difficult, and can severely limit the information content of the data, particularly when calculating Fourier transforms (see also our discussion of total scattering techniques from disordered materials in Section 1.4.3).
- For a given element, X-ray form factors are practically insensitive to A, and the scattering is therefore coherent. Neutron scattering can probe both coherent and incoherent (single-particle) correlations, as covered in Section 1.2.5.

- X-ray form factors display a strong energy dependence around absorption edges, described by both real (dispersive) and imaginary (dissipative) components. This behavior is termed *anomalous scattering* and can be exploited to distinguish and isolate specific elements in the sample, akin to isotope substitution and resonant scattering with neutrons—see also below.
- X-rays do not have a magnetic moment and so the interaction with magnetic materials is quite weak. X-ray polarization techniques such as circular dichroism can still be exploited to this end.

An important limitation is that neutrons are generally sensitive to all elements in a sample, compared with element-selective techniques using X-ray absorption edges [19–21]. In X-ray absorption, beam attenuation by the sample is measured as a function of incident energy, or sometimes it is more convenient to probe the resulting (background-free) fluorescence. In materials science, the important energy regions are near the edge (X-ray absorption near-edge spectroscopy, XANES) as well as the energy range above the edge (extended X-ray absorption fine-structure spectroscopy, EXAFS). Since the absorption edge is associated with an electronic transition from a core level to a free state, a detailed energy dependence of the XANES spectrum gives information on the electronic structure of valence and conduction electrons. EXAFS, on the other hand, is essentially a diffraction phenomenon whereby the photoelectron is scattered back from neighboring atoms such that the backscattered wave interferes with that of the primary photoelectron leading to a change in absorption probability. The higher the X-ray energy above the absorption edge, the larger the wave vector k of the photoelectron and, hence, the scattering wave vector $2k$ characterizing the diffraction process. The absorption spectrum as a function of the magnitude of the scattering vector then shows oscillations similar to that of the structure factor $S(Q)$ in X-ray or neutron diffraction experiments (see Section 1.4). EXAFS data therefore provide element-specific structural information by tuning to different absorption edges, yet it can be difficult to gain insight beyond nearest neighbors. Another approach possible with the advent of high-intensity synchrotron sources involves the use of anomalous X-ray scattering techniques. In this case, however, there is a correlation between the energy of the absorption edge and the maximum momentum transfer (Q_{max}), ultimately dictating the spatial resolution of the measurement. For the case of light elements (e.g., those lighter than germanium with a K-edge energy of $E_k = 11.1$ keV, $Q_{max} = 11.25$ Å^{-1}), other methods must be used including EXAFS, neutron diffraction (including isotopic substitution) or a combination of these and possibly other techniques. Other advantages of EXAFS include access to three-body particle correlations [22,23]. Also, measurements are relatively fast and, therefore, structural changes can be followed during rapid variations of temperature, pressure, or other external stimuli.

Element specificity with neutron-scattering techniques is primarily achieved using isotope-substitution techniques, hydrogen being the most notorious case.

Other methods include the use of nuclear resonances leading to anomalous dispersion (see, e.g., Ref. [24]), recoil scattering by epithermal neutrons (introduced in Section 1.4.3.2), and nuclear-spin alignment (see Section 1.7).

1.1.5 Adding Motion: Dynamics and Spectroscopy

In the general case where both momentum $\hbar\mathbf{Q}$ and energy E are exchanged with the sample, the conservation laws for these quantities are still given by Equations (1.5) and (1.6). The energy transferred to (or from) the sample may be taken up (or given off) by a single elementary excitation, for example, a quantum of vibration, in which case the variable E is often replaced by $\hbar\omega$, where ω is the mode frequency (or by $\hbar v$, $\hbar c \tilde{v}$, etc., as illustrated in Table 1.2). In other cases, the scattering event may involve multiple excitations and it is more convenient to refer to the total energy transfer E—note that E and $\hbar\omega$ will be used interchangeably throughout this work. It is worth emphasizing at this stage that the primary objective of inelastic neutron-scattering experiments is to measure scattered intensities as a function of both $\mathbf{Q}$ (a vector quantity) and E. As such, the physics underlying any inelastic-neutron-scattering event can be viewed as a Doppler shift in neutron velocities following their interaction with a moving target atom in the laboratory frame—we recall that the neutron speed in the center-of-mass scattering frame must remain the same before and after the collision so as to conserve linear momentum during scattering. Thus, these processes are not inelastic *per se* as far as the neutron is concerned, as such a possibility would necessarily imply the excitation of its internal degrees of freedom (quarks!).

Denoting as before the scattering angle by ϕ, Equation (1.5) can be rewritten as

$$k_i^2 - 2k_i k_f \cos\phi + k_f^2 = Q^2, \qquad (1.13)$$

Combining this expression with Equation (1.6) and eliminating k_f, we obtain

$$2\left[1 - (1 - E/E_i)^{1/2} \cos\phi\right] - E/E_i = Q^2/k_i^2 \qquad (1.14)$$

which generates a family of curves of E/E_i versus Q/k_i, with ϕ as a parameter. To illustrate these relationships, Figure 1.6 shows curves for fixed incident energies $E_i = 5$, 25, and 100 meV. It can be seen that, for any value of E_i, it is possible to make measurements over a range of values of Q and E by varying the scattering angle ϕ. The accessible ranges for both of these variables increase if E_i is increased. However, the resolution in Q and E is generally compromised as k_i, and thus E_i, are increased. The values of E_i and ϕ should therefore be optimized for the particular experiment at hand.

An inelastic-scattering experiment involves the determination of the magnitudes of $\mathbf{k}_i$ and $\mathbf{k}_f$ and the scattering angle ϕ for each detected neutron. For

Chapter 1 An Introduction to Neutron Scattering

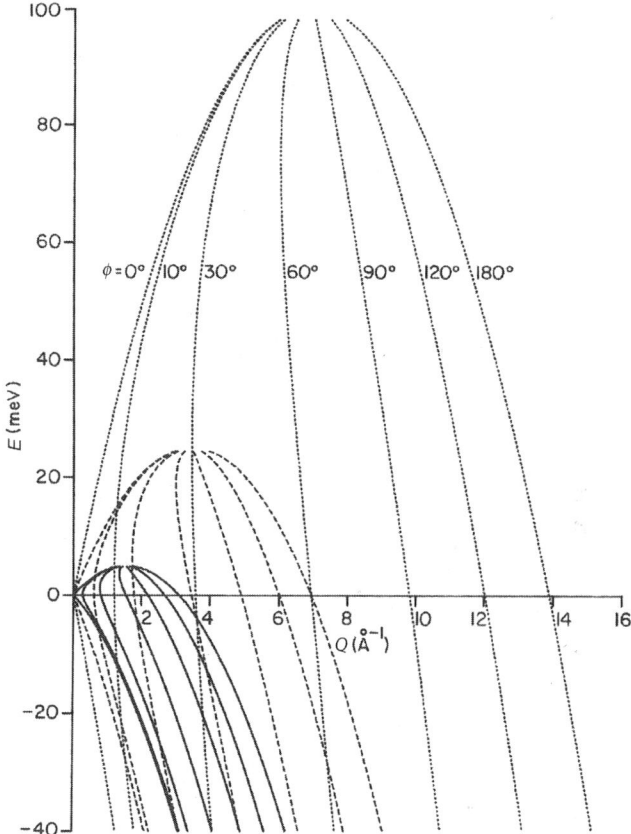

FIGURE 1.6 Relationship between energy transfer E and wave vector Q for three different incident energies $E_i = 5$ meV (solid curves), 25 meV (dashed curves), and 100 meV (dotted curves). For each E_i, loci are shown for different values of the scattering angle ϕ.

single-crystal measurements, the direction relative to the crystal lattice, that is, the direction of Q in the crystal coordinate system, must also be known. As discussed in Chapter 3, the magnitudes of $\mathbf{k}_i$ and $\mathbf{k}_f$ may be determined by measuring either wavelength or velocity, while the directions are determined from the geometry of the experimental arrangement.

Mathematically, inelastic experiments are described by a *double differential cross section* given by

$$\frac{\mathrm{d}^2\sigma}{\mathrm{d}\Omega\mathrm{d}E_\mathrm{f}} = \frac{C}{\Phi N(\Delta\Omega)\eta\Delta E}, \tag{1.15}$$

where C represents the count rate for a given (typically small) interval of energy transfers ΔE and solid angles $\Delta\Omega$. The double differential cross section (derived more rigorously in Section 1.2.1) is a function of the magnitudes and directions of both $\mathbf{k}_i$ and $\mathbf{k}_f$. It is customarily defined in terms of final-state (measurable) variables for the projectile, for example, Ω and E_f, but not E as sometimes written [15]. Integration of this observable over all energy transfers E at a fixed k_i and fixed direction $\hat{\mathbf{k}}_f$, gives the (single) differential cross section introduced in Equation (1.2):

$$\frac{d\sigma}{d\Omega} = \int_{-\infty}^{E_0} \frac{d^2\sigma}{d\Omega dE_f} dE. \qquad (1.16)$$

Thus, $d\sigma/d\Omega$ depends in an integral fashion on the dynamics of the system but does not reveal detailed information about the underlying excitations. Similarly, we may define a total scattering cross section for the system

$$\sigma_t = \int_{4\pi} \frac{d\sigma}{d\Omega} d\Omega, \qquad (1.17)$$

which can be measured in a simple transmission experiment. This observable depends only on $\mathbf{k}_i$, and obviously contains even less information about the structure and dynamics of the sample than $d\sigma/d\Omega$. Nevertheless, the ease with which it can be measured is sometimes a decisive advantage, particularly in technological applications.

Figure 1.7 illustrates the exquisite level of detail currently possible with inelastic-neutron-scattering techniques in the study of low-energy excitations in quantum matter. The data are represented in the form of (Q, E) color maps, making it self-evident that the observed spectral features are dispersive in nature. The ability to explore such low energy transfers with high resolution over wide regions of (Q, E) space remains a distinct advantage of neutron scattering relative to other techniques. The analysis and interpretation of this type of data is the subject of later sections.

1.1.5.1 Comparison with Other Spectroscopic Techniques

The count rate in a typical inelastic neutron-scattering experiment can be estimated along the same lines as in Section 1.1.4.1, except that we must account for a factor of order $\sim \Delta k_f/k_f$ because of the need to determine k_f as well as k_i for each scattering event. This additional requirement will lower the count rate by a factor of $\sim 10^{-2}$ if, for example, an energy-transfer resolution $\Delta E \approx 0.02 E_i$ is required and other experimental parameters remain the same. Also, scans must be made over a range of energy transfers E by either fixing incident or final energies. As a result, sample sizes are typically in the range of grams to tens of grams, and experiments can take from hours to several days to complete. Often, the limiting factor is the time available to an experimenter at a particular spectrometer.

As pointed out above, the energy transfers that can be measured with this technique are often well matched to the characteristic time and length scales of excitations in condensed matter. This very important point is brought to

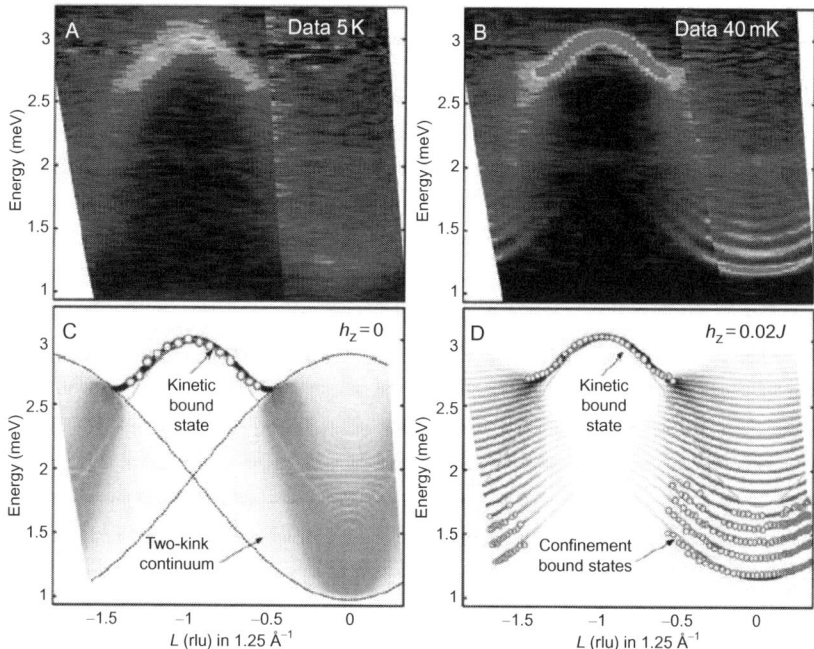

FIGURE 1.7 (a) and (b) Inelastic-neutron-scattering (Q, E) contour maps of spin excitations in $CoNb_2O_6$. (b) A progression of closely spaced Zeeman levels associated with an ordered spin phase in the material. (c) and (d) Model calculations to be compared with (a) and (b), respectively. Q is given in terms of reciprocal lattice units L of $2\pi/c$. *Reprinted with permission from Ref. [25].* (For color version of this figure, the reader is referred to the online version of this chapter.)

the fore in Figure 1.8, showing the wave-vector and energy ranges accessible with different experimental techniques. It is apparent that the range for neutron scattering covers many dynamic phenomena of interest to condensed-matter research. In particular, the important range of wave vectors centered at 1–2 Å^{-1}, corresponding to the Brillouin zone boundary in crystals and the first diffraction peak in glasses and liquids, is spanned by energy transfers extending several orders of magnitude. The extensive coverage of (Q, E) space possible with neutron-scattering techniques constitutes the main driver for the construction and operation of large centralized neutron facilities for condensed-matter research across the globe.

An exhaustive comparison between neutron scattering and the landscape of experimental techniques shown in Figure 1.8 transcends the scope of this introductory chapter, yet a few general remarks are in order. Inelastic X-ray scattering (IXS) can cover wide regions of (Q, E) space, yet it requires energies typically above 10 keV to access an adequate Q range for structural and dynamical studies of both ordered and disordered matter [27,28]. Given these high incident energies, it is also quite challenging to attain an energy

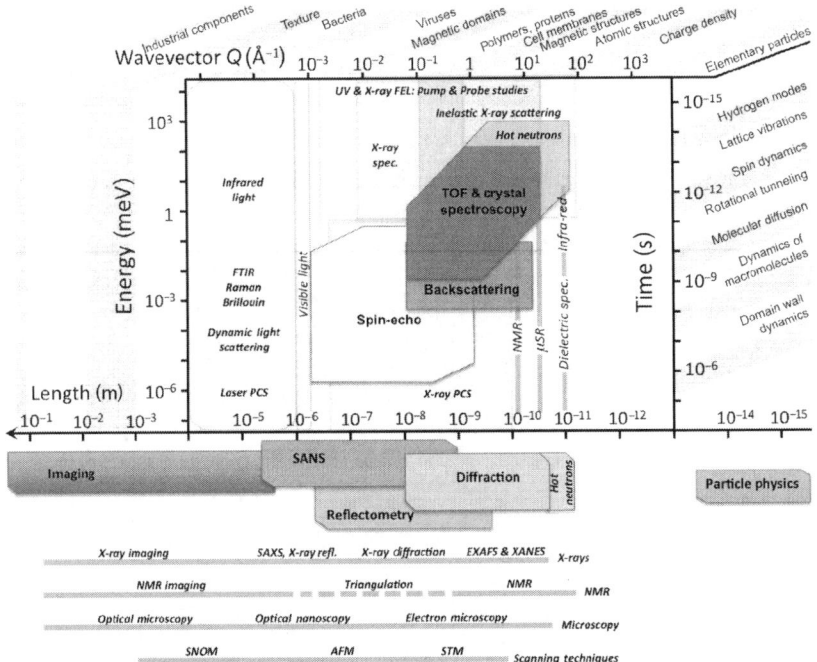

FIGURE 1.8 Regions of (Q, E) space accessible to neutrons and comparison with other experimental probes. For ease of comparison, associated length and time scales are given by the bottom and right axes, respectively. Neutron-scattering techniques include the use of epithermal (hot neutrons), thermal (diffraction as well as time-of-flight and crystal spectroscopy), and cold/ultracold wavelengths (imaging, SANS, reflectometry, backscattering, and spin-echo). Text entries around the main figure illustrate areas of scientific and technological application. *Reprinted with permission from Ref. [26].* (For color version of this figure, the reader is referred to the online version of this chapter.)

resolution comparable to low-energy inelastic neutron scattering. The present limit for IXS is ca. 1 meV [29,30], whereas neutron spectrometers can routinely access energy transfers which are orders of magnitude below this value. A compensating advantage of IXS resides in that the velocity of X-rays is orders of magnitude higher than sound velocities in materials. This feature avoids the kinematic restrictions that make it hard to perform spectroscopic measurements at low Q and high E with inelastic neutron scattering [30]. The resonant variant of IXS (or RIXS) exploits anomalous X-ray scattering and has gained much popularity in the past decade, particularly in the study of excitations in hard-condensed-matter systems such as semiconductors, insulators, transition-metal and rare-earth metals and oxides, or superconductors [31,32]. To gain access to slower (sub-meV) processes with photon-based probes over momentum transfers similar to those accessible to neutrons, X-ray photon correlation spectroscopy (XPCS) stands as the method of choice with the advent of third-generation synchrotrons [33,34]. XPCS measures the speckle pattern produced when coherent light is scattered off the sample and therefore allows a direct measurement of temporal correlations—in a sense, it is the X-ray analog of neutron-spin-echo techniques, discussed in more detail in Chapter 3. More recently, the unprecedented brightness afforded by X-ray free-electron lasers and the implementation of novel techniques such as, for example, ultrafast pump-probe diffraction schemes will certainly provide a plethora of yet-to-be-explored scientific opportunities using temporally and spatially coherent X-ray pulses [35].

Light sources in the visible and infrared regions span a broad energy-transfer range yet at the expense of very small momentum transfers ($Q \sim 10^{-5}$ Å^{-1}). Techniques include photocorrelation laser spectroscopy, dynamic light scattering, or Brillouin spectroscopy using primarily coherent light sources in the visible range [36], as well as the ubiquitous and well-established infrared and Raman spectroscopies to probe nuclear vibrations and magnetic phenomena at higher energy transfers [37,38]. In the far-infrared, the terahertz (THz ~ 4 meV) window is now routinely accessible with optical techniques [39], providing good complementarity with well-established inelastic neutron scattering in the study of low-energy excitations and stochastic phenomena, including those underpinning biological activity in proteins [40] or oligopeptides [41]. Above the visible range of the electromagnetic spectrum, photoelectron spectroscopy and, most notably, its angle-resolved variant ARPES can probe the electronic structure of solids, giving detailed information on band dispersion, the Fermi surface, and the nature of many-body correlations [42]. ARPES also finds a natural overlap with high-energy (eV) studies of strongly correlated electron systems using inelastic neutron scattering [43].

To explore short- and intermediate-range structure and relaxation, nuclear magnetic resonance (NMR) stands out as a mature and versatile tool applicable to a large number of nuclides across the periodic table [44]. It also offers a significant degree of overlap with neutron scattering, particularly after the development of pulsed-field-gradient (PFG-NMR) schemes to probe stochastic relaxation processes with spatial resolution [45,46]. Slow dynamics on surfaces can also be probed with ^{3}He spin-echo spectroscopy [47,48], a technique developed over the past decade to access timescales of picoseconds and length scales of nanometers—a recent review highlighting its overlap with neutron spectroscopy is given in Ref. [49]. Muon-based spectroscopies [50–53] also span a complementary dynamic range in time and length scales to NMR and, in some cases such as open-shell systems like metalloproteins, to electron paramagnetic resonance. The technique is based on the implantation of spin-polarized muons in matter followed by the subsequent detection of the influence of the surrounding medium (be it an atom, molecule, or unpaired electron) on their spin dynamics. Dedicated muon facilities for condensed matter research require the use of a proton accelerator for muon production, and these are often collocated with a spallation neutron source [54–57].

Last but not least, the simplicity and unrivalled dynamic range of dielectric (or impedance) spectroscopy techniques can be used to survey dynamical processes over many orders of magnitude in the frequency domain, from 10^{-4} all the way up to $>10^{+10}$ Hz [58]. Motions that can be followed via the measurement of the frequency-dependent dielectric constant (a complex quantity with both real and imaginary components) include fast rotations in dynamically disordered solids and liquids, all the way to sluggish molecular and supramolecular reorientations in glasses and biological systems. Although the technique has been used extensively for well over a century, a detailed theoretical description of the frequency-dependent dielectric susceptibility over such a wide frequency range remains a theoretical challenge, and closed-form expressions are mostly restricted to idealized cases.

The use of "computational experiments" to predict the structural and dynamical properties of materials overarches the space–time diagram shown in Figure 1.8. The field has experienced an explosive growth [59], from its early days of molecular dynamics and Monte Carlo simulations deeply rooted in the exploration of simple and paradigmatic condensed-matter systems in the 1960s, to the development of first-principles methods (most notoriously those based on density functional theory) [60] and multiscale techniques to probe disparate length and time scales [61]. In this respect, neutron scattering stands out as the experimental tool of choice for a direct comparison with computational and theoretical predictions, a recurring theme throughout this work.

In closing this section, and for reference purposes, Table 1.3 summarizes some of the *pros* and *cons* of neutron scattering in structural and dynamical investigations.

1.2 SCATTERING FOUNDATIONS

We proceed to establish a quantitative basis for the concepts introduced in Section 1.1. For a rigorous development of the subject, interested readers are encouraged to consult the excellent monographs by Squires [62] or Lovesey [63].

1.2.1 The Master Formula and Fermi's Golden Rule

Recalling the scattering experiment in Figure 1.2 and the concept of the differential cross section introduced in Section 1.1.2, we now allow explicitly for a change in state of the target (sample) from a state τ_i to a state τ_f, and also a possible change in neutron spin from σ_i to σ_f as a result of the scattering process. For an ideal detector of unit efficiency, Equation (1.2) can be written as

$$\left(\frac{d\sigma}{d\Omega}\right)_{k_i\sigma_i\tau_i \to k_f\sigma_f\tau_f} = \frac{1}{N\Phi\Delta\Omega} W_{k_i\sigma_i\tau_i \to k_f\sigma_f\tau_f}, \qquad (1.18)$$

where $W_{k_i\sigma_i\tau_i \to k_f\sigma_f\tau_f}$ ($=C/\eta$ in Equation 1.2) is the number of transitions per second from an initial state $k_i\sigma_i\tau_i$ of the combined neutron-target system to

TABLE 1.3 *Pro et Contra* of Neutron-Scattering Techniques for Condensed Matter Research

Pro

1. Neutrons carry no charge and therefore have high penetrating power into bulk samples

2. Thermal neutron wavelengths and energies match interatomic distances and excitation energies in condensed matter

3. Neutron-scattering lengths vary irregularly with Z—generally good for light atoms (especially hydrogen) and for discriminating between nearby elements in the periodic table

4. Scattering lengths vary irregularly with A—isotopic substitution can be used to provide information about a particular element, most notably hydrogen. Resonant and recoil scattering have also been used to achieve element specificity. See Section 1.4.3.2 and Chapters 4–6

5. Scattering intensities can be *quantitatively* related to the structural and dynamical properties of the sample—the main subject of later sections in this chapter

6. Scattering from nuclei can be both incoherent and coherent, thus sensitive to single- and two-particle spatiotemporal correlations in condensed matter. See Section 1.2.5

7. Scattering from unpaired electrons constitutes a unique probe of magnetism. See Sections 1.5 and 1.6, as well as

8. Scattering from nuclei with nonzero spin can be used to probe nuclear magnetism. See Section 1.7

9. Neutron-scattering experiments offer a direct link to the predictions of computational experiments—a recurring theme throughout the present and future volumes in this series

Contra

1. High-intensity neutron beams can only be produced at central facilities, i.e., you cannot (yet!) perform neutron-scattering experiments on your lab bench. See Chapter 2 for the most recent developments

2. Beam intensities may be too low to measure very small specimens, although current neutron instrumentation has improved significantly over the past two decades. See Chapter 3

3. Being neutral, neutrons are difficult to manipulate, e.g., focus into small spot sizes. See Chapter 3

a final state $k_f \sigma_f \tau_f$. The transition rate $W_{\mathbf{k}_i \sigma_i \tau_i \to \mathbf{k}_f \sigma_f \tau_f}$ on the right-hand side of Equation (1.18) may be evaluated within first-order perturbation theory using Fermi's Golden Rule:

$$W_{\mathbf{k}_i \sigma_i \tau_i \to \mathbf{k}_f \sigma_f \tau_f} = \frac{2\pi}{\hbar} |\langle \mathbf{k}_f \sigma_f \tau_f | V | \mathbf{k}_i \sigma_i \tau_i \rangle|^2 \rho_{\mathbf{k}_f \sigma_f}(E_f), \quad (1.19)$$

where V is the interaction potential between the neutron and the target and $\rho_{\mathbf{k}_f\sigma_f}(E_f)$ is the density of final neutron-scattering states per unit energy interval. To arrive at this expression, we note that first-order perturbation theory is valid for nuclear scattering because the neutron–nucleus interaction potential is short range and thermal energies are sufficiently low such that only s-wave (zero orbital angular momentum) scattering applies provided we are far away from resonances. Consequently, the Fermi pseudopotential (cf. Equation 1.27) used in conjunction with the Golden Rule gives the correct result. The magnetic scattering potential is not short range but it is weak and the usual conditions for the validity of first-order perturbation theory still apply. A more detailed discussion is given in Ref. [63].

If we consider the neutron and target to be in a large box of volume V_0, the incident and scattered neutron wave functions are the product of the spatial (plane-wave) and spin functions, that is, $V_0^{-1/2}e^{i\mathbf{k}_i\cdot\mathbf{r}}|\sigma_i\rangle$ and $V_0^{-1/2}e^{i\mathbf{k}_f\cdot\mathbf{r}}|\sigma_f\rangle$, respectively. Then the number of states over a final-energy interval dE_f is

$$\rho_{\mathbf{k}_f\sigma_f}(E_f)dE_f = \frac{V_0}{8\pi^3}d\mathbf{k}_f = \frac{V_0}{8\pi^3}k_f^2 dk_f \Delta\Omega. \tag{1.20}$$

Considering the relationship between kinetic energy and its associated linear momentum $dE_f = \hbar^2 k_f dk_f / m_n$ we can write

$$\rho_{\mathbf{k}_f\sigma_f}(E_f) = \frac{V_0}{8\pi^3}\frac{m_n k_f}{\hbar^2}\Delta\Omega. \tag{1.21}$$

The incident beam flux is the normalized number density times velocity v_i:

$$\Phi = \frac{1}{V_0}v_i = \frac{\hbar k_i}{V_0 m_n} \tag{1.22}$$

Using Equations (1.19)–(1.22) to rewrite Equation (1.18), we find

$$\left(\frac{d\sigma}{d\Omega}\right)_{\mathbf{k}_i\sigma_i\tau_i\to\mathbf{k}_f\sigma_f\tau_f} = \left(\frac{1}{N}\right)\frac{k_f}{k_i}\left(\frac{m_n V_0}{2\pi\hbar^2}\right)^2 |\langle\mathbf{k}_f\sigma_f\tau_f|V|\mathbf{k}_i\sigma_i\tau_i\rangle|^2, \tag{1.23}$$

where the ratio of final to incident neutron wave vectors k ensures flux conservation before and after scattering. Furthermore, energy conservation applies to both neutron and target

$$E = E_i - E_f = E_{\tau_f} - E_{\tau_i}, \tag{1.24}$$

resulting in a delta-function condition that must be fulfilled by the double differential scattering cross section

$$\left(\frac{d^2\sigma}{d\Omega dE_f}\right)_{\mathbf{k}_i\sigma_i\tau_i\to\mathbf{k}_f\sigma_f\tau_f} = \left(\frac{1}{N}\right)\frac{k_f}{k_i}\left(\frac{m_n V_0}{2\pi\hbar^2}\right)^2 |\langle\mathbf{k}_f\sigma_f\tau_f|V|\mathbf{k}_i\sigma_i\tau_i\rangle|^2 \delta(E+E_{\tau_i}-E_{\tau_f}). \tag{1.25}$$

Finally, we sum over all final states of the sample τ_f and final neutron-spin states σ_f as well as average over all initial states of the target occurring with probability p_{τ_i}, and over the initial spin states of the neutron (probability p_{σ_i}) to derive the *double differential cross section*:

$$\left(\frac{d^2\sigma}{d\Omega dE_f}\right)_{\mathbf{k}_i \to \mathbf{k}_f} = \left(\frac{1}{N}\right)\frac{k_f}{k_i}\left(\frac{m_n V_0}{2\pi\hbar^2}\right)^2 \sum_{\tau_i \sigma_i} p_{\tau_i} p_{\sigma_i} \sum_{\tau_f \sigma_f} |\langle \mathbf{k}_f \sigma_f \tau_f | V | \mathbf{k}_i \sigma_i \tau_i \rangle|^2 \delta(E + E_{\tau_i} - E_{\tau_f}). \tag{1.26}$$

Equation (1.26) has been termed the *master formula* [63] and it forms the basis for the interpretation of all neutron-scattering experiments.

1.2.2 Nuclear Scattering

The spatial range of the nuclear potential (fm) is several orders of magnitude smaller than typical thermal neutron wavelengths ($\text{Å} = 10^{+5}$ fm). It is, therefore, appropriate to describe it in terms of the *Fermi pseudopotential* involving a delta function:

$$\frac{2\pi\hbar^2}{m_n} b_{\sigma\tau} \delta(\mathbf{r} - \mathbf{R}), \tag{1.27}$$

where b is the scattering length introduced in Section 1.1.2.1 (also dependent on the spin state of the neutron-nucleus system), and $\mathbf{r}$ and $\mathbf{R}$ represent the instantaneous positions of the neutron and nucleus, respectively. The total interaction potential between the neutron and the sample is then obtained by summing over all atoms in the sample:

$$V(\mathbf{r}) = \frac{2\pi\hbar^2}{m_n} \sum_j b_j \delta(\mathbf{r} - \mathbf{R}_j). \tag{1.28}$$

Averaging Equation (1.28) over incident and final (translational) neutron wave functions we get

$$\langle k_f | V | k_i \rangle = \frac{2\pi\hbar^2}{m_n V_0} \sum_j b_j \int e^{-i\mathbf{k}_f \cdot \mathbf{r}} \delta(\mathbf{r} - \mathbf{R}_j) e^{i\mathbf{k}_i \cdot \mathbf{r}} d\mathbf{r} = \frac{2\pi\hbar^2}{m_n} \sum_j b_j e^{i\mathbf{Q} \cdot \mathbf{R}_j}, \tag{1.29}$$

where $\mathbf{Q} = \mathbf{k}_i - \mathbf{k}_f$ as in Section 1.1.3. The corresponding master formula (cf. Equation 1.26) then reads

$$\frac{d^2\sigma}{d\Omega dE_f} = \left(\frac{1}{N}\right)\frac{k_f}{k_i}\sum_{\tau_i \sigma_i} p_{\tau_i} p_{\sigma_i} \left| \sum_{\tau_f \sigma_f} \sum_j b_j \langle \sigma_f \tau_f | e^{i\mathbf{Q} \cdot \mathbf{R}_j} | \sigma_i \tau_i \rangle \right|^2 \delta(E + E_{\tau_i} - E_{\tau_f}), \tag{1.30}$$

where, for convenience, the subscript $\mathbf{k}_i \to \mathbf{k}_f$ has been omitted on the left-hand side.

The majority of applications of neutron–nuclear scattering use unpolarized beams where both neutron-spin states are equally probable for both incident and scattered beams. This situation requires summation over σ_i and σ_f to give

$$\frac{d^2\sigma}{d\Omega dE_f} = \frac{1}{N}\frac{k_f}{k_i}\sum_{\tau_i} p_{\tau_i} |\sum_{\tau_f}\sum_j b_j \langle \tau_f | e^{i\mathbf{Q}\cdot\mathbf{R}_j}|\tau_i\rangle|^2 \delta(E + E_{\tau_i} - E_{\tau_f}). \tag{1.31}$$

We shall return to Equation (1.30) when dealing with polarized neutrons for studies of electronic (Section 1.5.2) and nuclear magnetism (Section 1.7).

1.2.3 The Double Differential Cross Section in the Time Domain

Equation (1.31) gives the *double differential cross section* in terms of matrix elements of the operators $e^{i\mathbf{Q}\cdot\mathbf{R}_j}$ between initial and final target states τ_i and τ_f. Although this observable is not a time-dependent quantity as it describes an asymptotic scattering experiment, its evaluation becomes more transparent if we consider the explicit time evolution of particle positions $\mathbf{R}_j$. To achieve this, we replace the delta function in Equation (1.31) by an integral over time using its well-known integral representation:

$$\delta(E + E_{\tau_i} - E_{\tau_f}) = \frac{1}{2\pi\hbar}\int_{-\infty}^{\infty} \exp\left[-i\frac{t}{\hbar}(E + E_{\tau_i} - E_{\tau_f})\right] dt, \tag{1.32}$$

so that Equation (1.31) becomes

$$\frac{d^2\sigma}{d\Omega dE_f} = \frac{1}{N}\left(\frac{k_f}{k_i}\right)\frac{1}{2\pi\hbar}\sum_{\tau_i} p_{\tau_i}\sum_{\tau_f}\sum_{jj'} b_j^* b_{j'} \int_{-\infty}^{\infty} \langle \tau_i | e^{i\mathbf{Q}\cdot\mathbf{R}_j}|\tau_f\rangle$$
$$\langle \tau_f | e^{iHt/\hbar} e^{i\mathbf{Q}\cdot\mathbf{R}_{j'}} e^{-iHt/\hbar}|\tau_i\rangle e^{-iEt/\hbar} dt, \tag{1.33}$$

where H denotes a Hamiltonian (time evolution) operator such that

$$e^{iHt/\hbar}|\tau_i\rangle = e^{iE_i t/\hbar}|\tau_i\rangle, \tag{1.34}$$

in addition to an analogous expression for the final target states τ_f. From the closure relation $\sum_{\tau_f}|\tau_f\rangle\langle\tau_f| = 1$, the expression above simplifies to:

$$\frac{d^2\sigma}{d\Omega dE_f} = \frac{1}{N}\left(\frac{k_f}{k_i}\right)\frac{1}{2\pi\hbar}\int_{-\infty}^{\infty}\sum_{jj'} e^{-iEt/\hbar}\left\langle b_j^* b_{j'} e^{-i\mathbf{Q}\cdot\mathbf{R}_j} e^{iHt/\hbar} e^{i\mathbf{Q}\cdot\mathbf{R}_{j'}} e^{-iHt}/\hbar\right\rangle_{\tau_i} dt, \tag{1.35}$$

where the angular brackets represent a thermal average over the initial states τ_i. Noting that the term in brackets contains time-propagation terms of the form,

$$e^{i\mathbf{Q}\cdot\mathbf{R}_j(t)} \equiv e^{iHt/\hbar}e^{i\mathbf{Q}\cdot\mathbf{R}_j}e^{-iHt/\hbar} \qquad (1.36)$$

we finally get

$$\frac{d^2\sigma}{d\Omega dE_f} = \frac{1}{N}\left(\frac{k_f}{k_i}\right)\frac{1}{2\pi\hbar}\int_{-\infty}^{\infty}\sum_{jj'}\left\langle b_j^* b_{j'} e^{-i\mathbf{Q}\cdot\mathbf{R}_j(0)}e^{i\mathbf{Q}\cdot\mathbf{R}_{j'}(t)}\right\rangle e^{-iEt/\hbar}dt, \qquad (1.37)$$

which is a real-time representation of the double differential cross section in terms of thermal averages of a product of two operators, one involving a particle j at time 0, and the other a particle j' at time t. We are justified in putting the origin at time $t=0$ in Equation (1.37) since the time (Heisenberg) propagator depends only on time (or energy) differences between initial and final states. In a classical picture, $\mathbf{R}_j(t)$ can be equated to the actual (instantaneous) position of a given particle at a given time, averaged over configuration space. In other words, Equation (1.37) tells us that the double differential cross section (what we measure) contains information on the spatial (R) and temporal (t) evolution of the target system, weighted by pair products of nuclear-scattering lengths b. In the language of classical mechanics, such a result implies that Equation (1.37) may be used to calculate neutron-scattering cross sections from a history of the temporal and spatial development of the system, as obtained, for example, from a computer simulation. Much of the progress of neutron science over the past few decades has relied on such a synergy between experimental and computational techniques.

1.2.4 Farewell to Nuclear Physics

Despite the insights brought forward by Equation (1.37), it still represents a sum of averages for each pair of atoms where the nuclear states enter explicitly through their respective scattering lengths b. These quantities depend on the isotope and spin state of the jth nucleus, as presented earlier in Section 1.1.2.1 and, for all practical purposes, they are deeply rooted in nuclear physics, not condensed matter.

In cases where the atomic state is independent of nuclear variables (i.e., when nuclear spins and isotopes for a given element are randomly distributed), it is possible to represent the scattering process in terms of a function only dependent on atomic coordinates $\mathbf{R}_j(t)$. In most cases (bar a few exceptions like molecular hydrogen), the energy differences associated with different nuclear-spin orientations are $\sim 10^{-4}$ K per atom and, therefore, these spins are effectively random for temperatures above 1 mK—nuclear-spin correlations at ultralow temperatures is a special subject introduced in Section 1.7. Moreover, fractional energy differences associated with different isotopic masses are $\sim \Delta A/A$ and so the isotope distribution is generally random except for the lightest atoms at low temperatures, for example ^{3}He/^{4}He mixtures, which again must be treated as a special case. In general, therefore, we may average over initial nuclear-spin and isotope distributions in Equation (1.37) and write

$$\frac{d^2\sigma}{d\Omega dE_f} = \frac{1}{N}\left(\frac{k_f}{k_i}\right)\frac{1}{2\pi\hbar}\int_{-\infty}^{\infty}\sum_{dd'}\sum_{j\in d, j'\in d'}\overline{b_j^* b_{j'}}\left\langle e^{-i\mathbf{Q}\cdot\mathbf{R}_j(0)}e^{i\mathbf{Q}\cdot\mathbf{R}_{j'}(t)}\right\rangle e^{-iEt/\hbar}dt, \quad (1.38)$$

where d and d' refer to different elements and the bar represents an average over spin and isotope distributions for the corresponding pair of elements. In the case of isotope substitution experiments, we note that the index d may also refer to a particular isotope of an element, for example, if some hydrogen atoms in the sample are systematically exchanged for deuterium and the others left as hydrogen. For this reason, we will refer to "atom type" rather than "element" in the following sections.

1.2.5 Coherent and Incoherent Scattering

For conciseness, we define

$$S_{jj'}(\mathbf{Q}, E) = \frac{1}{2\pi\hbar}\int_{-\infty}^{\infty}\left\langle e^{-i\mathbf{Q}\cdot\mathbf{R}_j(0)}e^{i\mathbf{Q}\cdot\mathbf{R}_{j'}(t)}\right\rangle e^{-iEt/\hbar}dt, \quad (1.39)$$

and rewrite the double differential cross section in Equation (1.38) as

$$\frac{d^2\sigma}{d\Omega dE_f} = \frac{1}{N}\left(\frac{k_f}{k_i}\right)\sum_{dd'}\sum_{j\in d, j'\in d'}\overline{b_j^* b_{j'}} S_{jj'}(\mathbf{Q}, E). \quad (1.40)$$

If the nuclear spins and isotopes are uncorrelated, then

$$\overline{b_j^* b_{j'}} = \begin{cases} \overline{b_d^*}\,\overline{b_{d'}}, & j\neq j', \\ \overline{\left|b_d^2\right|}, & j = j', \end{cases} \quad (1.41)$$

or more succinctly using tensor notation

$$\overline{b_j^* b_{j'}} = \overline{b_d^*}\,\overline{b_{d'}} + \left(\overline{\left|b_d^2\right|} - \left|\overline{b_d}\right|^2\right)\delta_{jj'}\delta_{dd'}. \quad (1.42)$$

Substituting into Equation (1.40) we obtain

$$\frac{d^2\sigma}{d\Omega dE_f} = \frac{1}{N}\left(\frac{k_f}{k_i}\right)\sum_{dd'}\sum_{j\in d, j'\in d'}\left[\overline{b_d^*}\,\overline{b_{d'}} + \left(\overline{\left|b_d^2\right|} - \left|\overline{b_d}\right|^2\right)\delta_{jj'}\delta_{dd'}\right]S_{jj'}(\mathbf{Q}, E)$$

$$= \frac{1}{N}\left(\frac{k_f}{k_i}\right)\sum_{dd'}\overline{b_d^*}\,\overline{b_{d'}}\sum_{j\in d, j'\in d'}S_{jj'}(\mathbf{Q}, E) + \frac{1}{N}\left(\frac{k_f}{k_i}\right)\sum_d\left(\overline{\left|b_d^2\right|} - \left|\overline{b_d}\right|^2\right)\sum_{j\in d}S_{jj}(\mathbf{Q}, E)$$

$$(1.43)$$

The first term represents a sum over all possible pairs of atoms (j, j'), each term representing a two-body correlation between the position of atom

j at time 0 and that of atom j' at time t; it therefore contains the interference terms in the total scattering and gives rise to *coherent scattering*. Coherent scattering is therefore sensitive to the relative arrangement of atoms (structure).

The second term represents a sum over all atoms j one at a time, each term representing a correlation between the position of atom j at time 0 and that of the same atom at time t, that is, *incoherent scattering*, sensitive to single-particle spatiotemporal correlations. If for a particular atom type d there is only one possible scattering length b_d for all possible nuclear-spin states, then $\left|\overline{b_d^2}\right| = \left|\overline{b_d}\right|^2$ and there is no incoherent scattering from that atom type. Incoherent scattering arises from the random distribution of scattering lengths about the mean. Important examples that conform to this rule include all nuclides with zero nuclear spin such as ^{12}C and ^{16}O. We also note that these specific cases (must!) also coincide with NMR-inactive nuclides, all of which must also be pure coherent neutron scatterers.

Since isotopic compositions are generally uniform, coherent, incoherent, and total scattering cross sections may be regarded as universal properties of an atom type. These are defined as (see also the Appendix)

$$\sigma_c^d = 4\pi \left|\overline{b_d}\right|^2 \text{ (coherent cross section)},$$
$$\sigma_i^d = 4\pi \left[\left|\overline{b_d^2}\right| - \left|\overline{b_d}\right|^2\right] \text{ (incoherent cross section)}, \quad (1.44)$$
$$\sigma_t^d = 4\pi \left[\left|\overline{b_d^2}\right| - \left|\overline{b_d}\right|^2\right] \text{ (total scattering cross section)}.$$

To avoid confusion, we note that σ_i above no longer relates to the initial spin-state of the neutron. With these definitions, Equation (1.43) is then written

$$\frac{d^2\sigma}{d\Omega dE_f} = \frac{1}{N}\left(\frac{k_f}{k_i}\right)\sum_{dd'}\overline{b_d^*}\overline{b_{d'}}\sum_{j\in d, j'\in d'} S_{jj'}(\mathbf{Q}, E) + \frac{1}{N}\left(\frac{k_f}{k_i}\right)\sum_d \frac{\sigma_i^d}{4\pi}\sum_{j\in d} S_{jj}(\mathbf{Q}, E),$$

(1.45)

where for a single atom type $\overline{b_d^*}\overline{b_{d'}}$ may be replaced by $\sigma_c^d/4\pi$.

For a particular atom type that consists of different isotopes a with abundance c_a and nuclear spin I_a, the cross sections can be calculated form

$$\sigma_c = 4\pi\left\{\sum_a \frac{c_a}{2I_a+1}\left[(I_a+1)b_a^+ + I_a b_a^-\right]\right\}^2,$$
$$\sigma_s = 4\pi\sum_a \frac{c_a}{2I_a+1}\left[(I_a+1)\left(b_a^+\right)^2 + I_a\left(b_a^-\right)^2\right], \quad (1.46)$$
$$\sigma_i = \sigma_s - \sigma_c.$$

Values of σ_c and σ_i for most nuclides are listed in the Appendix. For non-standard compositions as those used for isotopic-substitution experiments, the appropriate cross sections can be calculated if the values for the individual isotopes are known.

1.2.6 Scattering Functions

To represent the sums over the $S_{jj'}(\mathbf{Q}, E)$ in Equation (1.45), we define the *coherent dynamic structure factor* or *scattering function*

$$S_c^{dd'}(\mathbf{Q}, E) = \frac{1}{(N_d N_{d'})^{1/2}} \sum_{j \in d, j' \in d'} S_{jj'}(\mathbf{Q}, E)$$
$$= \frac{1}{(N_d N_{d'})^{1/2}} \left(\frac{1}{2\pi\hbar}\right) \sum_{j \in d, j' \in d'} \int_{-\infty}^{\infty} \left\langle e^{-i\mathbf{Q}\cdot\mathbf{R}_j(0)} e^{i\mathbf{Q}\cdot\mathbf{R}_{j'}(t)} \right\rangle e^{-iEt/\hbar} dt,$$
(1.47)

and the *incoherent dynamic structure factor* or *scattering function*

$$S_i^d(\mathbf{Q}, E) = \frac{1}{N_d} \sum_{i \in d} S_{ii}(\mathbf{Q}, E) = \frac{1}{N_d} \left(\frac{1}{2\pi\hbar}\right) \sum_{j \in d} \int_{-\infty}^{\infty} \left\langle e^{-i\mathbf{Q}\cdot\mathbf{R}_j(0)} e^{i\mathbf{Q}\cdot\mathbf{R}_j(t)} \right\rangle e^{-iEt/\hbar} dt.$$
(1.48)

In terms of these functions, the double differential cross section has the form

$$\frac{d^2\sigma}{d\Omega dE_f} = \frac{k_f}{k_i} \left(\sum_{dd'} \left[c_d^{1/2} c_{d'}^{1/2} \bar{b}_d^* \bar{b}_{d'} \right] S_c^{dd'}(\mathbf{Q}, E) + \sum_{d} \left[c_d \frac{\sigma_i^d}{4\pi} \right] S_{dd}(\mathbf{Q}, E) \right), \quad (1.49)$$

where $c_d = N_d/N$ is the concentration of atom type d.

The time-dependent functions under the integral signs in Equations (1.47) and (1.48) are known as the *intermediate scattering function*

$$I^{dd'}(\mathbf{Q}, t) = \frac{1}{(N_d N_{d'})^{1/2}} \sum_{j \in d, j' \in d'} \left\langle e^{-i\mathbf{Q}\cdot\mathbf{R}_j(0)} e^{i\mathbf{Q}\cdot\mathbf{R}_{j'}(t)} \right\rangle, \quad (1.50)$$

and the *self-intermediate scattering function*

$$I_s^d(\mathbf{Q}, t) = \frac{1}{N_d} \sum_{j \in d} \left\langle e^{-i\mathbf{Q}\cdot\mathbf{R}_j(0)} e^{i\mathbf{Q}\cdot\mathbf{R}_j(t)} \right\rangle. \quad (1.51)$$

The intermediate scattering functions are dimensionless, while the scattering functions themselves are of dimension E^{-1}, as defined via Equations (1.38) and (1.40). As shown explicitly by Equation (1.49), scattering functions are obtained via the measurement of double differential cross sections. Intermediate scattering functions are directly accessible with neutron-spin-echo techniques, first introduced by Mezei in the early 1970s [64] and described in more detail in Chapter 3.

1.3 CANONICAL SOLIDS

The formalism developed in Section 1.2 finds its most natural and immediate application in the study of the solid state. At this point, it would be customary to consider the neutron-scattering response of a high-symmetry single crystal with just one atom per unit cell (Bravais lattice), as treated in depth in well-established monographs [62,63]. Notwithstanding the merits of this approach, much of the current interest focuses on disordered and amorphous systems, thus we choose to start out with a consideration of a general solid without the assumption of long-range order or periodicity. To this end, a solid is defined, operationally speaking, as a system in which each atom has a well-defined (and fixed) equilibrium position over the duration of the measurement. This somewhat restricted definition excludes quantum solids such as the condensed phases of helium, where atoms undergo quantum-mechanical exchange and are inherently delocalized in space. In this case, special methods are required to deal with a truly many-body problem, as described by Glyde [65]. Nor does it include solids in which atoms or ions undergo translational diffusion. The treatment of these cases requires considering the conceptual framework underpinning the scattering response of liquids, as introduced later on in Section 1.4.

1.3.1 Normal Modes of Vibration

According to our operational definition of a solid, each atom can be uniquely associated with a specific equilibrium site at position $\mathbf{j}$. The position of each atom around this site can be expressed in terms of the instantaneous displacement from equilibrium:

$$\mathbf{R}_j(t) = \mathbf{j} + \mathbf{u}_j(t), \, j = 1, \ldots, n. \tag{1.52}$$

Using Equations (1.47) and (1.48), the corresponding coherent and incoherent scattering functions considering only one type of atom in the lattice become

$$S_c(\mathbf{Q}, E) = \frac{1}{N}\left(\frac{1}{2\pi\hbar}\right) \sum_{jj'} \int_{-\infty}^{\infty} \left\langle e^{-i\mathbf{Q}\cdot\mathbf{u}_{j'}(0)} e^{i\mathbf{Q}\cdot\mathbf{u}_j(t)} \right\rangle e^{i\mathbf{Q}\cdot(\mathbf{j}'-\mathbf{j})} e^{-iEt/\hbar} dt \tag{1.53}$$

and

$$S_i(\mathbf{Q}, E) = \frac{1}{N}\left(\frac{1}{2\pi\hbar}\right) \sum_{j} \int_{-\infty}^{\infty} \left\langle e^{-i\mathbf{Q}\cdot\mathbf{u}_j(0)} e^{i\mathbf{Q}\cdot\mathbf{u}_j(t)} \right\rangle e^{-iEt/\hbar} dt. \tag{1.54}$$

To proceed, we assume that the forces in the solid are *harmonic*, that is, the force that acts to return each atom to its equilibrium position is a linear function of the displacement. In this case, the displacements $\mathbf{u}_j$ can be expressed as a superposition of *3N* normal modes [66],

$$\mathbf{u}_j(t) = \sum_k \left(\frac{\hbar}{2M_k\omega_k}\right)^{1/2} \left[\mathbf{e}_j^k e^{-\omega_k t} a_k + \mathbf{e}_j^{k*} e^{\omega_k t} a_k^+\right], \quad (1.55)$$

where ω_k is the frequency of the kth mode and $\mathbf{e}_j^k$ is the mode polarization vector for the jth atom in this mode; the $\mathbf{e}_j^k$ s are conventionally defined to be an orthonormal vector set. The frequencies and polarization vectors are eigenvalues and eigenvectors of the dynamical matrix $\mathbf{D}$, respectively, which is given by

$$\mathbf{D}_{jj'} = \frac{1}{(M_j M_{j'})^{1/2}} f_{jj'}, \quad (1.56)$$

where $f_{jj'}$ are force constants representing the coefficients of the displacements in the expansion of the potential energy of the solid, that is,

$$U - U_0 = \frac{1}{2}\mathbf{u}_j f_{jj'}\mathbf{u}_{j'}. \quad (1.57)$$

Thus, the dynamical matrix and hence the frequencies and polarization vectors can be calculated from the interatomic potential of the solid; a_j and a_j^+ are quantum-mechanical operators expressing, respectively, the annihilation and creation of one quantum of vibrational energy.

1.3.2 Scattering Under the Harmonic Approximation

To evaluate the thermal averages in Equations (1.53) and (1.54), we make recourse to the main results of the quantum-mechanical theory of a simple harmonic oscillator [62,67,68]. In particular, we recall that

$$\left\langle e^{-i\mathbf{Q}\cdot\mathbf{u}_j(0)} e^{i\mathbf{Q}\cdot\mathbf{u}_{j'}(t)} \right\rangle = e^{-[W_j(\mathbf{Q})+W_{j'}(\mathbf{Q})]} e^{\langle \mathbf{Q}\cdot\mathbf{u}_j(0)\mathbf{Q}\cdot\mathbf{u}_{j'}(t)\rangle}, \quad (1.58)$$

where the first (time-independent) factor is a function of $\mathbf{Q}$, usually referred to as the *Debye–Waller factor*:

$$e^{-W_j(\mathbf{Q})} = e^{-1/2\left\langle[\mathbf{Q}\cdot\mathbf{u}_j(0)]^2\right\rangle}. \quad (1.59)$$

For small (harmonic) displacements, the second factor in Equation (1.58) can be expanded to read [69,70]

$$e^{\langle \mathbf{Q}\cdot\mathbf{u}_j(0)\mathbf{Q}\cdot\mathbf{u}_{j'}(t)\rangle} = 1 + \langle \mathbf{Q}\cdot\mathbf{u}_j(0)\mathbf{Q}\cdot\mathbf{u}_{j'}(t)\rangle + \frac{1}{2!}\langle \mathbf{Q}\cdot\mathbf{u}_j(0)\mathbf{Q}\cdot\mathbf{u}_{j'}(t)\rangle^2 + \cdots. \quad (1.60)$$

The Debye–Waller factor acts as a multiplying factor for each term in the expansion of the scattering cross section. Using the normal-mode expansion of Equation (1.55), it can be shown that

$$W_j(\mathbf{Q}) = \frac{\hbar}{3M_j} \sum_k \frac{\left|\mathbf{Q}\cdot\mathbf{e}_j^k\right|^2}{\omega_k} \langle 2n_k + 1 \rangle, \quad (1.61)$$

where $\langle n_k \rangle$ is the Bose–Einstein population number for the kth mode, given by

$$\langle n_k \rangle = \frac{1}{e^{\hbar \omega_k / k_B T} - 1}. \tag{1.62}$$

The reader is referred to the work of Sears and Shelley [71] describing a simple and insightful model for the calculation of Debye–Waller factors. For isotropic systems (e.g., polycrystalline samples), integration over all orientations leads to a Q-dependent Debye–Waller factor of the form

$$2W(Q) = \frac{1}{3} Q^2 \langle u^2 \rangle, \tag{1.63}$$

where $\langle u^2 \rangle$ is the mean-square displacement. Equation (1.63) is used extensively in the study of disordered materials including glasses, polymers, and biological systems. It is also a quantity readily available from computer simulations using molecular dynamics methodologies. For example, in the particular case of incoherent (proton-dominated) neutron scattering from proteins, it has been used to quantify protein flexibility and how it may relate to biological function and activity [72].

1.3.3 Purely Elastic Events

Neglecting all time-dependent terms in the expansion shown by Equation (1.60), the coherent scattering function (Equation 1.53) becomes

$$\begin{aligned} S_{c,el}(\mathbf{Q}, E) &= \frac{1}{N} \left(\frac{1}{2\pi\hbar} \right) \sum_{jj'} \int_{-\infty}^{+\infty} e^{-[W_j(\mathbf{Q}) + W_{j'}(\mathbf{Q})]} e^{i\mathbf{Q} \cdot (\mathbf{j}' - \mathbf{j})} e^{-iEt/\hbar} dt \\ &= \left[\frac{1}{N} \sum_{jj'} e^{-[W_j(\mathbf{Q}) + W_{j'}(\mathbf{Q})]} e^{i\mathbf{Q} \cdot (\mathbf{j}' - \mathbf{j})} \right] \delta(E), \end{aligned} \tag{1.64}$$

where we have exploited the integral representation of the Dirac delta-function introduced in Equation (1.32). The presence of $\delta(E)$ implies purely elastic scattering and justifies the intuitive introduction of this concept given in Section 1.1.3. We note that this delta function in energy is the result of the fact that the argument in the correlation function of Equation (1.58) remains finite at infinite times, as implied by the concept of equilibrium sites (i.e., the existence of a well-defined time-averaged structure).

Integration of Equation (1.64) over energy gives the *elastic structure factor*

$$S_{el}(\mathbf{Q}) = \frac{1}{N} \sum_{jj'} e^{-[W_j(\mathbf{Q}) + W_{j'}(\mathbf{Q})]} e^{i\mathbf{Q} \cdot (\mathbf{j}' - \mathbf{j})}. \tag{1.65}$$

An analogous procedure using Equation (1.54) leads to similar expressions for the incoherent scattering function:

$$S_{i,el}(\mathbf{Q}, E) = \left[\frac{1}{N}\sum_j e^{-2W_j(\mathbf{Q})}\right]\delta(E), \quad (1.66)$$

and integration over energy gives

$$S_{i,el}(\mathbf{Q}) = \int S_{i,el}(\mathbf{Q}, E)dE = \frac{1}{N}\sum_j e^{-2W_j(\mathbf{Q})}, \quad (1.67)$$

which contains the average Debye–Waller factor for all atoms in the target system.

For a crystalline solid with long-range order, equilibrium lattice sites **j** may be described by the following vector relation

$$\mathbf{j} = \mathbf{l} + \mathbf{d}, \quad (1.68)$$

where **l** refers to the center of the unit cell and **d** to the relative position of a lattice site within the cell. Allowing for the presence of atom type d at a given position **d** in the unit cell, the set of vectors **l** makes up a crystal lattice with perfect long-range order. In this case the elastic structure factor becomes

$$S(\mathbf{Q})_{el}^{dd'} = \frac{(2\pi)^3}{v_0} e^{-[W_d(\mathbf{Q})+W_{d'}(\mathbf{Q})]} e^{i\mathbf{Q}\cdot(\mathbf{d'}-\mathbf{d})} \sum_{\boldsymbol{\tau}} \delta(\mathbf{Q}-\boldsymbol{\tau}), \quad (1.69)$$

where the vectors $\boldsymbol{\tau}$ are the reciprocal lattice vectors of the crystal and v_0 is the unit cell volume. Equation (1.69) renders itself to a simple physical interpretation whereby a series of (infinitely sharp) peaks appear every time **Q** is equal to a reciprocal lattice vector $\boldsymbol{\tau}$ of the crystal. These are the *Bragg peaks*, and Equation (1.69) is the exact formulation of the Bragg condition introduced in Equation (1.11).

From Equations (1.69) and (1.49) we may derive the *elastic coherent scattering cross section*

$$\left(\frac{d\sigma}{d\Omega}\right)_{c,el} = \frac{(2\pi)^3}{v_0} \sum_{\boldsymbol{\tau}} |F(\boldsymbol{\tau})|^2 \delta(\mathbf{Q}-\boldsymbol{\tau}), \quad (1.70)$$

where

$$F(\boldsymbol{\tau}) = \sum_{\mathbf{d}} \bar{b}_{\mathbf{d}} e^{-W_{\mathbf{d}}(\boldsymbol{\tau})} e^{i\boldsymbol{\tau}\cdot\mathbf{d}} \quad (1.71)$$

is the *unit-cell structure factor* for the crystal. Equations (1.70) and (1.71) form the basis of neutron crystallography, covered in Chapter 4.

Similarly, the incoherent elastic differential cross section is given by

$$\left(\frac{d\sigma}{d\Omega}\right)_{i,el} = \sum_{\mathbf{d}} c_{\mathbf{d}} \sigma_i^{\mathbf{d}} e^{-2W_{\mathbf{d}}(\mathbf{Q})}. \quad (1.72)$$

In the interpretation of neutron-diffraction experiments from crystalline materials, it is crucial to note that Equation (1.68) assumes a single value of **d** relative to the center of the unit cell. If (as in most cases!), there is a variation in the values of **d** through the crystal, translational invariance is no longer guaranteed and one needs to reconsider the general expression given by Equation (1.65). An immediate consequence of this loss of translational invariance is the emergence of scattering intensity (so-called *elastic diffuse scattering*) at all values of **Q**, not just at the reciprocal lattice points as implied by Equation (1.69). Section 1.4.4 discusses diffuse scattering in more depth. For systems with no long-range order, like glasses and other amorphous solids, all the elastic scattering is diffuse and one must (in principle) know all the time-averaged atomic positions in order to calculate $S_{el}(\mathbf{Q})$. This task can be efficiently achieved via the calculation of the differential cross section from computer models of the material under investigation [73,74].

1.3.4 Inelastic (One-Phonon) Scattering

Returning to Equation (1.53), straight substitution of the second term in the expansion given by Equation (1.60) gives

$$S_{c,1}(Q,E) = \frac{1}{N}\left(\frac{1}{2\pi\hbar}\right)\sum_{jj'}\int_{-\infty}^{+\infty} e^{-[W_j(\mathbf{Q})+W_{j'}(\mathbf{Q})]}\langle\mathbf{Q}\cdot u_j(0)\mathbf{Q}\cdot u_{j'}(t)\rangle e^{i\mathbf{Q}\cdot(\mathbf{j'}-\mathbf{j})}e^{-iEt/\hbar}dt \quad (1.73)$$

for the coherent scattering function, hereafter $S_{c,1}(\mathbf{Q},E)$. Dimensionally speaking, we note that this second-order term (describing harmonic vibrations) is quadratic in Q. Using the definition of normal modes given by Equation (1.55), we can rewrite it as

$$S_{c,1}(\mathbf{Q},E) = \frac{1}{N}\left(\frac{1}{4\pi}\right)\sum_{jj'}\int_{-\infty}^{+\infty} e^{-[W_j(\mathbf{Q})+W_{j'}(\mathbf{Q})]}\sum_k \frac{\left(\mathbf{Q}\cdot\mathbf{e}_j^k\right)^*\left(\mathbf{Q}\cdot\mathbf{e}_{j'}^k\right)}{(M_j M_{j'})^{1/2}\omega_k}\times \quad (1.74)$$
$$\times\left[e^{-i\omega_k t}\langle n_k+1\rangle + e^{i\omega_k t}\langle n_k\rangle\right]e^{i\mathbf{Q}\cdot(\mathbf{j'}-\mathbf{j})}e^{-iEt/\hbar}dt.$$

We express once more the time integral as a delta function in the energy domain to write this expression as the sum of two time-independent terms $S_{c,\pm 1}$ given by

$$S_{c,+1}(\mathbf{Q},E) = \frac{1}{2N}\sum_{jj'} e^{-[W_j(\mathbf{Q})+W_{j'}(\mathbf{Q})]}e^{i\mathbf{Q}\cdot(\mathbf{j'}-\mathbf{j})}\times$$
$$\times\sum_k \frac{\hbar^2\left(\mathbf{Q}\cdot\mathbf{e}_j^k\right)^*\left(\mathbf{Q}\cdot\mathbf{e}_{j'}^k\right)}{(M_j M_{j'})^{1/2}E_k}\langle n_k+1\rangle\delta(E-E_k) \quad (1.75)$$

and

$$S_{c,-1}(\mathbf{Q},E) = \frac{1}{2N}\sum_{jj'} e^{-[W_j(\mathbf{Q})+W_{j'}(\mathbf{Q})]} e^{i\mathbf{Q}\cdot(\mathbf{j}'-\mathbf{j})} \times$$
$$\times \sum_k \frac{\hbar^2 \left(\mathbf{Q}\cdot\mathbf{e}_j^k\right)^* \left(\mathbf{Q}\cdot\mathbf{e}_{j'}^k\right)}{(M_j M_{j'})^{1/2} E_k} \langle n_k \rangle \delta(E+E_k) \quad (1.76)$$

where $E_k = \hbar\omega_k$ corresponds to normal-mode energy eigenvalues. These two scattering functions predict peaks in the scattering function whenever E corresponds to $\pm E_k$, that is, to the creation (+1) or annihilation (−1) of a phonon from the scattering event, leading to the loss or gain of energy by the scattered neutron, respectively.

The corresponding *incoherent scattering function* also contains one-phonon emission and absorption peaks and it has a sensibly simpler functional form:

$$S_{i,+1}(\mathbf{Q},E) = \frac{1}{2N}\sum_j e^{-W_j(\mathbf{Q})} \sum_k \frac{\hbar^2 \left|\mathbf{Q}\cdot\mathbf{e}_j^k\right|^2}{M_j E_k} \langle n_k+1 \rangle \delta(E-E_k) \quad (1.77)$$

and

$$S_{i,-1}(\mathbf{Q},E) = \frac{1}{2N}\sum_j e^{-W_j(\mathbf{Q})} \sum_k \frac{\hbar^2 \left|\mathbf{Q}\cdot\mathbf{e}_j^k\right|^2}{M_j E_k} \langle n_k \rangle \delta(E+E_k). \quad (1.78)$$

If the solid has long-range order, $\mathbf{j}=\mathbf{l}+\mathbf{d}$ as before, the normal modes can be described as plane waves characterized by a reduced wave vector $\mathbf{q}$ and polarization vectors given by

$$\mathbf{e}_\mathbf{d}^k = \frac{1}{N_c^{1/2}} \sum_\mathbf{q} \mathbf{e}_\mathbf{d}^k(\mathbf{q}) e^{i\mathbf{q}\cdot\mathbf{l}}, \quad (1.79)$$

where N_c is the number of unit cells in the crystal. Then, for example, the one-phonon contribution to the coherent scattering function (cf. Equation 1.47) is given by

$$S_{c,+1}^{\mathbf{dd'}}(\mathbf{Q},E) = \frac{1}{2N_c}(2\pi)^3 e^{-[W_\mathbf{d}(\mathbf{Q})+W_{\mathbf{d'}}(\mathbf{Q})]} \sum_{\mathbf{q}k} \frac{\hbar^2 \left(\mathbf{Q}\cdot\mathbf{e}_\mathbf{d}^k(\mathbf{q})\right)^* \left(\mathbf{Q}\cdot\mathbf{e}_{\mathbf{d'}}^k(\mathbf{q})\right)}{(M_\mathbf{d} M_{\mathbf{d'}})^{1/2} E_k(\mathbf{q})} \times$$
$$\times e^{i\mathbf{Q}\cdot(\mathbf{d'}-\mathbf{d})} \langle n_{\mathbf{q}k}+1 \rangle \sum_\mathbf{\tau} \delta(\mathbf{Q}-\mathbf{q}-\mathbf{\tau})\delta(E-E_k(\mathbf{q})) \quad (1.80)$$

and the corresponding double differential cross section

$$\left(\frac{d^2\sigma}{d\Omega dE_f}\right)_{c,+1} = \frac{k_f}{k_i}\left[\frac{(2\pi)^3}{2N_c v_0}\right] \sum_\mathbf{\tau} \sum_{\mathbf{q}k} |F_1(\mathbf{Q},\mathbf{q}k)|^2 \frac{\hbar^2 \langle n_{\mathbf{q}k}+1 \rangle}{E_k(\mathbf{q})}$$
$$\delta(\mathbf{Q}-\mathbf{q}-\mathbf{\tau})\delta(E-E_k(\mathbf{q})), \quad (1.81)$$

where it is implied that the double differential cross section is both a function of **Q** and E. In analogy with Equation (1.71), we have also defined a one-phonon structure factor

$$F_1(\mathbf{Q},\mathbf{q}k) = \sum_\mathbf{d} \frac{\overline{b}_\mathbf{d}}{M_\mathbf{d}^{1/2}} e^{-W_\mathbf{d}(\mathbf{Q})} e^{i\mathbf{Q}\cdot\mathbf{d}} \left[\mathbf{Q}\cdot\mathbf{e}_\mathbf{d}^k(\mathbf{q})\right]. \quad (1.82)$$

The expression for one-phonon annihilation (neutron-energy gain) is analogous to Equation (1.81) with $\langle n_k \rangle$ instead of $\langle n_k+1 \rangle$ and the signs of **q** and E_k reversed. The delta-functions in Equation (1.81) imply that, for a single crystal, one-phonon scattering is only possible for values of **Q** and E satisfying

$$\mathbf{Q} = \boldsymbol{\tau} \pm \mathbf{q}, \ E = E_k(\mathbf{q}), \quad (1.83)$$

for some set of values of **q** and k. These restrictions lead to well-defined spectral features in the inelastic scattering observed from such crystals. Measurements of this kind are illustrated in Figure 1.9 and can be used to derive the *phonon dispersion relations* for a crystalline lattice, $E_k(\mathbf{q})$. In comparison with optical probes like infrared and Raman spectroscopy, neutron scattering can explore wide regions of the Brillouin zone, thus providing stringent tests to theoretical predictions.

Deviations in the structure from a given equilibrium position will give rise to diffuse scattering as the selection rule $\mathbf{Q} = \boldsymbol{\tau} \pm \mathbf{q}$ no longer applies. For a solid without long-range order, all the inelastic scattering is diffuse and we need to revert to the general expressions in Equations (1.75) and (1.76). It

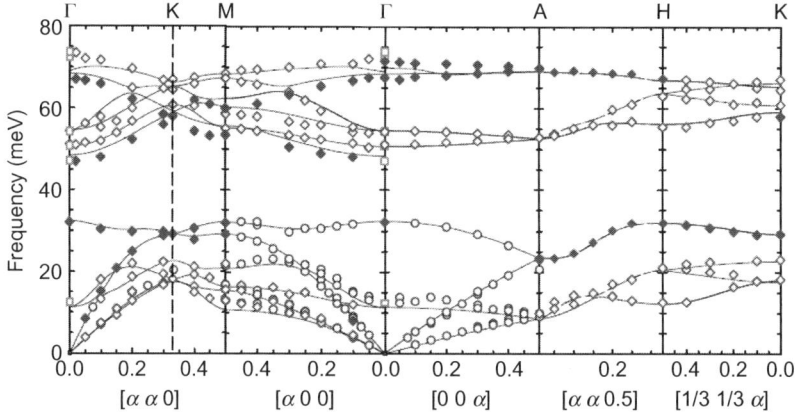

FIGURE 1.9 Phonon-dispersion relations for zinc oxide along the main symmetry directions. Diamonds and circles (red and blue) correspond to inelastic neutron-scattering data. Raman data at the Γ point ($Q=0$) are depicted by squares (green). *Ab initio* calculations are shown by the solid curves. *Reprinted with permission from Ref. [75].* (For interpretation of the references to color in this figure legend, the reader is referred to the online version of this chapter.)

should be noted, however, that as long as the harmonic approximation applies, the phonon expansion is still valid, but the normal modes are no longer described by plane waves with a well-defined value of **q**.

Inelastic incoherent scattering for a crystal is given by

$$\left(\frac{d^2\sigma}{d\Omega dE_f}\right)_{i,+1} = \frac{k_f}{k_i}\left(\frac{1}{2N}\right)\sum_\mathbf{d}\frac{1}{M_\mathbf{d}}\frac{\sigma_i^\mathbf{d}}{4\pi}e^{-2W_\mathbf{d}(\mathbf{Q})}\sum_{qk}\frac{\hbar^2\left|\mathbf{Q}\cdot\mathbf{e}_\mathbf{d}^k(\mathbf{q})\right|^2}{E_k(\mathbf{q})}\langle n_{\mathbf{q}k}+1\rangle\delta(E-E_k(\mathbf{q})), \tag{1.84}$$

with an analogous expression for phonon annihilation. Equation (1.84) represents a continuous function of **Q** and E since only one selection rule applies, namely

$$E = \pm E_k(\mathbf{q}). \tag{1.85}$$

We can therefore express the cross section in terms of a *one-phonon vibrational density of states* $Z(E)$, where $Z(E)dE$ is the fraction of normal modes with energies in the range $(E, E+dE)$. For neutron-energy loss, this becomes

$$\left(\frac{d^2\sigma}{d\Omega dE_f}\right)_{i,+1} = \frac{k_f}{k_i}\sum_\mathbf{d}\frac{3\sigma_i^\mathbf{d}}{24\pi}e^{-2W_\mathbf{d}(\mathbf{Q})}\frac{\hbar^2\overline{\left|\mathbf{Q}\cdot\mathbf{e}_\mathbf{d}^k(\mathbf{q})\right|^2}}{M_\mathbf{d}E}\langle n+1\rangle Z(E), \tag{1.86}$$

where the bar represents an average over all the modes at frequency E and

$$\langle n+1\rangle = \frac{1}{2}\left[1+\coth\left(\frac{E}{2k_\mathrm{B}T}\right)\right]. \tag{1.87}$$

For a one-component system of mass M in which the anisotropy in the Debye–Waller factor can be neglected, Equation (1.86) takes a simpler form

$$\left(\frac{d^2\sigma}{d\Omega dE_f}\right)_{i,+1} = \frac{k_f}{k_i}\left(\frac{\hbar^2 Q^2}{2M}\right)e^{-2W(Q)}\frac{\langle n+1\rangle}{E}Z(E), \tag{1.88}$$

so that the vibrational density of states $Z(E)$ can be deduced directly from the incoherent one-phonon scattering. This result holds for a one-component disordered system as well. For multicomponent systems, it is no longer the case since the scattering also depends on the polarization vectors of the different atom types through the factor $|\mathbf{Q}\cdot\mathbf{e}_\mathbf{d}^k(\mathbf{q})|^2$, which in general will be frequency-dependent. A more detailed discussion of vibrational densities of states and how these can be obtained from neutron-scattering experiments is given by Taraskin and Elliott [76]. Equation (1.88) also constitutes the starting point for the analysis of inelastic neutron-scattering experiments on proton-containing materials where incoherent scattering from hydrogen dominates the spectral response, as detailed in Ref. [77]. Readers interested in the use of inelastic neutron-scattering techniques for the study of hydrogen-containing materials are also referred to Ref. [78] for an extensive compilation of publicly available data.

1.3.5 Multiphonon Scattering

Higher-order terms in Equation (1.60) formally represent scattering processes in which two or more phonons are involved. The n-phonon contribution scales as $|\mathbf{Q} \cdot \mathbf{u}|^{2n}$, and for values of $\mathbf{Q}$ commonly encountered in thermal neutron scattering, it generally appears as a relatively small and smooth background under the one-phonon peaks; it tends to a continuous function for coherent as well as for incoherent scattering, particularly at high momentum transfers.

Using Equations (1.53), (1.54), and (1.58), calculations of the inelastic cross section can (in principle) be carried out to all orders in terms of multi-phonon contributions provided the phonon-dispersion relations are known. For the coherent cross section, this is a tedious calculation and it is usual to assume that multiphonon terms are equal to the incoherent cross section (*incoherent approximation*). For crystals with only one type of atom, an expression similar to Equation (1.88) can be used to give the exponent in Equations (1.53) and (1.54) as a weighted integral of the one-phonon density of states; alternatively, approximations such as that of Sjölander [63,79] can also be used.

In the language of the spectroscopist, multiphonon scattering corresponds to the appearance of overtone and combination bands associated with the simultaneous excitation of more than one quantum of vibration. These effects are best illustrated by considering the inelastic neutron scattering response from molecular systems where intermolecular (lattice) and intramolecular vibrational modes are typically well separated from each other in energy transfer. The color contour plot in Figure 1.10 shows inelastic neutron-scattering data for Rb_2PtH_6, a metal-hydride complex with an unusually high hydrogen content. The scattering is dominated by the strong incoherent response from hydrogen and the observed (nondispersive) bands at well-defined energy transfers correspond to either fundamental or overtone/combination (multiphonon) transitions from the ground vibrational state. The intensity of a given band as a function of momentum transfer (i.e., the *inelastic form factor*) scales as $Q^{2n} e^{-2W(Q)}$ with $n=1$ for a fundamental transition (cf. Equation 1.88), $n=2$ for a first overtone or binary combination, $n=3$ for a third overtone or ternary combination, etc. The specific value of n for a given transition can then be obtained directly from the experimental data by locating the position of $Q_{\max}$ corresponding to the maximum in the inelastic form factor, as within the harmonic approximation we can always write $n = Q_{\max}^2 \langle u^2 \rangle / 3$ for a polycrystalline system (cf. Equation 1.63). These simple considerations provide a unique method of discriminating between fundamentals and higher-order vibrational transitions using inelastic neutron

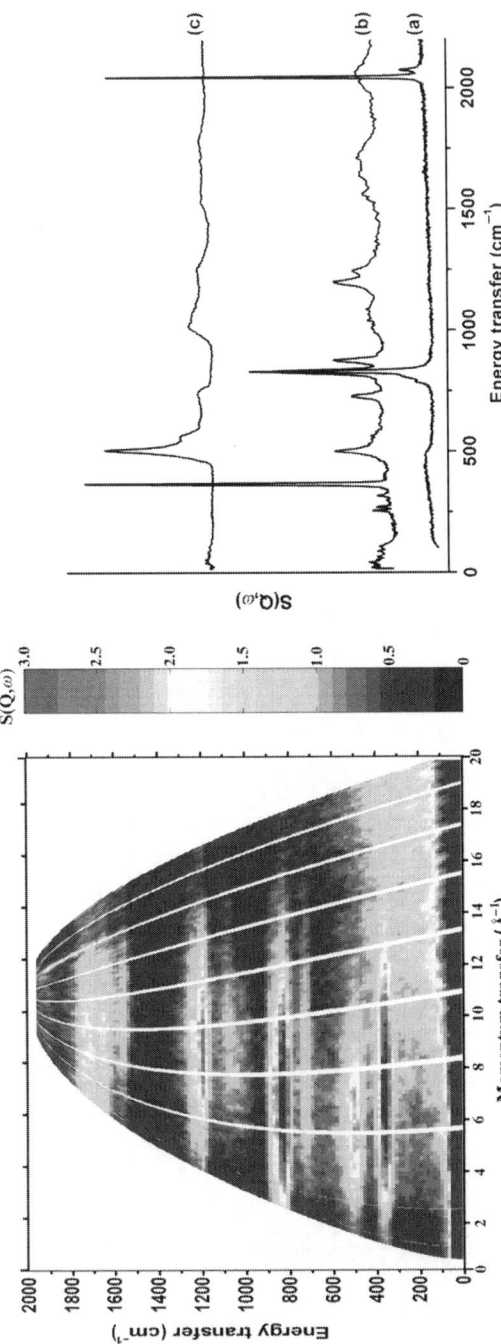

FIGURE 1.10 The contour color plot on the left shows inelastic neutron-scattering data for the metal-hydride complex Rb_2PtH_6, using an incident neutron energy of $2000\ cm^{-1}$ (~250 meV). The figure on the right compares vibrational spectra using (a) Raman scattering and (b) inelastic neutron scattering along the proton-recoil line for the same compound. The top trace (c) corresponds to the inelastic neutron-scattering response from the reference material RbH, also present as a reaction by-product in (b). The broad energy features in the Raman data below ca. $500\ cm^{-1}$ arise from fluorescence backgrounds. *Reprinted with permission from Ref. [80].* (For color version of this figure, the reader is referred to the online version of this chapter.)

scattering. The relative simplicity of these data on the (Q, E) plane arises from the predominance of incoherent scattering from hydrogen. In this situation, it is possible to focus data collection preferentially along the proton-recoil line so as to obtain a hydrogen-projected vibrational density of states (see spectra on the right-hand side of Figure 1.10). This approach provides a powerful means of boosting count rates by orders of magnitude with a much-improved spectral resolution, comparable to infrared spectroscopy and Raman scattering. It is also noteworthy that the neutron data contains far more spectral features than the corresponding optical spectrum, as the former is not subjected to hard selection rules and it can be directly related to the vibrational density of states. This feature of inelastic neutron scattering makes comparison with theoretical predictions relatively straightforward.

When $\mathbf{Q}$ becomes large, the expansion shown by Equation (1.60) is no longer meaningful, and a more useful limiting case is the *impulse approximation* whereby the target atom is assumed to recoil as an effectively free particle. Andreani *et al.* have discussed the transition between these two limits for the case of molecular hydrogen [81]. We will also return to this case in Section 1.4.3 in connection with the spectral moments of the scattering function.

1.3.6 Beyond Harmonic Vibrations

At higher temperatures (or even at the lowest temperatures for quantum solids), atomic displacements can be such a large fraction of the interatomic distance that the harmonic approximation is no longer valid (for reviews of anharmonic effects in crystals see, e.g., Refs. [82–84]). This departure from harmonic behavior has two primary consequences. First, higher-order terms appear in the expansion of $\left\langle e^{-i\mathbf{Q}\cdot\mathbf{u}_i(0)} e^{i\mathbf{Q}\cdot\mathbf{u}_{i'}(t)} \right\rangle$ in Equation (1.58), as recently revisited in the context of the analysis of diffraction data [85]. Numerical estimates [86] have shown that these effects are generally small in normal crystals at high temperatures. In quantum crystals, they appear to be more pronounced and lead to interference terms between one- and multiphonon scattering [65,87]. The second consequence of anharmonicity involves corrections to the thermal averages given in Section 1.3.4. One immediate effect is that the delta functions in Equations (1.75) and (1.76) are replaced by spectral distributions with temperature-dependent energies and widths that arise from phonon–phonon scattering. A formal separation into elastic, one-phonon, and multiphonon processes is however still valid. Direct experimental access to these (typically tiny) spectral shifts and line widths (excitation lifetimes) of phonons [88], magnetic excitations [89], and molecular rotations [90] has been made possible via the use of inelastic-spin-echo neutron scattering [91]. These data can now be compared with first-principles predictions, particularly those within the framework of density functional theory [92,93].

1.4 BEYOND CANONICAL SOLIDS

1.4.1 Space–Time (Van Hove) Correlation Functions

For systems exhibiting particle diffusion within experimental timescales (e.g., gases, liquids, and some solids like plastic crystals and battery materials), the operational definition of a solid introduced in Section 1.3 based on the concept of well-defined equilibrium sites is no longer applicable. We must then return to the general expressions for the scattering functions given in Section 1.2.6 and, in particular, to the definition of the intermediate scattering functions $I(\mathbf{Q}, t)$ and $I_s(\mathbf{Q}, t)$ introduced in Equations (1.50) and (1.51) at the end of that section. These two functions are intimately related to space–time correlation functions $G(\mathbf{r},t)$, first introduced by Van Hove [94], and often named after him:

$$G^{dd'}(\mathbf{r}, t) = \frac{1}{(2\pi)^3} \int I^{dd'}(\mathbf{Q}, t) e^{-i\mathbf{Q}\cdot\mathbf{r}} d\mathbf{Q}, \tag{1.89}$$

$$G_s^d(\mathbf{r}, t) = \frac{1}{(2\pi)^3} \int I_s^d(\mathbf{Q}, t) e^{-i\mathbf{Q}\cdot\mathbf{r}} d\mathbf{Q}. \tag{1.90}$$

Using again the integral representation of the delta function for spatial variables, and the definitions of $I(\mathbf{Q}, t)$ and $I_s(\mathbf{Q}, t)$ (Equations 1.50 and 1.51), these spatial Fourier transforms of the intermediate scattering functions are equal to

$$G^{dd'}(\mathbf{r}, t) = \frac{1}{(N_d N_{d'})^{1/2}} \sum_{\substack{j \in d \\ j' \in d'}} \int \langle \delta[\mathbf{r}' - \mathbf{R}_j(0)] \delta[\mathbf{r}' + \mathbf{r} - \mathbf{R}_{j'}(t)] \rangle d\mathbf{r}', \tag{1.91}$$

$$G_s^d(\mathbf{r}, t) = \frac{1}{N_d} \sum_{j \in d} \int \langle \delta[\mathbf{r}' - \mathbf{R}_j(0)] \delta[\mathbf{r}' + \mathbf{r} - \mathbf{R}_j(t)] \rangle d\mathbf{r}'. \tag{1.92}$$

The physical meaning of Equation (1.91) becomes more transparent when written in terms of particle-density operators $\rho(\mathbf{r},t)$ of the form

$$\rho_d(\mathbf{r}, t) = \sum_{j \in d} \delta[\mathbf{r} - \mathbf{R}_j(t)]. \tag{1.93}$$

so that the total scattering function reads

$$G^{dd'}(\mathbf{r}, t) = \frac{1}{(N_d N_{d'})^{1/2}} \int \langle \rho_d(\mathbf{r}', t) \rho_{d'}(\mathbf{r}' + \mathbf{r}, t) \rangle d\mathbf{r}'. \tag{1.94}$$

Similarly, we introduce particle-density operators in momentum space

$$\rho_d(\mathbf{Q}, t) = \int \rho_d(\mathbf{r}, t) e^{i\mathbf{Q}\cdot\mathbf{r}} d\mathbf{r} = \sum_{j \in d} e^{i\mathbf{Q}\cdot\mathbf{R}_j(t)}, \tag{1.95}$$

so that the intermediate scattering function becomes

$$I^{dd'}(\mathbf{Q},t) = \frac{1}{(N_d N_{d'})^{1/2}} \langle \rho_d(-\mathbf{Q},0)\rho_{d'}(\mathbf{Q},t) \rangle. \tag{1.96}$$

From the above, it becomes clear that $G_{dd'}(\mathbf{r}, t)$ and $I_{dd'}(\mathbf{Q}, t)$ represent spatiotemporal particle-density correlations in the system. As such, they embody all the information accessible via inelastic neutron-scattering experiments from atomic nuclei.

1.4.2 Pair Distribution Functions

Using the property that at $t=0$ the operators $\mathbf{R}_j(t)$ and $\mathbf{R}_{j'}(t)$ commute for a pair of distinguishable particles, Equation (1.91) simplifies to

$$G^{dd'}(\mathbf{r},0) = \left(\frac{N_d}{N_{d'}}\right) \sum_{\substack{j' \in d' \\ (0 \in d)}} \langle \delta[\mathbf{r}' - \mathbf{R}_0(0) + \mathbf{R}_{j'}(0)] \rangle, \tag{1.97}$$

where we neglect surface effects and assume that all d atoms are equivalent to the one at the origin. *Pair-density functions* are defined such that

$$\rho^{dd'}(\mathbf{r}) = \left(\frac{N_d}{N_{d'}}\right) \sum_{\substack{j' \in d' \\ (0 \in d)}} \langle \delta[\mathbf{r}' - \mathbf{R}_0(0) + \mathbf{R}_{j'}(0)] \rangle, \tag{1.98}$$

which give the average instantaneous density of particles of type d' with respect to one atom of type d sitting at the origin. $G(\mathbf{r}, 0)$ then becomes

$$G^{dd'}(\mathbf{r},0) = \delta_{dd'}\delta(\mathbf{r}) + \left(\frac{N_d}{N_{d'}}\right)\rho^{dd'}(\mathbf{r}) \tag{1.99}$$

and

$$G_s^d(\mathbf{r},0) = \delta(\mathbf{r}). \tag{1.100}$$

If we now define structure factors representing the energy integrals of the dynamic structure factor $S(\mathbf{Q}, E)$:

$$S^{dd'}(\mathbf{Q}) = \int_{-\infty}^{\infty} S_c^{dd'}(\mathbf{Q},E) dE = I^{dd'}(\mathbf{Q},0), \tag{1.101}$$

then Equations (1.89) and (1.90) lead to

$$\int S^{dd'}(\mathbf{Q})e^{-i\mathbf{Q}\cdot\mathbf{r}}d\mathbf{Q} = \int I^{dd'}(\mathbf{Q},0)e^{-i\mathbf{Q}\cdot\mathbf{r}}d\mathbf{Q}$$
$$= (2\pi)^3\left[\delta_{dd'}\delta(\mathbf{r}) + \left(\frac{N_d}{N_{d'}}\right)^{1/2}\rho^{dd'}(\mathbf{r})\right] \quad (1.102)$$

or equivalently

$$\frac{1}{(2\pi)^3}\int\left[S^{dd'}(\mathbf{Q}) - \delta_{dd'}\right]e^{-i\mathbf{Q}\cdot\mathbf{r}}d\mathbf{Q} = \left(\frac{N_d}{N_{d'}}\right)^{1/2}\rho^{dd'}(\mathbf{r}). \quad (1.103)$$

As shown by Equation (1.103), the pair-density function and its associated structure factor constitute the most fundamental quantities in a detailed analysis of both ordered and disordered solid-state structure using pair-distribution-function (PDF) methods [95]. Systems amenable to PDF analysis fall in the category of "canonical solids" as discussed in Section 1.3.

For incoherent scattering, we note from Equation (1.51) that

$$\int_{-\infty}^{\infty} S_i^d(\mathbf{Q},E)dE = I_s^d(\mathbf{Q},0) = 1. \quad (1.104)$$

The structure factors $S^{dd'}(\mathbf{Q})$ are measured in *a total scattering* experiment in which the energy-integrated scattering at constant $\mathbf{Q}$ is determined according to

$$\left.\frac{d\sigma}{d\Omega}\right|_\mathbf{Q} = \int_{-\infty}^{E_i}\frac{d^2\sigma}{d\Omega dE_f}dE_f \approx \sum_{dd'}c_d^{1/2}c_{d'}^{1/2}\overline{b_d^*b_{d'}}S^{dd'}(\mathbf{Q}) + \sum_d c_d\frac{\sigma_t^d}{4\pi}$$
$$= \sum_{dd'}c_d^{1/2}c_{d'}^{1/2}\overline{b_d^*b_{d'}}\left[S^{dd'}(\mathbf{Q}) - \delta_{dd'}\right] + \sum_d c_d\frac{\sigma^d}{4\pi}, \quad (1.105)$$

where the second term incorporates the total scattering cross section σ^d and the first term represents the "distinct" scattering excluding the self term. Equation (1.105), relating the differential cross section $d\sigma/d\Omega$ to the partial structure factors $[S^{dd'}(\mathbf{Q}) - \delta_{dd'}]$, and Equation (1.103), relating the latter to the pair-density functions, provide the basis for the measurement of the instantaneous structure of dense fluids with neutron diffraction. Of course, scattering measurements observe an ensemble and time average of such instantaneous structures.

A series of challenges arise in such experiments:

(1) The condition of integration over energy at constant $\mathbf{Q}$ is not easy to achieve. In a typical experimental arrangement as shown in Figure 1.2, measurements at constant scattering angle ϕ and energy transfer E involve a curved locus in $(\mathbf{Q}, E)$ (cf. Figure 1.6). Conversion of these data

to "constant-**Q**" cuts might not be a trivial task. Also, Equation (1.105) is not exact unless E_i is replaced by ∞ in the upper limit of the integral and a (k_f/k_i) factor is included in the integrand. The so-called static approximation neglects these limitations (see Section 1.4.4).

(2) The incoherent scattering cross sections σ_i^d must be known to sufficient accuracy, as well as their associated scattering lengths $\bar{b}_d$.

(3) A single measurement for a multicomponent system gives a linear combination of the various partial structure factors $S^{dd'}$, and multiple measurements with different contrasts (i.e., values of $\bar{b}_d$), for example, by varying isotopic compositions, are required to separate these out.

In spite of these hurdles, neutron diffraction on liquids and dense gases has become a mature tool for the study of an increasing number of complex disordered materials, including liquids and glasses. A classic result, for liquid argon, is shown in Figure 1.11. We may also measure the structure factors for glasses, subject to the same limitations. Note that these measurements give information about the instantaneous positions of the atoms in the glass as defined by Equation (1.98), compared with measurements of the purely elastic scattering (Equation 1.65), which only yield information about the equilibrium sites. The differences between the two depend on the details of the atomic vibrations. For fluids, there is no purely elastic scattering and the functions $S_c^{dd'}(\mathbf{Q},E)$ and $S_i^d(\mathbf{Q},E)$ continuously approach well-defined limits as $E \to 0$. To illustrate their typical behavior, these two scattering functions are shown in Figure 1.12 for liquid argon. For $E=0$, $S_i^d(\mathbf{Q},E)$ peaks at $Q=0$

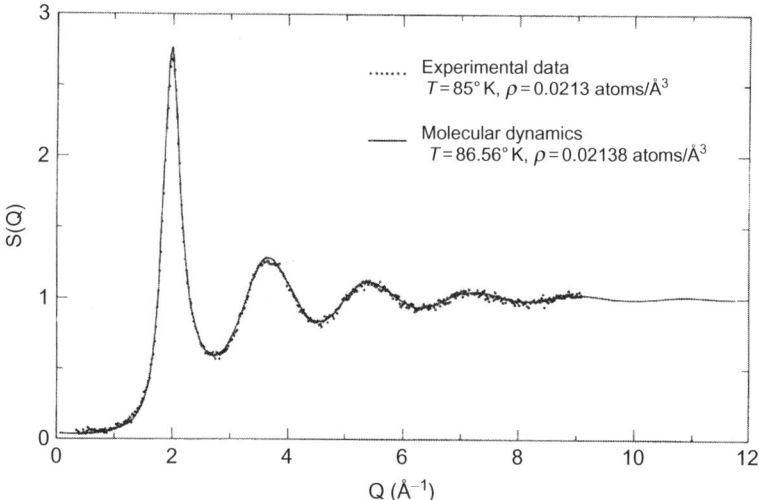

FIGURE 1.11 Experimental (black circles) structure factor for liquid ^{36}Ar at 85 K, and comparison with the predictions of molecular dynamics simulations (solid line) for a Lennard–Jones fluid. *Reprinted with permission from Ref. [96]*.

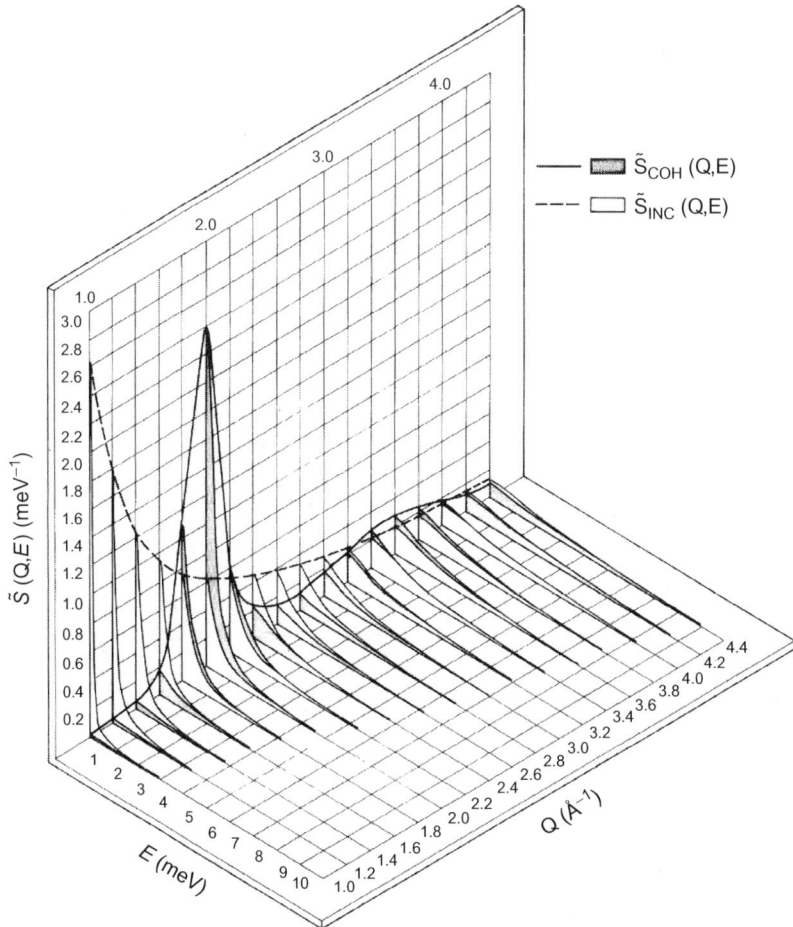

FIGURE 1.12 Coherent (solid line) and incoherent (dashed line) scattering functions for liquid argon. *Reprinted with permission from Ref. [97].*

and it decreases monotonically with Q. This behavior is offset by a concomitant increase in spectral widths, such that the total intensity remains constant when integrated over all energies (cf. Equation 1.112). We shall see in Section 1.4.5.1 that the Q dependence of the energy widths provides a direct measure of the self-diffusion coefficient at low momentum transfers. The coherent scattering function $S_c^{dd'}(\mathbf{Q}, E)$, on the other hand, displays a strong maximum at $Q \sim 2\ \text{Å}^{-1}$, followed by lower-amplitude oscillations at higher momentum transfers. This behavior mimics the one shown earlier in Figure 1.11 for the same system. These intensity oscillations are related to characteristic interatomic distances in the liquid, as implied by the Fourier relationship given by Equation (1.103). The attentive reader will also note that the spectral width around the first maximum in $S_c^{dd'}(\mathbf{Q},E)$ is narrower than adjacent energy spectra shown at lower and higher momentum transfers. This behavior is termed *de Gennes narrowing*, and necessarily arises from the physical properties of the coherent dynamic structure factor, as we shall see below. Intuitively, it might be rationalized as the arrest of atomic motions in the vicinity of characteristic distances associated with the solvation cage around a given atom. For further details, we strongly encourage a careful reading of the seminal (and very elegant) work of de Gennes, applicable to both nuclear [98] and magnetic [99] neutron scattering in liquids and paramagnets, respectively. A quantitative interpretation of $S_c^{dd'}(\mathbf{Q},E)$ is not as straightforward as that given above for $S_i^d(\mathbf{Q},E)$ and is deferred to our discussion of stochastic processes in Section 1.4.5.

In closing this discussion, we refer the reader to Ref. [100] for an online collection of examples of neutron-derived structure factors of disordered materials (both glasses and liquids) using total-scattering methods.

1.4.3 Properties of the Dynamic Structure Factor

The dynamic structure factors $S_c(\mathbf{Q}, E)$ and $S_i(\mathbf{Q}, E)$, and by implication the intermediate and space–time correlation functions, must satisfy certain general relations regardless of the details of the system under consideration. These relations are important for several reasons as they provide:

(1) A consistency check on experiment data.
(2) A necessary constraint on theoretical models.
(3) An indication as to how a classically based calculation, for example, using computer simulation, can be related to the experimentally observed neutron scattering where quantum effects might be at play.

Although these relations are valid for all types of systems, it is in the field of neutron scattering from liquids that they are most often invoked and it is therefore appropriate to introduce them here.

(a) *S(**Q**,E) is real (i.e., an experimental observable)*

$$S_c^{dd'}(\mathbf{Q},E) = S_c^{dd'}{}^*(\mathbf{Q},E), \tag{1.106}$$

$$S_i^d(\mathbf{Q},E) = S_i^d{}^*(\mathbf{Q},E). \tag{1.107}$$

These relations follow from the requirement that arbitrary linear combinations of the partial scattering functions with real coefficients lead to cross sections that must be real and positive; from the same reasoning, S_c^{dd} and S_i^d must be everywhere positive, but not necessarily $S_c^{dd'}$ when $d \neq d'$.

(b) *S(**Q**,E) must satisfy detailed balance*

$$S_c^{dd'}(\mathbf{Q}, E) = e^{E/k_B T} S_c^{dd'}(-\mathbf{Q}, -E), \qquad (1.108)$$

$$S_i^d(\mathbf{Q}, E) = e^{E/k_B T} S_i^d(-\mathbf{Q}, -E). \qquad (1.109)$$

Here we allow for the fact that the scattering functions may not be symmetric with respect to time reversal, for example in the presence of external (e.g., magnetic) fields. The condition of detailed balance governs the ratio of the scattering probabilities for two transitions proceeding in opposite directions. The matrix element is the same in both directions, but the probability of the sample being initially in the lower state is a factor $e^{E/k_B T}$ higher than that of it being initially in the higher state. Thus, the scattering is more intense when the neutron loses energy E than when it gains it.

The detailed-balance condition can also be expressed in terms of a relation between the real and imaginary parts of the correlation functions:

$$\operatorname{Im} G(r, t) = -\tanh\left(\frac{\hbar}{2k_B T}\frac{\partial}{\partial t}\right) \operatorname{Re} G(\mathbf{r}, t), \qquad (1.110)$$

which is a special case of the *fluctuation–dissipation theorem* expressing the relation between the dissipative response of the system to a perturbation and the equilibrium fluctuations of the system in the absence of the perturbation [101].

(c) *Zeroth moment*

The relations between total and dynamic structure factor introduced in Section 1.4.2 can be rewritten as

$$\int_{-\infty}^{\infty} S_c^{dd'}(\mathbf{Q}, E) dE = S^{dd'}(\mathbf{Q}) \qquad (1.111)$$

and

$$\int_{-\infty}^{\infty} S_i^d(\mathbf{Q}, E) dE = 1. \qquad (1.112)$$

The limit of $S^{dd'}(\mathbf{Q})$ as $|\mathbf{Q}| \to 0$ is related to the statistical fluctuation in the number of particles N_d in a fixed volume of dimension $V^{1/3} \gg 2\pi/|\mathbf{Q}|$:

$$S^{dd'}(0) = \frac{\overline{\Delta N_d \Delta N_{d'}}}{\overline{N_d N_{d'}}}. \qquad (1.113)$$

This limit is also related to the macroscopic isothermal compressibility for fluids (including their mixtures, see p. 77 in Ref. [103]) such that

$$S(0) = n\left(\frac{\partial V}{\partial P}\right)_T k_B T. \qquad (1.114)$$

(d) *First moment*

In the case of a system with velocity-independent interatomic forces, the first energy moments of the scattering functions are given by

$$\int_{-\infty}^{\infty} S_c^{dd'}(\mathbf{Q},E) E \, dE = \frac{\hbar^2 Q^2}{2M_d} \delta_{dd'}, \tag{1.115}$$

$$\int_{-\infty}^{\infty} S_i^d(\mathbf{Q},E) E \, dE = \frac{\hbar^2 Q^2}{2M_d}, \tag{1.116}$$

where the right-hand side of these expressions correspond to the recoil energy of an atom of mass M_d.

(e) *Second moment*

The second moment of the incoherent scattering function is

$$\int_{-\infty}^{\infty} S_i^d(\mathbf{Q},E) E^2 \, dE = \left(\frac{\hbar^2 Q^2}{2M_d}\right)^2 + \hbar^2 Q^2 \left\langle \left(\mathbf{v}\cdot\hat{\mathbf{Q}}\right)^2 \right\rangle, \tag{1.117}$$

where $\mathbf{v}\cdot\hat{\mathbf{Q}}$ is the velocity in the direction of $\mathbf{Q}$. The second moment of $S_i^d(\mathbf{Q},E)$ and the higher moments of both scattering functions depend on the interatomic potential. Explicit expressions valid for quantum systems have been given by Rahman et al. [104].

The significance of these moment relations is best illustrated by discussing two limiting cases, either fulfilling or violating these fundamental properties of the dynamic structure factor.

1.4.3.1 Total Scattering and the Static Approximation

If energy changes in the system are negligible compared with the incident neutron energy E_i, $(E+E_{\tau_i}-E_{\tau_f}) \sim E$, the delta function $\delta(E+E_{\tau_i}-E_{\tau_f})$ in Equation (1.31) can be replaced by $\delta(E)$. Similarly, $|\mathbf{k}_i|=|\mathbf{k}_f|$ and the double differential cross section simplifies to

$$\frac{d^2\sigma}{d\Omega dE_f} = \frac{1}{N}\sum_{\tau_i} p_{\tau_i} \sum_{\tau_f} \left| \sum_j b_j \langle \tau_f | e^{i\mathbf{Q}_0 \cdot \mathbf{R}_j} | \tau_i \rangle \right|^2 \delta(E), \tag{1.118}$$

where $\mathbf{Q}_0$ is the value of $\mathbf{Q}$ when $k_f=k_i$ for a given k_i and scattering angle ϕ. Integration over the final energy E_f gives

$$\begin{aligned}\frac{d\sigma}{d\Omega} &= \int_{-\infty}^{E_i} \frac{d^2\sigma}{d\Omega dE_f} dE_f = \frac{1}{N}\sum_{\tau_i} p_{\tau_i} \sum_{j\in d, j'\in d'} \overline{b_j^* b_{j'}} \left\langle e^{-i\mathbf{Q}_0 \cdot (\mathbf{R}_{j'}-\mathbf{R}_j)} \right\rangle \\ &= \sum_{dd'} c_d^{1/2} c_{d'}^{1/2} \overline{b_d^* b_{d'}} S^{dd'}(\mathbf{Q}_0) + \sum_d c_d \frac{\sigma_i^d}{4\pi}.\end{aligned} \tag{1.119}$$

It is noteworthy that Equation (1.119) is equivalent to Equation (1.105), the approximate result for the total scattering at constant $\mathbf{Q}$, with $\mathbf{Q}=\mathbf{Q}_0$. Diffraction experiments from liquids and glasses are often interpreted on the assumption that Equation (1.119) is valid in the case of measurements at constant E_i and ϕ. However, there is no inelastic scattering in this model and the first moment of the scattering function, for example, is zero instead of the exact result given by Equation (1.115). We may therefore expect this approximation to be valid when $E_i \gg \hbar^2 Q^2/2M$ or $k_i \gg Q/A^{1/2}$. For diffraction measurements out to $Q \approx 10\,\text{Å}^{-1}$, this condition may imply $k_i \approx 50\,\text{Å}^{-1}$ and neutron energies of several eV, readily available at pulsed spallation neutron sources. For measurements with thermal neutrons, attempts are usually made to correct for the errors in the treatment of elastic scattering using the incoherent approximation and an expansion in powers of $(1/A)$, as originally introduced by Placzek [105–107] and further discussed in Ref. [108].

1.4.3.2 Free Particles and the Impulse Approximation

The scattering function for identical (noninteracting and structureless) particles can be calculated analytically and provides important insights into the expected behavior of dynamic structure factors. To this end, we consider an ensemble of N such particles in a box of volume V with momentum distribution $n(\mathbf{p})$. The associated translational wave function for the jth particle in the ensemble is simply given by

$$|\tau\rangle = \frac{1}{V^{1/2}} e^{i\mathbf{p}\cdot\mathbf{R}/\hbar}. \tag{1.120}$$

Since the particles are independent, it suffices to consider the single-particle $S(\mathbf{Q}, E)$. Using

$$\langle \tau_f | e^{i\mathbf{Q}\cdot\mathbf{R}} | \tau_i \rangle = \delta_{\mathbf{Q},(\mathbf{p}_f - \mathbf{p}_i)/\hbar} \tag{1.121}$$

and

$$E_{\tau_f} - E_{\tau_i} = \frac{\hbar^2}{2M}\left(p_f^2 - p_i^2\right) = \frac{\hbar^2}{2M}\left(Q^2 + \frac{2\mathbf{Q}\cdot\mathbf{p}_i}{\hbar}\right). \tag{1.122}$$

The dynamic structure factor is then given by

$$\begin{aligned} S(\mathbf{Q}, E) &= \frac{1}{N}\sum_{\tau_i} p_{\tau_i} \sum_{\tau_f} \left\langle \tau_f | e^{i\mathbf{Q}\cdot\mathbf{R}} | \tau_i \right\rangle^2 \delta(E + E_{\tau_i} - E_{\tau_f}) \\ &= \sum_{\mathbf{p}_i} n(\mathbf{p}_i) \delta\left(E - \frac{\hbar^2 Q^2}{2M} - \frac{\hbar \mathbf{Q}\cdot\mathbf{p}_i}{M}\right). \end{aligned} \tag{1.123}$$

This is the well-known *impulse approximation* for scattering from a collection of independent particles. The scattering function is centered at the recoil energy

$$E_R = \frac{\hbar^2 Q^2}{2M} \tag{1.124}$$

and its zeroth moment is $\sum_{\mathbf{p}_i} n(\mathbf{p}_i) = 1$. Furthermore, its first moment is E_R and the second moment is given by

$$E_R^2 + 2E_R \sum_{\mathbf{p}_i} n(\mathbf{p}_i) \frac{\left(\mathbf{p}_i \cdot \hat{\mathbf{Q}}\right)^2}{M} = E_R^2 + \frac{2E_R \langle p_i^2 \rangle}{3M}. \tag{1.125}$$

In this particular case, we see that first three moment relations are fully satisfied. Furthermore, the coherent and incoherent scattering functions are identical since the particles are assumed to be uncorrelated. The impulse approximation is generally considered to represent the high-Q limit of the scattering functions for *any* system. When the recoil energy is much larger than the energies of the natural excitations in the system, Equation (1.125) should be a good approximation. Systematic deviations from this exact result have been the subject of much theoretical and experimental work to date [81,109–111].

The precise shape of the dynamic structure factor depends on the type of statistics obeyed by the particle ensemble. For a Boltzmann distribution at temperature T, it can be written as

$$S(\mathbf{Q}, E) = \frac{1}{\sqrt{2\pi}\sigma_Q} e^{-(E-E_R)^2/2\sigma_Q^2}, \tag{1.126}$$

where

$$\sigma_Q^2 = 2E_R k_B T = \hbar^2 Q^2 k_B T / M. \tag{1.127}$$

The intermediate scattering function is given by

$$I(\mathbf{Q}, t) = e^{-\sigma_t^2 Q^2/2} \tag{1.128}$$

and the space–time correlation function by

$$G(\mathbf{r}, t) = \frac{1}{(2\pi)^{3/2} \sigma_t^3} e^{-r^2/2\sigma_t^2}, \tag{1.129}$$

where

$$\sigma_t^2 = t(t - i\hbar/k_B T) k_B T / M. \tag{1.130}$$

We see that $G(\mathbf{r}, t)$ is given by a Gaussian whose spatial width increases at a rate given by

$$\frac{\sigma_t}{t} = \left\langle \frac{v^2}{3} \right\rangle^{1/2} \left(1 - \frac{i\hbar}{k_B T t}\right)^{1/2}. \tag{1.131}$$

For a classical system, we would clearly expect a rate given simply by $\langle v^2/3 \rangle^{1/2}$. Even though we have used classical statistics to describe the dynamics of the scattering system, there is an additional factor involving $\hbar$ that describes the quantum mechanics of the interaction process. This leads to a scattering function centered at E_R instead of $E=0$, which satisfies the detailed balance condition. The double differential cross section corresponding to Equation (1.126) can be integrated explicitly to give a total scattering cross section (see pp. 93–95 in Vol. I of Ref. [63])

$$\sigma_{\text{tot}} = \frac{\sigma}{(1+m_n/M)^2}\left[\left(1+\frac{1}{2\varepsilon}\right)\Phi(\varepsilon^{1/2}) + \frac{e^{-\varepsilon}}{(\pi\varepsilon)^{1/2}}\right], \quad (1.132)$$

where $\varepsilon = E_i M / m_n k_B T$ and $\Phi(x) = 2\pi^{-1/2}\int_0^x e^{-t^2} dt$. For $\varepsilon \ll 1$ this gives

$$\sigma_{\text{tot}} \propto \frac{1}{E^{1/2}}, \quad (1.133)$$

due to the kinematic factor (k_f/k_i) in the inelastic cross section, and for $\varepsilon \gg 1$ we get

$$\sigma_{\text{tot}} \approx \sigma\left(\frac{A}{1+A}\right)^2, \quad (1.134)$$

which is the free-atom cross section, as required.

For quantum systems, $n(\mathbf{p})$ will no longer have the Maxwellian form of Equations (1.126) and (1.127) since it is affected by delocalization and particle-exchange effects. In quantum solids and nondegenerate quantum liquids like ^{4}He and ^{3}He, early neutron experiments [112,113] and computer simulations [114] have established that $n(\mathbf{p})$ (as well as $S(\mathbf{Q}, E)$ by virtue of Equation 1.123) is, to a first-order approximation, Gaussian but with the actual temperature in Equation (1.127) replaced by an effective temperature representing the kinetic energy associated with zero-point motions. Figure 1.13 shows a series of recoil-scattering inelastic spectra for solid ^{4}He using neutrons with energies in the electron-volt range and available at spallation sources (cf. Chapter 2). In this high-energy regime, it is customary to represent the neutron data in terms of the so-called *neutron Compton profile* $J(\mathbf{Q},y)$ [111,116], where

$$y = \frac{M}{\hbar^2 Q}(E - E_R) \quad (1.135)$$

is a scaling variable which reflects the relationship between E and Q in the impulse limit. Within the impulse approximation, it is related to the dynamic structure factor $S_{\text{IA}}(\mathbf{Q},E)$ by

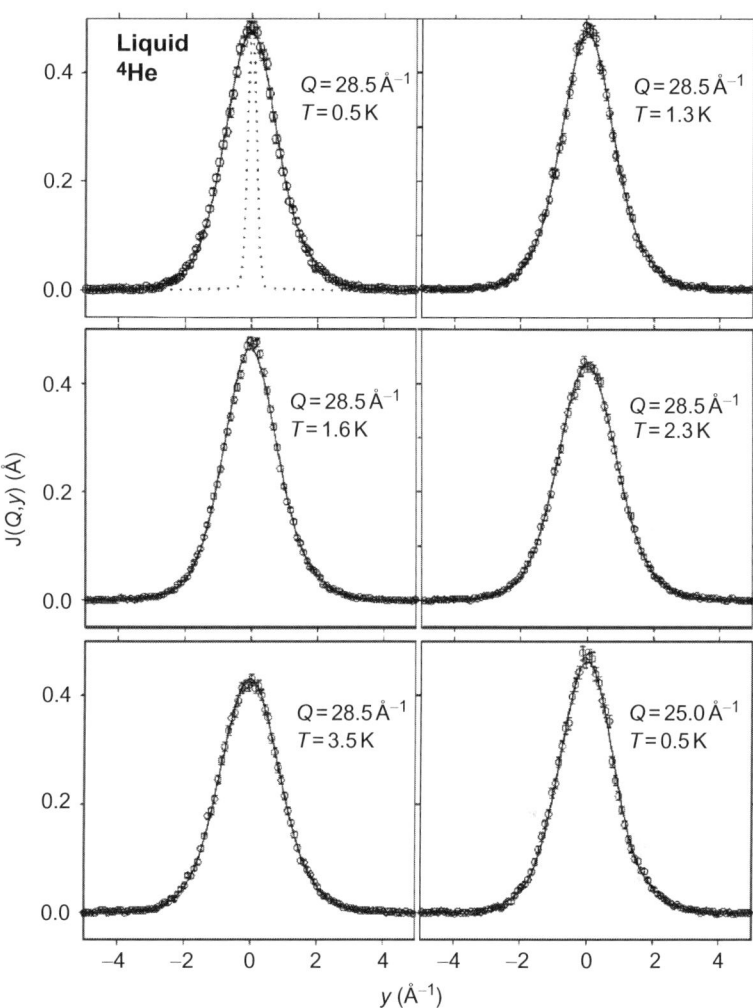

FIGURE 1.13 Recoil scattering of 0.75-eV neutrons from liquid ^{4}He above and below $T_\lambda = 2.17$ K. The dotted line denotes the instrumental resolution and the solid lines are fits to the experimental data using a model that accounts for the presence of a Bose–Einstein condensate fraction and departures from the impulse approximation. *Reprinted with permission from Ref. [115].*

$$J(\hat{\mathbf{Q}}, y) = \frac{\hbar Q}{M} S_{IA}(\mathbf{Q}, E). \quad (1.136)$$

Use of Equation (1.136) to represent experimental results necessarily assumes that it is possible to isolate the response from a given mass M in the neutron data. Also, measurements of this observable at different absolute Q values can be used to ascertain the validity and deviations from this limit – we also recall that at high momentum transfers Q, the static structure factor $S(Q)$ tends to unity and, therefore, the observed (total) dynamic structure factor is entirely incoherent. Conversely, the different recoil lines followed by different masses M provides a means of isolating the response of different elements in the sample (particularly light atoms). This task requires a judicious choice of the kinematic trajectories to "cut through" the mass-recoil lines in (Q, E) space with sufficient mass discrimination, as recently demonstrated in simultaneous measurements of the mean kinetic energies of several nuclides in lithium-containing materials [117,118].

More generally, the neutron Compton profile is related to the momentum distribution $n(\mathbf{p})$ via the integral relation

$$J\left(\hat{\mathbf{Q}}, y\right) = \hbar \int n(\mathbf{p}) \delta\left(\hbar y - \mathbf{p} \cdot \hat{\mathbf{Q}}\right) d\mathbf{p}. \tag{1.137}$$

Formally speaking, Equation (1.137) corresponds to the Radon transform of the momentum distribution $n(\mathbf{p})$ and it defines the probability for a given nucleus to have a momentum within an interval $\hbar(y, y+dy)$ parallel to $\hat{\mathbf{Q}}$. Within the impulse approximation, the neutron Compton profile should be independent of the magnitude of momentum transfer $\mathbf{Q}$, a result which is experimentally confirmed by the data shown in Figure 1.13. From these experiments, it is possible to extract quantitative estimates of the condensate fraction in this fundamental benchmark system. This experimental strategy has been used recently to establish whether claims of supersolidity in ^{4}He obtained from torsional-oscillator measurements could be traced back to microscopic mechanisms [119,120]. These studies have not detected any changes in the $n(\mathbf{p})$ of ^{4}He upon crossing the critical temperature, suggesting that the phenomenon is of a very different nature to the well-known case of superfluidity.

The use of epithermal neutrons to measure momentum distributions also transcends a purely academic interest in exotic quantum phases of matter. In the case of the hydrogen-bonded ferroelectric potassium dihydrogen phosphate (KH_2PO_4 or KDP), neutron Compton scattering has been used to probe the momentum distribution $n(\mathbf{p})$ of protons above and below the ferroelectric transition at $T = 124$ K [121]. These momentum distributions are shown in Figure 1.14 for a single crystal, from which it is possible to extract an effective Born–Oppenheimer potential for the hydrogen atoms in the material. These data also show that the proton remains coherently delocalized across two sites in the high-temperature phase, strongly suggesting the inapplicability of the commonly accepted order–disorder character of this transition.

As in the case of inelastic neutron-scattering experiments probing phonon structure at lower energy and momentum transfers, first-principles calculations can be used to predict the outcome of neutron Compton scattering measurements. Within the harmonic approximation, the neutron Compton profile for a given nuclide n may be described by a Gaussian line shape along a given crystallographic direction i by writing

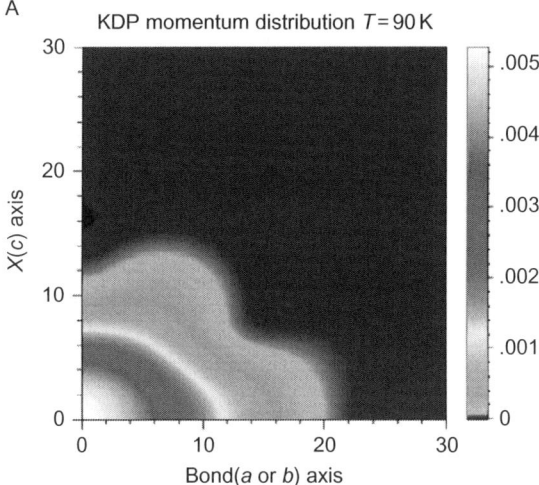

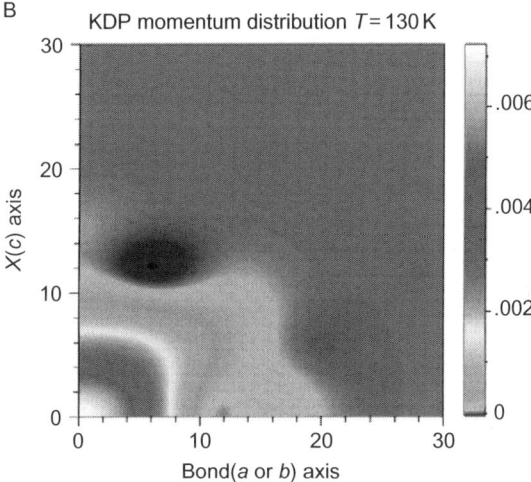

FIGURE 1.14 Proton momentum distributions in KDP below (a) and above (b) the ferroelectric transition. The green bulge in (b) ca. 30° away from the bond axis is ascribed to the repulsion experienced by the proton due to the neighboring phosphorous atoms in the lattice. *Reprinted with permission from Ref. [121].* (For interpretation of the references to color in this figure legend, the reader is referred to the online version of this chapter.)

$$J_{n,i}(y) = \frac{1}{\sqrt{2\pi}\sigma_{n,i}} e^{-y^2/2\sigma_{n,i}^2}, \quad (1.138)$$

where we have assumed that the impulse approximation holds (i.e., no implicit Q dependence) and $\sigma_{n,i}$ describes the second moment of the distribution for nucleus n along direction i. Using the formalism presented in

Section 1.3, the neutron Compton profile widths $\sigma_n(\hat{\mathbf{q}})$ along a given direction $\hat{\mathbf{q}}$ can be computed from the phonon-dispersion relations using [122]

$$\sigma_n^2(\hat{\mathbf{q}}) = \frac{M_n}{\hbar^2 N_\mathbf{q}} \sum_{\mathbf{q}\in BZ} \sum_k [\hat{\mathbf{q}}\cdot\mathbf{e}_n^k(\mathbf{q})]^2 \frac{E_k(\mathbf{q})}{2} \coth\left[\frac{E_k(\mathbf{q})}{2k_B T}\right], \quad (1.139)$$

where the sums run over all q-points sampled within the first Brillouin zone and over all phonon branches k. The *quasiharmonic* approximation [68] has been assumed in order to allow predictions at finite temperatures. Equation (1.139) provides a convenient starting point to discuss the various contributions to the Compton profile using the harmonic Born–Oppenheimer approximation, and has been shown to provide reliable estimates for a number of materials, including light hydrides [118,122], fluorides [117], and molecular hydrogen [123]. In the latter case, such an analysis of computational results has allowed the dissection of the experimental neutron data into various intra- and intermolecular contributions associated with the adsorbed state of this molecule, including the contributions from adsorbate–adsorbate interactions.

Beyond the harmonic approximation, path-integral molecular dynamics calculations allow for an exact evaluation of the nuclear momentum distribution $n(\mathbf{p})$ in terms of the density matrix $\rho(\mathbf{r},\mathbf{r}') = \langle\mathbf{r}|\exp(-\hat{H}/k_B T)|\mathbf{r}'\rangle$ [124]:

$$n(\mathbf{p}) = \frac{1}{2\pi\hbar^3} \int d\mathbf{r}d\mathbf{r}' e^{i/\hbar(\mathbf{p}\cdot(\mathbf{r}-\mathbf{r}'))} \frac{\rho(\mathbf{r},\mathbf{r}')}{Z}, \quad (1.140)$$

where Z is the partition function defined as

$$Z = \int d\mathbf{r}\rho(\mathbf{r},\mathbf{r}). \quad (1.141)$$

Whereas the partition function Z is an integral over diagonal terms of the particle density matrix, the momentum distribution $n(\mathbf{p})$ requires the inclusion of off-diagonal elements. Its accurate evaluation is, therefore, far from being a trivial task. In the context of neutron Compton experiments, Lin et al. [125] have recently proposed an efficient and at the same time insightful method to evaluate the open-path distribution represented by Equation (1.140). In their formulation, free-particle and environmental contributions factorize and the latter can be likened to the free-energy surface, therefore facilitating the interpretation of experimental data. The calculation of these observables using *ab initio* computational methods also provides solid ground for the understanding of nuclear quantum effects in condensed matter beyond those captured by the harmonic approximation. These include tunneling and particle delocalization [126], the only exception at present being the full inclusion of particle exchange.

Recent examples of the use of epithermal neutrons and state-of-the-art computational techniques in a growing number of materials include the study of proton momentum distributions and mean kinetic energies in supercritical and supercooled water [127,128], relaxation mechanisms in proton glasses [129], the potential energy landscape of protons in the superprotonic conductor $Rb_3H(SO_4)_2$ [130], nuclear quantum effects in the hydrogen-storage material lithium imide [131], and the determination of the spatial orientation of molecular hydrogen adsorbed in metal-graphite intercalates [123].

1.4.3.3 Classical Limit

Our final aim in this section is to establish a connection between the actual scattering function and that of a classical system interacting with the same potential. This is an important exercise because computer simulations using the molecular-dynamics method [132,133] play an invaluable role in furthering our understanding of classical liquids and dense fluids, and generally these calculations are done within the framework of classical mechanics. What is calculated is a classical intermediate scattering function (see Equation 1.96),

$$I_{cl}(\mathbf{Q},t) = \frac{1}{N}\langle \rho_d(-\mathbf{Q},0)\rho_{d'}(-\mathbf{Q},0)\rangle_{cl}, \qquad (1.142)$$

where the averages are now classical thermal averages over the history of the model system. In this case $I_{cl}(\mathbf{Q}, t)$, $G_{cl}(\mathbf{r}, t)$, and $S_{cl}(\mathbf{Q}, E)$ are all symmetric with respect to simultaneous inversion of both variables, and the detailed balance condition, Equation (1.108), is not satisfied. A solution to this problem was suggested by Schofield [134] who noted that substitution of

$$\tau = t - \frac{i\hbar}{2k_BT} \qquad (1.143)$$

in the classically derived correlation functions, yields functions

$$I(\mathbf{Q},t) = I_{cl}(\mathbf{Q},\tau) \qquad (1.144)$$

in accord with the detailed-balance condition. For example, the classical solution of the noninteracting gas problem discussed in the previous section is

$$I_{cl}(\mathbf{Q},t) = e^{-t^2 k_B T Q^2/2M}. \qquad (1.145)$$

The above procedure leads to the intermediate scattering function

$$I(\mathbf{Q},t) = e^{-\sigma_{t'}^2 Q^2/2} \qquad (1.146)$$

with

$$\sigma_{t'}^2 = \frac{k_BT}{M}\tau^2 = \frac{k_BT}{M}\left(t - \frac{i\hbar}{2k_BT}\right)^2. \qquad (1.147)$$

This result is identical to Equation (1.128) apart from a constant factor $\exp(E_R/4k_BT)$, which is generally close to unity for conditions under which measurements on classical liquids are performed. Similarly, the reverse procedure can be used to obtain "symmetrized" dynamic structure factors from experimental data:

$$\widetilde{S}(\mathbf{Q}, E) = e^{-E/2k_B T} S(\mathbf{Q}, E), \qquad (1.148)$$

corresponding to

$$\widetilde{I}(\mathbf{Q}, t) = I\left(\mathbf{Q}, t + \frac{i\hbar}{2k_B T}\right), \qquad (1.149)$$

which can then be directly compared with molecular dynamics results.

1.4.4 From Order to Disorder: Diffuse Scattering

Most materials of practical interest display some form of disorder, with profound consequences on their physical and chemical properties [135]. In certain cases, these properties are intentionally modified by introducing controlled amounts of disorder, as in the case of the alloying of metals for technological applications. In other cases, disorder arises as a result of the material's function, as in the cases of radiation damage of structural materials in nuclear reactors, or aging and corrosion in industrial applications. To gain an appreciation for the importance and technological relevance of disorder in our daily lives, the estimated global cost associated with corrosion prevention and mitigation *just* for the oil and gas sector has been valued at ca. $19 billion in 2011 [136].

Macroscopic features of disorder, such as dislocation patterns and grain structures, and their relationship to the material's properties are often understood at an empirical and phenomenological level. To advance beyond this level, static (within the timescale of the measurement) and dynamic disorder phenomena must be studied at the atomic scale. On this scale, disorder can be characterized in terms of vacancies and interstitials in the ideal lattice or by local fluctuations in chemical composition, all of which might also display a characteristic time dependence. The strains induced by these defects and the interaction and dynamical phenomena associated with these defects through their associated strain fields are important ingredients for a sound microscopic picture of disordered solids.

To proceed, we recall our preliminary discussion of diffuse scattering in Section 1.3.3. In a static picture (still conforming to our definition of a canonical solid), the loss of translational invariance in a solid gave way to diffuse scattering away from reciprocal lattice points satisfying the Bragg condition. In the extreme case of complete loss of translational disorder, all scattering becomes diffuse and one must abandon all together a description of the material in terms of well-defined Bragg reflections. This section extends these ideas and presents an outline of the use of diffuse nuclear scattering to study the emergence of disorder in solids. Magnetic diffuse scattering is discussed in Section 1.5.4. Beyond the classic text by Krivoglaz [137], more recent monographs [95,103,138–140] discuss the subject well beyond the scope of this chapter, including diffuse X-ray scattering.

For the intermediate case of solids with long-range order but with various types of local disorder, one observes a combination of Bragg scattering and diffuse scattering away from Bragg points. A detailed analysis of Bragg versus diffuse intensities provides a wealth of information about the underlying (static and dynamic) disorder. To illustrate the methodology on more quantitative grounds, let us consider a solid in which a periodic lattice can be used as a basis for indexing the atomic positions. In this case, the unit cells in the crystal are positioned at $\mathbf{l}^0$ and atomic positions within the unit cell are denoted by $\mathbf{d}^0$. The superscript "0" is used to denote the ideal (undistorted) lattice points. Deviations from the lattice points are given by the vectors $\mathbf{u}$. We also define an occupational probability P_{njk}, which is equal to one if lattice site n, j is occupied by an atom of type k and 0 otherwise. The elastic coherent differential neutron cross section is then given by

$$\left(\frac{d\sigma}{d\Omega}\right)_{c,el} = \frac{1}{N_c} \left| \sum_{n=0}^{N_c-1} \sum_{j=0}^{J-1} \sum_{k=0}^{K-1} P_{njk} \overline{b_k} e^{-W_{njk}} e^{i\mathbf{Q}\cdot\left(\mathbf{l}_n^0 + \mathbf{d}_j^0 + \mathbf{u}_{njk}\right)} \right|^2, \quad (1.150)$$

where N_c is the number of unit cells, J is the number of atomic positions within the unit cells, K is the number of different atom types, $\overline{b_k}$ is the coherent scattering length for atoms of type k, and W is the Debye–Waller exponent. The displacement $\mathbf{u}_{ijk}$ represents a linear superposition of displacements of atom n, j, k induced by all other atoms:

$$\mathbf{u}_{njk} = \sum_{n'j'k'} P_{n'j'k'} \mathbf{u}_{njk}^{n'j'k'}. \quad (1.151)$$

If we separate the occupation numbers into the average concentration of species k on site j, c_{jk}, and local fluctuations ΔP_{njk}, the displacement can be written as

$$\mathbf{u}_{njk} = \sum_{n'j'k'} c_{j'k'} \mathbf{u}_{njk}^{n'j'k'} + \sum_{n'j'k'} \Delta P_{n'j'k'} \mathbf{u}_{njk}^{n'j'k'}. \quad (1.152)$$

The first sum in Equation (1.152) can be seen to define a new reference lattice,

$$\mathbf{l}_n + \mathbf{d}_j = \mathbf{l}_n^0 + \mathbf{d}_j^0 + \overline{\sum_{n'j'k'} c_{j'k'} \mathbf{u}_{njk}^{n'j'k'}} \quad (1.153)$$

where the average of the sum is over atomic species.

The second sum in Equation (1.152) accounts for the fluctuations in the occupation number about the average. As such, it describes distortions relative to the new reference lattice. From Equations (1.152) and (1.153) we obtain

$$\left(\frac{d\sigma}{d\Omega}\right)_{c,el} = \frac{1}{N_c}\left|\sum_{njk} P_{njk}\langle\overline{b_k}\rangle e^{i\mathbf{Q}\cdot(\mathbf{l}_n+\mathbf{d}_j)}\exp\left(i\mathbf{Q}\cdot\sum_{n'j'k'}\Delta P_{n'j'k'}\mathbf{u}_{njk}^{n'j'k'}\right)\right|^2, \quad (1.154)$$

where $\langle\overline{b_k}\rangle$ is the average of $\overline{b_k}\exp(-W_{ij k})$ over the lattice indices n, j.

Further insight into the physical meaning of Equation (1.154) comes from considering the limiting case of small distortions. In this case, the second exponential in Equation (1.154) can be expanded as a power series. Keeping only the linear term in this expansion, the differential cross section becomes

$$\left(\frac{d\sigma}{d\Omega}\right)_{c,el} = \frac{1}{N_c}\left|\sum_{njk}[c_{jk}+P_{njk}]\langle\overline{b_k}\rangle e^{i\mathbf{Q}\cdot(\mathbf{l}_n+\mathbf{d}_j)}\left[1+i\mathbf{Q}\cdot\sum_{n'j'k'}\Delta P_{n'j'k'}\mathbf{u}_{njk}^{n'j'k'}\right]\right|^2.$$
(1.155)

By considering the expression resulting from the multiplication of the two square-bracketed terms in Equation (1.155), we can gain a first-order picture of the information content of the coherent elastic cross section for a solid with compositional disorder and small distortions:

- The term proportional to c_{jk} describes the Bragg conditions of the new reference lattice described by the vectors $\mathbf{l}$ and $\mathbf{d}$.
- The mixed terms involving c_{jk} times the sum of the other three terms in the square brackets will also affect the intensity in the Bragg peaks.
- The remaining terms give rise to diffuse scattering between the Bragg peaks.

The average distortion of the lattice due to disorder thus leads to a shift in the position of the Bragg peaks, while local distortions and compositional fluctuations are manifested both in the intensity of Bragg reflections and in the diffuse scattering between them. We also note that in our derivation of the differential cross section given by Equation (1.155), we have not assumed any implicit time dependence of the displacement terms. To treat this case, it becomes necessary to follow a similar methodology using the double differential cross section, followed by integration over final neutron energy. We shall return to this case in Section 1.4.5.

As a specific example of the information obtained from the results of a diffuse scattering experiment, we consider the case in which lattice distortions can be neglected. The differential cross section for diffuse scattering is then obtained from Equation (1.155) to read

$$\left(\frac{d\sigma}{d\Omega}\right)_{c,el}^{\text{diffuse}} = \frac{1}{N}\left|\sum_{njk}P_{njk}\langle\overline{b_k}\rangle e^{i\mathbf{Q}\cdot(\mathbf{l}_n\mathbf{d}_j)}\right|^2. \quad (1.156)$$

Representing compositional fluctuations by their Fourier transform in reciprocal space

$$\Delta P_{jk}(\mathbf{Q}) = \frac{1}{N_c}\sum_n \Delta P_{njk} e^{i\mathbf{Q}\cdot\mathbf{l}_n} \quad (1.157)$$

leads to

$$\left(\frac{d\sigma}{d\Omega}\right)^{\text{diffuse}}_{c,el} = \frac{N_c}{J}\left|\sum_{jk}\langle\overline{b_k}\rangle\Delta P_{jk}(\mathbf{q})e^{i\mathbf{Q}\cdot\mathbf{d}_j}\right|^2, \quad (1.158)$$

where ΔP is given as function of the reduced wave vector $\mathbf{q} = \mathbf{Q} - \boldsymbol{\tau}$, where $\boldsymbol{\tau}$ is a reciprocal vector of the lattice. Thus, the diffuse scattering in this case is entirely defined by the spectrum of compositional fluctuations in reciprocal space.

On a practical front, it becomes clear that the study of diffuse scattering is best accomplished using single-crystal specimens. Discerning static versus dynamic contributions to the observed scattering patterns may be accomplished by performing experiments at different incident wavelengths and temperatures. The former task is easily achievable at pulsed spallation sources using a broad range of incident energies. Figure 1.15 shows neutron-scattering data for deuterated benzil, and illustrates the wealth of information afforded by diffuse neutron-scattering experiments on hydrogen-containing molecular crystals. Similar to what is observed with X-rays and indicative of strong longitudinal-displacement correlations in this crystal, the middle image in this figure contains a diffuse hexagon connecting the set of intense reflections (400), (040), ($\bar{4}$40), (0$\bar{4}$0), ($\bar{4}$00), and ($\bar{4}$40) arising from Bragg scattering by 1.169 Å (59.8 meV) neutrons. As shown in the top image, this feature disappears from the data when the incident neutron wavelength is increased to 3.445 Å (6.9 meV), signaling a strong contribution of inelastic events to the former dataset. The bottom image shows broad regions of diffuse scattering, quite prominent at high momentum transfers and indicative of intramolecular motions—cf. Figure 1.10 and our previous discussion of inelastic form factors for one- and multiphonon scattering in Section 1.3.5. This particular case therefore represents a situation where analysis of scattered intensities away from Bragg reflections gives partial access to the double differential cross section. In some cases like polycrystalline MgO, it has been shown that detailed information on lattice dynamics for harmonic solids can also be obtained from total scattering data in a model-independent manner [142,143].

1.4.5 Stochastic Diffusion

Our treatment of the canonical solid presented in Section 1.3 made the tacit assumption that all atoms occupy well-defined equilibrium positions. In this case, the double differential neutron-scattering cross section is conveniently discussed in terms of *inelastic* scattering (i.e., exchange of energy and

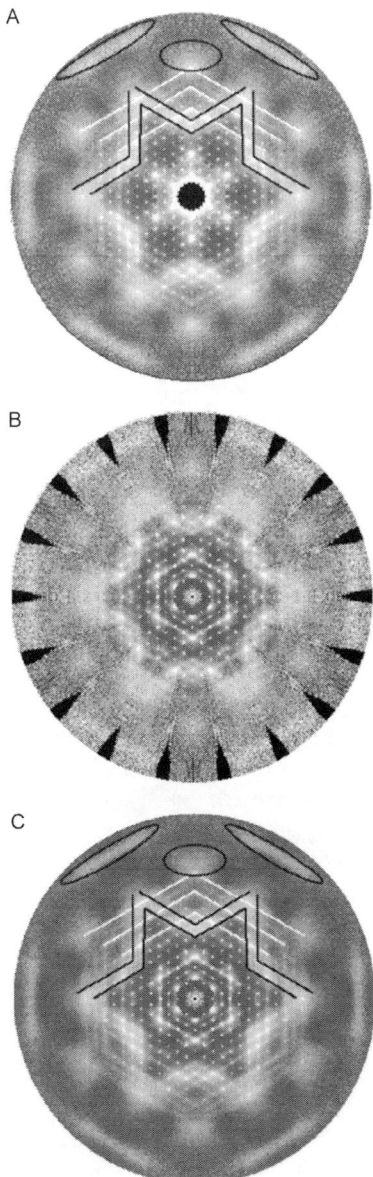

FIGURE 1.15 (*hk*0) reciprocal sections of deuterated benzil (1,2-diphenylethane-1,2-dione; $C_{14}D_{10}O_2$) measured on a pulsed-source single-crystal diffractometer at several scattering angles: (a) $+142.5°$; (b) $+37.5°$; and the sum over detectors located at $\pm 37.5°$, $\pm 90°$, and $\pm 142.5°$. In (c), only incident neutron energies above 27 meV have been considered in the data analysis. The dark lines in (a) and (c) emphasize prominent regions of diffuse intensity. *Reprinted with permission from Ref. [141].* (For color version of this figure, the reader is referred to the online version of this chapter.)

momentum with well-defined excitations in the solid), and *elastic* scattering (exchange of momentum with the entire solid recoiling as a rigid unit). Physically speaking, elastic scattering describes the response from particles at their equilibrium sites over the duration of the experiment (usually taken to be infinitely long relative to other characteristic timescales intrinsic to the material under investigation). In real solids, this is not generally the case. In addition to thermal motion around equilibrium sites (giving rise to a Debye–Waller factor, see Equation (1.59) in Section 1.3.2), atoms may occasionally move or diffuse between such sites. The ability to define a characteristic timescale for these diffusive motions immediately translates into the emergence of a spectral broadening of the elastic line in the energy domain. The energy widths of these *quasielastic* features constitute a direct measure of diffusion phenomena. If these processes are sufficiently fast, quasielastic spectral widths can be resolved experimentally and analyzed in terms of its details at the atomic scale. Recalling the space–time diagram shown by Figure 1.8 in Section 1.1.5.1, neutron scattering is quite unique as it provides simultaneous access to the spatial and temporal characteristics of diffusive motions via a study of the momentum and energy-transfer dependence of the double differential cross section.

Illustrations of this phenomenon in the solid state include metal hydrides, characterized by rapid translational diffusion of hydrogen atoms within the lattice, or fast-ion conductors, where the conducting ions are highly mobile. Lithium diffusion in LiC_6 and $LiCoO_2$ (materials sitting at the heart of modern lithium-ion-battery technology) are good examples of the latter. Likewise, in certain crystals displaying orientational disorder (so-called "plastic crystals") [144], molecules undergo rapid reorientations between a finite number of quasi-equilibrium configurations. This phenomenon corresponds to the rapid transport of individual atoms between quasi-equilibrium sites and, thus, also gives rise to quasielastic features in the neutron-scattering response. On a more fundamental front, molecular plastic crystals represent an interesting intermediate situation between liquids and canonical solids, where specific intermolecular normal modes of vibration become unstable and stochastic in nature prior to the onset of the liquid state *per se* [135,145,146]. In this case, quasielastic features in the neutron-scattering response are best understood as the transition between well-defined (underdamped) oscillatory motions characterized by an energy E much larger than its associated spectral width Γ such that $E \gg \Gamma$, to overdamped modes with $E \ll \Gamma$.

Liquids represent an extreme case where the concept of equilibrium sites breaks down altogether. In this situation, translational diffusion of the center of mass of a given particle happens in a more or less continuous manner, and quasielastic scattering is always observed. Moreover, vibrational excitations are strongly damped and the quasielastic and the inelastic regions of the dynamic structure factor can no longer be separated unambiguously. In certain cases and, in particular, at small values of the neutron momentum transfer, the quasielastic component can be isolated and used to derive distinct information about the diffusion process.

Chapter 6 provides a series of recent examples on the use of quasielastic neutron-scattering techniques to investigate stochastic dynamics and relaxation in solids and liquids. Detailed treatments can also be found in the monographs by Bee [147] and Hempelmann [148].

1.4.5.1 Liquids and Dense Gases

To describe particle diffusion, it is convenient to use the space–time formalism outlined in Section 1.4, that is, the space–time (van Hove) correlation functions [94]. In this case, the solution to the diffusion equation is obtained in $(\mathbf{r}, t)$-space and the dynamics of the diffusing particle is treated as a one-particle process. The corresponding Van Hove self-correlation function is often written within the so-called Gaussian approximation [149]

$$G_s(\mathbf{r}, t) = [4\pi \gamma(t)]^{-3/2} \exp[-r^2/4\gamma(t)], \quad (1.159)$$

where, for a classical system, $\gamma(t)$ is the mean-square displacement of a particle relative to its position at $t = 0$:

$$\gamma(t) = \frac{1}{6}\langle r^2(t) \rangle = \frac{1}{6}\int r^2 G_s(\mathbf{r}, t) d\mathbf{r}. \quad (1.160)$$

Recalling the properties of Gaussian distributions, the intermediate scattering function corresponding to Equation (1.159) is also a Gaussian (this time in Q space):

$$I_s(\mathbf{Q}, t) = \exp[-Q^2 \gamma(t)]. \quad (1.161)$$

In this approximation, $I_s(\mathbf{Q}, t)$, and thus $S_i(\mathbf{Q}, E)$, are completely specified by the time dependence of the mean-square displacement. We note that the Gaussian approximation is rigorously valid in certain cases only, namely, for a harmonic solid, for an ideal gas, and for the *long-time* behavior of an atom diffusing in a liquid [149,150]. It is, nonetheless, used extensively for other cases as well. To illustrate the formalism presented above, we consider two simple cases, namely, the free-flight motion of a particle in a dilute gas and the Brownian motion of an atom in a dense fluid.

For particles of mass M in a dilute gas, the mean-square displacement is given by

$$\gamma(t) = \frac{1}{6}\langle v^2 \rangle t^2 = \frac{k_B T}{2M} t^2, \quad (1.162)$$

and the scattering function is

$$S_i(\mathbf{Q}, E) = \left[\frac{M}{4\pi k_B T \hbar^2 Q^2}\right]^{1/2} \exp\left[-\frac{ME^2}{2k_B T \hbar^2 Q^2}\right], \quad (1.163)$$

that is, a Gaussian profile in energy transfer with a full-width-at-half-maximum given by

$$\Gamma(Q) = 2\sqrt{2 \ln 2}\ \hbar Q \left(\frac{k_B T}{M}\right)^{1/2}. \quad (1.164)$$

In dense fluids, such as in the liquid state, the atoms are in continuous interaction with their neighbors. As such, these interactions can be represented as an average fluctuating force on the particle. In the classical Langevin theory of Brownian motion, the corresponding mean-square displacement reads:

$$\gamma(t) = D\{t - \tau[1 - \exp(-t/\tau)]\}, \quad (1.165)$$

where

$$D = \frac{k_B T}{\eta M} = \frac{k_B T}{M}\tau \quad (1.166)$$

is the self-diffusion coefficient, η is the viscosity, and $\tau = 1/\eta$ is a measure of the rate of relaxation of velocity correlations in the fluid. From an expansion of the exponential function in Equation (1.165) and using Equation (1.166), it can be shown that Equation (1.165) is equal to the corresponding result for a particle in a dilute gas in the limit $t \ll \tau$. In the other limit, $t \gg \tau$, we obtain

$$\gamma(t) = D(t - \tau), \quad (1.167)$$

which should be compared to the result obtained from the solution of the macroscopic diffusion equation, namely

$$\gamma(t) = Dt, \quad (1.168)$$

that is, the inclusion of the friction term introduces a delay in the onset of purely diffusive behavior.

When inserted into Equation (1.161), $\gamma(t)$ in Equations (1.167) and (1.168) both yield the same result for the scattering function, namely

$$S_i(\mathbf{Q}, E) = \frac{1}{\pi}\frac{\hbar DQ^2}{E^2 + (\hbar DQ^2)^2}, \quad (1.169)$$

which is a Lorentzian with a half-width-at-half-maximum

$$\Gamma(Q) = \hbar DQ^2. \quad (1.170)$$

At small values of Q, where the above models are valid, the width of the quasielastic line is thus proportional to the diffusion coefficient. Typical diffusion coefficients in liquids lie in the range ~ 0.1 Å^2 ps^{-1}, resulting in a spectral half-width-at-half-maximum Γ of ~ 60 μeV for $Q \approx 1$ Å^{-1}. These spectral widths are routinely (and quite uniquely) accessible in scattering experiments using cold-neutron spectrometers.

Strictly speaking, the attentive reader will note that the incoherent scattering function shown in Equation (1.169) does not have a well-defined second moment and, therefore, is in clear violation of the moment rule given by Equation (1.117) in Section 1.4.3. Although this deficiency has little consequence in most practical applications, we note that it arises from considering only the dynamics at long times. To illustrate the origin of this behavior, we recall that the diffusion coefficient D can be expressed in terms of a time integral of the velocity autocorrelation function [151]

$$D = \int_0^\infty \langle \mathbf{v}(0) \cdot \mathbf{v}(t) \rangle \mathrm{d}t = \int_0^\infty Z(t) \mathrm{d}t, \qquad (1.171)$$

where for times much longer than the correlation time of stochastic forces we can write

$$Z(t) = D\left(\frac{\Gamma_v}{\hbar}\right) e^{-t(\Gamma_v/\hbar)}, \qquad (1.172)$$

where Γ_v is a friction or damping constant expressed in energy units. The corresponding spectral density in the energy domain is then given by

$$Z(E) = D\frac{1}{\pi}\frac{\Gamma_v}{E^2 + \Gamma_v^2}. \qquad (1.173)$$

As in the case of the dynamic structure factor, this spectral density is clearly ill-defined in terms of its second moment and therefore can only be valid for small energies E. One way to overcome this deficiency is to allow for an energy-dependent damping constant $\Gamma_v(E)$ that satisfies the fluctuation–dissipation theorem [152]. Using analytic-continuation techniques onto the complex plane, a commonly used functional form with a well-defined second (but not fourth!) moment reads

$$\Gamma_v(E) = \Gamma_{v0}\frac{\Gamma_{v1}}{\Gamma_{v1} + \mathrm{i}E}, \qquad (1.174)$$

where we note that $\Gamma_v(E) \to \Gamma_{v0}$ as $E/\Gamma_{v1} \to 0$. Therefore, only experiments seeking to probe *both* short- and long-time regimes will require such a generalization of the damping term. One such case is the extraction of the spectral density of the velocity autocorrelation function $Z(E)$ from neutron data using the relation [153]

$$E^2 \left[\frac{S_\mathrm{i}(Q,E)}{Q^2}\right]_{Q \to 0} \to Z(E) \qquad (1.175)$$

which, in combination with Equation (1.174), leads to a spectral density of the form

$$\left[\frac{E^2}{Q^2}S_\mathrm{i}(Q,E)\right]_{Q\to 0} \to \frac{\Gamma_{v0}\Gamma_{v1}^2}{\left(E^2+\Gamma_{v1}^2\right)E^2+\Gamma_0^2\Gamma_{v1}^2} \qquad (1.176)$$

More elaborate treatments of diffusion in liquids than what is captured by Equations (1.169) and (1.170) take into account a stepwise transport between quasi-equilibrium sites, including a description of pseudo-oscillatory behavior and residence time of a given atom temporarily trapped in a solvation cage [102,154]. In general, these models lead to Lorentzian-shaped scattering functions with a model-dependent $\Gamma(Q)$.

For bulk liquids, a commonly used extension of the Fickian limit described by Equation (1.170) considers a first-order correction to the high-Q limit describing a residence time τ_0 between translational jumps. In this jump-diffusion model of translational stochastic motions $\Gamma_\mathrm{T}(Q)$ is given by [153]

$$\Gamma_\mathrm{T}(Q) = \hbar\left[\frac{D_\mathrm{T}Q^2}{1+D_\mathrm{T}Q^2\tau_0}\right] \qquad (1.177)$$

where it is easily verified that $\Gamma_\mathrm{T}(Q)$ tends asymptotically to either $\hbar D_\mathrm{T}Q^2$ or $\hbar/\tau_0$ at low and high momentum transfers Q, respectively. An experimental determination of energy widths as a function of Q can then be used to test the validity of these theoretical models for translational diffusion. Figure 1.16 shows the momentum-transfer dependence of quasielastic spectral widths for incoherent (proton-dominated) scattering from liquid hydrogen fluoride as a function of temperature. The spectroscopic data are represented by recourse to a linearized form of Equation (1.177) reading

$$\Gamma_\mathrm{T}^{-1}(Q) = \frac{1}{\hbar}\left[D_\mathrm{T}^{-1}Q^{-2}+\tau_0\right] \qquad (1.178)$$

In this form, the jump-diffusion correction to Fick's law becomes an additive constant, and the slope as a function of Q^{-2} is inversely proportional to the *absolute value* of the microscopic self-diffusion coefficient D_T.

In molecular liquids such as in the example given above, in addition to the diffusive transport of the centre of mass, molecules can also rotate—the specific case of stochastic rotational motions in solids is also discussed in the next section. In the liquid phase, both translational and rotational motions determine the observed broadening. A first-order theoretical treatment of the resulting dynamic structure factor under the assumption of no couplings between translational and rotational modes in the time domain leads to the successive convolution (or folding) of the corresponding spectral functions in the energy domain. We can illustrate the approach by considering in more detail the case of hydrogen fluoride presented above.

As a heteronuclear diatomic molecule containing a single hydrogen atom, the incoherent self-intermediate scattering function in the time domain must include the contribution from center-of-mass translations of the whole molecule (assumed rigid) and rotation of the hydrogen atom about the molecular

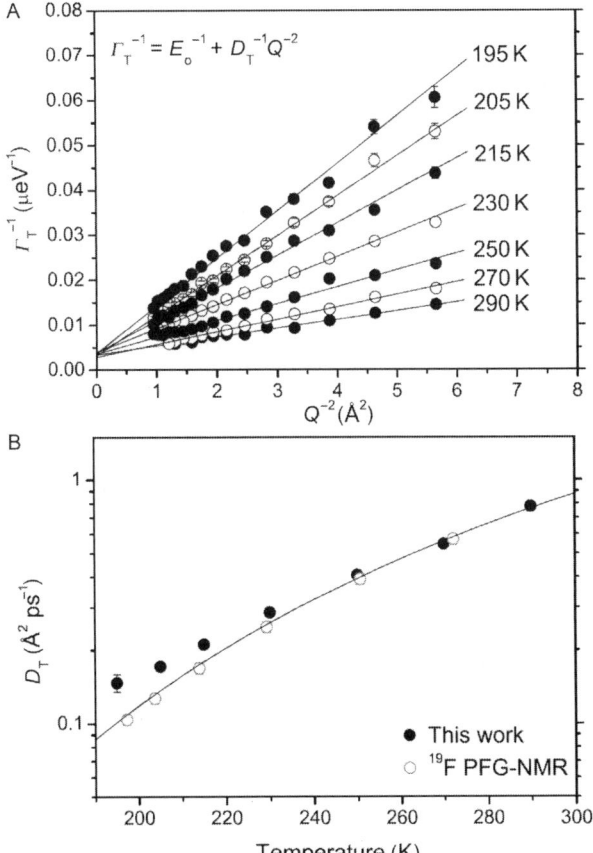

FIGURE 1.16 Quasielastic neutron-scattering data for liquid hydrogen fluoride as a function of temperature: (a) inverse spectral widths $\Gamma_T^{-1}(Q)$ for translational diffusion as a function of Q^{-2}, as described in the text; (b) derived translational self-diffusion coefficients and comparison with ^{19}F PFG-NMR results. *Reprinted with permission from Ref. [155].* (For color version of this figure, the reader is referred to the online version of this chapter.)

center of mass. If these two motions are decoupled from each other, we can then write the following self-intermediate scattering function for hydrogen:

$$I_S(Q,t) = I_T(Q,t) I_R(Q,t) \tag{1.179}$$

or in the energy domain

$$S_i(Q,E) = S_T(Q,E) \otimes S_R(Q,E), \tag{1.180}$$

where $\otimes$ denotes a convolution product. $S_T(Q,E)$ is given by Equation (1.177), and describes translational motions of the *entire* molecule, as well as the possibility of high-frequency vibrational motions as introduced in Equation (1.63)

$$S_T(Q,E) = e^{-Q^2\langle u^2\rangle/3}\left[\frac{1}{\pi}\frac{\Gamma_T(Q)}{E^2+\Gamma_T(Q)^2}\right]. \tag{1.181}$$

We note that both $\langle u^2\rangle$ and $\Gamma_T(Q)$ are accessible to experimental scrutiny via a measurement of intensities and spectral widths as a function of momentum transfer Q, respectively.

To describe stochastic rotational diffusion about the molecular centre of mass, a common expression to describe the isotropic case in terms of partial (angular momentum) waves is [156–158]

$$S_R(Q,E) = j_0^2(QR)\delta(E) + \sum_{l>0}(2l+1)j_l^2(QR)\left[\frac{1}{\pi}\frac{l(l+1)\Gamma_R}{E^2+(l(l+1)\Gamma_R)^2}\right]. \tag{1.182}$$

where R is radius of gyration and j_l is a spherical Bessel function of order l. The rotational diffusion constant D_R is defined such that $\Gamma_R = \hbar D_R$. Inserting Equations (1.181) and (1.182) into (1.180), the incoherent dynamic structure factor becomes

$$S_i(Q,E) = e^{-Q^2\langle u^2\rangle/3}\left\{j_0^2(QR)\left[\frac{1}{\pi}\frac{\Gamma_T(Q)}{E^2+\Gamma_T(Q)^2}\right] + \sum_{l>0}(2l+1)j_l^2(QR)\left[\frac{1}{\pi}\frac{\Gamma_{RT}(l;Q)}{E^2+\Gamma_{RT}^2(l;Q)}\right]\right\} \tag{1.183}$$

with $\Gamma_{RT}(l;Q) = \Gamma_T(Q) + l(l+1)\Gamma_R$. The first term in this expression corresponds to a purely translational contribution to the incoherent scattering function, and its form factor contains a $j_0^2(QR)$ term. In the particular case of hydrogen fluoride, it leads to the spectral widths shown in Figure 1.16a. We note, however, that *all* spectral components in Equation (1.183) have an intrinsic momentum-transfer dependence. For simple molecular liquids $\Gamma_R \gg \Gamma_T(Q)$ (particularly at low Q), and this dependence is far less marked in $\Gamma_{RT}(Q;l)$, usually leading to the approximation that it is Q independent. Moreover, the contribution of higher-order partial-wave terms beyond $l=1$ over typical momentum transfers is sufficiently small and it is generally omitted in the data analysis.

Equation (1.183) is also a convenient starting point to explore the effects of freezing translational motions, naturally leading to the formation of a plastic crystal (or an orientationally disordered glass). As $\Gamma_T(Q) \to 0$, the energy-dependent term in Equation (1.181) can be replaced by a delta function in E

$$S_T(Q,E) \to e^{-Q^2\langle u^2\rangle/3}\delta(E) \tag{1.184}$$

and the total incoherent scattering law reduces to

$$S_i(Q,E) = e^{-Q^2\langle u^2\rangle/3}\left\{j_0^2(QR)\delta(E) + \sum_{l>0}(2l+1)j_l^2(QR)\left[\frac{1}{\pi}\frac{l(l+1)\Gamma_R}{E^2+(l(l+1)\Gamma_R)^2}\right]\right\}. \tag{1.185}$$

The first term within the curly brackets represents a purely elastic term reflecting a vibrating molecule about its center of mass, that is, translationally speaking, the system behaves as a canonical solid, as introduced in Section 1.3. Higher-order terms in the sum shown by Equation (1.185) describe rotational stochastic motions, and their associated spectral widths are now Q independent. A multitude of models for the dynamic structure factor are discussed in Refs. [147,148]. Diffusion in bound and confined media lead to important changes to the momentum-transfer dependence of quasielastic neutron-scattering spectra particularly at low energy transfers, as originally discussed in Refs. [159,160]. For a recent treatment of this particular case, see Refs. [161,162]. As covered in detail in Chapter 6, computer simulation has become a key tool to interpret quasielastic neutron-scattering experiments, particularly for complex systems. Nonetheless, we emphasize that a physical interpretation of either experimental or computational data ultimately requires recourse to simplified models in order to gain the requisite level of insight.

In the preceding discussion, only the incoherent component to the scattering is considered and the results are directly related to self-correlation functions. Coherent scattering, on the other hand, is related to the full space–time correlation function. In this case, the theory is much less developed [153,154], yet still quasielastic neutron-scattering experiments have provided much-needed input to competing theories of the liquid state, see, for example, the elegant neutron experiments of Morkel *et al.* [163] to test the predictions of mode-coupling theories on liquid sodium. In certain cases, it is possible to separate the coherent and the incoherent scattering functions from measurements on samples with different isotopic compositions or via the use of polarization techniques (see Section 1.5.2). This has, for example, been done for liquid argon [97] as shown in Figure 1.12 in Section 1.4.2. In other cases, the Vineyard convolution approximation [149],

$$S_c(\mathbf{Q}, E) = S(\mathbf{Q})S_i(\mathbf{Q}, E), \qquad (1.186)$$

or an *ad hoc* modification of the convolution approximation [164]

$$S_c(\mathbf{Q}, E) = S(\mathbf{Q})S_i(\mathbf{Q}', E), \qquad (1.187)$$

with $\mathbf{Q}' = \mathbf{Q}\sqrt{S(\mathbf{Q})}$ is used to express the coherent scattering function in terms of the incoherent scattering function and the liquid structure factor. The observed scattering function can then again be analyzed for information about single-particle motions.

1.4.5.2 Disordered Solids

In certain solids, particle-diffusion rates are sufficiently large to enable neutron-scattering studies. For conventional spectrometers the condition is that $D \geq 0.01$ Å^2 ps^{-1}. However, backscattering or spin-echo spectrometers can offer spectral resolutions below 1 μeV, and measurable diffusion coefficients can be as low as 10^{-3} Å^2 ps^{-1}. Diffusion rates in these ranges are observed in fast-ion conductors and in certain metal hydrides, to name a few.

In analogy with some models for the liquid state, diffusion in the solid state may be regarded as taking place via discontinuous jumps between available quasi-equilibrium sites in the lattice. While residing at a given site, the particle undergoes high-frequency vibrations (describable via an appropriate Debye–Waller factor), and occasionally undergoes a transition to a nearby site. The appropriate rate equation for stepwise translational diffusion was first formulated by Chudley and Elliott [165]. Assuming instantaneous and statistically independent jumps across sites, the probability $P(\mathbf{r}, t)$ that the particle is at site $\mathbf{r}$ at time t is obtained from the following differential equation:

$$\frac{\partial}{\partial t}P(\mathbf{r},t) = \frac{1}{n\tau}\sum_{k=1}^{n}[P(\mathbf{r}+\mathbf{d_k},t) - P(\mathbf{r},t)], \tag{1.188}$$

where τ is the residence time at a given site and $\mathbf{d}_k$ is the distance to the kth site in the lattice. It is assumed that the available sites form a Bravais lattice so that the set of jump vectors $\mathbf{d}_k$ is the same for all sites. With the condition that the particle starts at the origin at time zero

$$P(\mathbf{r},0) = \delta(\mathbf{r}), \tag{1.189}$$

we obtain

$$G_s^D(\mathbf{r},t) = P(\mathbf{r},t), \tag{1.190}$$

where $G_s^D(\mathbf{r},t)$ is the Van Hove correlation function that describes the diffusion process. We will further assume that the diffusion process is decoupled from the thermal vibrations so that the scattering function corresponding to G_s^D may be derived independently.

Using an intermediate scattering function of the form

$$I_s^D(\mathbf{Q},t) = \exp\left[-\frac{\Gamma(\mathbf{Q})}{\hbar}t\right], \tag{1.191}$$

and the integral relationship between I_s and G_s given in Section 1.4.1, the incoherent dynamic structure factor is of the form

$$S_i^D(\mathbf{Q},E) = \frac{1}{\pi}\frac{\Gamma(\mathbf{Q})}{E^2 + \Gamma^2(\mathbf{Q})}, \tag{1.192}$$

with

$$\Gamma(\mathbf{Q}) = \frac{\hbar}{\tau n}\sum_{k=1}^{n}\left[1 - e^{-i\mathbf{Q}\cdot\mathbf{d}_k}\right]. \tag{1.193}$$

Equations (1.192) and (1.193) render themselves to a simple physical interpretation, as the scattering function is a Lorentzian with a width that is determined by the jump distance **d** and jump rate τ. Nonetheless, it is at first sight somewhat surprising that the incoherent scattering should contain structural information. However, we note that this information is not obtained from wave interference from different particles (as in the case of coherent scattering), but rather from the interference of waves scattered from the same particle at different times. To illustrate the use of Equation (1.193), the width function for a simple cubic lattice with lattice parameter a is given by

$$\Gamma(\mathbf{Q}) = \frac{\hbar}{3\tau}\left(3 - \cos Q_x a - \cos Q_y a - \cos Q_z a\right), \quad (1.194)$$

where we note that for $|\mathbf{Q}| \to 0$, the half-width-at-half-maximum becomes

$$\Gamma(Q) = \hbar \frac{Q^2 a^2}{6\tau} \equiv \hbar D Q^2, \quad (1.195)$$

that is, the result derived above for continuous (Fickian) diffusion in a fluid is recovered in this limit. For systems in which the sites do not form a Bravais lattice, the scattering function is obtained from a coupled set of differential equations, each one representing a Bravais subset of lattice points, and the scattering function is then a sum of Lorentzians. An in-depth discussion of the mathematical formalism and physical models applicable to a wide range of solid-state systems can be found in the monograph by Hempelmann [148].

To illustrate the above concepts, Figure 1.17 shows neutron-spin-echo for the hydrated perovskite $BaZr_{0.90}Y_{0.10}O_{2.95}$, a proton-conducting material [166,167]. As mentioned in Section 1.2.6, spin-echo methods probe the intermediate scattering functions and can extend the dynamic range of neutron spectroscopy well into the nanosecond time domain. Notwithstanding the experimental challenges associated with the isolation of the spectral response associated with proton motions using spin-echo techniques, these data provide a direct measurement of stochastic diffusion over timescales associated with atomic transport at the mesoscale. Ongoing developments in neutron sources and instrumentation aim to make these challenging measurements routine in next-generation neutron facilities.

The theoretical approach outlined thus far for solid-state diffusion has only considered the incoherent scattering function. In the case of scattering from hydrogen (e.g., hydrides, proton conductors), the incoherent scattering dominates and only $S_i(\mathbf{Q}, E)$ is of practical interest. In the case of other fast-ion conductors, the mobile atoms often display significant coherent scattering and the treatment above is then not applicable. In some cases, notably Ag diffusion in the well-studied fast-ion conductor AgI [168], the Vineyard convolution approximation [149], or an *ad hoc* modification of the Vineyard approximation [164], has been used to evaluate the coherent scattering function [169]. When possible, comparison of incoherent and coherent experiments also provide information on the differences between self and collective diffusion coefficients and the underlying short- and long-range

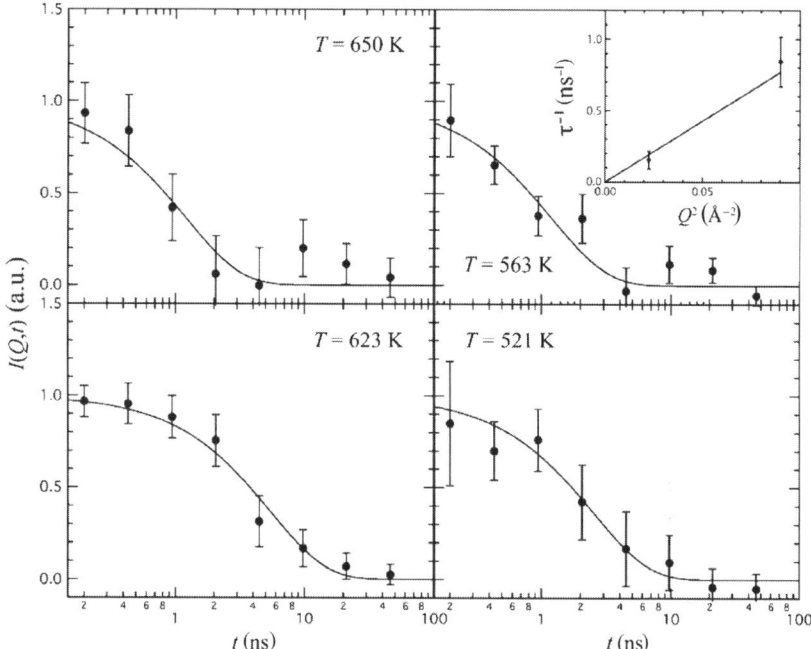

FIGURE 1.17 Intermediate scattering functions for hydrated $BaZr_{0.90}Y_{0.10}O_{2.95}$ at several temperatures and momentum transfers: $T=650$ K (0.20 Å^{-1}), 623 K (0.20 Å^{-1}), 563 K (0.30 Å^{-1}), and 521 K (0.30 Å^{-1}). Solid lines are fits to an exponentially decay with a Q-dependent relaxation time τ. The inset illustrates the Q^2 dependence of $\tau(Q)$ at $T=563$ K. *Reprinted with permission from Ref. [166].*

order associated with inter-species (also called "chemical") diffusion. For a detailed discussion, see Chapter 6 in Ref. [148]. As example, quasielastic scattering data from diffusing ions in yttria-stabilized zirconia at elevated temperatures are shown in Figure 1.18. In this case, the observed spectral broadenings are related to the diffusion of O^{2-} ions in the lattice. Studies of this sort at elevated temperatures have been greatly facilitated by ongoing developments in levitated (containerless) techniques [171].

There is one more important class of materials that show quasielastic scattering, namely, certain molecular solids in which specific functional groups (particularly methyl groups) are free to reorient on the time scale of picoseconds. As reorientations take place, individual atoms will move about a given molecular axis occupying a sequence of quasi-equilibrium orientations. Rate equations such as the one given in Equation (1.188) can again be used to solve the dynamics of these reorientational jumps. In this case, however, there is a finite probability that the particle will visit a given site more than once and, therefore, the total probability density will remain finite in the limit $t \to \infty$. As shown by Equation (1.185), this feature gives rise to a truly elastic component to the scattering, in addition to a broad quasielastic line corresponding to

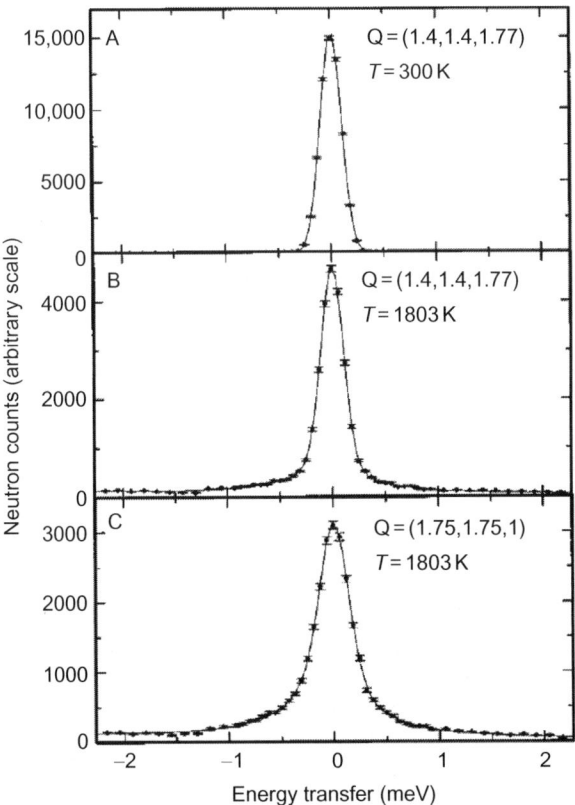

FIGURE 1.18 Quasielastic scattering spectra at constant Q from a single crystal of $(ZrO_2)_{0.87}(Y_2O_3)_{0.13}$ at room temperature and 1800 K. *Reprinted with permission from Ref. [170].*

the temporal redistribution of probability density between the sites. At sufficiently low temperatures, these reorientational motions can give rise to well-defined spectroscopic transitions associated with quantum-mechanical particle delocalization across sites, also called "tunneling transitions" [172,173]. These interesting dynamical effects are not only present in simple molecular solids involving a high-symmetry species such as solid methane or ammonium salts [173] but have also been observed in complex materials like polymers [174]. Interest in understanding low-energy excitations (uniquely accessible with neutron scattering) transcends a purely academic interest in solid-state spectroscopy, as it constitutes the basis for a growing number of neutron studies probing the interaction of technologically relevant molecules for gas storage and catalysis. These include methane [175], molecular hydrogen [176,177], or carbon dioxide [178] with surfaces and nanoporous media. Chapter 6 covers these in more detail in the context of the dynamics of atoms and molecules.

1.4.6 Beyond Atoms and Molecules: Large-Scale Structures

Neutron-scattering techniques can also be used to probe the properties of so-called "large-scale structures," supramolecular collections of atoms of sizes up to several thousand times interatomic distances. Examples of such systems are found in a plethora of scientific and technological applications including precipitates and voids in solids, macro, and biomolecules in solution, polymer chains in solid and liquid solutions and thin films, food, and consumer products, etc. In principle, scattering experiments yield information on both size and shape of the relevant structural units (form factor), as well as on the relative arrangement of these units (structure factor).

The use of neutron scattering to study large-scale structures has experienced an impressive growth over the past three decades. The origin of the discipline dates back to the 1970s, when position-sensitive detectors became generally available for small-angle neutron-scattering (SANS) experiments [179]. Further progress in instrumentation in the 1980s led to the development of neutron-reflectometry (NR) techniques for the study of interfaces and layered materials. In both cases, the increased availability of cold moderators at major neutron sources (both reactor- and spallation-based) has greatly stimulated this area of neutron science. As a brief complement to Chapter 5, we limit ourselves to a general overview of the overarching principles, as well as to how these can be related to our earlier discussion of neutron scattering by atomic and molecular systems.

In ordinary Bragg diffraction (Section 1.1.4), intense and coherent elastic scattering is observed whenever the Bragg condition is satisfied:

$$\lambda_i = 2d \sin \frac{\phi}{2}. \tag{1.196}$$

This condition can also be written

$$Q = 2\pi/d, \tag{1.197}$$

where Q is the wave vector at the Bragg peak, namely,

$$Q = \frac{4\pi}{\lambda_i} \sin \frac{\phi}{2}. \tag{1.198}$$

We also recall that the Bragg condition results from the constructive interference of neutron waves scattered from individual atomic planes. These planes define a periodically modulated neutron-scattering-length density. Operationally speaking, a minimum scattering angle corresponding to the maximum plane spacing can always be derived from Equation (1.196). In principle, there are no physical restrictions on the length scales associated with variations in scattering-length density, and these may very well vary over distances that are significantly larger than typical interatomic distances. As implied by Equation (1.197), the scattering vector Q is inversely proportional to the characteristic distance over which the scattering-length density fluctuates. Large objects thus give rise to scattering at small wave vectors and hence, in general, to scattering at small angles.

1.4.6.1 Small-Angle Scattering

The differential cross section for neutron scattering from individual atoms was introduced in Section 1.3 and reads

$$\frac{d\sigma}{d\Omega} = \frac{1}{N}\left|\sum_{\mathbf{R}} \overline{b_{\mathbf{R}}} e^{i\mathbf{Q}\cdot\mathbf{R}}\right|^2. \tag{1.199}$$

For scattering at small wave vectors, $Q \ll \pi/d$, individual atoms are not resolved and the scattering process is effectively governed by the interference of neutron waves scattered from regions of linear dimensions l such that

$$l \approx \pi/Q. \tag{1.200}$$

In this case, the summation over discrete atoms in Equation (1.199) can be written as a spatial integral of the form

$$\frac{d\sigma}{d\Omega} = \frac{1}{N}\left|\int_V \rho_b(\mathbf{r}) e^{i\mathbf{Q}\cdot\mathbf{r}} d\mathbf{r}\right|^2, \tag{1.201}$$

where $\rho_b(\mathbf{r})$ is the local scattering-length density, that is, a continuum approximation to the actual scattering-length density at a particular point. The spatial integration is over the entire volume of the target. We also express $\rho_b(\mathbf{r})$ in terms of its average value plus fluctuations around the average, namely,

$$\rho_b(\mathbf{r}) = \overline{\rho_b} + \delta\rho(\mathbf{r}). \tag{1.202}$$

The constant term $\bar{\rho}_b$ gives a contribution to the differential cross section at $Q=0$ only. For $Q>0$, the scattering is given by

$$\frac{d\sigma}{d\Omega} = \frac{1}{N}\left|\int_V \delta\rho(\mathbf{r}) e^{i\mathbf{Q}\cdot\mathbf{r}} d\mathbf{r}\right|^2. \tag{1.203}$$

For the case of N particles of homogeneous scattering-length density ρ_p imbedded in a matrix of homogeneous scattering length density ρ_m, Equation (1.203) gives

$$\frac{d\sigma}{d\Omega} = \frac{1}{N}\left(\rho_p - \rho_m\right)^2 \left|\int_{V_p} e^{i\mathbf{Q}\cdot\mathbf{r}} d\mathbf{r}\right|^2, \tag{1.204}$$

where the integral is over all particles in the system. If all the N_p particles are identical, the differential cross section becomes

$$\frac{d\sigma}{d\Omega} = \frac{V_p^2 N_p}{N}(\rho_p - \rho_m)^2 |F_p(\mathbf{Q})|^2, \qquad (1.205)$$

where the last term is the single-particle form factor given by

$$P(\mathbf{Q}) = |F_p(\mathbf{Q})|^2 = \left|\frac{1}{V_p}\int_{V_p} e^{i\mathbf{Q}\cdot\mathbf{r}} d\mathbf{r}\right|^2, \qquad (1.206)$$

Since $F_p(0) = 1$ in the limit $Q \to 0$, we can write:

$$\left(\frac{d\sigma}{d\Omega}\right)_{Q\to 0} = \frac{V_p^2 N_p}{N}(\rho_p - \rho_m)^2. \qquad (1.207)$$

For a sphere of radius R, the single-particle form factor $P(\mathbf{Q})$ displays the following momentum-transfer dependence:

$$P(\mathbf{Q}) = \{3[\sin QR - QR\cos QR]/Q^3 R^3\}^2. \qquad (1.208)$$

The differential cross section thus shows characteristic maxima and minima, a result which is not only restricted to spherically shaped objects, as shown in Figure 1.19.

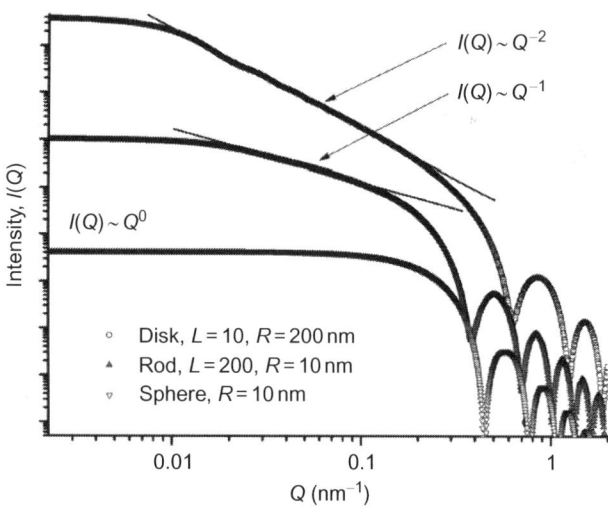

FIGURE 1.19 Simulated SANS profiles for monodisperse and dilute ensembles of disks (top trace), rods (middle), and spheres (bottom). The strong decay of the oscillations at high momentum transfers is the result of Porod's law. Disks (rods) exhibit a region where the integrated intensity $I(Q)$ scales as $Q^{-2}(Q^{-1})$. The spherical case (the smallest object of the trio) most clearly shows a flat plateau extending to low momentum transfers—this feature is actually present in all cases as long as $RQ < 1$. Characteristic radii of gyration can be obtained from a Guinier analysis of these data, as described in the main text. *Reprinted with permission from Ref. [180].*

In the general case of a single particle of arbitrary shape and orientation, it is not possible to express the scattering function in a simple closed form. Notwithstanding these methodological limitations, there are two limiting cases which can still provide useful information from the general behavior of the scattering function: the *Guinier approximation* [181], which applies at small values of Q such that $Ql \ll 1$, where l is the characteristic length over which the scattering-length density varies (the particle size); and the *Porod approximation* [182], which is valid at sufficiently large Q values such that $Ql \gg 1$.

When the Guinier approximation is applied to randomly oriented particles, the form factor is given by

$$P(\mathbf{Q}) = \exp(-Q^2 R_G^2/3), \tag{1.209}$$

where R_G, the radius of gyration, is obtained from

$$R_G^2 = \frac{1}{V_p} \int_{V_p} r^2 d\mathbf{r}, \tag{1.210}$$

and the integration is over the entire volume of the particle. Spherical particles of radius R give $R_G = \sqrt{3/5} R$ and the Guinier approximation is valid for $QR_G \leq 1.2$.

For particles of surface area A_p and homogeneous scattering length density, the Porod approximation predicts a scattering law proportional to Q^{-4}:

$$S(\mathbf{Q}) = 2\pi \left(\frac{A_p}{V_p^2}\right) \frac{1}{Q^4}. \tag{1.211}$$

Additional information can be obtained from integrated intensities. From Equations (1.202) and (1.203) we obtain

$$I_{\text{TOT}} = \frac{N}{V_p} \int \frac{d\sigma}{d\Omega} d\mathbf{Q} = (2\pi)^3 \left[\delta \rho^2(\mathbf{r})\right]_{\text{av}}, \tag{1.212}$$

where the average is taken over the whole system. For homogeneous particles imbedded in a homogeneous matrix, the integrated intensity is given by a slightly modified expression

$$I_{\text{TOT}} = (2\pi)^3 c_p (1 - c_p) \left(\rho_p - \rho_m\right)^2, \tag{1.213}$$

where $c_p = N_p V_p / V$. Thus, total scattered intensities provide a direct measure of volume fractions. Extrapolation to $Q = 0$ (see Equation 1.207) gives $V_p^2 N_p / N = V_p c_p$. In combination with c_p obtained from the integrated intensity, it is then possible to determine the average particle volume V_p. An example of the determination of particle volumes by SANS in ionic-liquid-in-oil microemulsions is shown in Figure 1.20.

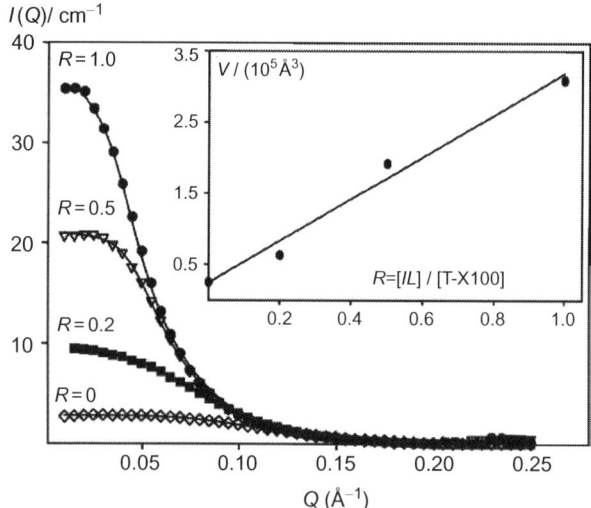

FIGURE 1.20 SANS from microemulsions of the ionic liquid [h-bmin]BF$_4$ in cyclohexane-d_{12}, stabilized by the nonionic surfactant T-X100 at 55 °C. R is defined as the ratio of ionic liquid to nonionic surfactant, as shown in the inset. Fits to the SANS data have been performed using a form factor for homogeneous ellipsoids. The inset shows the particle volume as a function of R, indicating that substantial micelle swelling takes places as R is increased. *Adapted with permission from Ref. [183].*

Up until now, we have explicitly assumed low particle concentrations so that interparticle correlations can be safely neglected. In many cases, the relative arrangement of large structural units is important—for example, in colloidal solutions at high concentrations and in biological molecules where the arrangement of the subunits is often of vital importance for biological function (e.g., quaternary protein structure or DNA pairing). To obtain the differential cross section in the more general case, the system can be subdivided into N_p cells such that each cell contains exactly one particle. The position of the centre of the jth cell is given by $\mathbf{R}_j$ and the position of the kth particle in the cell relative to the centre of the cell is denoted by $\mathbf{d}_k$. With these definitions, the macroscopic differential cross section becomes

$$\frac{d\Sigma}{d\Omega} = \frac{1}{V}\left\langle \left| \sum_{j=1}^{N_p} e^{i\mathbf{Q}\cdot\mathbf{R}_j} \sum_{k=1}^{N_i} \overline{b_{jk}} e^{i\mathbf{Q}\cdot\mathbf{d}_k} \right|^2 \right\rangle, \quad (1.214)$$

where $\overline{b_{ij}}$ is the scattering length for particle j in cell I, and N_i is the number of atoms in the cell. Defining form factors for the individual cells as

$$F_j(\mathbf{Q}) = \sum_{k=1}^{N_i} \overline{b_{jk}} e^{i\mathbf{Q}\cdot\mathbf{d}_k}, \quad (1.215)$$

allows us to rewrite Equation (1.214) in the following form

$$\frac{d\Sigma}{d\Omega} = \frac{1}{V}\left\langle \sum_{j=1}^{N_p}\sum_{k=1}^{N_p} F_j(\mathbf{Q})F_k(\mathbf{Q})e^{i\mathbf{Q}\cdot(\mathbf{R}_j-\mathbf{R}_k)} \right\rangle. \quad (1.216)$$

For a system of identical particles with the same orientation, or spherical particles, the form factors are the same for all particles and Equation (1.216) can be written as

$$\frac{d\Sigma}{d\Omega} = \frac{N_p}{V}|F(\mathbf{Q})|^2 S(\mathbf{Q}), \quad (1.217)$$

where $S(\mathbf{Q})$ is now used to denote the *interparticle structure factor*

$$S(\mathbf{Q}) = \frac{1}{N_p}\left\langle \sum_{j=1}^{N_p}\sum_{k=1}^{N_p} e^{i\mathbf{Q}\cdot(\mathbf{R}_j-\mathbf{R}_k)} \right\rangle. \quad (1.218)$$

Equation (1.217) forms the basis for the determination of the relative arrangement of large structural units and it is used extensively in the study of the architecture underlying supramolecular chemical and biological systems. By selective deuteration, any chosen subunit can be made to match the scattering-length density of the matrix, which often is an aqueous solvent. If all but two subunits in a structure are made to match the solvent, $S(\mathbf{Q})$ will contain a term $S'(Q)$ describing the interference between scattering-length density centers of the two subunits separated by a distance l. This term is approximately given by

$$S'(Q) \approx j_0(Ql), \quad (1.219)$$

where $j_0(Ql) = \sin(Ql)/Ql$ is the already-familiar spherical Bessel function of the first kind with $l=0$. The zeros of this function occur at $Q=2\pi n/l$ for any nonzero integer n, from which it is possible to obtain the distance between subunits. This approach has been used to determine the mesoscale structure of ribosomes in the seminal studies of Engelman *et al.* [184].

SANS techniques for the study of biological structure have evolved significantly in the past two decades [185]. Owing to the inherent complexity of these systems (e.g., presence of tertiary and quaternary structure in proteins, etc.), a common strategy involves the simultaneous analysis of neutron and X-ray small-angle data, as illustrated in Figure 1.21.

1.4.6.2 Reflection from Surfaces and Interfaces

Neutron reflectivity (NR) has become a powerful technique for the study of density fluctuations in surfaces and interfaces. As in optics, we can define a neutron refractive index n describing the change of neutron wave vector k_n in a medium relative to that *in vacuo*:

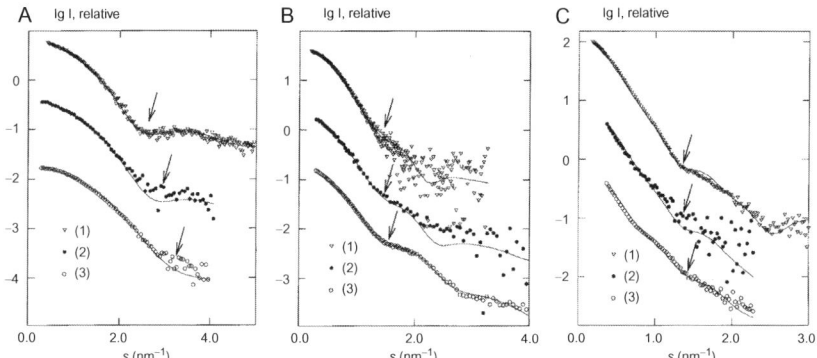

FIGURE 1.21 Small-angle scattering from proteins: (a) lysozyme; (b) *Escherichia coli* thioredoxin reductase; (c) and *Escherichia coli* ribonucleotide reductase protein *R1*. Solid lines are fits using the CRYSOL and CRYSON programs. In each figure, labels (1–3) correspond to X-rays, neutrons with H$_2$O, and neutrons with D$_2$O, respectively. *Reprinted with permission from Ref. [186].*

$$n = 1 - \frac{1}{2\pi}\lambda_n^2 \rho \overline{b}, \qquad (1.220)$$

where ρ is the number of scatterers per unit volume. As the coherent neutron-scattering length $\overline{b}$ for most materials is positive (hydrogen being an important exception), the associated neutron refractive index n is generally smaller than one and neutrons will be externally reflected by most materials (see also Section 1.5.5). This property is exploited in neutron guides to provide long experimental flight paths without reducing the acceptance angle and hence the outgoing intensity, as discussed in Chapter 3.

Total reflection from a surface or interface will take place at grazing angles of incidence θ below the critical angle θ_c. Values of θ_c are, however, generally small. In the case of total reflection, diffraction measurements can also be made by varying the azimuthal angle, giving information about the in-plane structure of the atoms at the surface. These *grazing-* or *glancing-incidence* measurements (sometimes called GISANS) open up the possibility of investigating the surface properties of nanostructured materials, including information about surface roughness and lateral correlations [187].

For values of $\theta > \theta_c$, information about the surface or interface structure can be obtained, including the study of buried interfaces not accessible using optical or electron microscopies. The specular component to the reflectivity is characterized by equal incident and reflected angles relative to the surface normal z and can be conveniently described within the kinematic approximation [188]. As in photon-based optics, this component to the reflectivity profile is confined to the plane defined by incident and reflected neutron beams. As such, it is related to the square of the Fourier transform of the scattering length density profile, $\rho(z)$, along z:

$$R(Q_z) = \frac{16\pi^2}{Q_z^2} |\int \rho(z) e^{-iQ_z z} dz|^2, \qquad (1.221)$$

where Q_z is the wave-vector transfer perpendicular to the surface ($Q_z = (4\pi/\lambda)$ sin θ, where θ is the angle of incidence).

The vast majority of NR studies performed to date have focused on an analysis of the specular reflectivity described by Equation (1.221), and many examples are given in Chapter 5. Gaining information about in-plane structure (on the xy plane) requires the detection of scattered intensities away from the specular plane so as to access nonzero momentum transfers away from the surface normal. So-called off-specular NR can also be of a diffuse character when it is associated with disorder and surface roughness [189,190]. A comprehensive account of the formalism to describe neutron and X-ray reflectivity is given in Ref. [191], and more recent applications of nonspecular NR in soft matter have been summarized by Dalgliesh [192]. The study of magnetic large-scale structures is also an active and growing area of research, and it is discussed below in Section 1.5.5.

In closing our discussion on large-scale structures, we emphasize that, in reality, there is no sharp distinction between SANS and the total-scattering techniques to probe shorter length scales (Section 1.4.2). Profiting from the availability of a wide range of incident neutron energies at spallation sources, a number of last-generation instruments [193–195] have been purposely designed to bridge the gap between microscopic and mesoscopic length scales, all within a single measurement.

1.5 MAGNETIC STRUCTURE AND POLARIZED NEUTRONS

This section covers the use of neutron scattering to study the spatial distribution of unpaired electron density in magnetic materials. This task is primarily achieved by means of neutron diffraction experiments. Such studies not only tell us about magnetism *per se* but also provide insights into the electronic structure of the material. Polarized neutrons are also included in this discussion, as they have been used extensively to discriminate between nuclear and magnetic neutron scattering.

1.5.1 Basic Principles

We outline the theory of neutron–magnetic interactions and associated scattering cross sections, so as to provide a suitable framework for the discussion of specific results in the sections to follow.

1.5.1.1 Magnetic Interactions and Cross Sections

In analogy to our discussion of the master formula and nuclear scattering in Section 1.2, we now consider the interaction of a neutron (in spin state σ) with a moving electron of momentum **p** and spin state **s**—note that Pauli operators **σ** are used to define the spin state of the neutron [196]. As introduced in Section 1.1.1, their corresponding magnetic moments are $\gamma\mu_N\boldsymbol{\sigma}$ and $g_e\mu_B\mathbf{s}$, where $\gamma = -1.913$, $g_e = -2.002319$ is the electron g-factor, and μ_N and μ_B are the nuclear and Bohr magnetons, respectively. In our treatment below, we shall make the implicit assumption that $g_e = -2$, an approximation of a similar nature to assuming m_n to be equal to the unified atomic-mass unit for nuclear scattering. While not conventional choices in the neutron literature, these definitions of the neutron-magnetic-moment-to-nuclear-magneton ratio γ and electron g-factor g_e conform to the latest recommendations on the use of units and physical constants [2]. Convincing physical arguments behind this sign convention have been presented in Ref. [197] for both atomic and molecular systems.

With these definitions in mind, the neutron–electron interaction potential is a distance- and momentum-dependent function of the form

$$V(\mathbf{r}) = -(2\gamma\mu_N\mu_B)\boldsymbol{\sigma} \cdot \left(\nabla \times \left(\frac{\mathbf{s} \times \hat{\mathbf{r}}}{r^2} \right) + \frac{\mathbf{p} \times \hat{\mathbf{r}}}{\hbar r^2} \right). \quad (1.222)$$

Physically speaking, this interaction potential is sensibly more intricate than the corresponding one for nuclear scattering involving the Fermi pseudo-potential, and given explicitly by Equation (1.27).

The double differential scattering cross section for a target containing a collection of unpaired electrons is obtained by substituting Equation (1.222) in the master formula, Equation (1.26). In the evaluation of matrix elements of the form $\langle \mathbf{k}_f \sigma_f \tau_f | V | \mathbf{k}_i \sigma_i \tau_i \rangle$, we first consider expectation values associated with the neutron plane-wave states $|\mathbf{k}_i\rangle$ and $|\mathbf{k}_f\rangle$, namely,

$$\langle \mathbf{k}_f | V | \mathbf{k}_i \rangle = 8\pi(\gamma\mu_N\mu_B)\boldsymbol{\sigma} \cdot \mathbf{D}_\perp (\mathbf{k}_i - \mathbf{k}_f), \quad (1.223)$$

where $\mathbf{D}_\perp(\mathbf{Q})$ is the *magnetic interaction operator*

$$\mathbf{D}_\perp(\mathbf{Q}) = \sum_j \left(\hat{\mathbf{Q}} \times \left(\mathbf{s}_j \times \hat{\mathbf{Q}} \right) + \frac{i}{\hbar Q} \mathbf{p} \times \hat{\mathbf{Q}} \right) e^{i\mathbf{Q}\cdot\mathbf{r}_j}. \quad (1.224)$$

To evaluate the matrix elements over neutron spin states $|\sigma_i\rangle$ and $|\sigma_f\rangle$, we first consider the case of *unpolarized* neutrons, where the corresponding matrix elements of products involving the neutron-spin operator **σ** satisfy the conditions

$$\sum_{\sigma_i} p_{\sigma_i} \sum_{\sigma_f} |\langle \sigma_f | \sigma_\alpha \sigma_\beta | \sigma_i \rangle|^2 = \delta_{\alpha\beta}, \quad (1.225)$$

and so the master formula becomes

$$\frac{d^2\sigma}{d\Omega dE_f} = \frac{1}{N_m}\left(\frac{k_f}{k_i}\right)(\gamma r_0)^2 \sum_\alpha \sum_{\tau_i} p_{\tau_i} \sum_{\tau_f} \langle \tau_i | \mathbf{D}^+_{\perp\alpha} | \tau_f \rangle \langle \tau_f | \mathbf{D}_{\perp\alpha} | \tau_i \rangle \delta(E + E_{\tau_i} - E_{\tau_f}),$$

$$(1.226)$$

where

$$r_0 = \frac{e^2}{4\pi\varepsilon_0 m_e c^2} = 2.8179 \text{ fm} \quad (1.227)$$

is the classical electron radius (of a similar magnitude as the range of nuclear interactions), and N_m represents the total number of magnetic centers. As in Section 1.2.3, it is convenient to switch to a real-time representation of the double differential cross section

$$\frac{d^2\sigma}{d\Omega dE_f} = \frac{1}{N_m}\left(\frac{k_f}{k_i}\right)(\gamma r_0)^2 \sum_\alpha \frac{1}{2\pi\hbar}\int \langle \mathbf{D}_{\perp\alpha}(-\mathbf{Q},0)\mathbf{D}_{\perp\alpha}(\mathbf{Q},t)\rangle e^{-iEt/\hbar}dt, \quad (1.228)$$

where the $\mathbf{D}_{\perp\alpha}(\mathbf{Q},t)$ terms are magnetic interaction operations in the Heisenberg representation. Comparison of Equation (1.228) with the nuclear-scattering counterpart (cf. Equation 1.37) shows that the characteristic magnetic scattering cross section per electron is $4\pi(\gamma r_0)^2 = 3.652$ b, and under suitable conditions magnetic scattering can therefore be of the same order of magnitude as nuclear scattering.

The magnetic interaction operators $\mathbf{D}_\perp(\mathbf{Q},t)$ are inextricably linked to the magnetic properties of real systems, as they are projections of a generalized vector operator $\mathbf{D}(\mathbf{Q},t)$ on a plane perpendicular to the neutron-momentum transfer $\mathbf{Q}$. We can always write:

$$\mathbf{D}_\perp = \hat{\mathbf{Q}} \times \mathbf{D} \times \hat{\mathbf{Q}} = \mathbf{D} - \left(\mathbf{D}\cdot\hat{\mathbf{Q}}\right)\left(\hat{\mathbf{Q}}\right), \quad (1.229)$$

where $\mathbf{D}$ is related to the Fourier transform of the magnetization operator $\mathbf{M}$ via:

$$\mathbf{D}(\mathbf{Q},t) = -\frac{\mathbf{M}(\mathbf{Q},t)}{2\mu_B} = -\frac{1}{2\mu_B}\int \mathbf{M}(\mathbf{r},t)e^{i\mathbf{Q}\cdot\mathbf{r}}d\mathbf{r}. \quad (1.230)$$

Using Equation (1.229), Equation (1.230) can be reexpressed in terms of $\mathbf{D}(\mathbf{Q},t)$ as

$$\frac{d^2\sigma}{d\Omega dE_f} = \frac{1}{N_m}\left(\frac{k_f}{k_i}\right)(\gamma r_0)^2 \sum_{\alpha\beta}\left(\delta_{\alpha\beta} - \hat{Q}_\alpha\hat{Q}_\beta\right)\frac{1}{2\pi\hbar}\int \langle D_\alpha(-\mathbf{Q},0)D_\beta(\mathbf{Q},t)\rangle e^{-iEt/\hbar}dt.$$

$$(1.231)$$

Therefore, scattering probabilities depend on the magnetic fluctuations in the system, an entirely analogous situation to the way in which nuclear scattering depends on density fluctuations and discussed in Section 1.4.1.

The directional dependence through the tensor $\left(\delta_{\alpha\beta} - \hat{Q}_\alpha\hat{Q}_\beta\right)$ becomes a crucial and distinct feature of magnetic neutron scattering, as it "picks out"

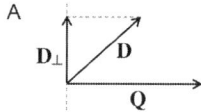

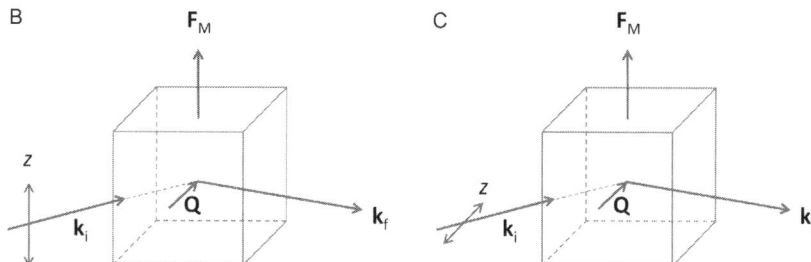

FIGURE 1.22 (a) Definition of the magnetic interaction operator $\mathbf{D}_\perp$. (b) Scattering geometry with the direction of neutron polarization z orthogonal to $\mathbf{Q}$. (c) Scattering geometry with z parallel to $\mathbf{Q}$. (For color version of this figure, the reader is referred to the online version of this chapter.)

the components of the magnetization orthogonal to the momentum transfer $\mathbf{Q}$, as shown in Figure 1.22a. For a spatially uniform magnetization, it is convenient to use a unit vector $\hat{\eta}$ ($\hat{\eta}$ is a vector quantity) in which case the directional dependence implied in Equation (1.231) is described by the factor

$$1 - \left(\hat{\mathbf{Q}} \cdot \hat{\eta}\right)^2. \qquad (1.232)$$

This factor makes it possible to deduce the orientation of the magnetic moments in a crystalline material relative to a given crystallographic direction.

The evaluation of $\mathbf{D}$ in terms of microscopic properties depends entirely on the nature of the magnetic electrons—for a detailed discussion, see Chapter 6 in Ref. [108]. For electrons of intrinsic spin and orbital angular momentum $\mathbf{s}$ and $\mathbf{l}$, respectively, localized on atoms at positions $\mathbf{R}_j$, and with positions $\mathbf{r}_n$ relative to their corresponding atomic centers,

$$\mathbf{D}(\mathbf{Q}, t) = \sum_j e^{i\mathbf{Q} \cdot \mathbf{R}_j} \sum_{n \in j} e^{i\mathbf{Q} \cdot \mathbf{r}_n} (\mathbf{s}_n + \mathbf{l}_n). \qquad (1.233)$$

For electrons obeying Russell–Saunders (or *LS*) coupling, we can write

$$\mathbf{D}(\mathbf{Q}, t) = \sum_j f_j(\mathbf{Q}) \boldsymbol{\mu}_j(t) e^{i\mathbf{Q} \cdot \mathbf{R}_j(t)}, \qquad (1.234)$$

where we define $\boldsymbol{\mu}_j = \frac{1}{2} g_j \mathbf{S}_j$ as the magnetic moment of the jth atom (in Bohr magnetons μ_B), and $\mathbf{S}_j$ its atomic-spin operator. Use of the symbol $\mathbf{S}$ above adheres to convention, yet we note that it can be used (rather interchangeably!) to denote intrinsic spin, total electronic angular momentum $\mathbf{J}$, or (more generally) an effective angular-momentum operator for the case of partially quenched orbital angular momenta, the latter case requiring magnetization measurements on the system under investigation. With these definitions, $f_j(\mathbf{Q})$ is the corresponding form factor defined by

$$f(\mathbf{Q}) = \frac{g_S}{g}\bar{j}_0(\mathbf{Q}) + \frac{g_L}{g}[\bar{j}_0(\mathbf{Q}) + \bar{j}_2(\mathbf{Q})] \qquad (1.235)$$

with

$$\bar{j}_n(\mathbf{Q}) = \int j_n(Qr)|\psi(r)|^2 d\mathbf{r}, \qquad (1.236)$$

where $\psi(r)$ is the electronic wave function, j_n's are spherical Bessel functions, and g, g_L, and g_S correspond to g-factors associated with total, orbital, and intrinsic electronic spins, respectively. We also note that $f_j(\mathbf{Q})$ is normalized so that $f_j(0) = 1$. Thus, $\boldsymbol{\mu}_j(t)$ represents the magnitude and direction of the magnetic moment and $f_j(\mathbf{Q})$ the Fourier transform of its spatial distribution about the atom center.

Assuming that the magnetic moments associated with the electrons have a negligible effect on interatomic forces, the sums over electronic and nuclear coordinates of $\mathbf{D}$ in Equation (1.234) can be carried out separately and Equation (1.231) becomes

$$\frac{d^2\sigma}{d\Omega dE_f} = \frac{1}{N_m}\left(\frac{k_f}{k_i}\right)(\gamma r_0)^2 \sum_{\alpha\beta}\left(\delta_{\alpha\beta} - \hat{Q}_\alpha\hat{Q}_\beta\right)\sum_{jj'} f_j^*(\mathbf{Q})f_{j'}(\mathbf{Q}) \times$$
$$\times \frac{1}{2\pi\hbar}\int \langle \mu_{j\alpha}(0)\mu_{j'\beta}(t)\rangle \langle e^{-i\mathbf{Q}\cdot\mathbf{R}_j(0)}e^{i\mathbf{Q}\cdot\mathbf{R}_{j'}(t)}\rangle e^{-iEt/\hbar}dt, \qquad (1.237)$$

which is the magnetic analog of Equation (1.38). It is important to note that the double differential scattering cross section shown in Equation (1.237) still contains correlation functions for the atomic positions, so that magnetic scattering is still sensitive to the structure and dynamics of these, as well as to magnetic properties.

1.5.1.2 Random Spins

The simplest case to apply the formalism introduced in the previous section corresponds to a paramagnetic solid with localized electrons at zero magnetic field. In this case, there are no correlations between spin orientations on different atoms, and therefore

$$\langle \mu_{j\alpha}(0)\mu_{j'\beta}(t)\rangle = \langle \mu_{j\alpha}(0)^2\rangle \delta_{jj'}\delta_{\alpha\beta} = \frac{1}{12}g_j^2 S_j(S_j+1)\delta_{jj'}\delta_{\alpha\beta}. \qquad (1.238)$$

Since $\sum_{\alpha\beta}\left(\delta_{\alpha\beta} - \hat{Q}_\alpha\hat{Q}_\beta\right) = 2$, Equation (1.237) gives

$$\frac{d^2\sigma}{d\Omega dE_f} = \frac{1}{N_m}\left(\frac{k_f}{k_i}\right)(\gamma r_0)^2 \sum_j |f_j(Q)|^2 \frac{1}{6}g_j^2 S_j(S_j+1) \frac{1}{2\pi\hbar} \int \left\langle e^{-i\mathbf{Q}\cdot\mathbf{R}_j(0)} e^{i\mathbf{Q}\cdot\mathbf{R}_j(t)}\right\rangle e^{-iEt/\hbar} dt, \quad (1.239)$$

which is equivalent to Equation (1.48) for incoherent nuclear scattering with the cross section σ_i replaced by

$$4\pi(\gamma r_0)^2 |f_j(Q)|^2 \frac{1}{6}g_j^2 S_j(S_j+1). \quad (1.240)$$

Apart from the magnetic form factors, there are no new features associated with the magnetic scattering from such a system. Equation (1.235) no longer applies in the presence of exchange interactions between different ions, or if there is an electrostatic interaction between magnetic and neighboring ions giving rise to a *crystal field* that lifts the M-sublevel degeneracy of the electronic ground state. Another way to lift this angular-momentum degeneracy is via the application of an external magnetic field B leading to characteristic Zeeman splitting of the magnetic sublevels. If B is applied in the z-direction, it will induce a nonzero average spin S_z, leading to coherent scattering with an effective scattering length of

$$(\gamma r_0) f_j(Q) \frac{1}{2} g_j \bar{S}_z. \quad (1.241)$$

We note that this quantity depends on the magnitude of $\mathbf{Q}$, not its direction.

1.5.1.3 Elastic Scattering from Spin Order

Spin-ordering phenomena imply nonzero correlations between magnetic moments on different atoms in the material. If these persist for long times (relative to the duration of the measurement), they necessarily lead to purely elastic scattering, just as the static order in nuclear positions discussed in Section 1.3. In this case the correlation function in Equation (1.228) becomes a time-independent quantity and the differential scattering cross section reads

$$\left(\frac{d\sigma}{d\Omega}\right)_{el} = \frac{1}{N_m}(\gamma r_0)^2 |\langle \mathbf{D}_\perp(\mathbf{Q})\rangle|^2. \quad (1.242)$$

For a periodic structure, we define a time-dependent position vector for the jth spin as $\mathbf{R}_j(t) = \mathbf{l} + \mathbf{d} + \mathbf{u}_j(t)$, where the $\mathbf{l}$ vectors now have a periodicity corresponding to the *magnetic structure*. After averaging the electronic and nuclear parts independently, Equation (1.234) becomes

$$\langle \mathbf{D}(\mathbf{Q})\rangle = \sum_l \sum_d f_d(\mathbf{Q})\langle \boldsymbol{\mu}_d\rangle e^{i\mathbf{Q}\cdot(\mathbf{l}+\mathbf{d})} e^{-W_d(\mathbf{Q})} = \frac{1}{|\gamma r_0|} \sum_l \mathbf{F}_M(\mathbf{Q}) e^{i\mathbf{Q}\cdot\mathbf{l}}, \quad (1.243)$$

with the *magnetic unit-cell structure factor* $\mathbf{F}_M(\mathbf{Q})$ is defined as

$$\mathbf{F}_M(\mathbf{Q}) = |\gamma r_0| \sum_d f_d(\mathbf{Q}) \langle \mu_d \rangle e^{i\mathbf{Q}\cdot\mathbf{d}} e^{-W_d(\mathbf{Q})}. \tag{1.244}$$

Summation over lattice vectors gives

$$\left(\frac{d\sigma}{d\Omega}\right)_{el} = \frac{1}{N_m} \frac{(2\pi)^3}{v_0} \sum_{\tau_M} \delta(\mathbf{Q} - \tau_M) |\mathbf{F}_{M\perp}(\tau_M)|^2, \tag{1.245}$$

where

$$\mathbf{F}_{M\perp} = \hat{\mathbf{Q}} \times \mathbf{F}_M \times \hat{\mathbf{Q}} \tag{1.246}$$

is sometimes called the *magnetic structure factor*, τ_M represent the reciprocal lattice vectors of the magnetic structure, and v_0 is the magnetic unit-cell volume. For the special case of itinerant electrons (e.g., magnetic metals), Equation (1.245) still holds provided that the unit-cell structure factor is replaced by

$$\mathbf{F}_M(\mathbf{Q}) = \hat{\boldsymbol{\eta}} |\gamma r_0| \left(\int_{V_0} |\varphi(\mathbf{r})|^2 e^{i\mathbf{Q}\cdot\mathbf{r}} d\mathbf{r} \right) \left(\frac{n_\uparrow - n_\downarrow}{2} \right) e^{-W(\mathbf{Q})}, \tag{1.247}$$

where $|\phi(\mathbf{r})|^2$ is the spin density in each cell and $n_\uparrow$ and $n_\downarrow$ are the average numbers of up and down spins. In general, $\hat{\boldsymbol{\eta}}$ will itself be a function of $\mathbf{Q}$ if the moments do not all point in the same direction. A detailed discussion of the physics behind Equation (1.247) is given in Sections 5.5 and 7.2 of Ref. [108].

Comparison of the nuclear-scattering structure factor with Equation (1.237) or (1.244) indicates that the magnetic analog provides information on both the magnitude and direction of magnetic moments on each site, and the form factors reflect the distribution of unpaired electron density in the material. Both of these features provide a strong motivation for carrying out magnetic diffraction studies.

1.5.2 Polarized Neutrons

In our discussion up until this point, the spin state (or polarization) of both incident and scattered neutron beams has been assumed to be evenly distributed between the two possible $\pm 1/2$ eigenstates for a spin-1/2 particle. Furthermore, Section 1.2.2 and 1.5.1 have presented differential scattering cross sections by performing an explicit average over all possible spin states. For many problems involving magnetic scattering (and some involving nuclear scattering), it is nonetheless advantageous to have the additional variable of neutron polarization at one's disposal so as to distinguish between different types of scattering. To this end, it becomes necessary to retrace the formalism of the previous sections taking explicit account of initial and final polarizations.

Chapter | 1 An Introduction to Neutron Scattering

1.5.2.1 Nuclear Scattering

The dependence of nuclear scattering on neutron spin requires defining scattering-length operators of the form:

$$\hat{b}_j = \bar{b}_j + \frac{1}{2}b_{Nj}(\sigma \cdot \mathbf{I}_j) \tag{1.248}$$

where $\mathbf{I}$ and $1/2\sigma$ are the spin angular momentum operators for the jth nucleus and neutron, respectively. If the eigenvalues of $\hat{b}_j$ are $\hat{b}_j^{\pm}$ for the compound neutron–nucleus system of total angular momentum $I_j \pm 1/2$, respectively, the corresponding (spin-dependent) scattering lengths are given by

$$\begin{aligned}\bar{b}_j &= \frac{1}{2I_j+1}\left[(I_j+1)b_j^+ + I_j b_j^-\right], \\ b_{Nj} &= \frac{2}{2I_j+1}\left[b_j^+ - b_j^-\right].\end{aligned} \tag{1.249}$$

The *spin-dependent scattering length* b_N defined above is proportional to the so-called *incoherent scattering length* (see the Appendix). Expectation values of the scattering-length operators over the possible polarization states of incoming and outgoing neutrons can be obtained by recourse to angular-momentum algebra:

$$\begin{aligned}\langle \uparrow |\hat{b}_j| \uparrow \rangle &= \bar{b}_j + \frac{1}{2}b_{Nj}I_{jz}, \\ \langle \downarrow |\hat{b}_j| \downarrow \rangle &= \bar{b}_j - \frac{1}{2}b_{Nj}I_{jz}, \\ \langle \downarrow |\hat{b}_j| \uparrow \rangle &= \frac{1}{2}b_{Nj}(I_{jx} + iI_{jy}), \\ \langle \uparrow |\hat{b}_j| \downarrow \rangle &= \frac{1}{2}b_{Nj}(I_{jx} - iI_{jy}),\end{aligned} \tag{1.250}$$

where down/up arrows denote the state with spin anti/parallel to the quantization (z) axis. The upper two cases correspond to non-spin-flip (nsf) and the lower two to spin-flip (sf) events.

For a single-element specimen consisting of randomly mixed isotopes a with concentrations c_a and randomly oriented spins I_a, the corresponding coherent scattering lengths are given by

$$\begin{aligned}\bar{b}_{nsf} &= \sum_a c_a \bar{b}_a = \sum_a \frac{c_a}{2I_a+1}\left[(I_a+1)b_a^+ + I_a b_a^-\right] = \bar{b}, \\ \bar{b}_{sf} &= 0,\end{aligned} \tag{1.251}$$

where $\bar{b}$ is the mean scattering length for unpolarized neutrons (see Equation 1.44). Thus, coherent scattering takes place without a neutron-spin flip. The corresponding incoherent cross sections are given by

$$\sigma_{i,\text{nsf}} = 4\pi \sum_a c_a \overline{b_a^2} - 4\pi \left(\sum_a c_a \overline{b}_a\right)^2 + \frac{\pi}{3} \sum_a c_a I_a(I_a+1) b_{Na}^2 = \sigma_i - \sigma_{i,\text{sf}},$$

$$\text{with } \sigma_{i,\text{sf}} = \frac{2\pi}{3} \sum_a c_a I_a(I_a+1) b_{Na}^2 = \frac{8\pi}{3} \sum_a c_a \frac{I_a(I_a+1)}{(2I_a+1)^2} \left[\overline{b}_a^+ - \overline{b}_a^-\right]^2.$$

(1.252)

Thus, if there are isotopes with $b_a^+ \neq b_a^-$, there is a nonzero spin-flip scattering component equal to two-thirds of the spin-incoherent scattering. In this case, polarization analysis can be used to change the ratio of coherent to incoherent scattering and thus separate out both scattering functions. With the appropriate substitutions of Equations (1.251) and (1.252), the formalism in Section 1.2 can be taken over.

Figure 1.23 shows a recent example where neutron-polarization analysis has been used to separate the coherent and incoherent response in hemoglobin and myoglobin solutions. The data above $Q = 1\,\text{Å}^{-1}$ shown in Figure 1.23c display a similar behavior to that discussed earlier in relation to the incoherent and coherent response from simple liquids (cf. Section 1.4.2 and Figure 1.12), with a prominent peak at $2.0\,\text{Å}^{-1}$ associated with solvent–solvent correlations. At lower energy transfers, coherent-scattering features are ascribed to intramolecular helix–helix separations within the protein (0.6–$0.7\,\text{Å}^{-1}$) and protein–protein distances (0.10–$0.15\,\text{Å}^{-1}$). Also note the predominance of coherent scattering in the low-angle region below $0.05\,\text{Å}^{-1}$.

1.5.2.2 Magnetic Scattering

For magnetic scattering, it is necessary to evaluate the neutron-spin dependence of the matrix elements associated with the interaction potential (cf. Equation 1.223). In analogy with Equation (1.250) for spin-dependent nuclear scattering, we define

$$\begin{aligned}
\langle \uparrow | \sigma \cdot \mathbf{D}_\perp | \uparrow \rangle &= D_{\perp z}, \\
\langle \downarrow | \sigma \cdot \mathbf{D}_\perp | \downarrow \rangle &= -D_{\perp z}, \\
\langle \downarrow | \sigma \cdot \mathbf{D}_\perp | \downarrow \rangle &= D_{\perp x} + iD_{\perp y}, \\
\langle \uparrow | \sigma \cdot \mathbf{D}_\perp | \downarrow \rangle &= D_{\perp x} - iD_{\perp y}.
\end{aligned}$$

(1.253)

The matrix elements can then be used to compute differential cross sections using the master formula.

1.5.2.3 The Localized Paramagnet

As a first illustration of polarization analysis in magnetic systems, we use the double differential cross section for non-spin-flip magnetic scattering given by

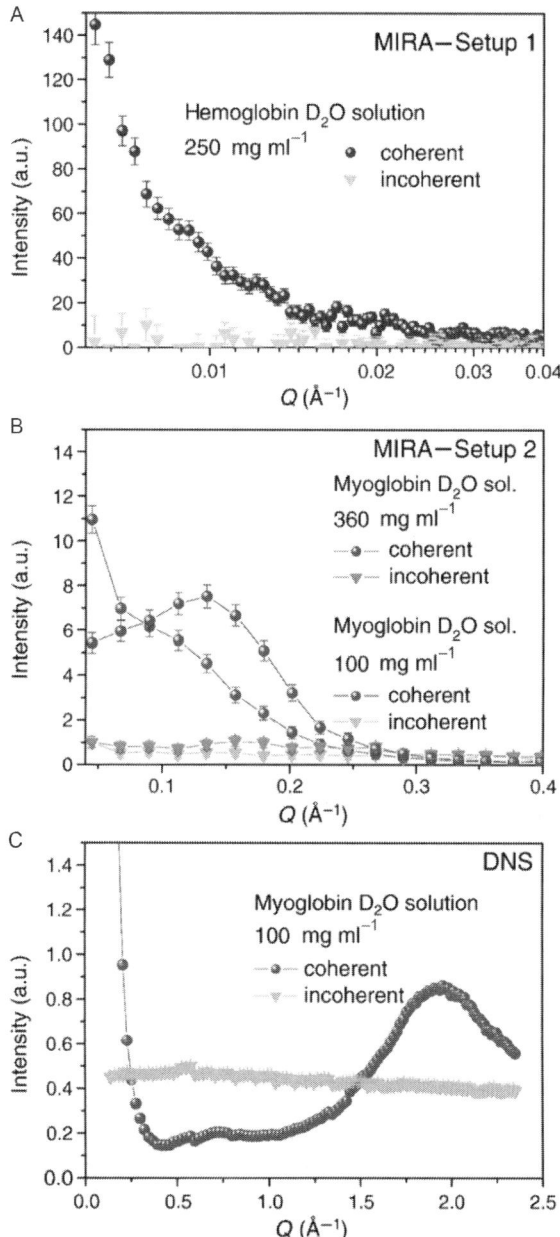

FIGURE 1.23 Incoherent- and coherent-scattering intensities of a series of protein samples: (a) low; (b) intermediate; and (c) intermediate and high momentum transfers Q. *Reprinted with permission from Ref. [198].* (For color version of this figure, the reader is referred to the online version of this chapter.)

$$\frac{d^2\sigma}{d\Omega dE_f} = \frac{1}{N_m}\left(\frac{k_f}{k_i}\right)(\gamma r_0)^2 \sum_{\tau_i} p_{\tau_i} \sum_{\tau_f} |\langle \tau_f | D_{\perp z} | \tau_i \rangle|^2 \delta(E + E_{\tau_i} - E_{\tau_f}). \quad (1.254)$$

For a paramagnetic solid with localized electrons in zero magnetic field, we may integrate over energy, and sum over initial (τ_i) and final target states (τ_f), to obtain

$$\left(\frac{d\sigma}{d\Omega}\right)_{nsf} = \frac{1}{N_m}(\gamma r_0)^2 \langle D^+_{\perp z} D_{\perp z}\rangle. \quad (1.255)$$

Using the results of Equations (1.234) and (1.248), this becomes

$$\left(\frac{d\sigma}{d\Omega}\right)_{nsf} = \frac{1}{N_m}(\gamma r_0)^2 \sum_j |f_j(Q)|^2 \frac{g_j^2}{12} S_j(S_j+1)\left(1 - \hat{Q}_z^2\right). \quad (1.256)$$

Comparison with Equation (1.240) shows that the spin-flip scattering must be

$$\left(\frac{d\sigma}{d\Omega}\right)_{sf} = \frac{1}{N_m}(\gamma r_0)^2 \sum_j |f_j(Q)|^2 \frac{g_j^2}{12} S_j(S_j+1)\left(1 + \hat{Q}_z^2\right). \quad (1.257)$$

Thus, if **Q** is parallel to the direction of neutron polarization, $\left(1 - \hat{Q}_z^2\right) = 0$ and all the scattering is spin-flip. Alternatively, if **Q** is perpendicular to the polarization direction, $\hat{Q}_z = 0$ and both spin-flip and non-spin-flip scattering cross sections are equal. The use of polarized neutrons makes it possible to separate paramagnetic from nuclear scattering by means of measurements using two distinct polarizations.

1.5.2.4 Probing Magnetic Order

In a second illustration of polarization analysis, we examine coherent and elastic neutron scattering from a system of magnetic atoms or ions, allowing for the presence of both nuclear and magnetic scattering. The interaction potential for nuclear scattering has matrix elements given by Equation (1.29) with the generalization to polarized neutrons (cf. Equation 1.248):

$$\langle \mathbf{k}_f | V | \mathbf{k}_i \rangle_{nucl} = \frac{2\pi \hbar^2}{m_n} \sum_j \hat{b}_j e^{i\mathbf{Q}\cdot\mathbf{R}_j}. \quad (1.258)$$

For magnetic scattering, we use Equations (1.223), (1.229), and (1.234) to write

$$\langle \mathbf{k}_f | V | \mathbf{k}_i \rangle_{mag} = \frac{2\pi \hbar^2}{m_n}(\gamma r_0)\sigma \cdot \mathbf{D}_\perp = \frac{2\pi \hbar^2}{m_n}\sum_j \sigma \cdot \mathbf{C}_j e^{i\mathbf{Q}\cdot\mathbf{R}_j}, \quad (1.259)$$

where the coefficient $\mathbf{C}_j$ is defined as

$$\mathbf{C}_j = (\gamma r_0) f_j(\mathbf{Q}) \left(\hat{\mathbf{Q}} \times \boldsymbol{\mu}_j \times \hat{\mathbf{Q}} \right). \quad (1.260)$$

The total matrix element is therefore

$$\langle \mathbf{k}_f | V | \mathbf{k}_i \rangle_{\text{mag}} = \frac{2\pi \hbar^2}{m_n} \sum_j (\hat{b}_j + \boldsymbol{\sigma} \cdot \mathbf{C}_j) e^{i\mathbf{Q} \cdot \mathbf{R}_j}, \quad (1.261)$$

and from Equation (1.25) the scattering cross section for an event with initial and final neutron spin states σ_i and σ_f and initial and final states of the system τ_i and τ_f is

$$\left(\frac{d^2\sigma}{d\Omega dE_f} \right)_{\sigma_i \tau_i \to \sigma_f \tau_f} = \frac{1}{N_m} \left(\frac{k_f}{k_i} \right) \left| \left\langle \sigma_f \tau_f \left| \sum_j (\hat{b}_j + \boldsymbol{\sigma} \cdot \mathbf{C}_j) e^{i\mathbf{Q} \cdot \mathbf{R}_j} \right| \sigma_i \tau_i \right\rangle \right|^2 \delta(E + E_{\tau_i} - E_{\tau_f}).$$

$$(1.262)$$

For coherent elastic scattering, we may use the results from Section 1.2.3 and generalize Equation (1.70) to the case of polarized-neutron analysis of combined magnetic and nuclear scattering:

$$\left(\frac{d\sigma}{d\Omega} \right)_{\sigma_i \to \sigma_f} = \frac{(2\pi)^3}{v_0} \sum_{\boldsymbol{\tau}} |\langle \sigma_f | \hat{F}(\boldsymbol{\tau}) | \sigma_i \rangle|^2 \delta(\mathbf{Q} - \boldsymbol{\tau}) \quad (1.263)$$

with a combined nuclear-magnetic form factor of the form

$$\hat{F}(\boldsymbol{\tau}) = \sum_d (\bar{b}_d + \boldsymbol{\sigma} \cdot \mathbf{C}_d) e^{-W_d} e^{i\boldsymbol{\tau} \cdot \mathbf{d}}. \quad (1.264)$$

For the case of coherent scattering, the neutron-spin-dependent matrix elements of $\hat{F}(\boldsymbol{\tau})$ can be obtained from Equations (1.250) and (1.253), and are given by

$$\begin{aligned}
\langle \uparrow | \hat{F} | \rangle \uparrow &= F_N - F_{M\perp z}, \\
\langle \downarrow | \hat{F} | \rangle \downarrow &= F_N + F_{M\perp z}, \\
\langle \downarrow | \hat{F} | \rangle \uparrow &= -\left(F_{M\perp x} + i F_{M\perp y} \right), \\
\langle \uparrow | \hat{F} | \rangle \downarrow &= -\left(F_{M\perp x} - i F_{M\perp y} \right),
\end{aligned} \quad (1.265)$$

where $F_N(\boldsymbol{\tau})$ is the unit-cell structure factor for nuclear scattering given by Equation (1.71), and $F_{M\perp}(\boldsymbol{\tau})$ is the magnetic structure factor given by Equations (1.244) and (1.246). In essence, Equation (1.263) with the matrix elements shown in Equation (1.265) replaces the expressions for unpolarized neutrons given earlier by Equations (1.70) and (1.245). The matrix elements given by Equation (1.265) also serve to illustrate that the scattering depends on the relative orientation of three vectors, namely: the neutron-spin quantization axis z, the magnetization of the system described by $\mathbf{F}_M$, and the wave-vector $\mathbf{Q}$ (governing the orientation and magnitude of $\mathbf{F}_{M\perp}$). The participation of these three distinct vector quantities suggests that there are several distinct ways of conducting magnetic-structure experiments with polarized neutrons.

When z is orthogonal to $\mathbf{Q}$ and parallel to $\mathbf{F}_M$ (see Figure 1.22b), Equation (1.265) shows that the spin-flip-scattering channel is zero and that the *flipping ratio* (ratio of cross sections for neutrons parallel and antiparallel to $\mathbf{F}_{M\perp}$) is

$$r = \left(\frac{F_N - F_{M\perp}}{F_N + F_{M\perp}}\right)^2 \approx 1 - \frac{4F_{M\perp}}{F_N}. \qquad (1.266)$$

When $F_{M\perp}$ is small compared with F_N, it is preferable to measure it through the flipping ratio, not the total scattering cross section as the latter is increased only by a fraction $(F_{M\perp}/F_N)^2$ arising from the presence of magnetic scattering. Furthermore, flipping ratios are amenable to accurate measurements owing to the cancellation of common systematic errors. From these considerations, it should come as little surprise that much of the study of magnetic structures with polarized neutrons has involved extensive use of Equation (1.266).

The flipping ratio as defined by Equation (1.266) also provides the most commonly used method for producing and analyzing polarized beams, arranging for $\mathbf{F}_M$ in a magnetic crystal to be perpendicular to $\mathbf{Q}$, so as to fulfill the condition $\mathbf{F}_{M\perp} = \mathbf{F}_N$, and choosing crystals such that $\mathbf{F}_N \approx \mathbf{F}_{M\perp}$. In this case, the outgoing neutron beam becomes polarized either parallel or antiparallel to $\mathbf{F}_M$. For a Bravais lattice, the required condition is equivalent to

$$|b| = f(\mathbf{Q})|\gamma r_0 \mu|, \qquad (1.267)$$

a requirement which holds reasonably well, for example, for the (1.200) reflection of the Heussler alloy $Co_{0.92}Fe_{0.08}$. The output beam in this case is both monochromatic and polarized. For the production of broadband polarized beams, other methods must be used such as: neutron guides (see Refs. [199–201] and Section 1.5.5 below); ^{4}He-based spin filters [202,203]; the spin dependence of nuclear absorption resonances such as that from ^{149}Sm [204]; or the scattering from polarized nuclei (see Ref. [205] and Section 1.7).

An alternative geometry to the one described above involves arranging $\mathbf{Q}$ parallel to the neutron quantization axis z, so that $\mathbf{F}_{M\perp}$ becomes perpendicular to z (see Figure 1.22c). As shown by Equation (1.265), magnetic scattering becomes all spin-flip, while nuclear scattering remains non-spin-flip. For a ferromagnet, a magnetic field must be applied to align the domains, and this will define the z direction, so that normally there will be no component of $\mathbf{F}_{M\perp}$ perpendicular to z. For noncollinear magnetism (both ferro and antiferro), however, such components are nonvanishing and can be identified by measuring the spin-flip and non-spin-flip cross sections separately.

1.5.3 Magnetic Bragg Scattering

1.5.3.1 Ferromagnets

Ferromagnetic crystals are characterized by identical magnetic and nuclear unit cells and, therefore, magnetic and nuclear scattering take place at identical positions in reciprocal space. The simplest (and most widely used) method to separate these two contributions involves taking the difference between datasets above and below the ordering (Curie) temperature. Another approach involves applying a saturating magnetic field along the momentum-transfer vector $\mathbf{Q}$. The effect of the field is to remove magnetic scattering since in this case $\mathbf{F}_{M\perp} \equiv 0$. At zero field, on the other hand, the domains will be randomly oriented with a scattered intensity proportional to

$$1 - \overline{(\hat{\boldsymbol{\tau}} \cdot \hat{\boldsymbol{\eta}})^2}, \tag{1.268}$$

where the second term represents an average over domains. For randomly oriented powders and cubic crystals this term is equal to 2/3, and in the general case it is greater than zero. For uniaxial or orthorhombic structures, the average will depend on the specific orientation of the reciprocal-lattice vector $\hat{\boldsymbol{\tau}}$. As such, it can be used to deduce the inclination of the spin direction relative to the principal axes. For powders, it is generally hard to saturate the sample parallel to $\mathbf{Q}$, and the temperature dependence is frequently used to separate magnetic and nuclear scattering.

The most powerful and least ambiguous method for measuring ferromagnetic structures is diffraction from single crystals with polarized neutrons using the configuration $\mathbf{F}_M \| z \perp \boldsymbol{\tau}$ (Figure 1.22b). In this case $\mathbf{F}_{M\perp} \approx \mathbf{F}_M$ and the flipping ratios give $\mathbf{F}_M(\boldsymbol{\tau})$ in terms of $\mathbf{F}_N(\boldsymbol{\tau})$. The results may be interpreted in terms of magnetic moments on individual atoms as in Equation (1.244), or $\mathbf{F}_M(\boldsymbol{\tau})$ can be Fourier transformed to give the spin density as a function of position in the unit cell. Figure 1.24 shows the spin density obtained for the (001) plane in iron by Shull and Yamada in the early days of polarized neutron scattering [13]. The higher densities along the direction of the cube edges show that the 3d electronic orbitals are preferentially aligned along these directions—see also Figure 1.4 in Section 1.1.

1.5.3.2 Antiferromagnets

Antiferromagnets represent a departure from the previous case, as they possess distinctly different nuclear and magnetic lattices leading to purely magnetic Bragg reflections. In this case, it is therefore not necessary to use polarized neutrons to isolate the magnetic response—most antiferromagnetic structures have been determined with unpolarized beams, and often with powder samples. If the structure has sufficiently low symmetry, it may be possible to obtain the orientation of the moments in the magnetic structure, as already discussed in Section 1.5.3.1.

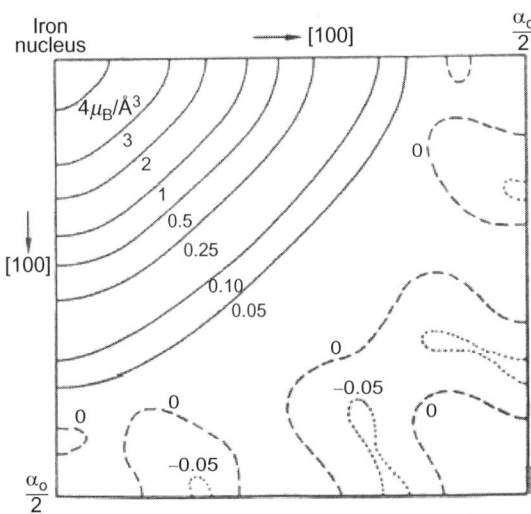

FIGURE 1.24 Spin density for iron on the (001) plane. *Reprinted with permission from Ref. [13].*

An instructive illustration of antiferromagnetic ordering is provided by compounds whose magnetic structure is characterized by a spin direction that changes with position in the lattice. When the repeat distance is not an exact multiple of the repeat distance of the nuclear structure (a rather common situation), the magnetic structure is then called incommensurate. If the magnitude and direction of the moments on each magnetic atom are described by a function $\boldsymbol{\mu}(\mathbf{r})$, the periodicity of $\boldsymbol{\mu}(\mathbf{r})$ can be described with propagation vectors $\boldsymbol{\lambda}$:

$$\boldsymbol{\mu}(\mathbf{r}) = \sum_n \boldsymbol{\mu}_n e^{in\boldsymbol{\lambda}\cdot\mathbf{r}}. \qquad (1.269)$$

And the magnetic unit-cell structure factor, Equation (1.244), for the nth component is

$$\mathbf{F}_{Mn}(\mathbf{Q}) = -(\gamma r_0) \sum_d \boldsymbol{\mu}_n f_d(\mathbf{Q}) e^{-W_d(\mathbf{Q})} e^{i(n\boldsymbol{\lambda}+\mathbf{Q})\cdot\mathbf{d}}. \qquad (1.270)$$

The differential cross section in Equation (1.245) becomes

$$\left(\frac{d\sigma}{d\Omega}\right)_{el} = \frac{(2\pi)^3}{v_0} \sum_n \sum_\tau \delta(\mathbf{Q}+n\boldsymbol{\lambda}-\boldsymbol{\tau}) |\mathbf{F}_{Mn\perp}(\boldsymbol{\tau})|^2. \qquad (1.271)$$

The most salient feature of Equation (1.271) is the emergence of satellite peaks in the magnetic diffraction pattern, displaced from nuclear Bragg peaks by multiples of the propagation vector $\boldsymbol{\lambda}$. The intensity of these satellites depends on the Fourier components $\boldsymbol{\mu}_n$.

The experiments on holmium by Koehler *et al.* [206] provide a beautiful and legendary example of satellite reflections in a magnetic structure. In the temperature region below the Néel temperature (133 K), the magnetic moments form a simple helical structure

$$\mu_x(\mathbf{r}) = \mu\cos(\boldsymbol{\lambda}\cdot\mathbf{r}), \quad \mu_y(\mathbf{r}) = \mu\sin(\boldsymbol{\lambda}\cdot\mathbf{r}), \quad \mu_z(\mathbf{r}) = 0, \qquad (1.272)$$

which has just one Fourier component, so there is a pair of satellites $\boldsymbol{\tau}+\boldsymbol{\lambda}$ about each Bragg reflection. Figure 1.25a shows the crystal rocking curve about the (100) reflection at 77 K; at this temperature λ is about 0.46 π/c, where c is the lattice parameter. As the temperature is lowered, λ decreases and at the same time new peaks arise in the diffraction pattern. At 4.2 K (Figure 1.25b), $\lambda = \pi/(1.3c)$, that is, the helical structure is now commensurate with the crystal lattice, and peaks are now observed at $\boldsymbol{\tau}\pm 5\boldsymbol{\lambda}$ (labeled 5, 7 in the figure) and also at $\boldsymbol{\tau}\pm\boldsymbol{\lambda}$ (labeled 1, 11). This is due to slight changes in the orientation of the moments, in opposite directions on successive planes, which produce a lower energy configuration on account of crystal-anisotropy effects. The new magnetic structure is thus described by

$$\mu_x(\mathbf{r}^\pm) = \mu\sin\gamma\cos(\boldsymbol{\lambda}\cdot\mathbf{r}\pm\delta), \quad \mu_y(\mathbf{r}^\pm) = \mu\sin\gamma\sin(\boldsymbol{\lambda}\cdot\mathbf{r}\pm\delta), \quad \mu_z(\mathbf{r}^\pm) = \mu\cos\gamma, \qquad (1.273)$$

where the plus and minus signs refer to alternating planes normal to the c axis, and the z component represents a small ferromagnetic contribution. The occurrence of the fifth- and seventh-harmonic satellites can be qualitatively understood by considering the structure of Equation (1.273) as a doubling of the magnetic unit cell structure in the c direction. This structure leads to new reciprocal points $\boldsymbol{\tau}'$, such as (001), leading to new satellites at $\boldsymbol{\tau}'+\boldsymbol{\lambda}$, for example, (005/6) and (007/6). Comparison of the measured intensities of these satellites with those calculated from the magnetic unit-cell structure factor (Equation 1.270) shows that δ is about 6° at 4.2 K.

1.5.4 Diffuse Scattering from Magnetic Disorder

As in the nuclear case, when the magnetic moments do not follow a periodic pattern, the elastic magnetic scattering becomes a continuous function of $\mathbf{Q}$—diffuse scattering. Assuming that the moments lie along a unique direction $\hat{\boldsymbol{\eta}}$, the differential cross section given by Equation (1.242) is

$$\frac{d\sigma}{d\Omega} = \frac{1}{N_m}(\gamma r_0)^2\left(1 - \hat{\mathbf{Q}}\cdot\hat{\boldsymbol{\eta}}\right)^2|\langle D(\mathbf{Q})\rangle|^2 \qquad (1.274)$$

with

$$|\langle D(\mathbf{Q})\rangle|^2 = \sum_{jj'} f_j(\mathbf{Q})f_{j'}(\mathbf{Q})\langle\mu_j\rangle\langle\mu_{j'}\rangle e^{i\mathbf{Q}\cdot(\mathbf{j}'-\mathbf{j})}e^{-W_j-W_{j'}}. \qquad (1.275)$$

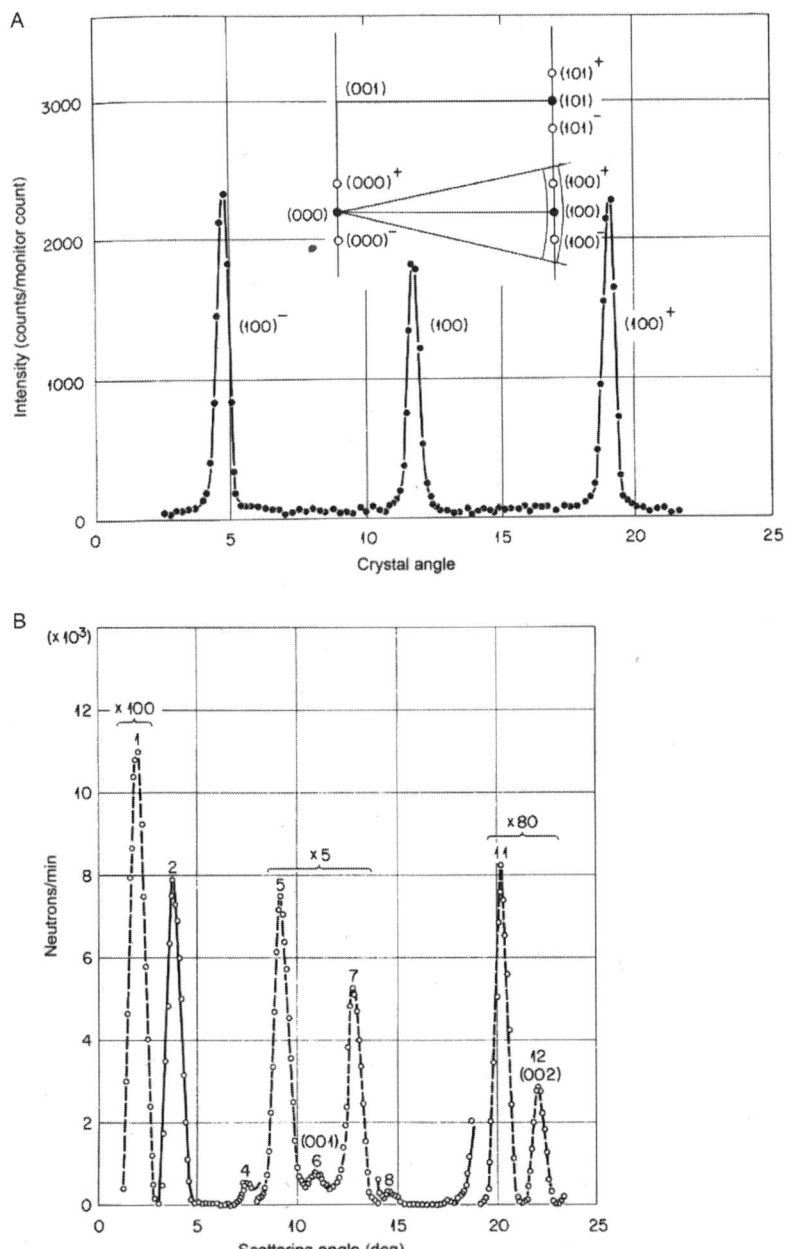

FIGURE 1.25 (a) Crystal rocking curve about the (100) reflection in holmium at 77 K and (b) at 4.2 K. Reprinted with permission from Ref. [206].

Considering an average spin of atoms of type d as

$$\bar{\mu}_d = \frac{1}{N_d} \sum_{j \in d} \langle \mu_j \rangle, \qquad (1.276)$$

this average value in combination with Equation (1.275) leads to Bragg scattering as defined in the previous section. Deviations from the average, given by

$$\langle \mu_j \rangle \approx \bar{\mu}_d + \Delta \mu_j, \qquad (1.277)$$

give rise to diffuse scattering proportional to

$$|\langle D(\mathbf{Q}) \rangle|^2_{\text{diff}} = \sum_{\substack{j \in d \\ j' \in d'}} f_d(\mathbf{Q}) f_{d'}(\mathbf{Q}) \Delta \mu_j \Delta \mu_{j'} e^{i\mathbf{Q} \cdot (\mathbf{j}' - \mathbf{j})} e^{-W_d - W_{d'}}. \qquad (1.278)$$

These considerations are entirely analogous to the diffuse nuclear scattering discussed in Section 1.4.4. As illustration of this formalism, a disordered alloy with concentration c of impurity atoms is characterized by deviations of the form

$$\Delta \mu_j = \Delta \mu \cdot p_j + \sum_{j' \neq j} \phi(\mathbf{j} - \mathbf{j}') p_{j'}, \qquad (1.279)$$

where $\Delta \mu$ is the difference between the magnetic moment of the impurity and that of the host atoms. Furthermore, $p_j = 1$ if the site is occupied by an impurity atom and 0 otherwise, and $\phi(\mathbf{j} - \mathbf{j}')$ represents the effect of an impurity atom on the spin of the host. Ignoring the effect of the impurity atom on $f(\mathbf{Q})$, one gets

$$|\langle D(\mathbf{Q}) \rangle|^2_{\text{diff}} = f^2(\mathbf{Q}) c(1-c) [\Delta \mu + \phi(\mathbf{Q})]^2 e^{-2W}, \qquad (1.280)$$

where $\phi(\mathbf{Q})$ is the Fourier transform of $\phi(\mathbf{j} - \mathbf{j}')$. This equation assumes that the magnetic defects associated with each impurity atom are independent of each other, a reasonable assumption in the dilute limit. Also, $\phi(\mathbf{Q} = 0)$ represents the total magnetic-defect moment per impurity atom other than that at the impurity site itself.

A classic set of measurements of dilute magnetic alloys was carried out by Low and Collins in the 1960s [207]. They used a simple time-of-flight diffractometer with a saturating magnetic field that could be rotated parallel and perpendicular to the scattering vector $\mathbf{Q}$ in order to isolate the magnetic scattering. Some results on iron alloys are shown in Figure 1.26. An interesting feature is the behavior shown by several alloys in which the cross section passes through zero near $Q \approx 0.5$ Å^{-1}. This result implies that $\phi(\mathbf{Q} = 0)$ is opposite in sign to $\Delta \mu$ so that the sum of the two crosses zero when $|\phi| = |\Delta \mu|$. Thus, the change in moment at the impurity site itself is opposite in sign to the net moment due to the impurity. The fact that the scattering goes

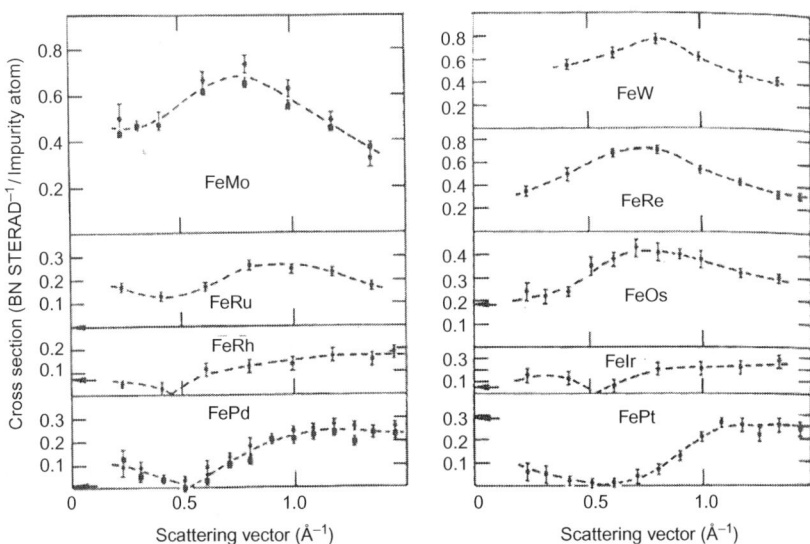

FIGURE 1.26 Diffuse magnetic scattering from iron alloys. *Reprinted with permission from Ref. [207].*

precisely to zero at a certain value of Q implies that the distribution of $\phi(\mathbf{Q})$ with respect to direction must be nearly isotropic. More precise treatments must take account of the effects of short-range order (which can be inferred from the *nuclear* diffuse scattering) and interactions between the defects ignored in the treatment leading to Equation (1.280). In this case, departures from the dilute limit in defect concentration can be examined. A comprehensive review in given in Ref. [208].

In certain classes of materials, there are static moments which do not exhibit long-range order either due to disorder or spatial constraints (e.g., nanoparticles). In this case, the magnetic scattering is diffuse in character. One class is the spin glasses, where the moments exhibit short-range correlations and there is no net magnetic moment. Another is that of amorphous magnetic materials and magnetic liquids, where the atomic positions as well as the moments only have short-range order. Relevant reviews on magnetically disordered systems can be found in Refs. [209–211].

To conclude this section, we return to a more fundamental example of the use of polarized neutron diffraction to probe diffuse magnetic scattering in the spin ice $Ho_2Ti_2O_7$, a material that has been predicted to support magnetic-monopole excitations [212]. Figure 1.27 shows diffuse-scattering data using an experimental setup configured to measure two distinct scattering functions associated with two-spin correlation functions along the y (spin-flip) and z (non-spin-flip) directions. The characteristic pinch points observed in the

Chapter 1 An Introduction to Neutron Scattering

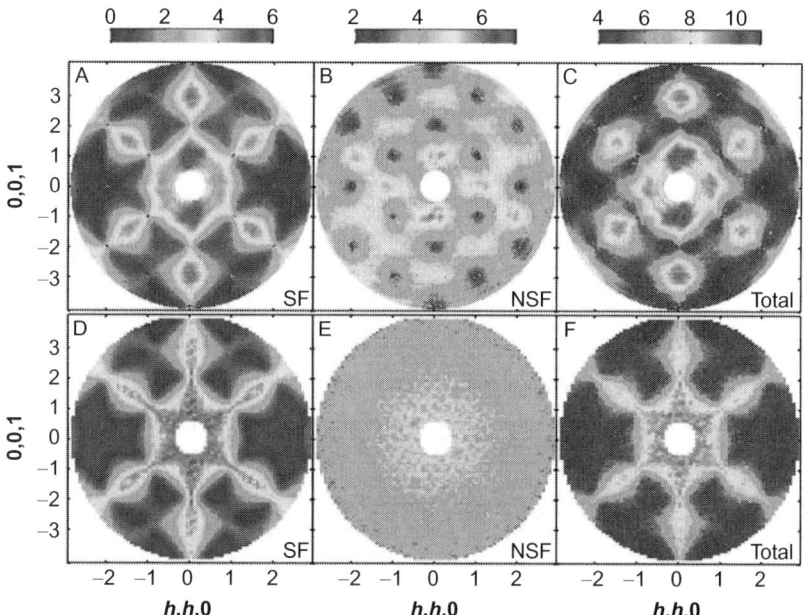

FIGURE 1.27 Diffuse scattering using polarized neutron diffraction techniques (a–c) compared against theoretical predictions using a near-neighbor model (d–f) for a single crystal of the spin-ice compound $Ho_2Ti_2O_7$: (a) spin-flip scattering at $T = 1.7$ K; (b) non-spin-flip scattering at the same temperature; (c) sum of spin-flip and non-spin-flip scattering intensities. *Reprinted with permission from Ref. [212].* (For color version of this figure, the reader is referred to the online version of this chapter.)

spin-flip data at the Brillouin zone centers (0,0,2), (1,1,1), and (2,2,2) in Figure 1.27a provide an unambiguous signature of the dipolar form of the spin correlation function expected for a Coulomb phase, the quasiparticle vacuum for magnetic monopoles. These characteristic features in the spin-flip data are not at all evident in the unpolarized data shown in Figure 1.27c.

1.5.5 Large-Scale Magnetic Structures

We recall the expression for the differential cross section for elastic magnetic scattering (cf. Equation 1.242)

$$\left(\frac{d\sigma}{d\Omega}\right)_{el} = \frac{1}{N_m}(\gamma r_0)^2 |\langle \mathbf{D}_\perp(\mathbf{Q}) \rangle|^2, \quad (1.281)$$

where $\mathbf{D}(\mathbf{Q}) = -\mathbf{M}(\mathbf{Q})/2\mu_B$, and $\mathbf{M}(\mathbf{Q})$ is the Fourier transform of the magnetization density

$$\mathbf{M}(\mathbf{Q}) = \int \mathbf{M}(\mathbf{r}) e^{i\mathbf{Q}\cdot\mathbf{r}} d\mathbf{r}. \quad (1.282)$$

Long-range fluctuations in the magnetization lead to small-angle scattering, analogous to the small-angle nuclear scattering caused by density fluctuations (Section 1.4.6). In real systems, these fluctuations can be caused, for example, by dislocations in a metal as a result of magnetoelastic coupling between the elastic strain field of the dislocations and the magnetization density. The resulting small-angle scattering can be much larger than the nuclear small-angle scattering caused by the dislocations.

An elegant use of small-angle magnetic neutron scattering in the study of emergent properties and exotic phases of matter is shown in Figure 1.28. At

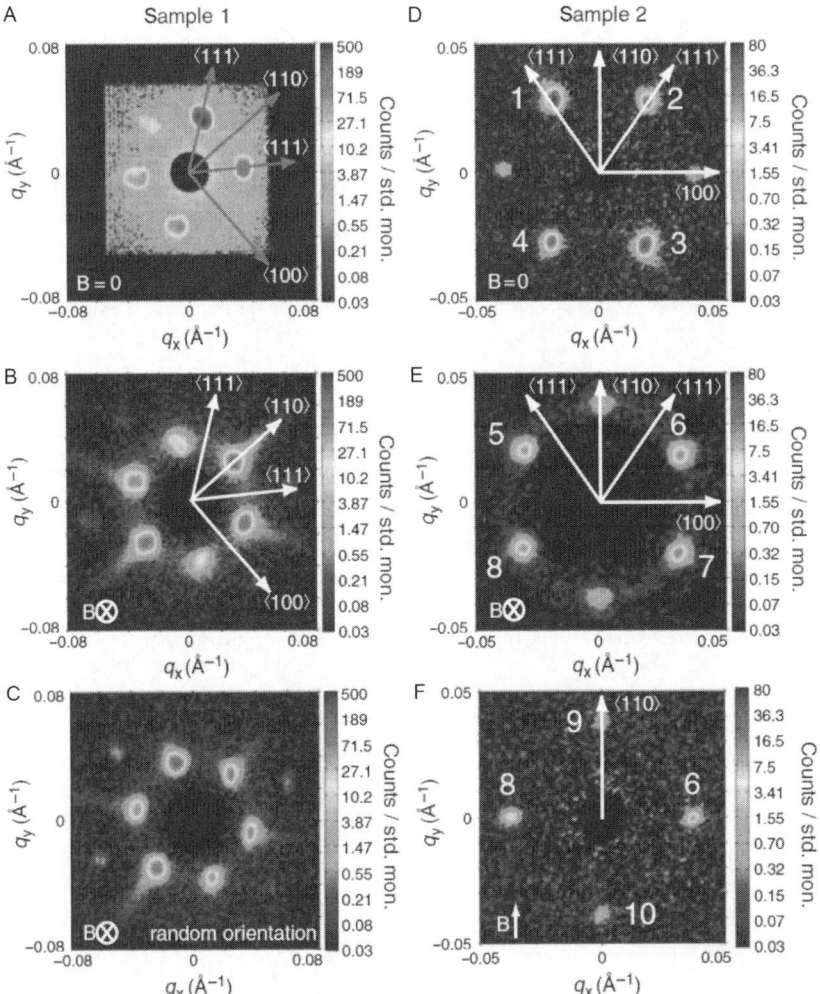

FIGURE 1.28 Small-angle neutron-scattering data showing the formation of a skyrmion lattice in two separate MnSi specimens, as labeled in the figure. *Reprinted with permission from Ref. [213].* (For color version of this figure, the reader is referred to the online version of this chapter.)

zero applied magnetic field, the small-angle scattering pattern for the chiral itinerant-electron magnet MnSi evinces the presence of helical order in the zero-field-cooled state. Sixfold symmetry emerges when a small field ($B \sim 0.15$ T) is applied to the sample, as shown in (b) of this figure. Remarkably, the same structure is still present for a randomly oriented sample (c). The observed magnetic structure is orthogonal to the applied magnetic field and entirely independent of the underlying crystalline order. These beautiful neutron-scattering data have been interpreted in terms of the spontaneous formation of a two-dimensional lattice of skyrmion lines—a topologically stable field configuration with particle-like properties, in this case involving electron-spin states.

As discussed earlier in the context of NR (Section 1.4.6.2), neutron transport through materials can be described in terms of a refractive index:

$$n = 1 - \frac{\rho}{2\pi} \lambda_n^2 \bar{b}, \quad (1.283)$$

where ρ is the number density of scattering centers characterized by a mean scattering length $\bar{b}$. In a magnetic medium, the latter necessarily contains magnetic contributions. If the spin direction is along the direction of magnetization, Equation (1.283) reads

$$n^\pm = 1 - \frac{\rho}{2\pi} \lambda_n^2 (\bar{b} \pm C), \quad (1.284)$$

for the magnetization parallel (+) and antiparallel (−) to the neutron-spin direction. For a neutron travelling in air and encountering the surface of such a medium, total reflection takes place at glancing angles less than the critical angles defined by

$$\theta_c^\mp = \lambda_n \left[\frac{\rho}{\pi} (\bar{b} \mp C) \right]^{1/2}. \quad (1.285)$$

For angles larger than θ_c, reflection is still possible but falls off over an angular range $\sim \theta_c$. This feature constitutes the basis for a magnetic neutron mirror. For $b \approx C$, it will reflect neutrons of the down-spin state only. Mirrors are used as polarizing devices where relatively long-wavelength neutrons are required and small angular divergences can be tolerated.

Equation (1.284) also provides a means for studying domain distributions in magnetically ordered materials, since refraction at a domain boundary is also governed by the refractive index n. An early experiment in this area was carried out by Hughes et al. [214] who measured the angular divergence of a neutron beam passed through an iron slab in magnetized and unmagnetized states. The average deflection at each boundary inferred from the estimated domain size and sample thickness agreed with that calculated from Equation (1.284) within 10%.

An important application of neutron-optical effects in magnetic scattering is in the field of surface magnetism, where the profile of the magnetization near the surface can be derived from measurements of the reflected intensity as a function of grazing angle and neutron polarization. In a pioneering experiment by Felcher et al. [215], the penetration of magnetic field into the surface of superconducting niobium was measured from the reflection of polarized neutrons. In this case Equation (1.284) takes the form

$$n^{\pm}(z) = 1 - \frac{1}{2\pi}[\rho\bar{b} \mp cH(z)], \qquad (1.286)$$

where c is a material-dependent constant. It is also assumed that the magnetic-field penetration into the material is of the form

$$H(z) = He^{-z/\Lambda}, \qquad (1.287)$$

where Λ is a characteristic penetration depth. An external field H is applied parallel to the surface and it is less than the critical entry field for magnetic flux, H_{c1}. Using a polarized beam with a broadband wavelength spectrum (from a pulsed neutron source), and polarized by a mirror of the type described above in conjunction with a spin flipper in front of the sample, the flipping ratio of the reflectivities for the two spin states, R^+/R^-, as a function of neutron wavelength can be measured. Once again, measurement of a flipping ratio provides a sensitive measurement since most sources of systematic error are eliminated. The expression for the reflectivities is more complicated than Equation (1.284) for the refractive indices, but an approximate expression can be written down explicitly for the field dependence given above. From the flipping ratio and an extrapolation of the temperature dependence to $T=0$, a value of $\Lambda(0) = 410 \pm 40$ Å can be derived. This figure agrees well with estimates based on local Ginzburg–Landau theory and the measured temperature dependence of the critical fields.

Progress in the use of polarized neutrons to study magnetic interfaces and superlattices has moved well beyond these seminal studies. Figure 1.29 shows NR data for a heterostructure made of thin (nm) layers of the half-metallic ferromagnet $La_{2/3}Ca_{1/3}MnO_3$ and the superconductor $YBa_2Cu_3O_7$. In this figure, the line across $Q_x=0$ corresponds to specular reflectivity (both nuclear and magnetic) and remains the dominant contribution to the signal well below the onset of ferromagnetism at $T=180$ K. The absence of diffuse scattering under these conditions also tells us that the in-plane magnetic roughness in this multi-layered material is negligible. Below the structural transition at $T=105$ K, there is a broadening of the main Bragg feature along the off-specular direction, thought to be related to the distortion of the substrate surface. Below the superconducting temperature, the off-specular scattering shows a dramatic change in terms of in-plane structure, characteristic of the presence of stripe-like magnetic domains. From the total range and inter-peak separation of these Bragg features along Q_x, it becomes possible to quantify both domain size (d) and domain-wall width (w). These properties are quite

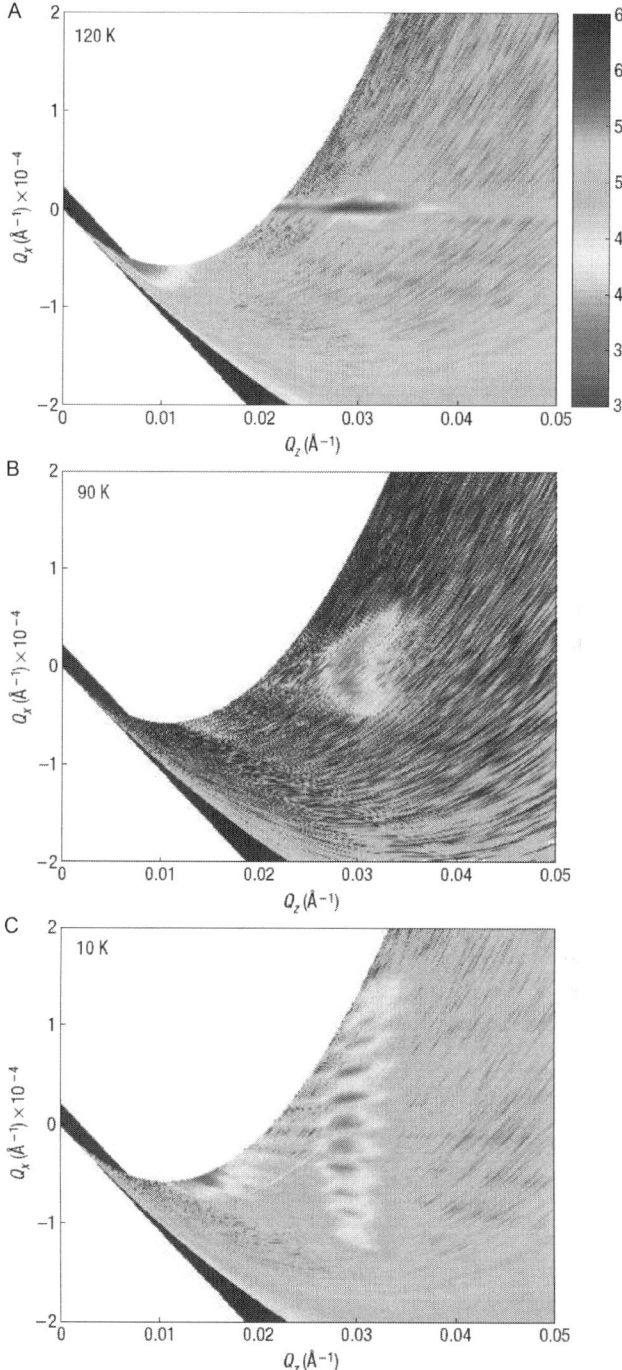

FIGURE 1.29 Temperature dependence of specular and off-specular neutron reflectivity around the first superlattice Bragg reflection in $La_{2/3}Ca_{1/3}MnO_3/YBa_2Cu_3O_7$ multilayers: (a) below the onset of ferromagnetism; (b) below the structural transition in the substrate at $T = 105$ K; and (c) below the superconducting transition temperature at 78 K. Note the large difference in momentum-transfer scales along the specular (bottom) and off-specular directions (left). *Reprinted with permission from Ref. [216].* (For color version of this figure, the reader is referred to the online version of this chapter.)

sensitive to temperature, with $d = 24(1.41)\,\mu\text{m}$ and $w = 2(1.3)\,\mu\text{m}$ at $T = 10$ (1.63) K. In addition to providing information on the spin and orbital electronic structure at the interfaces, these data thus demonstrate the possibility of extensive rearrangements of magnetic-domain structure at the superconducting transition temperature in artificial nanostructured materials.

1.6 SPIN DYNAMICS

1.6.1 Generalized Susceptibility

As in the case of inelastic nuclear scattering presented in Section 1.3, magnetic spins can sustain single-particle and collective excitations (spin waves). Their theoretical treatment requires going back to the double differential scattering cross section presented in Equation (1.237) of Section 1.5.1:

$$\frac{d^2\sigma}{d\Omega dE_f} = \frac{1}{N_m}\left(\frac{k_f}{k_i}\right)(\gamma r_0)^2 \sum_{\alpha\beta}\left(\delta_{\alpha\beta} - \hat{Q}_\alpha \hat{Q}_\beta\right)\sum_{jj'} f_j^*(\mathbf{Q})f_{j'}(\mathbf{Q}) \times$$
$$\times \frac{1}{2\pi\hbar}\int \langle \mu_{j\alpha}(0)\mu_{j'\beta}(t)\rangle \langle e^{-i\mathbf{Q}\cdot\mathbf{R}_j(0)}e^{i\mathbf{Q}\cdot\mathbf{R}_{j'}(t)}\rangle e^{-iEt/\hbar}dt. \quad (1.288)$$

The two correlation functions shown under the integrand above can be expressed as the sum of time-independent and time-dependent parts:

$$J_{jj'}^{\alpha\beta}(t) \equiv \langle \mu_{j\alpha}(0)\mu_{j'\beta}(t)\rangle = J_{jj'}^{\alpha\beta}(\infty) + \widetilde{J}_{jj'}^{\alpha\beta}(t),$$
$$I_{jj'}(\mathbf{Q}, t) = \langle e^{-i\mathbf{Q}\cdot\mathbf{R}_j(0)}e^{i\mathbf{Q}\cdot\mathbf{R}_{j'}(t)}\rangle = I_{jj'}(\mathbf{Q}, \infty) + \widetilde{I}_{jj'}(\mathbf{Q}, t). \quad (1.289)$$

With these definitions, the Fourier time integral in Equation (1.288) can be written as

$$\int \left[J_{jj'}^{\alpha\beta}(\infty) + \widetilde{J}_{jj'}^{\alpha\beta}(t)\right]\left[I_{jj'}(\mathbf{Q}, \infty) + \widetilde{I}_{jj'}(\mathbf{Q}, t)\right]e^{-iEt/\hbar}dt. \quad (1.290)$$

From the four distinct terms arising from this expression, the time-independent term $J_{jj'}^{\alpha\beta}(\infty)I_{jj'}(\mathbf{Q}, \infty)$ corresponds to elastic scattering, as discussed in Section 1.5. The *magneto-vibrational* term $J_{jj'}^{\alpha\beta}(\infty)I_{jj'}(\mathbf{Q}, t)$ has the same energy dependence as the nuclear scattering but the interaction of the neutron with the system is through the magnetic potential. For ferromagnets with identical spins aligned parallel to the z direction, we can write

$$J_{jj'}^{\alpha\beta}(\infty) = \delta_{\alpha\beta}\delta_{\alpha z}\langle\mu^z\rangle^2 \quad (1.291)$$

and the double differential scattering cross section for elastic and magneto-vibrational scattering is

$$\frac{d^2\sigma}{d\Omega dE_f} = \frac{1}{N_m}\left(\frac{k_f}{k_i}\right)(\gamma r_0)^2\left(1-\hat{Q}_z^2\right)f^2(\mathbf{Q})\langle\mu^z\rangle^2\sum_{jj'}\frac{1}{2\pi\hbar}$$
$$\int_{-\infty}^{\infty}\left\langle e^{-i\mathbf{Q}\cdot\mathbf{R}_j(0)}e^{i\mathbf{Q}\cdot\mathbf{R}_{j'}(t)}\right\rangle e^{-iEt/\hbar}dt \quad (1.292)$$

Comparison with Equation (1.47) shows that this expression corresponds to the coherent nuclear scattering function with the factor

$$(\gamma r_0)^2\left(1-\hat{Q}_z^2\right)f^2(\mathbf{Q})\langle\mu^z\rangle^2 \quad (1.293)$$

instead of $|\overline{b}|^2$ for nuclear scattering.

$J_{jj'}^{\alpha\beta}(t)I_{jj'}(\mathbf{Q},\infty)$ in the product above is responsible for inelastic magnetic scattering. For a solid with long-range order we may write (similar to Equation 1.68)

$$\mathbf{j} = \mathbf{l} + \mathbf{d} \quad (1.294)$$

and the results of Section 1.2.3 can be used to give the double differential magnetic scattering cross section (elastic plus inelastic) as

$$\frac{d^2\sigma}{d\Omega dE_f} = \frac{1}{N_m}\left(\frac{k_f}{k_i}\right)\frac{(\gamma r_0)^2}{4\pi\mu_B^2}\sum_{\alpha\beta}\left(\delta_{\alpha\beta}-\hat{Q}_\alpha\hat{Q}_\beta\right)\left(1-e^{-E/k_BT}\right)^{-1}$$
$$\sum_{dd'}e^{-[W_d(\mathbf{Q})+W_{d'}(\mathbf{Q})]}\left(\chi''\right)^{dd'}_{\alpha\beta}(\mathbf{Q},E), \quad (1.295)$$

where χ'' is the imaginary part of a *generalized susceptibility* function $\chi^{dd'}_{\alpha\beta}(\mathbf{Q},E)$, corresponding to

$$\left(\chi''\right)^{dd'}_{\alpha\beta}(\mathbf{Q},E) = 4\pi\mu_B^2 f_d^*(\mathbf{Q})f_{d'}(\mathbf{Q})\left(1-e^{-E/k_BT}\right)$$
$$\times\sum_l e^{i\mathbf{Q}\cdot(\mathbf{l}+\mathbf{d}'-\mathbf{d})}\frac{1}{2\pi\hbar}\int_{-\infty}^{\infty}\left\langle\mu_{0d\alpha}(0)\mu_{ld'\beta}(t)\right\rangle e^{-iEt/\hbar}dt. \quad (1.296)$$

The generalized susceptibility function is analogous to the dynamic structure factor $S(\mathbf{Q}, E)$ for nuclear inelastic scattering, as it contains detailed information about the dynamics of the magnetic system. The factor $\left(1-e^{-E/k_BT}\right)$ in Equation (1.296) ensures that detailed balance (see Section 1.4.3) is obeyed, also consistent with the condition

$$\chi''(\mathbf{Q},-E) = -\chi''(\mathbf{Q},E). \quad (1.297)$$

The function

$$\chi(\mathbf{Q}) \equiv \chi(\mathbf{Q},0) \quad (1.298)$$

is the *isothermal susceptibility* representing the **Q** dependence of the response to a static magnetic field. According to the Kramers–Kronig causality relations, $\chi(\mathbf{Q})$ is also given by

$$\chi(\mathbf{Q}) = \frac{1}{\pi} \int_{-\infty}^{\infty} \chi''(\mathbf{Q}, E) \frac{dE}{E}, \quad (1.299)$$

which represents a sum rule for $\chi''(\mathbf{Q},E)$ analogous to the zeroth-moment sum rule for $S(\mathbf{Q},E)$ (Equation 1.111). In what follows, we seek to examine $\chi''(\mathbf{Q},E)$ in some specific situations.

1.6.2 Spin Waves

Long-range magnetic order can sustain collective excitations known as spin waves. In essence, these excitations arise from sinusoidal deviations of the spin components along a particular direction as one moves across magnetic centers. Like phonons, spin waves are quantized, and a single quantum is called a magnon. As such, much of the treatment of spin waves using a second-quantization picture is analogous to what we have discussed for excitations associated with nuclear motion. For a Bravais lattice, the Heisenberg operators for the spins $\mathbf{S}_l(t)$ can be written in terms of the spin waves as

$$S_l^x(t) + S_l^y(t) = \left(\frac{2S}{N}\right)^{1/2} \sum_{\mathbf{q}} b(\mathbf{q}) e^{i[\mathbf{q}\cdot\mathbf{l}-\omega(\mathbf{q})t]},$$

$$S_l^x(t) - S_l^y(t) = \left(\frac{2S}{N}\right)^{1/2} \sum_{\mathbf{q}} b^+(\mathbf{q}) e^{-i[\mathbf{q}\cdot\mathbf{l}-\omega(\mathbf{q})t]}, \quad (1.300)$$

$$S_l^z(t) = S - \frac{1}{N} \sum_{\mathbf{q}\mathbf{q}'} b(\mathbf{q}) b^+(\mathbf{q}') e^{i\{(\mathbf{q}'-\mathbf{q})\cdot\mathbf{l}-[\omega(\mathbf{q}')-\omega(\mathbf{q})]t\}},$$

where $\omega(\mathbf{q})$ is the frequency of a spin wave of reduced wave vector $\mathbf{q}$ and the z axis is taken as the direction of spin quantization. The $\hbar\omega(\mathbf{q})$ terms represent eigenvalues of the spin Hamiltonian, which for a Heisenberg ferromagnet has the form

$$\mathrm{H} = -\sum_{ll'} J_{ll'} \mathbf{S}_l \cdot \mathbf{S}_{l'}, \quad (1.301)$$

where the coefficients $J_{ll'}$ are Heisenberg exchange integrals arising from electrostatic interactions between the electrons. The $b(\mathbf{q})$ and $b^+(\mathbf{q})$ are quantum-mechanical operators representing the annihilation or creation of a magnon, analogous to the phonon annihilation and creation operators introduced in Section 1.3.1. Under the so-called linear approximation (the magnetic analog of the harmonic approximation for phonons), deviations of the spin z-components from ground-state values are small relative to the S_l. In this case, the Hamiltonian in Equation (1.301) can be expressed in terms of these annihilation and creation operators

$$H = H^0 + \sum_{\mathbf{q}} \hbar\omega(\mathbf{q}) b^+(\mathbf{q}) b(\mathbf{q}), \tag{1.302}$$

where

$$H^0 = -S^2 N J(0), \tag{1.303}$$

$$\hbar\omega(\mathbf{q}) = 2S[J(0) - J(\mathbf{q})], \tag{1.304}$$

and

$$J(\mathbf{q}) = \sum_{ll'} J_{ll'} e^{i\mathbf{q}\cdot(\mathbf{l}-\mathbf{l}')}. \tag{1.305}$$

Spin-wave dispersion relations are explicitly given by Equation (1.304) in terms of Heisenberg exchange integrals. For a lattice with r atoms per unit cell, the dispersion relation has in general r branches with both acoustic and optic magnons. For reference, the case for a hexagonal-closed-packed ferromagnet is treated in detail in Section 9.5 of Ref. [108].

For a Bravais lattice, Equation (1.296) can be rewritten as

$$\chi''(\mathbf{Q}, E) = \pi g^2 \mu_B^2 \left(1 - e^{-E/k_B T}\right) \sum_{\mathbf{l}} e^{i\mathbf{Q}\cdot\mathbf{l}} \frac{1}{2\pi\hbar} \int_{-\infty}^{\infty} \langle S_{0\alpha}(0) S_{l\beta}(t) \rangle e^{-iEt/\hbar} dt. \tag{1.306}$$

From Equation (1.300), it can be shown that the only nonzero components of the time-correlation function $\langle S_{0\alpha}(0) S_{l\beta}(t) \rangle$ are

$$\langle S_{0x}(0) S_{lx}(t) \rangle = \langle S_{0y}(0) S_{ly}(t) \rangle$$
$$= \frac{S}{2N} \sum_{\mathbf{q}} \left[e^{-i[\mathbf{q}\cdot\mathbf{l}-\omega(\mathbf{q})t]} \langle n_\mathbf{q} + 1 \rangle + e^{i[\mathbf{q}\cdot\mathbf{l}-\omega(\mathbf{q})t]} \langle n_\mathbf{q} \rangle \right], \tag{1.307}$$

$$\langle S_{0x}(0) S_{ly}(t) \rangle = -\langle S_{0y}(0) S_{lx}(t) \rangle$$
$$= \frac{iS}{2N} \sum_{\mathbf{q}} \left[e^{-i[\mathbf{q}\cdot\mathbf{l}-\omega(\mathbf{q})t]} \langle n_\mathbf{q} + 1 \rangle + e^{i[\mathbf{q}\cdot\mathbf{l}-\omega(\mathbf{q})t]} \langle n_\mathbf{q} \rangle \right], \tag{1.308}$$

and

$$\langle S_{0z}(0) S_{lz}(t) \rangle = S^2 - \frac{S}{2N} \sum_{\mathbf{q}} \langle n_\mathbf{q} \rangle, \tag{1.309}$$

where

$$\langle n_\mathbf{q} \rangle = \frac{1}{\exp[\hbar\omega(\mathbf{q})/k_B T] - 1} \tag{1.310}$$

is the Bose population factor already encountered in our previous treatment of vibrations (both phonons and magnons are bosonic pseudoparticles). The time-independent correlation function shown in Equation (1.309) represents the elastic scattering discussed in Section 1.5. Furthermore, the xy and yx cross terms cancel in the expression for the cross section (cf. Equation 1.295), leaving only xx and yy contributions. The double differential inelastic scattering cross section is therefore reduced to a sum of terms for one-magnon creation and annihilation:

$$\left(\frac{d^2\sigma}{d\Omega dE_f}\right)_{m,+1} = \left(\frac{k_f}{k_i}\right)\frac{(2\pi)^3}{2N_m v_0}(\gamma r_0)^2\left(\frac{g^2 S}{4}\right)\left(1+\hat{Q}_z^2\right)f^2(\mathbf{Q})e^{-2W(\mathbf{Q})} \times$$
$$\times \sum_{\tau \mathbf{q}} \langle n_\mathbf{q}+1 \rangle \delta(\mathbf{Q}-\mathbf{q}-\boldsymbol{\tau})\delta[E-\hbar\omega(\mathbf{q})],$$

(1.311)

$$\left(\frac{d^2\sigma}{d\Omega dE_f}\right)_{m,-1} = \left(\frac{k_f}{k_i}\right)\frac{(2\pi)^3}{2N_m v_0}(\gamma r_0)^2\left(\frac{g^2 S}{4}\right)\left(1+\hat{Q}_z^2\right)f^2(\mathbf{Q})e^{-2W(\mathbf{Q})}$$
$$\times \sum_{\tau \mathbf{q}} \langle n_\mathbf{q} \rangle \delta(\mathbf{Q}+\mathbf{q}-\boldsymbol{\tau})\delta[E+\hbar\omega(\mathbf{q})],$$

(1.312)

These differential scattering cross sections have a similar structure as those encountered in our previous treatment phonon scattering, see, for example, Equation (1.81). As in the phonon case, strong scattering features ought to satisfy

$$\mathbf{Q} = \boldsymbol{\tau} \pm \mathbf{q}, \quad E = \hbar\omega(\mathbf{q}). \tag{1.313}$$

The geometrical factor $\left(1+\hat{Q}_z^2\right)$ averages to unity for a multidomain cubic crystal. For an applied magnetic field strong enough to orient the spin direction, this factor can vary between 1 and 2 depending on the direction of the field.

The Hamiltonian of Equation (1.301) is based on a localized picture of electronic band structure, which might not be appropriate in cases such as, for example, transition-metal ferromagnets with wide d bands. Nonetheless, it has been found to provide a reasonable description of spin-wave structure, perhaps because of the strong couplings among d electrons on a given atom compared with (typically) weaker electronic correlations with other atoms.

Figure 1.30 shows spin-wave maps for the magnetically ordered iron chalcogenide $Fe_{1.05}Te$. These data have been measured on state-of-the art spallation-based time-of-flight neutron spectrometers. These instruments access high energy transfers up to the electron-volt range and have become workhorses in the study of strongly correlated electron systems. Dissection of these experimental data along well-defined crystallographic directions allows for the determination of the spin-wave Hamiltonian via the use of Equations (1.302)–(1.305). This particular example is of particular relevance to

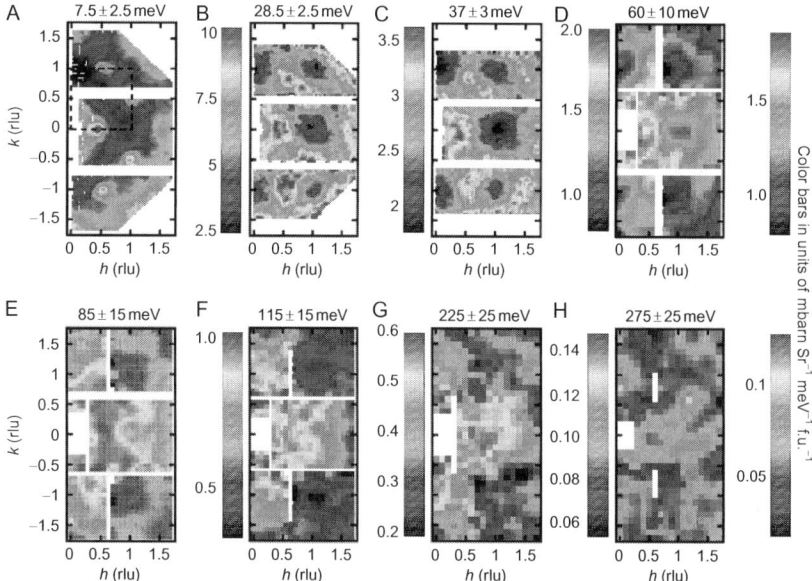

FIGURE 1.30 Constant energy-transfer slices showing the spin-wave structure as a function of increasing energy in $Fe_{1.05}Te$ at $T=10$ K. A crystallographic Brillouin zone is shown by the dashed line. Also note the absolute units of the color scale. *Reprinted with permission from Ref. [217].* (For color version of this figure, the reader is referred to the online version of this chapter.)

our understanding of superconductivity in the pnictides, a new class of materials with unusually high T_c's. The data shown in Figure 1.30 show that these spin-wave relations are distinctly different from first-principles predictions and parallel observations in the closely related compound $CaFe_2As_2$. Whereas nearest-neighbor exchange couplings are quite different in both systems, next-nearest-neighbor terms are remarkably similar. This finding suggests that superconductivity in the pnictides and chalcogenides share a common magnetic origin intimately related to next-nearest-neighbor energy couplings.

Antiferromagnetic materials with purely exchange interactions between ions on different sublattices lead to dispersion relations for the "acoustic" branches that are linear in q for small q. Anisotropic interactions, in which the spin direction is linked to the crystal lattice through the interaction of the orbital moment with the ligand field, usually introduce an extra term leading to a dispersion relation with finite frequency at $q=0$.

1.6.3 Crystal Fields and Magnetic Clusters

Couplings between nearby spins lead to collective spin-wave behavior. In the opposite limit, strong interactions with the electrostatic crystalline field of neighboring ions can lead to energy levels with zero or little dispersion.

Transitions between these levels are observed with inelastic neutron scattering provided that the symmetry is such so as to allow for magnetic-dipole-mediated transitions. These neutron experiments may be particularly useful when optical data are not available, for example, metallic systems.

The double differential cross section for such processes can be derived from Equations (1.226), (1.229), and (1.234):

$$\frac{d^2\sigma}{d\Omega dE_f} = \frac{1}{N_m}\left(\frac{k_f}{k_i}\right)(\gamma r_0)^2 \sum_{\alpha\beta}\left(\delta_{\alpha\beta} - \hat{Q}_\alpha\hat{Q}_\beta\right)|f(\mathbf{Q})|^2 e^{-2W} \times$$
$$\times \frac{1}{Z}\sum_{\tau_i\tau_f} e^{-E_{\tau_i}/k_BT}\langle\tau_i|\mu_\alpha^+|\tau_f\rangle\langle\tau_f|\mu_\beta|\tau_i\rangle\delta(E+E_i-E_f),$$
(1.314)

where Z is the partition function $\sum_{\tau_i} e^{-E_{\tau_i}/k_BT}$.

For rare-earth compounds, simple models for the crystal field based on charged point ions may provide a good description of experimental data if appropriate values are taken for the ionic charges and the moments of the radial wave functions of the magnetic ions. Figure 1.31 shows neutron spectra for PrBi for which these simple considerations give an excellent fit to the data [218]. Crystal-field levels in actinide compounds are usually less well defined because of strong exchange interactions—see the review by Pfleiderer [219].

Magnetic clusters involving interactions between neighboring magnetic centers represent an intermediate case between the limits of long-range order and isolated spins discussed so far. Interest in this regime (typically involving tens of magnetic ions) comes from their ubiquity in nature, from artificial nanostructures to metalloproteins. Small clusters also offer the opportunity to study fundamental magnetic interactions in a controlled manner and can display novel quantum phenomena such as quantum tunneling of the magnetization. Figure 1.32 shows inelastic neutron data for the magnetic cluster Mn_{12} acetate [220]. This compound is characterized by a total spin of $S=10$ in its ground state, giving rise to an unusually rich neutron spectrum at low energy transfers associated with the splitting of this multiplet structure. In a simplified picture of this cluster, an external ring of eight Mn(III) ions with $S=2$ surrounds a tetrahedron of four Mn(IV) ions ($S=3/2$), and the molecule adopts a plate-like shape with local tetragonal symmetry. At low temperatures, this cluster exhibits regular steps in the hysteresis cycle associated with the reversal of its magnetization, a phenomenon associated with a quantum-tunneling mechanism governed by small (higher-order) spin interactions. The spin Hamiltonian is of the form

$$H = D\left[S_z - \frac{1}{3}S(S+1)\right] + B_4^0 O_4^0 + B_4^4 O_4^4$$
(1.315)

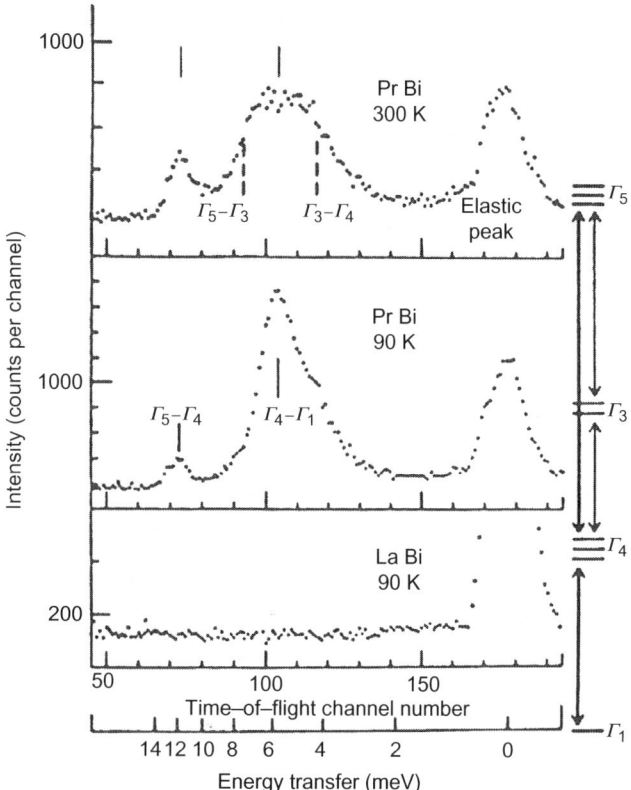

FIGURE 1.31 Neutron-scattering spectra of crystal-field excitations in PrBi. A schematic diagram of the underlying energy-level structure is shown on the right. *Reproduced with permission from Ref. [218].*

with

$$O_4^0 = 35 S_z^4 - [30 S(S+1) - 25] S_z^2 - 6 S(S+1) + 3 S^2 (S+1)^2 \quad (1.316)$$

and

$$O_4^4 = \frac{1}{2}[S_+^4 + S_-^4], \quad (1.317)$$

where $S_\pm$ are raising and lowering operators. The inelastic neutron data are shown in Figure 1.32, where only $\Delta M = \pm 1$ transitions are allowed by virtue of the character of magnetic neutron interactions. A decrease in energy spacings with decreasing energy transfer is primarily dominated by the B_4^0 term above. The best fit of the experimental data using the Hamiltonian given by Equation (1.315) yields $D = -56.7(2)\,\mu\text{eV}$ and $B_4^0 = -2.89(2) \times 10^{-3}\,\mu\text{eV}$. A nonzero B_4^4 will mix states with different M values and manifests itself most clearly for energy transfers below 0.3 meV, corresponding to energy levels near the top of the anisotropy barrier. Analysis of the inelastic neutron spectra

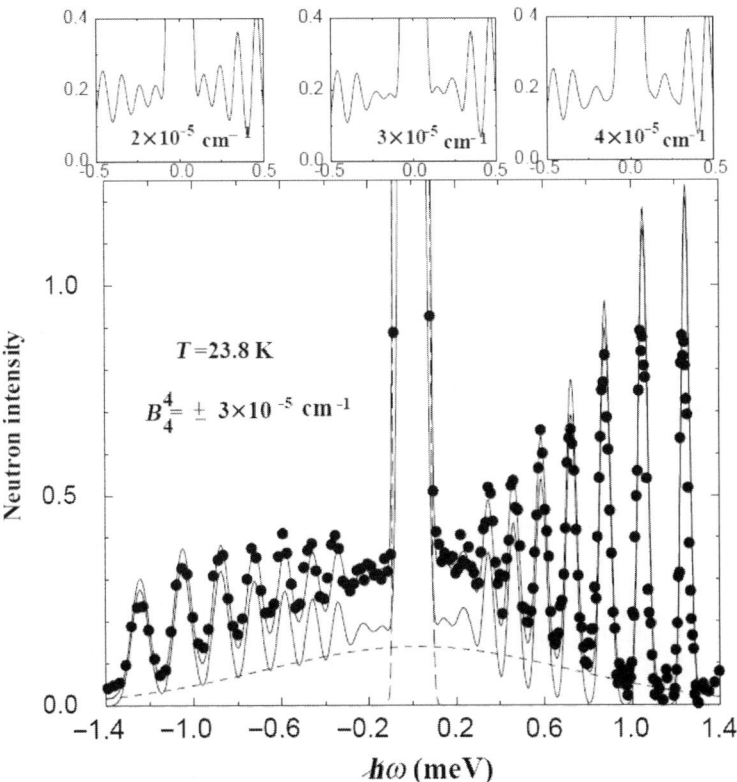

FIGURE 1.32 Inelastic neutron-scattering data for the magnetic cluster Mn_{12} acetate at cryogenic temperatures. The thin line in the main figure corresponds to a simulated spectrum. The top three figures show the sensitivity of the neutron spectra to different values of the coupling constant B_4^4 (see main text for details). *Reprinted with permission from Ref. [220].*

gives $|B_4^4| = 3.71(3) \times 10^{-3}\,\mu eV$, where the absolute sign of this parameter remains undetermined owing to the symmetry of the Hamiltonian. This term allows for tunneling subject to the selection rule $\Delta M = \pm 4$, as observed experimentally in magnetization measurements. Interested readers may also wish to consult a recent review of this growing area of research [221].

1.6.4 Spin Fluctuations

Section 1.5.1.2 presented a simple formula for the scattering from a paramagnetic system of localized spins, representing elastic scattering with a **Q** dependence given by $|f(\mathbf{Q})|^2$. The situation is more complicated if the spins interact, say, with ligand electric charges in the crystal (the case discussed in the last section) or with lattice vibrations and conduction electrons via Coulomb and exchange interactions. In this case, both elastic and inelastic crystal-field transitions will experience spectral broadening as a function of temperature.

A number of metallic compounds exhibit spectra with pronounced line broadenings. These appear to be related to hybridization with conduction electrons. These interactions lead to a decay in the spin correlations with a characteristic energy $\Gamma(T)$:

$$\langle \mu_\alpha(0)\mu_\beta(t)\rangle = \frac{1}{3}\delta_{\alpha\beta}\langle \mu^2(T)\rangle e^{-\Gamma(T)t/2\hbar}, \qquad (1.318)$$

Using Equation (1.296), this magnetic-moment correlation function gives rise to a susceptibility of the form

$$\chi''(\mathbf{Q},E) = \pi|f(\mathbf{Q})|^2 \chi_s(T) \frac{1}{\pi} \frac{\Gamma(T)/2}{E^2 + [\Gamma(T)/2]^2} E, \qquad (1.319)$$

where $\chi_s(T)$ is the static susceptibility. Substitution of this expression into the double differential scattering cross section (Equation 1.295) shows that the scattering spectrum is a Lorentzian function centered at $E=0$ with full-width-at-half-maximum $\Gamma(T)$.

Mixed-valence compounds such as $Ce_{0.74}Th_{0.26}$ exhibit large values of $\Gamma(T)$ and Figure 1.33 shows neutron-scattering spectra for this compound [222]. The widths are so large that it is difficult to measure the wings of the distributions with thermal neutron spectrometers, yet these have been confirmed in high-energy-transfer neutron-scattering experiments [223,224]. An extensive series of measurements in mixed-valence compounds is described by Holland-Moritz *et al.* [225]. The relevance of this class of materials in ferroelectromagnetic and colossal magnetoresistance phenomena have also been reviewed in Refs. [226] and [227], respectively.

In closing this discussion, we note that the temperature dependence of the widths in most mixed-valence compounds is often small or nonexistent, compared with the linear dependence expected for the usual Korring-type relaxation. Even more remarkable, large widths at low temperatures are observed in many actinide compounds such as UAs or UAl_2, even down to low enough temperatures for some compounds to become magnetically ordered [228]. These have been ascribed to interactions with the conduction electrons.

1.6.5 Interband Transitions

For ferromagnetic metals where magnetic properties derive from the presence of a band of itinerant electrons, the generalized susceptibility can be written down in terms of electron energies and occupation numbers as follows:

$$\chi_0(\mathbf{Q},E) = -\frac{(g\mu_B)^2}{N_m}\sum_{\mathbf{k}} \frac{n_{(\mathbf{k}+\mathbf{Q})\uparrow} - n_{\mathbf{k}\downarrow}}{E_{(\mathbf{k}+\mathbf{Q})\uparrow} - E_{\mathbf{k}\downarrow} + \Delta + E + i\delta}, \qquad (1.320)$$

where δ is a small real number enabling one to calculate $\chi''(\mathbf{Q}, E)$ as $\delta \to 0$. Equation (1.320) neglects interactions between electrons and cannot give collective excitations such as spin waves. If the electrons are allowed to interact

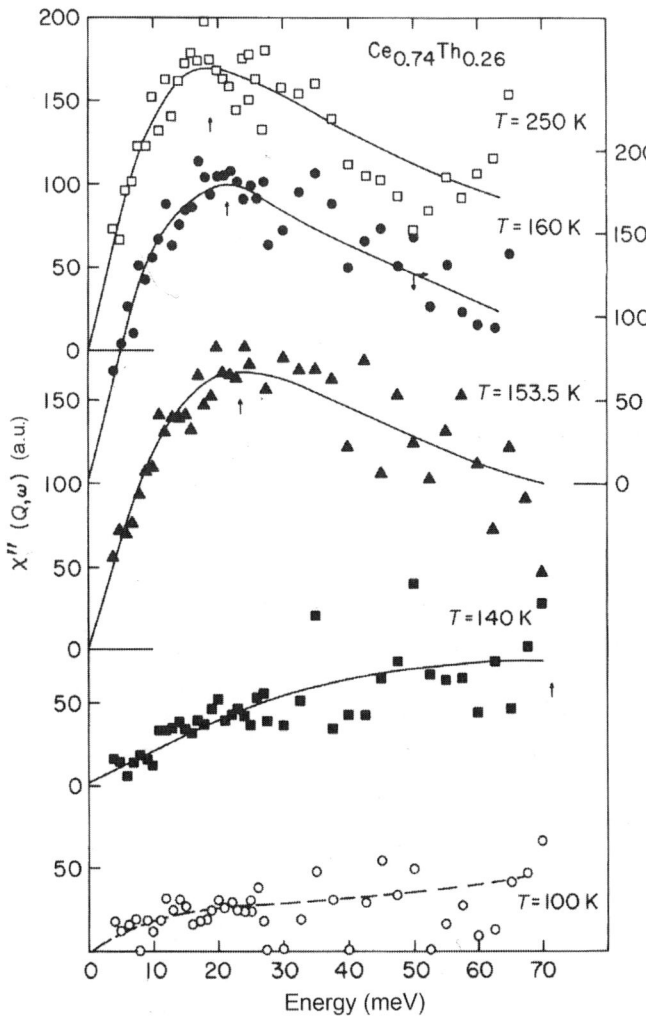

FIGURE 1.33 Susceptibilities obtained from inelastic neutron-scattering data for the mixed-valence compound $Ce_{0.74}Th_{0.26}$ at several temperatures. *Reprinted with permission from Ref. [222].*

electrostatically, with I representing an average value for the interaction, the resulting susceptibility can be written as

$$\chi(\mathbf{Q}, E) = \frac{\chi_0(\mathbf{Q}, E)}{1 - I\chi_0(\mathbf{Q}, E)}. \quad (1.321)$$

This susceptibility produces spin waves for $(\mathbf{Q}, E)$ values such that the denominator goes to zero. As shown in Section 9.3d of Ref. [108], this condition leads to an expression for the double differential cross section that can be related to Equations (1.311) and (1.312), originally derived for localized spins.

We also note that at zero momentum transfer $\mathbf{Q}$, Equation (1.320) simplifies to

$$\chi_0''(\mathbf{Q}, E) = \pi(g\mu_B)^2 \frac{\Delta}{I} \delta(E - \Delta), \qquad (1.322)$$

where

$$\Delta = I(n_\downarrow - n_\uparrow) \qquad (1.323)$$

is the up-down band splitting, and results in a peak in the scattering at $E = \Delta$. For $|\mathbf{Q}| > 0$, these modes broaden into a band of single-particle excitations known as Stoner modes, with characteristic energies of order several electron volts. Until now, these have not been conclusively identified in neutron experiments, although indirect evidence has been presented—see, for example, Refs. [229,230].

In the paramagnetic region, the up and down spin states are degenerate and Equation (1.320) becomes

$$\chi_0''(\mathbf{Q}, E) = -\pi \frac{(g\mu_B)^2}{N_m} \sum_{\mathbf{k}} n_{(\mathbf{k}+\mathbf{Q})}(1 - n_\mathbf{k}) \delta(E_{(\mathbf{k}+\mathbf{Q})} - E_\mathbf{k} - E). \qquad (1.324)$$

For a polycrystalline specimen, Equation (1.324) can be written in terms of the electronic density of states $g(\varepsilon)$ as

$$\chi_0''(\mathbf{Q}, E) = -\pi \frac{(g\mu_B)^2}{N_m} \iint d\varepsilon d\varepsilon' g(\varepsilon) g(\varepsilon') n(\varepsilon) [1 - n(\varepsilon')] \delta(\varepsilon' - \varepsilon - E). \qquad (1.325)$$

It becomes therefore possible to measure (in principle) the joint density of states within some distance of the Fermi surface, perhaps ~ 1 eV. Whilst such measurements may be possible at spallation neutron sources, numerical estimates indicate that the intensities will be extremely low except in exceptionally favorable cases [231].

1.6.6 Critical Scattering

So-called magnetic critical scattering is a particular case of the anomalous scattering arising from large fluctuations near a second-order phase transition. This phenomenon has been the subject of an extremely active field, and we limit ourselves to providing a brief account of some simple cases.

For a ferromagnetic Bravais lattice characterized by a Heisenberg Hamiltonian (cf. Equation 1.301), the susceptibility defined by Equation (1.306) can be written as

$$\chi''_{\alpha\alpha}(\mathbf{Q},E) = \pi g^2 \mu_B^2 \left(1 - e^{-E/k_B T}\right) \frac{1}{2\pi\hbar} \int_{-\infty}^{\infty} \langle S_\alpha(\mathbf{Q},0) S_\alpha(\mathbf{Q},t) \rangle e^{-iEt/\hbar} dt, \quad (1.326)$$

where the correlation functions are of the form

$$S_\alpha(\mathbf{Q}, t) = \sum_l e^{-i\mathbf{Q}\cdot\mathbf{l}} S_l(t). \quad (1.327)$$

Defining

$$\Delta S_\alpha(\mathbf{Q}, t) = \sum_l e^{-i\mathbf{Q}\cdot\mathbf{l}} [S_l(t) - \langle S_l(t) \rangle], \quad (1.328)$$

we can then rewrite the susceptibility as

$$\chi''_{\alpha\alpha}(Q,E) = \pi g^2 \mu_B^2 \left(1 - e^{-E/k_B T}\right) \left[N^2 \delta_{Q0} \langle S_\alpha \rangle^2 \delta(E) + \frac{1}{2\pi\hbar} \int_{-\infty}^{\infty} \langle \Delta S_\alpha(\mathbf{Q},0) \Delta S_\alpha(\mathbf{Q},t) \rangle e^{-iEt/\hbar} dt \right]. \quad (1.329)$$

Since $\langle S_\alpha \rangle \to 0$ as $T \to T_c$, the first term within square brackets (corresponding to Bragg scattering) is small near T_c. On the contrary, the second term (related to fluctuations in the magnetization) diverges as $T \to T_c$. This behavior is most apparent in the limits $\mathbf{Q} \to 0$ (or $\mathbf{Q} \to \tau$) and $t \to 0$ (static approximation), where the second term tends to

$$\chi''_{\alpha\alpha,cr} = \pi g^2 \mu_B^2 \left(1 - e^{-E/k_B T}\right) \langle \Delta S_\alpha(0,0)^2 \rangle = \pi \left(1 - e^{-E/k_B T}\right) \langle \Delta M_\alpha^2 \rangle, \quad (1.330)$$

where ΔM is the fluctuation in the magnetization and, therefore,

$$\chi''_{\alpha\alpha,cr} = \pi \left(1 - e^{-E/k_B T}\right) N k_B T \chi_{s,\alpha\alpha}, \quad (1.331)$$

where χ_s is the static susceptibility. Near the critical temperature T_c, magnetization fluctuations become large and χ_s diverges as $T \to T_c$. This divergent behavior leads to anomalously large scattering at $\mathbf{Q} = \tau$ known as *magnetic critical scattering*.

Away from Bragg reflections, the energy and wave vector dependence of critical scattering can be examined via the use of the isothermal susceptibility introduced in Equation (1.298):

$$\chi''_{\alpha\alpha,cr} = \pi N \chi_\alpha(\mathbf{Q}) E F_\alpha(\mathbf{Q}, E), \quad (1.332)$$

where $F_\alpha(\mathbf{Q},E)$ is a spectral-weight function having unit area when integrated over energy E (see Equation 1.299). By analogy with the classical theory of critical opalescence, Van Hove [232] postulated the existence of instantaneous (exponentially decaying) spatial correlations between spins. The isothermal susceptibility then has the form

$$\chi_\alpha(\mathbf{Q}) = \frac{\chi_0}{r_1^2(q^2 + \kappa_1^2)}, \quad (1.333)$$

where $q = |\mathbf{Q} - \boldsymbol{\tau}|$, χ_0 is the susceptibility of a noninteracting spin system, r_1 represents the range of the exchange interactions, and $\kappa_1^2 r_1^2 = (T/T_c - 1)$. To include an implicit time dependence, Van Hove assumed that the spin fluctuations $S_\alpha(\mathbf{Q}, t)$ decay according to stochastic translational diffusion, that is,

$$F_s(\mathbf{Q}, E) = \left(\frac{1}{\pi}\right) \frac{\Gamma(q)}{\Gamma^2(q) + E^2}. \quad (1.334)$$

with

$$\Gamma(q) = \hbar \Lambda q^2, \quad (1.335)$$

where Λ is a diffusion constant expected to vary with temperature as $\kappa_1^{1/2}$ or $\chi_\alpha(0)^{1/4}$. Dynamical scaling theory [233,234] also gives a Lorentzian line shape, but with the width functions given more generally by

$$\Gamma(q, \kappa_1) = c q^{5/2} f\left(\frac{\kappa_1}{q}\right), \quad (1.336)$$

which goes over to Equation (1.335) only in the hydrodynamic limit $\kappa_1/q \to 0$.

Figure 1.34 from Collins et al. [235] shows data for the critical scattering from iron as a function of temperature near T_c. Well below T_c, the scans at constant $Q = 0.05$ Å^{-1} show well-defined energy-loss and energy-gain magnon peaks. As a temperature is raised, the magnon line widths increase, and as T_c is approached only a single peak is observed, reaching maximum intensity at $T = T_c$. This critical scattering is observed to follow the behavior predicted by Equations (1.326)–(1.334). The temperature dependence of $\Gamma(q)$ was found to be much weaker than that predicted by Equation (1.335). It was pointed out by Als-Nielsen [236] that this is consistent with Equation (1.336), however, since the κ_1/q values covered in this experiment lie in a region where calculated scaling functions $f(\kappa_1/q)$ exhibit a minimum and thus vary quite slowly with temperature.

Critical scattering is also observed near the Néel temperature T_N in antiferromagnetic materials. In uniaxial crystals, two components of the susceptibility are involved, one parallel and one transverse to the anisotropy axis, with different values of κ_1 and r_1. In MnF$_2$, for example, $\chi_\parallel(\mathbf{Q})$ diverges at T_N, whereas $\chi_\perp(\mathbf{Q})$ remains finite at T_N but obeys critical behavior corresponding to a temperature $T_\perp < T_N$ [237]. In the planar antiferromagnetic K$_2$NiF$_4$, critical scattering is concentrated along ridges in the direction perpendicular to the NiF$_2$ planes, that is, no three-dimensional critical scattering is observed [238]. Thus, even though the actual transition is three-dimensional, the long-range order is achieved simultaneously within and between the planes, and it is the ordering within the planes that drives the phase transition—for further details, see the comprehensive monograph by Collins [239].

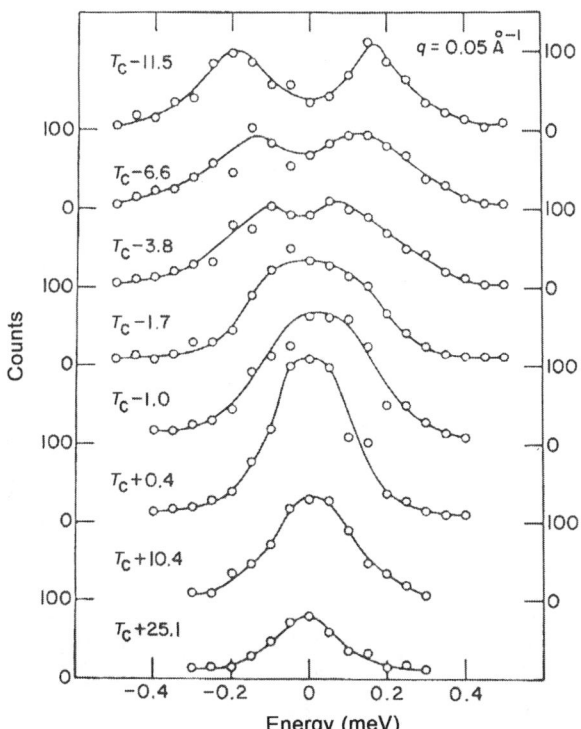

FIGURE 1.34 Magnetic critical scattering from iron. *Reprinted with permission from Ref. [235].*

1.7 NUCLEAR SPIN: ORDER AND DISORDER

As discussed in Section 1.5, neutron scattering from magnetic centers has an intrinsic dependence on spin orientation arising from the electromagnetic interaction between the magnetic moment of the neutron and that of the unpaired electron density, the latter giving rise to permanent magnetic moments on the order of a Bohr magneton μ_B. In the case of neutron scattering from nuclei, the magnetic moments are on the order of a nuclear magneton, $\mu_N = (m_e/m_p)\mu_B$, and the corresponding electromagnetic interaction is (while not necessarily zero) much weaker. Notwithstanding these limitations, we recall that nuclear scattering is mediated by the strong nuclear force, and for the case of non-zero-spin (i.e., NMR active nuclides), the scattering length is generally dependent on the relative orientation of neutron and nuclear spins during the collision. In cases where the nuclear spins are completely disordered, the spin dependence of the scattering process leads to the incoherent component of the scattering cross section. However, in cases where the nuclear spins are correlated, the spin dependence of the scattering amplitude can be exploited for the determination of nuclear-spin correlations. This section examines the basic features of the scattering of thermal neutrons from ensembles of nuclear spins. The formalism follows closely the one presented in the excellent monograph of Abragam and Goldman [240].

1.7.1 A Closer Look at Nuclear Spins

Away from nuclear resonances, the scattering of thermal neutrons by nuclei does not involve the participation of partial waves with orbital angular momentum $l > 0$, that is, in the language of scattering theory, nuclear scattering is entirely s-wave in character. Physically speaking, this condition arises from the orders-of-magnitude difference between neutron wavelengths (Å) and the size of nuclei (fm $= 10^{-5}$ Å). In this case, the neutron–nucleus scattering state has a total angular moment $\mathbf{J}$ given by the quantum-mechanical vector addition of their intrinsic spins, namely,

$$\mathbf{J} = \mathbf{I} + \frac{1}{2}\boldsymbol{\sigma}, \tag{1.337}$$

where $\mathbf{I}$ and $1/2\boldsymbol{\sigma}$ correspond to nuclear and neutron spins, respectively. Since the neutron has spin $1/2$, there are only two possible values for J:

$$J = I \pm \frac{1}{2}. \tag{1.338}$$

The strength of the scattering for each ($\pm$) angular-momentum state is characterized by their corresponding scattering lengths b^+ and b^-, respectively. In operator form, the general expression for the spin-dependent neutron-scattering lengths was already introduced in Equations (1.248) and (1.249) in the context of polarized neutrons. To facilitate the present discussion, we rewrite them here

$$\hat{b} = \bar{b} + \frac{1}{2} b_N \boldsymbol{\sigma} \cdot \mathbf{I}, \tag{1.339}$$

where

$$\begin{aligned} \bar{b} &= \frac{1}{(2I+1)}[(I+1)b^+ + Ib^-], \\ b_N &= \frac{2}{(2I+1)}[b^+ - b^-], \end{aligned} \tag{1.340}$$

and we recall that $(2I+1)$ corresponds to the M_I sublevel degeneracy of the nuclear angular momentum.

1.7.2 Scattering Cross Sections

The interactions between nuclear spins are very weak and, therefore, the energies involved in their dynamics are very small. Even in the case of solid ^{3}He (where the exchange interaction is several orders of magnitude larger than dipolar interactions), the energies involved are on the order of a fraction of a μeV. Information on the static correlations between nuclear spins can be obtained from elastic scattering. Following the early work by Hayter et al. on single crystals of lanthanum magnesium nitrate [241], several such experiments have been reported [242–257]. From a practical viewpoint, the approach offers the possibility of enhancing coherent Bragg intensities relative to incoherent scattering. Inelastic neutron-scattering studies are significantly more challenging, yet weak nuclear-spin excitations associated with hyperfine transitions in compounds such as the antiferromagnetic metal oxide CoO [258], the spin ice $Ho_2Ti_2O_7$ [259], or metallic $HoAl_2$ [260] have been observed at energy transfers in the μeV region using state-of-the-art neutron instrumentation. A notable exception to the considerations given above occurs in certain systems of high symmetry such as H_2, CH_4, and methyl (CH_3) functional groups (as well as their deuterated counterparts), where the Pauli Exclusion Principle ties the molecular and nuclear-spin wavefunctions in such a way that energy changes associated with atomic motions (primarily rotations) are tied to changes in total nuclear spin. H_2 represents the most extreme case, where the transition between $J=0$ and $J=1$ rotational levels at 15 meV is accompanied by a change in total nuclear spin from $I=0$ (*para* state) to $I=1$ (*ortho* state). While interesting in their own right, these particular cases will not be treated any further in the discussion that follows below (see discussion in Section 1.4.5.2 and Chapter 6).

To examine how neutron scattering can be used to probe nuclear-spin correlations, we recall the basic formula for the differential cross section:

$$\frac{d\sigma}{d\Omega} = \frac{1}{N}\left(\sum_j \langle |\hat{b}_j|^2 \rangle + \sum_{j \neq j'} \langle \hat{b}_j \hat{b}_{j'}^+ \rangle e^{-i\mathbf{Q}\cdot(\mathbf{R}_j - \mathbf{R}_{j'})}\right), \quad (1.341)$$

where the averages are over the spin variables. For nuclei with non-zero-spin, the scattering lengths given by Equations (1.339) and (1.340) give

$$|\hat{b}_j|^2 = \bar{b}^2 + \bar{b}b_N \sigma \cdot \mathbf{I}_j + \frac{1}{4}b_N^2 (\sigma \cdot \mathbf{I}_j)^2. \quad (1.342)$$

Using the general relation for the square of a scalar product between quantum-mechanical angular-momentum operators

$$(\sigma \cdot \mathbf{I})^2 = \mathbf{I}\cdot\mathbf{I} + i\sigma \cdot \mathbf{I} \times \mathbf{I} = I(I+1) - \sigma \cdot \mathbf{I}, \quad (1.343)$$

the two scattering-length products in Equation (1.341) can be expressed as

$$\begin{aligned}|\hat{b}_j|^2 &= \bar{b}^2 + \bar{b}b_N \sigma \cdot \mathbf{I}_j + \frac{1}{4}b_N^2[I(I+1) - \sigma \cdot \mathbf{I}_j], \\ \hat{b}_j\hat{b}_{j'}^\dagger &= \bar{b}^2 + \frac{1}{2}\bar{b}b_N [\sigma \cdot \mathbf{I}_j + \sigma \cdot \mathbf{I}_{j'}] + \frac{1}{4}b_N^2 [\mathbf{I}_j \cdot \mathbf{I}_{j'} + i\sigma \cdot (\mathbf{I}_j \times \mathbf{I}_{j'})].\end{aligned} \quad (1.344)$$

Substitution of these expressions into Equation (1.341) provides a general expression for the spin-dependent scattering differential cross section.

1.7.3 Uncorrelated and Correlated Spin Ensembles

The simplest case to consider involves random nuclear-spin orientations, as there is no correlation between the spin-dependent scattering amplitudes of two particles and their relative position. In this case, Equation (1.341) simplifies to

$$\frac{d\sigma}{d\Omega} = \frac{1}{N}\left(\sum_j \langle |\hat{b}_j|^2 \rangle + \sum_{j \neq j'} \langle \hat{b}_j \hat{b}_{j'}^+ \rangle_{j \neq j'} e^{-i\mathbf{Q}\cdot(\mathbf{R}_j - \mathbf{R}_{j'})}\right). \quad (1.345)$$

If on the right-hand side of Equation (1.345) we add and subtract $\langle \hat{b}_j \hat{b}_{j'}^+ \rangle_{j \neq j'}$ to the first and second term, respectively, we obtain

$$\frac{d\sigma}{d\Omega} = \frac{1}{N}\sum_j \left[\langle |\hat{b}_j|^2 \rangle - \langle \hat{b}_j \hat{b}_{j'}^+ \rangle_{j \neq j'}\right] + \frac{1}{N}\sum_{j,j'} \langle \hat{b}_j \hat{b}_{j'}^+ \rangle_{j \neq j'} e^{-i\mathbf{Q}\cdot(\mathbf{R}_j - \mathbf{R}_{j'})}. \quad (1.346)$$

Even in the absence of spin correlations in the above sense, the spin system can have a net macroscopic polarization and the beam can also be polarized. Denoting the polarization of the target spin system by P and the polarization of the neutron beam by p, both referred to the same quantization axis, we have

$$\langle \mathbf{I}_j \cdot \mathbf{I}_{j'} \rangle = \langle \mathbf{I}_j \rangle \cdot \langle \mathbf{I}_{j'} \rangle = P^2 I^2, \quad \langle \mathbf{I}_j \times \mathbf{I}_{j'} \rangle = 0, \quad \langle \sigma \cdot \mathbf{I}_j \rangle = pPI. \quad (1.347)$$

From Equation (1.344):

$$\langle |\hat{b}_j|^2 \rangle = \bar{b}^2 + \bar{b}b_N pPI + \frac{1}{4}b_N^2[I(I+1) - pPI],$$

$$\langle \hat{b}_j \hat{b}_{j'}^+ \rangle_{i \neq j} = \bar{b}^2 + \bar{b}b_N pPI + \frac{1}{4}b_N^2 P^2 I^2, \quad (1.348)$$

which gives a differential scattering cross section [cf. Equation (1.346)] of the form

$$\frac{d\sigma}{d\Omega} = \frac{1}{4}b_N^2[I(I+1) - pPI - P^2 I^2] \\ + \left(\bar{b}^2 + \bar{b}b_N pPI + \frac{1}{4}b_N^2 P^2 I^2\right)\frac{1}{N}\sum_{j,j'} e^{-i\mathbf{Q}\cdot(\mathbf{R}_j - \mathbf{R}_{j'})}. \quad (1.349)$$

For a crystal with one atom per unit cell, the double sum in Equation (1.349) simplifies to

$$\frac{1}{N}\sum_{j,j'} e^{-i\mathbf{Q}\cdot(\mathbf{R}_j - \mathbf{R}_{j'})} = \sum_j e^{-i\mathbf{Q}\cdot\mathbf{R}_j}, \quad (1.350)$$

where the sum is over all N atoms (or cells). Equation (1.349) then reads

$$\frac{d\sigma}{d\Omega} = \frac{1}{4}b_N^2[I(I+1) - pPI - P^2 I^2] \\ + \left(\bar{b}^2 + \bar{b}b_N pPI + \frac{1}{4}b_N^2 P^2 I^2\right)\sum_j e^{-i\mathbf{Q}\cdot\mathbf{R}_j}, \quad (1.351)$$

or in terms of cross sections

$$\frac{d\sigma}{d\Omega} = \frac{1}{4\pi}\left(\sigma_i + \sigma_c \frac{(2\pi)^3}{v_0}\sum_\tau \delta(\mathbf{Q}-\boldsymbol{\tau})\right), \qquad (1.352)$$

where the incoherent and coherent cross sections are given by

$$\sigma_i = \pi b_N^2 \left[I(I+1) - pPI - P^2 I^2\right] \qquad (1.353)$$

and

$$\sigma_c = 4\pi\left(\overline{b}^2 + \overline{b}b_N pPI + \frac{1}{4}b_N^2 P^2 I^2\right), \qquad (1.354)$$

respectively.

For an unpolarized target ($P=0$) and with $\overline{b}$ and b_N expressed in terms of the up and down scattering amplitudes [cf. Equation (1.340)], the cross sections can be written as

$$\sigma_i = \pi I(I+1)b_N^2 = 4\pi \frac{I(I+1)}{(2I+1)^2}(b^+ - b^-)^2 \qquad (1.355)$$

and

$$\sigma_c = 4\pi \overline{b}^2 = 4\pi \frac{1}{(2I+1)^2}[(I+1)b^+ + Ib^-]^2. \qquad (1.356)$$

These two expressions are the same as those derived in Section 1.2 [cf. Equation (1.46)]. From Equation (1.353), a fully polarized target ($P=1$) will have no incoherent scattering when $pP=1$, that is, when both target spins and neutrons are polarized along the same direction.

To treat correlations between nuclear-spin orientations and the relative positions of particle pairs, it becomes necessary to generalize the treatment given above. For example, if Equation (1.347) is replaced by

$$\langle \mathbf{I}_d \cdot \mathbf{I}_{d'}\rangle = \langle \mathbf{I}_d\rangle \cdot \langle \mathbf{I}_{d'}\rangle = P_d P_{d'} I^2, \qquad (1.357)$$

then we are implicitly assuming that there is only one kind of nucleus with spin I, and P_d is the polarization for a given nucleus d. The differential cross section in this case is given by

$$\frac{d\sigma}{d\Omega} = \frac{1}{4\pi}\left(\sigma_i + \sigma_c \frac{(2\pi)^3}{v_0}\sum_\tau |F(\boldsymbol{\tau})|^2 \delta(\mathbf{Q}-\boldsymbol{\tau})\right) \qquad (1.358)$$

with an incoherent cross section of the form

$$\sigma_i = \pi b_N^2 \sum_d \left[I(I+1) - pP_d I - P_d^2 I^2\right] \qquad (1.359)$$

summed over all nuclei in the unit cell. The new structure factor $F(\tau)$ is given by

$$|F(\tau)|^2 = \sum_{d,d'} \left[\bar{b}^2 + \frac{1}{2}\bar{b}b_N p(P_d + P_{d'})I + \frac{1}{4}b_N^2 P_d P_{d'} I^2 \right] e^{-i\mathbf{Q}\cdot(\mathbf{d}-\mathbf{d}')}. \quad (1.360)$$

In the case of ferromagnetic ordering, Bragg peaks will experience an intensity modulation arising from the spin-dependent terms in Equation (1.360), but no new peaks will appear. Also, the onset of polarization will affect the incoherent scattering, as explicitly shown by Equation (1.359).

For antiferromagnetic nuclear-spin ordering, new Bragg features will appear in addition to those observed in the same structure with no nuclear-spin order. For the antiferromagnetic peaks, the differential cross section reads

$$\left(\frac{d\sigma}{d\Omega}\right)_m = \frac{(2\pi)^3}{v_m} \frac{1}{4} \sum_{d,d'} b_N^2 P_d P_{d'} I^2 e^{-i\mathbf{Q}\cdot(\mathbf{d}-\mathbf{d}')} \sum_{\tau_m} \delta(\mathbf{Q}-\tau_m), \quad (1.361)$$

where the subscript "m" denotes summation over the antiferromagnetic unit cell. Figure 1.35 shows neutron-diffraction data from a ^{65}Cu single crystal at ultralow (milliKelvin) temperatures. These data evince the emergence of long-range nuclear-spin antiferromagnetic order upon the application of a small magnetic field. The time evolution of the nuclear-spin antiferromagnetic (100) peak depends strongly on the magnitude of the applied magnetic field between zero and its critical value at 0.25 mT.

Although these experiments still remain quite challenging, there is clearly scope for further work in this area. In particular, the implementation of *dynamic nuclear polarization* schemes [261] including their implementation at high magnetic fields [262] offers the exciting prospects of attaining a sufficiently large nuclear polarization so as to enable a number of important applications of neutron scattering where element-specific contrast variation would be highly desirable. Signal-enhancement factors of ca. 10^{+4} have already been demonstrated in liquid-state NMR studies [263] and promising neutron experiments have been reported recently using Laue diffraction [257] and small-angle techniques [254–256]. In this context, the scheme proposed by Buckingham [264] to modulate neutron-scattering intensities via the use of well-established pulsed neutron-magnetic-resonance techniques is most enticing, and still a challenge for its successful realization in the laboratory. Further improvements in neutron sources and instrumentation would certainly accelerate the above trend.

1.8 OUTLOOK

This introductory chapter has brought together the basic theory and formalism of thermal neutron scattering for condensed matter research, and given some examples to illustrate its present use in a variety of applications. Our choices

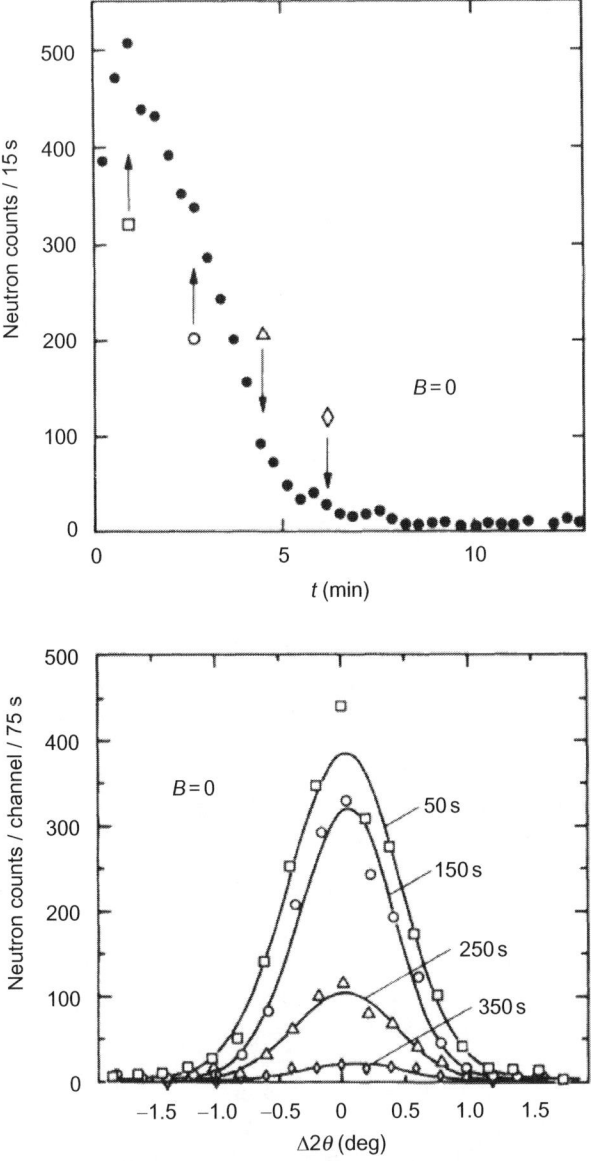

FIGURE 1.35 Top: time dependence of the integrated intensity of the (100) antiferromagnetic Bragg reflection from a ^{65}Cu single crystal following demagnetization. Bottom: (100) reflection at the points in time marked in the figure. *Reprinted with permission from Ref. [247].*

in the selection of this material are unavoidably colored by our personal preferences, appreciation, and knowledge of the vast and rapidly growing number of applications of neutron scattering. It is our hope that the in-depth treatments of specific research areas given in the subsequent chapters of this and future volumes will constitute a base reference for the assessment, planning, and successful execution of neutron-scattering experiments in all of these diverse applications.

The advances made in the quarter-century following the publication of the previous version of this volume [265] have been spectacular, and it is tempting, perhaps foolhardy, to speculate about what the next quarter-century will bring in this field. Some exciting and imaginative proposals have been suggested by Furrer [266]. We could add some more obvious possibilities as follows.

First, qualitative scientific advances generally follow developments in sources and instrumentation. New innovations in these areas are respectively described in Chapters 2 and 3. In particular, the third-generation spallation sources at Oak Ridge and Tokai are getting well underway, and the novel type of source represented by the European Spallation Source provides an exciting challenge for new kinds of instruments.

Instrumental developments such as innovative uses of polarized neutrons, new detectors, and the exploitation of the explosion in computing and data-handling capabilities will likewise pave the way for major scientific advances.

No less important are unabated developments in sample-environment capabilities. Of the many that we could mention, the access to ever-higher pressures, and larger samples at moderately high pressures, is making it possible to obtain new science on solids and liquids and relate it to important issues in geology. Recent advances in the various levitation techniques are making it possible to study materials at previously inaccessible temperatures and in a greater range of metastable states, while *in situ* measurements in real time and under technologically and industrially relevant conditions can connect neutron measurements with the important problems in battery and fuel-cell materials or catalysis.

Finally, the excitement of a neutron-scattering measurement derives from the materials and phenomena being investigated. As one example, recent advances in nanoscale science and technology provides a host of opportunities for neutron experiments, ranging from the study of atomic and magnetic structure as one goes to smaller particle dimensions and confinement, to the measurement of size distributions with small-angle scattering and the investigation of nm-thin films and coatings with reflectivity techniques.

We hope that the background to neutron-scattering techniques outlined in this volume will help our increasingly interdisciplinary scientific community to exploit these exciting opportunities beyond the imaginable at the present time.

REFERENCES

Section 1
[1] Wietfeldt FE, Greene GL. Rev Mod Phys 2011;83:1173.
[2] For recommended values of all constants defined in this chapter, see the *List of Commonly Used Symbols* and references therein.
[3] Chadwick J. Nature 1932;129:312.

[4] URL: www.nobelprize.org/nobel_prizes/physics/laureates/1935/ [last accessed on July 1, 2013].
[5] Segrè E. Rev Mod Phys 1955;27:257.
[6] Dubbers D, Schmidt MG. Rev Mod Phys 2011;83:1111.
[7] Alexandrox YuA. Fundamental properties of the neutron. Oxford: Oxford University Press; 1992.
[8] Magán M, Sordo F, Zanini L, Terrón S, Ghiglino A, Martínez F, et al. Nucl Instr Methods A 2013;729:417.
[9] See Chapter 4 in Windsor CG. Pulsed neutron scattering. London: Taylor and Francis; 1981.
[10] Price DL, Fu L, Bermejo FJ, Fernandez-Alonso F, Saboungi M-L. Chem Phys 2013; http://dx.doi.org/10.1016/j.chemphys.2013.02.020, in press.
[11] Enderby JE. Chem Soc Rev 1995;24:159.
[12] Sears VF. Neutron optics: an introduction to the theory of neutron optical phenomena and their applications. Oxford: Oxford University Press; 1989.
[13] Shull CG, Yamada Y. J Phys Soc Jpn 1962;17(Suppl. B-III):1.
[14] Word RE, Trammell GT. Phys Rev B 1981;24:2430.
[15] Newton RG. Scattering theory of waves and particles. New York: McGraw-Hill; 1966.
[16] Schimmel HG, Huot J, Chapon LC, Tichelaar FD, Mulder FM. J Am Chem Soc 2005;127:14348.
[17] Als-Nielsen J, McMorrow D. Elements of modern X-ray physics. 2nd ed. Chichester: John Wiley & Sons; 2011.
[18] Margaritondo G. Elements of synchrotron light for biology, chemistry, and medical physics. Oxford: Oxford University Press; 2002.
[19] Rehr JJ, Albers RC. Rev Mod Phys 2000;72:621.
[20] Parsons JG, Aldrich MV, Gardea-Torresdey JL. Appl Spectrosc Rev 2002;37:187.
[21] Rehr JJ, Amkudinov AL. Coord Chem Rev 2005;249:131.
[22] Filipponi A, Di Cicco A, Natoli CR. Phys Rev B 1995;52:15122.
[23] Filipponi A, Di Cicco A. Phys Rev B 1995;52:15135.
[24] Cole JM, Wright AC, Newport RJ, Sinclair RN, Fischer HE, Cuello GJ, et al. J Phys Condens Matter 2007;19:056002.
[25] Coldea R, Tennant DA, Wheeler EM, Wawrzynska E, Prabhakaran D, Telling M, et al. Science 2010;327:177.
[26] See Chapter 2 of the ESS Technical Design Report. URL: eval.esss.lu.se/cgi-bin/public/DocDB/ShowDocument?docid=274 [Last accessed on July 4, 2013].
[27] Burkel E. Rep Prog Phys 2000;63:171.
[28] Scopigno T, Ruocco G, Sette F. Rev Mod Phys 2005;77:881.
[29] Toellner TS, Alatas A, Said AH. J Synchrotron Radiat 2011;18:605.
[30] Sinn H. J Phys Condens Matter 2001;13:7525.
[31] Ament LJP, van Veenendaal M, Devereaux TP, Hill JP, van den Brink J. Rev Mod Phys 2011;83:705.
[32] Kotani A, Shin S. Rev Mod Phys 2001;73:203.
[33] Grubel G, Zontone F. J Alloys Compd 2011;363:3.
[34] Leheny RL. Curr Opin Colloid Interface Sci 2012;17:3.
[35] portal.slac.stanford.edu/sites/lcls_public/Pages/Default.aspx [last accessed on August 21, 2013].
[36] Berne BJ, Pecora R. Dynamic light scattering, with applications to chemistry, biology, and physics. Mineola: Dover Publications; 2000.
[37] Hayes W, Loudon R. Scattering of light by crystals. Mineola: Dover Publications; 2004.

[38] Harrys DC, Bertolucci MD. Symmetry and spectroscopy—an introduction to vibrational and electronic spectroscopy. Mineola: Dover Publications; 1989.
[39] Baxter JB, Guglietta GW. Anal Chem 2011;83:4342.
[40] Ebbinghaus S, Kim SK, Heyden M, Yu X, Heugen U, Gruebele M, et al. Proc Natl Acad Sci USA 2007;104:20749.
[41] Kutteruf MR, Brown CM, Iwaki LK, Campbell MB, Korter TM, Heilweil EJ. Chem Phys Lett 2003;375:337.
[42] Damascelli A. Phys Scr 2004;T109:61.
[43] Damascelli A, Hussain Z, Shen ZX. Rev Mod Phys 2003;75:473.
[44] Sanders JKM, Hunter BK. Modern NMR spectroscopy—a guide for chemists. Oxford: Oxford University Press; 1994.
[45] Price WS. Concept Magn Res 1998;10:197.
[46] Parella T. Magn Reson Chem 1998;36:467.
[47] Jardine AP, Hedgeland H, Alexandrowicz G, Allison W, Ellis J. Prog Surf Sci 2009;84:323.
[48] Jardine AP, Dworski S, Fouquet P, Alexandrowicz G, Riley DJ, Lee GYH, et al. Science 2004;304:1790.
[49] Fouquet P, Hedgeland H, Jardine AP. Z Phys Chem 2010;224:61.
[50] Yaouanc A, Dalmas de Reotier P. Muon spin rotation, relaxation, and resonance: applications to condensed matter. Oxford: Oxford University Press; 2011.
[51] Nagamine K. Introductory muon science. Cambridge: Cambridge University Press; 2003.
[52] Lee SL, Cywinski R, Kilcoyne SH, editors. Muon science: muons in physics, chemistry, and materials. New York: Taylor & Francis; 1999.
[53] Schenck AGE. Muon spin rotation spectroscopy: principles and applications in solid state physics. New York: Taylor & Francis; 1985.
[54] www.isis.stfc.ac.uk/groups/muons/ [last accessed on August 20, 2013].
[55] j-parc.jp/researcher/MatLife/en/instrumentation/ms.html [lLast accessed on August 20, 2013].
[56] lmu.web.psi.ch/ [last accessed on August 20, 2013].
[57] http:/musr.ca/ [last accessed on August 20, 2013].
[58] Kremer F, Schönhals A, editors. Broadband bielectric spectroscopy. Berlin: Springer; 2002.
[59] Frenkel D, Smit B. Understanding molecular simulation: from algorithms to applications. 2nd ed. San Diego: Academic Press; 2001.
[60] Martin RM. Electronic structure: basic theory and practical methods. Cambridge: Cambridge University Press; 2008.
[61] Attinger S, Koumoutsakos P, editors. Multiscale modelling and simulation. Berlin: Springer; 2004.

Section 2
[62] Squires GL. Introduction to the theory of thermal neutron scattering. 3rd ed. Cambridge: Cambridge University Press; 2012.
[63] Lovesey SW. Theory of neutron scattering from condensed matter, vols. I and II. Oxford: Oxford University Press; 1986.
[64] Mezei F. Z Phys 1972;255:146.

Section 3
[65] Glyde HR. Excitations in liquid and solid helium. Oxford: Oxford University Press; 1995.
[66] Born M, Huang K. Dynamical theory of crystal lattices. Oxford: Oxford University Press; 1988.
[67] Wallace DC. Thermodynamics of crystals. New York: John Wiley & Sons; 1972.

[68] Dove MT. Introduction to lattice dynamics. Cambridge: Cambridge University Press; 1993.
[69] Mezei F. In: Hansen JP, Levesque D, Zinn-Justin J, editors. Liquids, freezing, and the glass transition. Amsterdam: North Holland; 1991. p. 629–87.
[70] Willis BTM, Pryor AW. Thermal vibrations in crystallography. Cambridge: Cambridge University Press; 1975.
[71] Sears VF, Shelley SA. Acta Cryst A 1991;47:441.
[72] Zaccai G. Science 2000;288:1604.
[73] Dove MT, Tucker MG, Keen DA. Eur J Mineral 2002;14:331.
[74] Tucker MG, Keen DA, Dove MT, Goodwin AL, Hui Q. J Phys Condens Matter 2007;19:335218.
[75] Serrano J, Manjón FJ, Romero AH, Ivanov A, Cardona M, Lauck R, et al. Phys Rev B 2010;81:174304.
[76] Taraskin SN, Elliott SR. Phys Rev B 1997;55:117.
[77] Mitchell PCH, Parker SF, Ramirez-Cuesta AJ, Tomkinson J. Vibrational spectroscopy with neutrons, with applications in chemistry, biology, materials science and catalysis. Singapore: World Scientific; 2005.
[78] http:/www.isis.stfc.ac.uk/instruments/tosca/ins-database/ [last accessed on August 8, 2013].
[79] Sjölander A. Ark Fys 1958;14:315.
[80] Parker SF, Bennington SM, Ramirez-Cuesta AJ, Auffermann G, Bronger W, Herman H, et al. J Am Chem Soc 2003;125:11656.
[81] Andreani C, Colognesi D, Pace E. Phys Rev B 1999;60:10008.
[82] Fultz B. Prog Mater Sci 2010;55:247.
[83] Cowley RA. Rep Prog Phys 1968;31:123.
[84] Cowley RA. Adv Phys 1963;12:421.
[85] Safarik DJ, Llobet A, Lashley JC. Phys Rev B 2012;85:174105.
[86] Maradudin AA, Ambegaokar V. Phys Rev 1964;13:A1070. Cowley RA, Buyers WJL. J Phys C 1969;2:2262.
[87] Glyde HR. Can J Phys 1974;52:2281. Varma CM, Werthamer NR. In: Bennemann KH, Ketterson JB, editors. The physics of liquid and solid helium, part I. New York: Wiley; 1974. p. 503.
[88] Aynajian P, Keller T, Boeri L, Shapiro SM, Habicht K, Keimer B. Science 2008;319:1509.
[89] Bayrakci SP, Keller T, Habicht K, Keimer B. Science 2006;312:1926.
[90] Fernandez-Alonso F, Cabrillo C, Fernandez-Perea R, Bermejo FJ, Gonzalez MA, Mondelli C, et al. Phys Rev B 2012;86:144524.
[91] Mesot J. Science 2006;312:1888.
[92] Paulatto L, Mauri F, Lazzeri M. Phys Rev B 2013;87:214303.
[93] Baroni S, de Gironcoli S, Dal Corso A, Gianozzi P. Rev Mod Phys 2001;73:515.

Section 4
[94] Van Hove L. Phys Rev 1954;95:249.
[95] Egami T, Billinge SJL. Underneath the Bragg peaks: structural analysis of complex materials. Amsterdam: Pergamon; 2003.
[96] Yarnell JL, Katz MJ, Wenzel RG, Koenig SH. Phys Rev A 1973;7:2130.
[97] Sköld K, Rowe JM, Ostrowski GE, Randolph PD. Phys Rev A 1972;6:1107.
[98] de Gennes PG. Physica 1959;25:825.
[99] de Gennes PG. J Phys Chem Solids 1958;4:223.
[100] http:/www.isis.stfc.ac.uk/groups/disordered-materials/database/ [last accessed on August 10, 2013].

[101] Sjölander A. In: Egelstaff PA, editor. Inelastic neutron scattering. New York: Academic Press; 1965. p. 301–3, and references therein.
[102] Sjölander A. In: Egelstaff PA, editor. Thermal neutron scattering. New York: Academic Press; 1965. p. 291.
[103] Cusack NE. The physics of structurally disordered matter, an introduction. Bristol: Adam Hilger; 1988.
[104] Rahman A, Singwi KS, Sjölander A. Phys Rev 1962;126:986.
[105] Placzek G. Phys Rev 1952;86:377.
[106] Placzek G. Phys Rev 1954;93:895.
[107] Placzek G. Phys Rev 1957;105:1240.
[108] Marshall W, Lovesey SW. Theory of thermal neutron scattering. Oxford: Oxford University Press; 1971.
[109] Mayers J. Phys Rev B 1990;41:41.
[110] Sears VF. Phys Rev B 1984;30:44.
[111] Andreani C, Colognesi D, Mayers J, Reiter GF, Senesi R. Adv Phys 2005;54:377.
[112] Hilleke RO, Chaddah P, Simmons RO, Price DL, Sinha SK. Phys Rev Lett 1984;52:847. Sokol PE, Simmons RO, Price DL, Hilleke RO. Eckern U, Schmid A, Weber W, Wahl H, editors. Proceedings of the 17th international conference on low temperature physics. Amsterdam: North-Holland Publishing Company; 1984. p. 1213.
[113] Sokol PE, Sköld K, Price DL, Kleb R. Phys Rev Lett 1985;54:909. Mook H. Phys Rev Lett 1985;55:2452.
[114] Whitlock PA, Panoff RM. Proceedings of the 1984 workshop on high-energy excitations condensed matter. Rep. LA-10227, vol. II; 1984. p. 430, and references therein.
[115] Glyde HR, Azuah RT, Stirling WG. Phys Rev B 2000;62:14337.
[116] Watson GI. J Phys Condens Matter 1996;8:5955.
[117] Seel AG, Ceriotti M, Edwards PP, Mayers J. J Phys Condens Matter 2012;24:365401.
[118] Krzystyniak M, Richards SE, Seel AG, Fernandez-Alonso F. (submitted).
[119] Adams MA, Mayers J, Kirichek O, Down RBE. Phys Rev Lett 2007;98:085301.
[120] Diallo SO, Pearce JV, Azuah RT, Kirichek O, Taylor JW, Glyde HR. Phys Rev Lett 2007;98:205301.
[121] Reiter GF, Mayers J, Platzman P. Phys Rev Lett 2002;89:135505.
[122] Krzystyniak M, Fernandez-Alonso F. Phys Rev B 2011;83:134305.
[123] Krzystyniak M, Adams MA, Lovell A, Skipper NT, Bennington SM, Mayers J, et al. Faraday Discuss 2011;151:171.
[124] Berne BJ, Thirumalai D. Annu Rev Phys Chem 1986;37:401.
[125] Lin L, Morrone JA, Car R, Parrinello M. Phys Rev Lett 2010;105:110602.
[126] Ceriotti M, Markland TE. J Chem Phys 2013;138:014112.
[127] Pantalei C, Pietropaolo A, Senesi R, Imberti S, Andreani C, Mayers J, et al. Phys Rev Lett 2008;100:177801.
[128] Pietropaolo A, Senesi R, Andreani C, Botti A, Ricci MA, Bruni F. Phys Rev Lett 2008;100:127802.
[129] Feng YJ, Ancona-Torres C, Rosenbaum TF, Reiter GF, Price DL, Courtens E. Phys Rev Lett 2006;97:145501.
[130] Homouz D, Reiter G, Eckert J, Blinc R. Phys Rev Lett 2007;98:115502.
[131] Ceriotti M, Miceli G, Pietropaolo A, Colognesi D, Nale A, Catti M, et al. Phys Rev B 2010;82:174306.
[132] Rahman A. In: Woods Halley J, editor. Correlation functions and quasiparticle interactions in condensed matter. New York: Plenum; 1977. p. 417.

[133] Allen MP, Tildesley DJ. Computer simulation of liquids. Oxford: Clarendon Press; 1996.
[134] Schofield P. Phys Rev Lett 1960;4:239.

Section 4B

[135] Price DL, Bermejo FJ, Saboungi M-L. Rep Prog Phys 2003;66:407.
[136] www.visiongain.com/Report/721/The-Oil-Gas-Corrosion-Prevention-Market-2011-2021 [last accessed on August 23, 2013].
[137] Krivoglaz MA. Theory of X-ray and thermal-neutron scattering by real crystals. New York: Plenum; 1969.
[138] Ziman JM. Models of disorder: the theoretical physics of homogeneously disordered systems. Cambridge: Cambridge University Press; 1979.
[139] Nield VM, Keen DA. Diffuse neutron scattering from crystalline materials. Oxford: Oxford University Press; 2001.
[140] Welberry TR. Diffuse X-ray scattering and models of disorder. Oxford: Oxford University Press; 2004.
[141] Welberry TR, Goossens DJ, David WIF, Gutmann MJ, Bull MJ, Heerdegen AP. J Appl Cryst 2003;36:1440.
[142] Goodwin AL, Tucker MG, Dove MT, Keen DA. Phys Rev Lett 2004;93:075502. Goodwin AL, Tucker MG, Dove MT, Keen DA. Phys Rev Lett 2005;95:119901.
[143] Goodwin AL, Tucker MG, Cope ER, Dove MT, Keen DA. Physica B 2006;385–386:285.
[144] Parsonage NG, Staveley LAK. Disorder in crystals. Oxford: Clarendon Press; 1978.
[145] Angell CA. J Non-Cryst Solids 1991;131:13.
[146] Brand R, Lunkenheimer P, Loidl A. J Chem Phys 2002;116:10386.
[147] Bée M. Quasielastic neutron scattering, principles and applications in solid state chemistry, biology and materials science. Bristol: Adam Hilger; 1988.
[148] Hempelmann R. Quasielastic neutron scattering and solid state diffusion. Oxford: Oxford University Press; 2000.
[149] Vineyard G. Phys Rev 1959;110:999.
[150] Nijboer BRA, Rahman A. Phys Rev 1962;126:989.
[151] Hansen JP, McDonald IR. Theory of simple liquids. Boston: Academic Press; 1986.
[152] Kubo R. In: Meixner J, editor. Statistical mechanics of equilibrium and non-equilibrium. Amsterdam: North Holland; 1965.
[153] Egelstaff PA. An introduction to the liquid state. Oxford: Oxford University Press; 1992.
[154] Balucani U, Zoppi M. Dynamics of the liquid state. Oxford: Oxford University Press; 1994.
[155] Fernandez-Alonso F, Bermejo FJ, McLain SE, Turner JFC, Molaison JJ, Herwig KW. Phys Rev Lett 2007;98:077801.
[156] Sears VF. Can J Phys 1966;44:1279.
[157] Sears VF. Can J Phys 1966;44:1299.
[158] Sears VF. Can J Phys 1967;45:237.
[159] Volino F, Dianoux AJ. Mol Phys 1981;41:271.
[160] Hall PL, Ross DK. Mol Phys 1981;42:673.
[161] Bicout DJ. Phys Rev E 2000;62:261.
[162] Bicout DJ. Phys Rev E 2001;64:011910.
[163] Morkel Chr, Gronemeyer Chr, Gläser W, Bosse J. Phys Rev Lett 1987;58:1873, Phys Rev Lett 1987;58:2609.
[164] Sköld K. Phys Rev Lett 1967;19:1023.
[165] Chudley CT, Elliot RJ. Proc Phys Soc Lond 1961;77:353.

[166] Karlsson M, Engberg D, Björketun ME, Matic A, Wahnström G, Sundell PG, et al. Chem Mater 2010;22:740.
[167] Karlsson M. Dalton Trans 2013;42:317.
[168] Hull S. Rep Prog Phys 2004;67:1233.
[169] Höch A, Funke K, Lechner RE, Ohachi T. Solid State Ionics 1983;9/10:1353, and references therein.
[170] Goff JP, Hayes W, Hull S, Hutchings MT, Klausen KN. Phys Rev B 1999;59:14202.
[171] Price DL. High-temperature levitated materials. Cambridge: Cambridge University Press; 2010.
[172] Press W. Single-particle rotations in molecular crystals. Berlin: Springer-Verlag; 1981.
[173] Prager M, Heidemann A. Chem Rev 1997;97:2933.
[174] Colmenero J, Mukhopadhyay R, Alegría A, Frick B. Phys Rev Lett 1998;80:2350.
[175] Larese JZ, Martin y Marero D, Sivia DS, Carlile CJ. Phys Rev Lett 2002;87:206102.
[176] Lovell A, Fernandez-Alonso F, Skipper NT, Refson K, Bennington SM, Parker SF. Phys Rev Lett 2008;101:126101.
[177] Fernandez-Alonso F, Bermejo FJ, Cabrillo C, Loutfy RO, Leon V, Saboungi M-L. Phys Rev Lett 2007;98:215503.
[178] Yang S, Sun J, Ramirez-Cuesta AJ, Callear SK, David WIF, Anderson DP, et al. Nat Chem 2012;4:887.
[179] Feigin LA, Svergun DI. Structure analysis by small-angle X-Ray and neutron scattering. New York and London: Plenum Press; 1987.
[180] Hollanby MJ. Phys Chem Chem Phys 2013;15:10566.
[181] Guinier A. Ann Phys (Paris) 1939;12:161.
[182] Porod G. Kolloid Z 1951;124:83.
[183] Eastoe J, Gold S, Rogers SE, Paul A, Welton T, Heenan RK, et al. J Am Chem Soc 2005;127:7302.
[184] Engelman DM, Moore PB. Sci Am 1976;235:44.
[185] Petoukhov MV, Svergun DI. Biophys J 2005;89:1237.
[186] Svergun DI, Richard S, Koch MHJ, Sayers Z, Kuprin S, Zaccai G. Proc Natl Acad Sci USA 1998;95:2267.
[187] Müller-Buschbaum P. Polym J 2013;45:34.
[188] Penfold J, Thomas RK. J Phys Condens Matter 1990;2:1368.
[189] Sinha SK, Sirota EB, Garoff S, Stanley HB. Phys Rev B 1988;38:2297.
[190] Pynn R. Phys Rev B 1992;45:602.
[191] Dietrich S, Haase A. Phys Rep 1995;260:1.
[192] Dalgliesh R. Curr Opin Colloid Interface Sci 2002;7:244.
[193] Bowron DT, Soper AK, Jones K, Ansell S, Birch S, Norris J, et al. Rev Sci Instrum 2010;81:033905.
[194] Neuefeind J, Feygenson M, Carruth J, Hoffmann R, Chipley KN. Nucl Instr Methods B 2012;287:68.
[195] j-parc.jp/MatLife/en/instrumentation/ns_spec.html#bl21 [last accessed on August 21, 2013].

Section 5

[196] Merzbacher E. Quantum mechanics. 3rd ed. New York: John Wiley & Sons; 1998.
[197] Brown JM, Buenker RJ, Carrington A, di Lauro C, Dixon RN, Field RW, et al. Mol Phys 2000;98:1597.

[198] Gaspar AM, Busch S, Appavou M-S, Haeussler W, Georgii R, Su Y, et al. Biochim Biophys Acta 2010;1804:76.
[199] Penfold J, Webster J, Bucknall D. The use of polarisation analysis in polarised neutron reflection studies. Rutherford Appleton Laboratory Technical Report RAL-93-018, Didcot: Rutherford Appleton Laboratory; 1993.
[200] Böni P. Physica B 1997;234–236:1038.
[201] Dalgliesh RM, van Well AA, Boag S, Charlton TR, Frost CD, de Haan VO, et al. Physica B 2007;397:176.
[202] Andersen KH, Jullien D, Petoukhov AK, Mouveau P, Bordenave F, Thomas F, et al. Physica B 2009;404:2652.
[203] Surkau R, Becker J, Ebert M, Grossmann T, Heil W, Hofmann D, et al. Nucl Instr Methods A 1997;384:444.
[204] Freeman FF, Williams WG. J Phys E Sci Instrum 1978;11:459.
[205] Felcher GP, Kleb R, Hinks DG, Potter WH. Phys Rev B 1984;29:4843.
[206] Koehler WC, Cable JW, Wilkinson MK, Wollan EO. Phys Rev 1961;151:414.
[207] Low GGE, Collins MF. J Appl Phys 1963;34:1195. Collins MF, Low GG. Proc Phys Soc Lond 1965;86:535.
[208] Hicks JH. In: Kostorz G, editor. Neutron scattering. New York: Academic Press; 1979. p. 337.
[209] Binder K, Young AP. Rev Mod Phys 1985;58:801.
[210] Dormann JL, Fiorani D, Tronc E. Adv Chem Phys 1997;98:283.
[211] Holm C, Weiss JJ. Curr Opin Colloid Interface Sci 2005;10:133.
[212] Fennell T, Deen PP, Wildes AR, Schmalzl K, Prabhakaran D, Boothroyd AT, et al. Science 2009;326:415.
[213] Mühlbauer S, Binz B, Jonietz F, Pfleiderer C, Rosch A, Neubauer A, et al. Science 2009;323:915.
[214] Hughes J, Burgy MT, Helier RB, Wallace JW. Phys Rev 1949;75:565.
[215] Felcher GP, Kampwirth RT, Gray KE, Felici R. Phys Rev Lett 1984;52:1539.
[216] Chakhalian J, Feeland JW, Srajer G, Strempfer J, Khaliullin G, Cezar JC, et al. Nat Phys 2006;2:244.

Section 6
[217] Lipscombe OJ, Chen GF, Fang C, Perring TG, Abernathy DL, Christianson AD, et al. Phys Rev Lett 2011;106:057004.
[218] Birgeneau RJ, Bucher E, Passel L, Price DL, Turberfield KC. J Appl Phys 1970;41:900.
[219] Pfleiderer C. Rev Mod Phys 2009;81:1551.
[220] Mirebeau I, Hennion M, Casalta H, Andres H, Güdel HU, Irodova AV, et al. Phys Rev Lett 1999;83:628.
[221] Furrer A, Waldmann O. Rev Mod Phys 2013;85:367.
[222] Shapiro SM, Axe JD, Birgeneau RJ, Lawrence JM, Parks RD. Phys Rev B 1977;16:2275.
[223] Loong C-K. J Appl Phys 1985;57:3772.
[224] Loong C-K, Grier BH, Shapiro SM, Lawrence JM, Parks RD, Sinha SK. Phys Rev B 1987;35:3092.
[225] Holland-Moritz E, Wohlleben D, Loewenhaupt M. Phys Rev B 1982;25:7482.
[226] Chupis IE. Low Temp Phys 2010;36:477.
[227] Loktev VM, Pogorelov YuG. Low Temp Phys 2000;26:171.
[228] Loewenhaupt M, Lander GH, Murani AP, Murasik A. J Phys C 1982;15:5199.
[229] Matan K, Morinaga R, Iida K, Sato TJ. Phys Rev B 2009;79:054526.

[230] Hordequin C, Pierre J, Currat R. J Magn Magn Mater 1996;162:75.
[231] Sinha SK. Silver RN, editor. Proceedings of the 1984 workshop on high-energy excitations condensed matter, Rep. LA-10227, vol. 11. Los Alamos, New Mexico: Los Alamos Natl. Lab; 1984. p. 346, Cooke JB. ibid., p. 401.
[232] Van Hove L. Phys Rev 1954;95:1374.
[233] Halperin BI, Hohenberg PC. Phys Rev 1969;177:952.
[234] Foster D. Hydrodynamic fluctuations, broken symmetry, and correlation functions. Amsterdam: W.A. Benjamin Inc.; 1975.
[235] Collins MF, Minkiewicz VJ, Nathans R, Passel L, Shirane G. Phys Rev 1969;179:417, see also Minkiewicz VJ. Int J Magn 1971;1:149.
[236] Als-Nielsen J. Brookhaven Natl Lab Rep BNL 14793; 1971.
[237] Schulhof MP, Heller P, Nathans R, Linz A. Phys Rev B 1970;1:2304.
[238] Birgeneau RD, Skalyo J, Shirane G. Phys Rev B 1971;3:1736.
[239] Collins MF. Magnetic critical scattering. Oxford: Oxford University Press; 1989.

Section 7

[240] Abragam A, Goldman M. Nuclear magnetism: order and disorder. Oxford: Clarendon Press; 1982.
[241] Hayter JB, Jenkin GT, White JW. Phys Rev Lett 1974;33:696.
[242] Roinel Y, Bouffard V, Bacchella GL, Pinot M, Meriel P, Roubeau P, et al. Phys Rev Lett 1978;41:1572.
[243] Leslie M, Jenkin GT, Hayter JB, White JW, Cox S, Wagner G. Philos Trans R Soc Lond B 1980;290:497.
[244] Steiner M, Bevaart L, Ajiro Y, Millhause AJ, Ohlhoff K, Rahn G, et al. J Phys C 1981;14:L597.
[245] Suzuki H, Ohtsuka T, Kawarazaki S, Kunitomi N, Moon RM, Nicklow RM. Solid State Commun 1984;49:1157.
[246] Benoit A, Bossy J, Flouquet J, Schweizer J. J Phys Lett 1985;46:923.
[247] Jyrkkiö TA, Huiku MT, Lounasmaa OV, Siemensmeyer K, Kakurai K, Steiner M, et al. Phys Rev Lett 1988;60:2418.
[248] Steiner M. Phys Scr 1990;42:367.
[249] Tuoriniemi JT, Nummila KK, Vuorinen RT, Lounasmaa OV, Metz A, Siemensmeyer K, et al. Phys Rev Lett 1995;75:3744.
[250] Oja AS, Lounasmaa OV. Rev Mod Phys 1997;69:1.
[251] Siemensmeyer K, Clausen KN, Lefmann K, Lounasmaa OV, Metz A, Nummila KK, et al. Physica B 1998;241–243:506.
[252] van den Brandt B, Glättli H, Grillo I, Hautle P, Jouve H, Kohlbrecher J, et al. Europhys Lett 2002;59:62.
[253] Steiner M. J Low Temp Phys 2004;135:545.
[254] van den Brandt B, Glättli H, Hautle P, Kohlbrecher J, Konter JA, Michels A, et al. J Appl Cryst 2007;40:S106.
[255] Kumada T, Noda Y, Koizumi S, Hashimoto T. J Chem Phys 2010;133:054504.
[256] Noda Y, Kumada T, Hashimoto T, Koizumi S. J Appl Cryst 2011;44:503.
[257] Piegsa FM, Karlsson M, van den Brandt B, Carlile CJ, Forgan EM, Hautle P, et al. J Appl Cryst 2013;46:30.
[258] Chatterji T, Schneider GJ. Phys Rev B 2009;79:212409.
[259] Ehlers G, Mamontov E, Zamponi M, Kam KC, Gardner JS. Phys Rev Lett 2009;102:016405.

[260] Chatterji T, Jalarvo N, Szytula A. Solid State Commun 2013;161:42.
[261] Abragam A, Goldman M. Rep Prog Phys 1978;41:395.
[262] Maly T, Debelouchina GT, Baja VS, Hu KN, Joo CG, Mak-Jurkauskas ML, et al. J Chem Phys 2008;128:052211.
[263] Ardenkjaer-Larsen JH, Fridlund B, Gram A, Hansson G, Hansson L, Lerche MH, et al. Proc Natl Acad Sci U S A 2003;100:10158.
[264] Buckingham AD. Chem Phys Lett 2003;371:517.

Section 8

[265] Sköld K, Price DL, editors. Neutron scattering, experimental methods in the physical sciences, vol. 23A. New York: Academic Press; 1986.
[266] Furrer A. Neutron News 2007;18:2.

Chapter 2

Neutron Sources

Francisco J. Bermejo[*,†] and Fernando Sordo[‡,§]
[*]Consejo Superior de Investigaciones Científicas, C.S.I.C., Instituto de Estructura de la Materia, Madrid, Spain
[†]UAI Física Aplicable CSIC—UPV/EHU, Facultad de Ciencia y Tecnología, Campus de Leioa-Erandio, Bilbao, Spain
[‡]Consorcio ESS-Bilbao, Leioa, Spain
[§]Instituto de Fusión Nuclear, José Gutierrez Abascal 2, Madrid, Spain

Chapter Outline

2.1. Scope	138	
2.2. Useful Neutron Production Reactions	140	
2.2.1. Fission	141	
2.2.2. Direct and Stripping Reactions	142	
2.2.3. Bremsstrahlung	144	
2.2.4. Spallation Reactions	146	
2.3. Neutron Slowing Down and Moderators	153	
2.3.1. Moderators	158	
2.4. Basic Building Blocks of Accelerators to Drive Neutron Sources	166	
2.4.1. Beam Injectors	167	
2.4.2. Targets	178	
2.5. Accelerator-Driven Sources: Some Predecessors	197	
2.6. State-of-the-Art Accelerator Drivers for Neutron Sources	199	
2.6.1. Last-Generation Megawatt-Range Sources	199
2.6.2. Medium-Power (100 kW) Sources	205
2.6.3. Compact, Accelerator-Driven Sources	209
2.7. Research Reactors	212
2.7.1. Core Designs	213
2.7.2. Reactor Vessel	217
2.8. Future Prospects	219
2.8.1. Accelerators	219
2.8.2. Hybrid Systems	225
2.8.3. Reactors	226
2.9. Nonneutron-Scattering Uses of Neutron Sources	227
2.9.1. Isotope Production, In-Vessel Irradiation, γ-Radiation, and Neutron Activation Analysis	228

Experimental Methods in the Physical Sciences, Vol. 45. http://dx.doi.org/10.1016/B978-0-12-398374-9.00002-4
© 2013 Elsevier Inc. All rights reserved.

2.9.2. Nuclear Physics and Engineering: Astroparticle Physics, Nuclear Structure and Reactions, and Transmutation of Nuclear Waste 228
2.9.3. Hadron Physics: Neutrino-Related Phenomena 229
2.9.4. Fundamental Physics: Foundations of Quantum Mechanics, Effects of Gravity on Isolated Particles, Search for Dark Matter Using Ultracold Neutrons, Tests, and Validations of the Standard Model of Particle Physics 230
2.9.5. Use of Muon Beams for Condensed Matter and Fusion Research 231
Acknowledgments 231
Appendix A. Some Basic Relationships 232
Appendix B. The Transport Equation for Neutrons 234
References 237

2.1 SCOPE

Rather than aiming to be a self-contained document on neutron sources, this chapter is conceived as an update to that written by Carpenter and Yelon [1] back in 1986 for this very same book series. Proper acknowledgment should also be given to the first widely available book on the topic regarding pulsed neutron sources due to Windsor [2], which served as the reference manual for a few generations of experimentalists. A number of important developments have taken place since 1986 which have significantly changed our perspective about this field. First and foremost, some 30 years of operation of the intense pulsed neutron source (IPNS) at Argonne National Laboratory (ANL), KENS II at the High-Energy Accelerator Research Organization (KEK), and the Spallation Neutron Source (now ISIS) at the Rutherford Appleton Laboratory have shown that such sources can be fully competitive with traditional reactor-based installations as measured both by the output in science publications and by their relative impact, an example of which is given in Table 2.1. Second, advances in instrumentation for neutron applications together with newer neutron delivery systems that have taken place at reactor sources such as the Institut Laue Langevin (ILL) have resulted in large improvements in count rates of about 20 times at very modest costs. In addition, the impressive results evidenced as a consequence of refurbishment programs of known workhorses such as the reactor at the NIST Center for Neutron Research (NCNR at the National Institute of Standards and Technology, NIST, previously named as National Bureau of Standards) and the high-flux isotope reactor (HFIR, Oak Ridge National Laboratory) reactors show that steady, reactor-based sources are to stay for the coming decades. As a

TABLE 2.1 Publications Appeared in Fully Referred Journals Originated from the Most Relevant Neutron Facilities [3]

Year	ILL	NCNR (NIST) (NIST)	ISIS (STFC)	HFIR-SNS (ORNL)	MLSF (J-PARC)	SINQ (PSI)
2011	611 (141)	370 (105)	375 (139)	346 (100)	* (22)	140 (42)
2010	652 (179)	360 (85)	375 (135)	272 (64)	215 (20)	110 (30)
2009	635 (170)	400 (92)	345 (82)	189 (46)	149 (3)	120 (40)

Year	LANSCE (LANL)	Orphee (CEA-Saclay)	FRM-II (TU-Munich)	BER-II (HZB)	OPAL (ANSTO)
2011	16* (5)	* (38)	179 (31)	103 (29)	159 (26)
2010	166 (19)	161 (55)	206 (42)	78 (20)	62 (18)
2009		156 (52)	126 (24)	159 (23)	49 (17)

The acronyms within brackets are the institutions or laboratories hosting the facility. The numbers in parenthesis refer to papers published within a selected list of high-impact journals which is used as a benchmark for facility intercomparisons (*Chemistry of Materials, Europhysics J.E., JMB, JACS, Langmuir, Macromolecules, Nature, Nature: Materials, Nature: Physics, PRL, PRB, PRC, PRE, Science*). Such a list is obvious incomplete and somewhat biased toward the *Condensed Matter Sciences* and is to be taken as indicative of the productivity of the facility for mostly external users. The numbers correspond to articles signed by authors affiliated to a given facility. It does not necessarily imply that the reported results were obtained at the same premises. Data pertaining ORNL correspond to the combined output of both facilities. The numbers regarding MLSF include publications authored by JAERI scientists and mostly correspond to results obtained in other installations as well as at J-PARC. The asterisks refer to lack of publicly displayed information for the total published output.

matter of fact, the past decade witnessed the position of NCNR as the flagship of American neutron-scattering facilities, which together with the ILL in France and the spallation source ISIS at STFC-UK are considered as, up to now, the most productive and effective scientific neutron facilities in the world. Third and final, several sources of the last generation ushered in within the past three decades, such as the 20 MW OPAL reactor at the Australian Nuclear Science and Technology Organization (ANSTO), the quasi-steady-state spallation source SINQ at the Paul Scherrer Institut (PSI), and the two large spallation sources, the Spallation Neutron Source and the Materials and Life Sciences Facility (MLSF) at J-PARC, sited at Oak Ridge and at the Japan-Proton Accelerator Project Complex, Tokai-mura, Japan, respectively.

Within these, the 13 new neutron instruments at OPAL, managed by the Bragg Institute, represent the most remarkable effort for an installation of its size.

The saga of megawatt-range spallation facilities is now more than 20 years old, as witnessed by the first efforts, leading to the publication of a full feasibility study for the construction of the European Spallation Source, delivered in 1996. A dedicated 1 MW source named as the Spallation Neutron Source was built at Oak Ridge and completed in 2006. Full attainment of its specification in its operation was achieved in 2010 reaching full beam power. The Japanese MLSF is hosted by the KEK and JAEA within the J-PARC collaborative project together with facilities for hadron and neutrino physics as well as a facility for accelerator-driven transmutation of nuclear waste. Finally, the current European Spallation Source project, now in its design phase, constitutes a scaling down of those projects taken into consideration in 1996 and 2003 by the several European countries involved within the collaboration. The current baseline design retains the original goals of the collaboration, although its breadth has been shortened due to cost reduction issues.

A truly novel part of the present scenario of neutron sources with respect to the time when the previous monograph was written concerns the ongoing development of small accelerator-based neutron sources around the world, an initiative aimed to promote the exchange of information on emerging science and novel applications relevant to long-pulsed and/or medium-flux neutron sources. The potential of such installations has been reviewed in several IAEA technical documents [4], with special attention put on the role played by such installations in helping large, user-based facilities to achieve full performance.

At the time of writing, the most comprehensive list of neutron sources around the world open to external users is given by NIST in Ref. [5]. Such reference does not exhaust the catalog of neutron activities taking place around the world, and indeed, some examples will be given concerning the role played by national or regional centers as developers of important aspects of neutron instrumentation and techniques carried out in close collaboration with the larger facilities.

2.2 USEFUL NEUTRON PRODUCTION REACTIONS

Although there are a fairly large number of nuclear reactions and processes leading to emission of hadrons and light nuclei, only a few have been seriously considered as meriting due consideration to build usable, experiment-driven neutron sources. The separation of neutron production reactions into different groups is somewhat arbitrary since most reactions taking place at the higher energies also comprise characteristic processes which may take place at lower energies. With such proviso in mind, we can categorize such reactions into broadly three types, depending upon the type of nuclear excited state that such reactions generate. At low energies, typically below few megaelectron-volts, neutron production occurs as a consequence of the decay of an excited state of a compound nucleus. Above energies of the order of 15 MeV or so, nuclei reach well-defined resonance states which further decay ejecting neutrons. Finally, at substantially larger energies of the impinging particles, the nuclei are excited into a continuum of states where individual discrete states are not resolvable but some of the processes seen at lower energy are identifiable. In what follows, we describe the main neutron production reactions following such criteria.

2.2.1 Fission

The reaction, known since 1939, can either take place by thermal neutrons, that is, having energies of the order of 0.025 eV captured by fissile materials such as $^{235}_{92}U$ or by bombardment of high mass-number nuclei such as $^{238}_{92}U$ by higher energy neutrons with characteristic energies above 1 MeV or so. Fission by thermal neutrons involves the formation of a compound nucleus $^{236}_{92}U$ which becomes unstable against spontaneous fission and leads the nucleus to disintegrate into two unequal mass fragments within some 10^{-13} s. The initial fission reaction common to most reactors used as neutron sources for experimentation is as follows:

$$n + ^{235}_{92}U \rightarrow ^{236}_{92}U^* \rightarrow n + ^{96}_{39}Y^* + ^{139}_{53}I^*$$

where the star stands for unstable fragments which subsequently β-decay leading to some 72% $^{138}_{56}Ba$ and 16% $^{95}_{42}Mo$ as final decay products. The total energy released is about 193 MeV that is distributed between fission fragments (83.1%), neutrons (3.1%), β-particles (2.6%), neutrinos (5.6%), and γ-rays (5.6%) [6]. The energy spectrum of the neutrons generated by such a reaction is remarkably asymmetric, showing a distribution extending up to 17 MeV with mean value of 2 MeV and a mode about 0.75 MeV. Such a distribution is Maxwellian in the center-of-mass reference frame and can be represented on empirical grounds by [6]

$$p(E_n)dE_n = 0.775 E_n^{1/2} \exp[-0.775 E_n] dE_n \quad (2.1)$$

where E_n stands for the neutron energy. Usually, fission neutrons are classified into three groups, as fast ($E_n > 0.5$ MeV), epithermal or resonance ($0.2 < E_n < 0.5$ MeV), and thermal ($E_n < 0.2$ MeV), the latter being in almost thermodynamic equilibrium with the motion of atoms composing the moderating medium.

As the relevant figure of merit, one gets an average number of neutrons produced by the fission reaction of highly enriched (95%) uranium fuel (HEU) to be about 2.4 neutrons per absorbed thermal neutron.

The neutron economy of a fission device is governed by a multiplication factor k which relates the number of thermal neutrons in the $(j+1)$th generation to the n_j neutrons of the jth by $n_{j+1}=kn_j$ (see Appendix B). The quantity is given in terms of several parameters as $k=f\eta\varepsilon p$, where f is the fraction of neutrons absorbed by the fuel elements, η is the number of fast-fission neutrons produced per captured thermal neutron, ε stands for a fast-fission factor, and p is the resonance escape probability that measures the likelihood of a neutron escaping capture by $^{238}_{92}U$. A value just about (i.e., slightly below or above) unity corresponds to the reactor critical operation at any chosen power value set by the operator. Keeping such a value sets limits to the values of the four factors entering the above expression. In turn, the specific parameter values are governed by considerations pertaining the reactor design (i.e., its fuel cell dimensions and geometry, the chosen moderator, or the ratio of fuel to moderator) as well as the fuel composition (i.e., specific fuel and degree of enrichment with fissile material).

2.2.2 Direct and Stripping Reactions

Nuclear reactions at relatively low energies can be classified as either "direct" or "compound nucleus" reactions depending upon whether the reaction takes place at timescales comparable with those required for the impinging particles to transit across the nucleus (of the order of 10^{-22} s) or those where the particle impacting the nucleus is absorbed and the compound nucleus survives long enough (about 10^{-16} s or longer) before decaying to a ground state, to make the processes of absorption and decay statistically independent. This means that the kinetic energy of the incident particle is shared among all the nucleons, and all memory of the incident particle and target is lost. In terms of characteristic energies, one can set a somewhat diffuse boundary at 3–10 MeV, above which most reactions are predominantly of the direct type. Such reactions are triggered by a variety of incident particles, from γ-photons to multiply charged ions and energies extending up to the gigaelectron-volts range. They usually show a marked angular anisotropy and a relatively weak dependence on the energy of the incident particle beyond energies of a few megaelectron-volts.

The characteristics of direct nuclear reactions can be explained assuming that the particles emitted from the nucleus acquire energy and momentum in the process of direct interaction with the incident particle. In this respect, direct nuclear reactions are quite the opposite of nuclear reactions that pass through a stage of formation of a compound nucleus, where the energy introduced into the nucleus is statistically distributed among all the nucleons because of multiple collisions of the nucleons with each other. It is believed that [7] direct nuclear reactions occur at the periphery of the nucleus, where the nucleon density is low. Consequently, a nucleon that has acquired sufficient energy through interaction with an external agent has a considerable probability of leaving the nucleus without any collisions. The peripheral layer of the nucleus extends for ≈ 1 fm (1 fermi $= 10^{-15}$ m), a quantity to be compared with the radius of a medium-size, heavy nuclei which comes to be of the order of 10 fm. This class of reactions comprises a full range of phenomena such as elastic, quasielastic, and inelastic scattering, few-particle trapping such as stripping and pickup processes, depending on whether the incident particle has lost or acquired nucleons in the reaction, charge-exchange interactions, and, finally, nuclear knockout. As an illustrative example, let us consider proton-induced reactions, denoted as (p,n) processes. The ^{7}Li(p,n) and ^{9}Be(p,n) reactions are presently considered as good candidates to build quasi-monoenergetic neutron sources in the energy region up to 50 MeV. The reaction ^{9}Be(p,xn) with a thick target (i.e., thicker than the stopping power) is being exploited for the construction of medium-flux neutron sources for applied purposes owing to high neutron yields, good heat conduction, and high melting point of the target metal [8,9] (Figure 2.1).

On quantitative grounds, parametric forms giving the neutron yield for the reaction ^{9}Be(p,n) and proton energies below 23 MeV in units of neutrons

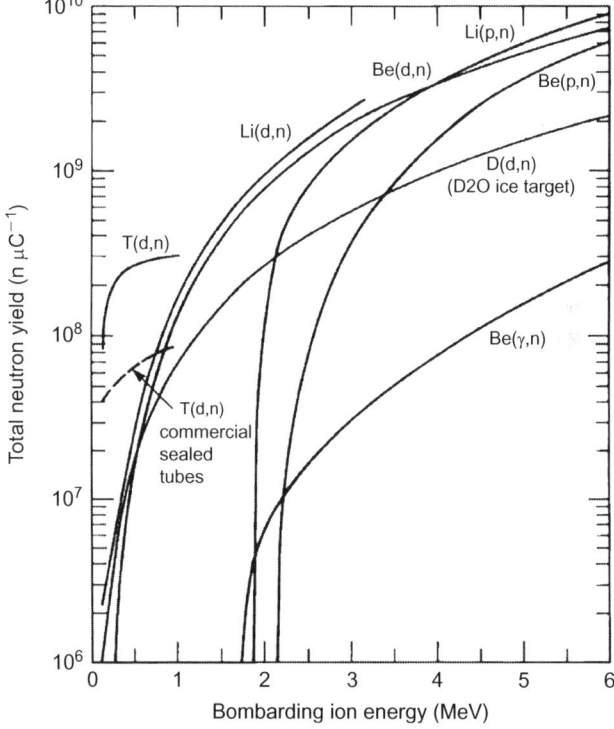

FIGURE 2.1 Neutron yield for low-energy particle beam reactions with a thick target. Data are taken from Ref. [1] which in turn reports on data originally compiled by L.D. Stevens and A.J. Miller, LLNL Report. URCL 19386 (1969). Data correspond to (γ,n), (p,n), (d,n) reactions on Li, Be, D$_2$O ice as well as the yield from commercially sealed generators. *Reprinted with permission.*

produced per micro-Coulomb of proton current (n/μC) as well as somewhat above that limit are reproduced below [8] as a function of the incoming proton energy,

$$Y_N(E_p) = 3.42 \times 10^8 (E_p - 1.87)^{2.05}, \text{ for } E_p < 23\,\text{MeV} \quad (2.2)$$

$$Y_N(E_p) = 5.58 \times 10^8 (E_p)^{1.69}, \text{ for } 23 \leq E_p \leq 80\,\text{MeV} \quad (2.3)$$

Nuclear stripping reactions mostly involve deuterons and correspond to processes where, because of the small binding energy of the deuterium nucleus (2.2 MeV), one of the particles is absorbed into the struck nucleus, whereas the other simply carries off the balance of momentum and energy. Reactions of this type are particularly relevant for the design of the International Fusion Materials Irradiation Facility (IFMIF) accelerator at Rokasho, Japan [10], intended for the study of fusion materials under intense radiation conditions [11]. The chosen reactions for such a purpose are those triggered by 40 MeV deuterons directed against a liquid lithium target and comprise ^{7}Li(d,2n)^{7}Be, ^{6}Li(d,n)^{7}Be, and ^{6}Li(d,t)^{4}He. In its current specification, the accelerator delivers 250 mA of deuteron current totalling 10 MW of beam power generating 10^{17} n s^{-1}. Such high-flux source results from the rather large beam current since the neutron production yield of such reactions is indeed modest (0.003 n/d) (see Table 2.6).

2.2.3 Bremsstrahlung

Although an energetic electron beam can produce neutrons as a consequence of its impact with a target material [12], the efficiency or cross section for such a process is too small to be considered as a useful mechanism for neutron production. However, a beam of accelerated electrons can generate large quantities of neutrons by means of a two-stage mechanism which involves the production of γ-ray photons as a consequence of rapid deceleration of such a beam when impinging on heavy target nuclei. The energy loss of incoming electrons with energy E_e can be parameterized as

$$\frac{-dE_e}{dx} = 4\alpha_h N_A \frac{Z^2}{A} \left(\frac{e}{4\pi\varepsilon_0 m_e c^2}\right) E_e \ln\left(\frac{183}{Z^{1/3}}\right) \quad (2.4)$$

where the symbols refer to the fine-structure constant α_h, which is the "coupling constant" or measure of the strength of the electromagnetic force that governs how electrically charged elementary particles (e.g., electron, muon) and light (photons) interact, and is defined as $\alpha_h = e^2\hbar c/4\pi\varepsilon_0$; Avogadro number, Z, and A are the atomic number and atomic weight (in g mol^{-1}) of the target material; ε_0 is the vacuum permittivity, and m_e is the electron mass.

The process known as bremsstrahlung or braking radiation has been used by a number of facilities to generate neutron yields of the order of 0.05

neutrons per impinging electron, leading to obtainable source terms of the order of 10^{14} n s^{-1} achieved at the now defunct Harwell electron linac [13]. There are at least three modes of neutron production by bremsstrahlung depending upon the energies involved. At low energies ($E \leq 5$ MeV), photons are absorbed by a mechanism known as the dipole interaction within which a compound nucleus is formed and later decays emitting protons and neutrons and other heavier particles. This mechanism involves the photon electric field which transfers its energy to the nucleus by inducing an oscillation, which leads to a relative displacement of tightly bound neutrons and protons inside the nucleus. The incident photons thus result in an excitation of the nucleus into a higher discrete energy state, and the extra energy is emitted in the form of neutrons. For heavy nuclei, the excited nucleus comes into ground state by emission of single neutrons by a (γ,n) reaction with some contribution expected for double neutron emission (γ,2n) at higher photon energies. On the other hand, the large Coulomb barrier hinders proton emission in heavy nuclei, whereas for $Z \leq 20$ the proton yield tends to be larger than the neutron yield. For energies larger than the threshold for nucleon emission, which comes to be about 8 MeV for most heavy nuclei, a giant resonance peak is observed about 14–18 MeV as well as subsidiary maxima at higher energies.

At higher energies ($E \geq 25$ MeV), the γ-photon interacts with the nucleus through absorption onto a quasi-deuteron state, which subsequently decays producing a neutron and a proton pair which can interact with the rest of the nucleus. Within such a process, the incident photon interacts with the dipole moment of a neutron–proton pair inside the nucleus rather than with the nucleus as a whole. At still higher energies, the photopion production becomes possible and competes with the quasi-deuteron process. The minimum energy needed to knock a neutron off nuclei by a γ-photon has to be above the binding energy of such neutron to the nuclei which in most cases is within the 7-to 15-MeV interval. The underlying physical mechanism of this kind of nuclear excitations in heavy nuclei brought forward by γ capture at low energies is understood in terms of a hydrodynamic model in which the nucleus is viewed as a mixture of proton and neutron fluids at constant density where the absorption of the incident photon leads to a separation of both fluids. Due to the strong binding of protons and neutrons inside the struck nucleus, the absorption of the γ-photon creates a high-frequency dipole which will ultimately decay by ejecting some nucleons.

For high enough photon energies $E \approx 50$–100 MeV photodisintegration channels open up, and finally for $E \geq 140$ MeV photopion absorption processes are enabled [14]. In fact, the cross section for pion production $\sigma(\gamma,\pi)$ is dominated by a strong peak at some 200 MeV of photon energy and a number of secondary maxima. At this stage, it is worth recalling that the triplet of pions, π^-, π^0, π^+, as well as the $\rho-$ and ω^-mesons are charged particles (mesons, that is having mass values intermediate between the electron and proton) whose primary role is to mediate the nuclear force, responsible for holding the nucleus together. Once such a meson is created, it will decay after a lapse of time of 10^{-16} to 10^{-8} s into different channels giving in some cases such as that for the neutral pion decay $\pi^0 \rightarrow 2\gamma$ additional γ-photons generating additional phenomena known as an *electromagnetic cascade* which in turn will be able to generate additional neutrons.

The energy distribution of the emitted neutrons is peaked about 1 MeV, although there is a high-energy tail extending up to tens of MeV generated by the higher energy processes just referred to. The low-energy part of the spectrum, referred to as the *fission spectrum*, can be approximated by [2]

$$\phi(E_n) = \frac{\sqrt{\pi}}{2E_{fis}^{3/2}} E_n^{1/2} \exp\left(\frac{-E_n}{E_{fis}}\right) \tag{2.5}$$

with a constant $E_{fis} \approx 1.4$ MeV. The quantity $\phi(E)$ stands for the neutron flux per unit energy. The distribution is basically isotropic. Above ≈ 2 MeV, the neutron flux mainly arises from *evaporation neutrons*, the origin of which are discussed below, and above 20 MeV, one still finds a tail originated from *direct processes* which have been discussed above and extends up to the energy of the incident particle.

As far as neutron yields of (γ,n) reactions, the production rate rises steeply with electron energy up to about 30–45 MeV. Above such figures, a plateau is reached and the yield produced by a variety of materials can be parameterized as [14]

$$N(Z) = 9.3 \times 10^{10} Z^{0.73} \tag{2.6}$$

where the units correspond to yields given in neutrons per kilowatt of beam power per second. The above equation serves to represent most materials examined, exception made of uranium for which the measured value turns out to be twice as large as the prediction.

Existing sources of this type are still in operation at Oak Ridge (ORELA) [15], at the Institute for Reference Materials and Measurements (IRMM-GELINA) at Geel, Belgium [16], Hokkaido, Kyoto, and Tohoku Universities in Japan or the CAB at Bariloche, Argentina. The most relevant installations for the purpose of this chapter are listed in Table 2.2.

The main bottlenecks limiting the development of the technique concern the large amount of heat generated by the process (2000 MeV per emitted neutron) and the relatively large γ backgrounds which impose severe requirements to cooling and shielding engineering.

2.2.4 Spallation Reactions

The origins and authorship of the term "nuclear spallation reaction" are somewhat unclear. Some of the basic phenomenology was already known in pre-World War II years, and in fact, suggestions by Bohr and Weisskopf [18]

TABLE 2.2 Electron Linacs Involved in Neutron Production Activities

Facility	Target materials	Pulse length (ns)	Repetition rate (Hz)	Electrical energy (MeV)	Average power (kW)	Neutron yield n s^{-1} × 10^{14}
ORELA(ORNL, TN, USA)	Ta	4–30	2/1000	180	50	1
Hokkaido-LINAC (Hokkaido, Japan)	Pb	10/3000	200	45	3	0.04
KURRI-LINAC(Osaka, Japan)	Ta	2/3000	100/300	30	6	0.08
Tohoku LINAC(Sendai, Japan)	W	3000	300	300		0.14
GELINA(Geel, Belgium)	U–Mo	<1/10	800	105	8.5	0.38
C.A. Bariloche (Bariloche, Argentina)	U	10/2000	100	25		0.2
RFNC-VNIIEF(Sarov, Russia)		2400	10	75		
DAΦNE-INFN(Frascati, Italy)	W	10	50	510		9 × 10^{-6}

All data reported above have been taken from sources open to the general public. Data for the DAΦNE-BTF source at Frascati have been elaborated from fluence values reported in Ref. [17].

on the possibility of neutron emission from excited nuclei were reported a few years before the discovery of nuclear fission. Fully fledged reports of experiments involving the analysis of products resulting from reactions denominated by such term appeared in the late 1940s [19,20]. Later on during the 1950s, a description of the spallation reaction as a two-step process involving energy deposition and later evaporation was finally postulated [21].

The basic tenet of those studies considered that for incident particles, say protons with mass m_p and energies E_p above some 100 MeV, their de Broglie wavelength,

$$\lambda_p = \frac{h}{\sqrt{2m_p E_p}} \qquad (2.7)$$

became of the order of 1 fm and thus allows the impinging particle to interact with individual nucleons. In this respect, it is worth recalling that nuclei have dimensions within 1–10 fm and that nucleons have also finite radii which for the proton come to be of 0.74 fm as measured by electron scattering experiments on hydrogen and deuterium. The nuclear force which binds protons and neutrons together leading to the formation of stable, bound states such as the deuteron when the two nucleon spins are aligned, is maximal for hadron separations of some 0.9 fm and vanishes for distances of 2.5 fm. The strength of such a force, which is nowadays considered as a *residual strong force* which binds the *quarks* together inside the nucleons by means of exchange of particles known as *gluons*, is of the order of 10^4 N for typical internucleon separations of about 1.3 fm.

It is also worth remembering that the binding energy per nucleon for most elements over the periodic table comes to be $\approx 8 \pm 1$ MeV, and therefore, it is expected that a nucleus if struck by protons, deuterons, or α-particles with energies beyond 100 MeV will result in the ejection of nucleons.

Our current understanding of processes originated by particles with GeV energies still is somewhat sketchy, even if some unified descriptions of the variety of phenomena underlying such processes are being attempted [22]. The main difficulties hindering further progress arise from the main obstacles nuclear physicists encounter for the development of *ab initio* approaches for nuclear structure calculations and have to do with the fact that most nuclei of interest are truly many-body systems, the description of which requires accurate knowledge of three-body and possibly higher order particle interactions. To the authors knowledge, such a goal has only been reached for nuclei with masses up to $A = 12$.

The present view of these reactions portrays them as taking place stepwise. Right after the impact with the incoming ion, a first process takes place leading to an *intranuclear cascade* where the incident particle triggers a series of reactions with characteristic energies from ≈ 20 MeV up to that of the incident particle, with nucleons leading to the creation of secondary particles. The characteristic time of such process is of the order of 10^{-22} s leaving the nucleus in a highly excited state. The particular characteristics of such cascade depend upon the number of target nucleons as well as on the energy of the incoming particles. At particle energies of ≈ 100 MeV, all interactions take place between nucleons as such. Increasing the incoming beam energy to a few hundred megaelectron-volts leads to pion production, which in turn will promote the nuclei into a higher lying excited state. The de-excitation of such a nucleus will proceed via the decay ejecting neutrons, protons, π-mesons as well as many other high-energy particles. Some of these projectiles, which are mainly emitted in the forward direction, may in turn trigger an additional spallation reaction in neighboring nuclei, thus giving rise to an *internuclear cascade*. The particles which result from this step of the reaction are mainly ejected in the forward direction with energies up to that of the incoming ion. A second stage then develops where nuclear de-excitation takes place by means of what is known as *evaporation*, a process that occurs at timescales of 10^{-16} s and yields additional neutrons, protons, α-particles, nuclear fragments, etc., which usually are some 15 atomic units lighter than the original target nucleus [23], this time all of them with energies below 20 MeV. Also, nuclear fission of the excited nucleus can take place if the energy involved in the evaporation process is high enough. In contrast with particles being ejected from nuclear cascades, nuclear evaporation yields particles isotropically distributed.

Because of reasons given above, most of our current knowledge on spallation reactions comes from model-building approaches [23], which are then scrutinized against experimental data. Such models rely on basic assumptions such that interactions between high-energy projectiles and the nucleus can be treated as free particle–particle collisions inside the nucleus. Even so, the level of complexity required for the purposes becomes remarkable indeed. As a matter of fact, some of the processes here lumped together as de-excitation and evaporation usually involve competition between fission and evaporation events which makes the physics of nuclear fragmentation by impact of energetic particles an active and interesting research field within studies on collective phenomena of nuclear matter.

In terms of the energy distribution of the spallation neutrons, Figure 2.2 compares the distribution in energy of fission neutrons to those resulting from spallation reactions. As expected, the latter are generated with energies up to that of the incoming proton, although in both cases, the energy distribution shows a distinctive peak at some 2 MeV.

The neutron yield (or neutron multiplicity) for typical proton energies, say 1 GeV, and target materials as Hg is about 30 n/p, whereas the energy deposited per emitted neutron is just 55 MeV. The first results on the neutron yield by spallation reactions on thick targets date back to times where the Canadian Intense Neutron Generation (ING) project was alive [25]. The results, valid for energies below 1.5 GeV, were parameterized in terms of neutron multiplicities per incident proton with energy E_p as,

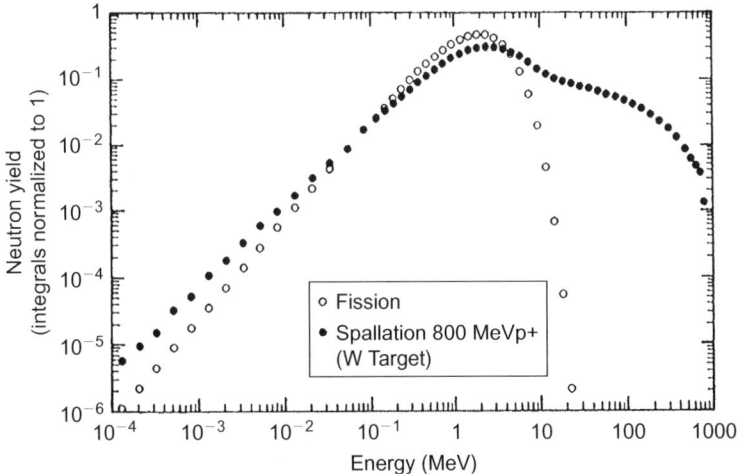

FIGURE 2.2 Energy spectra for fission of $_{92}^{235}$U compared to that resulting from spallation of 800 MeV protons impinging on a tungsten target of 10 cm diameter and 30 cm length [24]. *Reprinted with permission.*

TABLE 2.3 Approximate Neutron Multiplicities (n/p) for Different Target Materials for Thin Targets [26] and Three Different Values of the Incoming Proton Energy

Energy (GeV)	U	W	Ta	Pb	PbBi	Hg	Sn	SnPb
0.3	5.3	3.0	3.0	3.1	3.1	3.3	2.4	2.8
0.6	17	10	9.5	9.7	9.8	10.5	7	8
1.0	34	19	18	18	18	20	12	15

$$M(E_p) = 0.1(E_p - 0.12)(A + 20) \quad (2.8)$$

$$M(E_p) = 5.0(E_p - 0.12) \quad (2.9)$$

where the first equation accounts for nonfissile targets and the second for a target made of ^{238}U. Here A is the target mass number. The largest yield of about 57 n/p was found for a depleted uranium target of dimensions 10.2×61 cm.

A brief compilation of yields for different materials and for some selected values of the final proton energies is given in Table 2.3.

Studies on neutron yields per charge number for different ions (^{3}He, ^{4}He, and ^{12}C) were reported by Vassilkov and Yurevich [27]. The use of heavier ions with mass number A_i requires to be efficient, to increase the kinetic energy to keep the ion range inside the target long enough to allow nuclear reactions, since the collision cross section goes as

$$\sigma_{\text{coll}} \propto \left(A_i^{1/2} + A_{\text{target}}^{1/3}\right)^2 \quad (2.10)$$

Put into real numbers the above equation states that whereas for protons having a nuclear collision mean-free-path times the target density of some 200 g cm^{-2} and thus if one wants a proton range that will allow all protons to interact, that is basically three times the path, the required beam energy is about 1 GeV. In contrast, to meet the same conditions for an ion such as oxygen [23], a kinetic energy of about 30 GeV would be required. Choosing a lighter ion such as deuterons where stripping reactions result in an additional source of neutrons and protons yields increases in neutron production rates within 10–15% [28] only, at the expense of a significant increase of accelerator costs.

As regards the optimal incident energies for protons, it is worth delving at the following considerations,

- The proton range for beams above some 100 MeV follows a slightly sublinear trend with increasing proton energy. The nuclear collision probability, however, reaches a constant value above some 800–900 MeV since [29]

$$\begin{aligned} P_{\text{coll}} &= 1 - \exp\left[R/\left(33 A^{1/3}\right)\right] \\ R &= \rho^{-1}\, 233\, Z^{0.23} \left(E_p - 0.032\right)^{1.4} \end{aligned} \quad (2.11)$$

where ρ and Z stand for the target density and atomic number.

- The volume where neutron production takes place increases with increasing proton energy, which poses stringent demands on the reflector and moderator designs,
- Above a few hundred megaelectron-volts, the pion production cross sections become large enough to guarantee the operation of the existing sources as providers of muon beams produced by $\pi^{\pm} \to \mu^{\pm} + \nu_\mu$ or $\bar{\nu}_\mu$ decays. For incident proton energies significantly above 1 GeV, pion production becomes the dominant process [30]. While $\pi^{\pm}$-induced reactions lead to neutron production albeit less effectively than that generated by protons, neutral pions generated by proton collisions on a thick target decay as $\pi^0 \to 2\gamma$ and their mean lifetime of 10^{-16} s is too short to allow them to take part in the internuclear cascades.

In consequence, although the use of proton beam kinetic energies substantially above 1–2 GeV may be beneficial from the point of view of target radiation damage, the costs incurred in building the accelerator are not compensated by the neutron performance. In addition, experimental results carried out by the ASTE collaboration (The AGS Spallation Target Experiment, where AGS stands for the Brookhaven Alternating Gradient Synchrotron [31]) using 1.94, 12, and 24 GeV proton beams of 10^{12} particles per pulse impinging on a Hg target show that the maximum peak intensity over the Maxwellian peak region as measured in units of n cm^{-2} sr^{-1} W^{-1} linearly decreases as the proton energy increases.

To the best of the authors knowledge, the energy dependence of available neutron yield data for a wide range of incident proton energies reaching 12 GeV can be parameterized as [23]

$$M(E_p) = M_0 + M_1 E_p^x \qquad (2.12)$$

where for a thick Pb target the exponent gets down to $x = 0.75$, and the corresponding coefficient values are $M_0 = -8.2 \pm 1.6$ and $M_1 = 29.3 \pm 1.3$.

Contrary to the case of other neutron-producing reactions, the spallation products arising from reactions with most heavy-metal nuclei result in a wide distribution of unstable isotopes that decay generating a significant amount of radiation and afterheat in the target even after shutdown. The distribution of these spallation fragments shows a marked peak for atomic numbers close to the target mass number, together with a fission tail generated within the evaporation stage. Figure 2.3 shows the residuals generated for a Pb target [32].

From Figure 2.3, it is seen that elements with mass numbers up to that of the target material are formed as reaction products. For mercury targets, a number of radionuclides, mostly, H, I, Hg, Sr, Gd, Hf, and Au isotopes, are found in the target radiotoxic inventories which comprise a total of about 500 nuclides. From experience using solid targets during the past four decades, several unpleasant nuclides have been found which need to be tackled

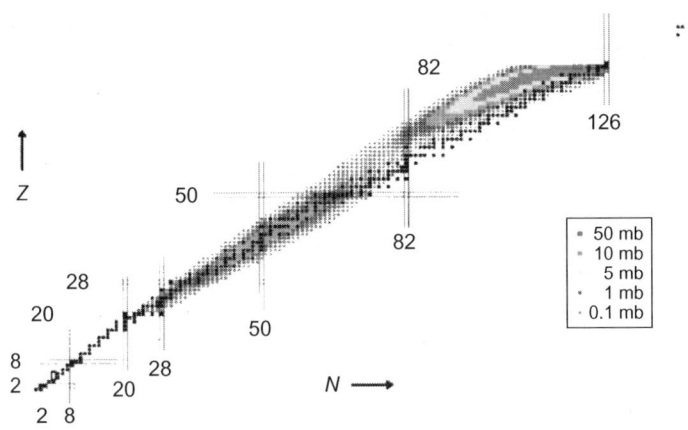

FIGURE 2.3 Measured distribution of spallation products for a ^{208}Pb target irradiated with 1 GeV protons. The data due to Enqvist *et al.*[32] are plotted as a nuclide chart where Z stands for the number of protons and N is the number of neutrons. Filled black squares correspond to the stable isotopes. Spallation and fission are separated by a minimum of cross sections at $Z = 58 \pm 3$. The color code depicts the cross-section values given in millibarns. The graph includes data for about 900 isotopes. *Reprinted with permission.* (For color version of this figure, the reader is referred to the online version of this chapter.)

with care. In particular, ^{7}Be is produced as a spallation product of water oxygen, has a half life of 53 days, and tends to accumulate on the walls of the cooling loops, leading to some local dose levels. Experience has shown that purification methods are efficient enough to remove the isotope from the cooling water. Other sources such as ^{11}C, ^{11}N, ^{13}N, and ^{15}O are short-lived positron emitters which are created in large quantities and may yield intense levels of ≈ 500 keV radiation. The fact needs to be taken care of to avoid high doses to accumulate in cooling elements within the loop.

Active research into spallation processes also takes place within the community involved in cosmic ray physics where the relevance of spallation reactions in astrophysics was realized about 40 years ago [33] and came as a consequence of the observed discrepancy between the chemical composition of cosmic rays which for some elements shows abundances well in excess of those found in the solar system. In fact, our current understanding of the excess by three to four orders of magnitude of elements with mass numbers within 20–25 relies on the interaction of galactic cosmic rays (mostly protons and α-particles) with C, N, and O within the interstellar medium. Research into the cosmogenic formation of nuclides, some of them easily observable in the earth atmosphere, has experienced as a consequence a significant boost in the past decades.

2.3 NEUTRON SLOWING DOWN AND MODERATORS

As shown in Figure 2.2, most neutrons produced by either fission or spallation reactions do show characteristic energies with a mean value of some 1–6 MeV. The use of such fast neutrons is however limited to a few applications in radiation physics of materials and devices, somewhat outside the scope of this review. The energies of neutron beams aiming at studies in condensed matter science are typically within the range 0.1 meV to 10 eV so that most of the neutrons produced by means of nuclear reactions have to be brought down or *moderated* to significantly lower energies. The processes of neutron moderation mostly involve the contact of the neutron ejected from a nucleus with a material kept at a given temperature, called the *moderator*, which by means of multiple collisions with its constituent atoms or molecules bring down the neutron energy to values about those corresponding to thermal energy of the moderating media. The moderation process mostly involves repeated inelastic collisions in which neutrons lose energy without being absorbed.

A broad classification of the usable neutron energy ranges is given in Table 2.4, together with some indications of typical applications.

To put ideas concerning neutron moderation on quantitative grounds, we write the basic equations giving the velocity change of a neutron entering a moderating media composed by atoms with mass A, which are initially at rest, that is, with velocities $v_b = 0$, with an initial velocity v_n (energy E_i) transferring energy down to E_f (velocity v'_n) to a moderator particle which then raises

TABLE 2.4 Commonly Accepted Neutron Energy Ranges Covering Those Characteristic of the Last-Generation Sources

Energy range	Neutron energy (meV)	Neutron wavelength (Å)	Moderator temperature (K)	Typical applications
Ultra cold	$<3 \times 10^{-4}$	≥ 500	$\approx 5^*$	Fundamental properties of the neutron
Very cold	$3 \times 10^{-4} – 0.1$	30–500		Ultra-small-angle scattering
Cold	0.1–5	4–30	5–20	Spectroscopy, supermolecular structures
Thermal	5–100	1–4	20–100	High-resolution spectroscopy
Epithermal	100–10^3	1–0.3	120–300	Spectroscopy, diffraction
Resonant	$10^3 – 10^5$	0.3–0.03	300	High-energy spectroscopy
Intermediate	$10^5 – 10^8$	$0.03 – 9 \times 10^{-4}$	300	Deep inelastic scattering
Fast	$10^8 – 10^{10}$	$9 \times 10^{-4} – 9 \times 10^{-5}$		Atomic physics, device irradiation
Ultrafast	$10^{10} – 3^{12}$	$9 \times 10^{-5} – 5 \times 10^{-6}$		Nuclear physics
				Hadron physics

The difference between very cold and ultracold neutrons is based upon the rather specific properties exhibited by the latter. In fact, by ultracold neutrons (UCNs), we refer to free neutrons which can be stored in traps made from certain materials which reflect the UCN. The kinetic energy of 300 neV corresponds to a maximum velocity of 7.6 m s^{-1} or a minimum wavelength of 520 Å. A cryogenic converter is usually placed after the cold moderator to achieve lower temperatures by means of inelastically downscattered cold neutrons. See Section 2.3.1.4.

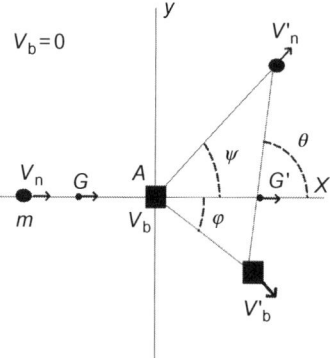

FIGURE 2.4 Collision parameters.

its velocity to v'_b. The relevant laboratory-fixed frame of reference corresponds to that depicted in Figure 2.4.

Since we assume that energy and momentum are conserved, the relevant equations become

$$m_n v_n = m_n v'_n \cos\psi + A v'_b \cos\varphi \tag{2.13}$$

$$m_n v'_n \sin\psi - A v'_b \sin\varphi = 0 \tag{2.14}$$

$$m_n v_n^2 = m_n v'^2_n + A v'^2_b \tag{2.15}$$

The fractional change in neutron energy after the collision is thus given by

$$X = \frac{E_f}{E_i} = \frac{v'^2_n}{v_n^2} = \frac{(1+A)^2 - 4A\cos^2\varphi}{(1+A)^2} \tag{2.16}$$

with

$$X = \frac{1 + A^2 + 2A\cos\theta}{1 + A^2 + 2A} \tag{2.17}$$

Rewriting now the previous expression using

$$\alpha = \left(\frac{A-1}{A+1}\right)^2 \tag{2.18}$$

$$E_f = \frac{E_i}{2}[1 + \alpha + (1-\alpha)\cos\theta] \tag{2.19}$$

we reach to a simple formula giving the minimum neutron energy that can be obtained in an collision in the center-of-mass frame as

$$E_{\min} = \alpha E_i \tag{2.20}$$

For the neutron energies of interest, one can assume that most elastic scattering events conform to s-wave processes so that the angular distribution of the outgoing neutron is isotropic, and therefore, all the output energies are equally probable (Figure 2.5).

A final integration over angles leads to the equation giving the average energy of the neutron after colliding with the moderator material as

$$\langle E_f \rangle = \int_{\alpha E_i}^{E_i} E_f p(E_i \to E_f) dE_f = \frac{E_i}{2}(1+\alpha) \quad (2.21)$$

A short compilation of α, A, and $\langle E_f \rangle$ values entering the above equations is given in Table 2.5 for some candidate materials. A glance to it shows

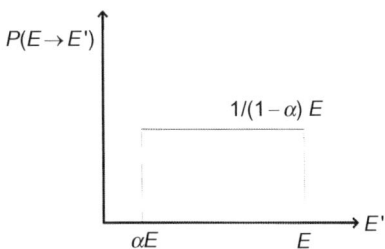

FIGURE 2.5 Collision law for a nucleus initially at rest.

TABLE 2.5 Neutronic Properties of Basic Moderator Materials

Atom	A	α	$\langle \frac{E_f}{E_i} \rangle$	$S_p(\text{cm}^{-1})$	$N_{MB}(\lambda)/N_{epi}(\lambda)$	$V \times t_s(\tau)$ cm	U_l
^{1}H	1	0.00	0.50	0.86	61	3.47	1
^{2}H	2	0.11	0.56	0.124		21.37	0.73
^{4}He	4	0.36	0.68	9×10^{-6}	51		0.427
^{9}Be	9.01	0.64	0.82	0.16	167	13.58	0.21
^{12}C	12.01	0.72	0.86	0.087	215	24.51	0.16
^{56}Fe	55.85	0.93	0.96	0.034	0.51	59.86	0.04
^{59}Ni	58.71	0.93	0.96	0.055	0.59	36.77	0.03
^{207}Pb	207.19	0.98	0.99	0.004	0.59	576.49	0.01
H$_2$O	18			1.43	62	2.15	0.926
D$_2$O	20			0.18	4830	14.25	0.51
CH$_4$	16			1.41	200	2.12	0.811
Polyethylene	(14)			1.61	122	1.77	0.913

The slowing-down power S_p, the moderating ratio $N_{MB}(\lambda)/N_{epi}(\lambda)$, the slowing-down length $V \times t_s(\tau)$, and the lethargy U_l are defined in Appendix A.

that hydrogen is best since neutrons may lose 100% of their energy in a single collision and, on average, they lose 50%. All other materials produce a much slower moderation process.

The moderation process usually involves repeated collisions until thermal equilibrium with the material atoms is attained. Under such conditions, the energy distribution of the constituent atoms of the moderating media E_a is given by the Maxwell–Boltzmann law at absolute temperature T as given below:

$$P(E_a)dE_a = a\left(\frac{E_a}{k_BT}\right)^{1/2}\exp\left(-\frac{E_a}{k_BT}\right)d\left(\frac{E_a}{k_BT}\right) \quad (2.22)$$

where k_B is the Boltzmann constant (8.617065×10^{-5} eV K^{-1} = 1.38054×10^{-23} J K^{-1}), T is the absolute temperature, and $p(E_a)$ is the probability distribution valid for gases in which molecules do not show large spatio-temporal correlations and so its validity excludes solids where atoms are embedded within a crystal lattice.

Once equilibrium is attained, the velocity distributions of neutrons and atoms forming the moderating material is given by

$$N_n(v_n)dv_n = N_n\left(\frac{m_n}{2\pi k_BT}\right)^{3/2}\exp\left(-\frac{m_nv_n^2}{2k_BT}\right)4\pi v_n^2 dv_n \quad (2.23)$$

$$N_b(v_b)dv_b = N_b\left(\frac{A}{2\pi k_BT}\right)^{3/2}\exp\left(-\frac{Mv_b^2}{2k_BT}\right)4\pi v_b^2 dv_b \quad (2.24)$$

where N_n and N_b are the density of neutrons and moderator atoms, m_n the neutron mass, and M the mass of the moderator atoms.

The most probable value for the energy for a thermalized neutron is given by k_BT, while the average value comes to be $3/2k_BT$. This put into real numbers gives the most probable energy for a neutron in thermal equilibrium with the environment (300 K) to be:

$$E_T = k_BT \sim 25\,\text{meV} \quad (2.25)$$

When neutron slowing-down processes take place in a medium composed by light nuclei such as hydrogen or deuterium, the equations written above tell that almost all its energy will be lost at a single collision and therefore there will be a negligible correlation between its energy and its age, namely, the time spent since the neutron was created. In contrast, if the nuclei scattering the neutrons are sufficiently heavy, energy loss takes place stepwise and the concept of *lethargy* or fractional energy loss defined in Appendix A becomes relevant. From statistical physics, slowing down within light media follows a Poisson process, that is the energy changes in discrete random jumps, whereas the heavier atom moderators obey to Gaussian statistics where energy loss is represented as a consequence of a large number of infinitesimal increments. The quantity of interest is here the slowing-down density $\phi(U_l)$ which may be defined as the number of neutrons rising above a given value for the lethargy. Mathematically, its main significance stems from its role as an effective source function in the diffusion equations that govern the time and space dependence of neutron moderation. The most useful form of theory of neutron moderation is cast within the "age equations," the beauty of which stands from their formal similarity to Fourier heat conduction equations whose solutions are known for most geometries and a whole variety of boundary conditions. The original formulations can be consulted in Ref. [34], and more recent versions of it can be consulted in several standard textbooks such as that of Ref. [35] and here we only sketch some of the results mostly within Appendices A and B.

2.3.1 Moderators

From the arguments given above, it is seen that the energy distribution of neutrons can be tailored by controlling the temperature of the moderating medium. The neutron output for any moderator geometry will, because its finite thickness, contain an additional epithermal or slowing-down component which, for steady-state sources follows the $1/E_n$ (or $1/\lambda_n$ if cast in neutron wavelengths) slowing-down spectrum. The strength of such a component can be brought down in reactor sources to a modest 1/100 but remains a well-visible component of the neutron pulse in accelerator-driven sources. Such component will contribute as a strong rise up at low neutron wavelengths on neutron spectra coming out from an accelerator source and can be modeled in terms of the Maxwellian distribution of neutron wavelengths, that is, Equation (2.22) cast in terms of the neutron wavelength λ_n, $N_{MB}(\lambda_n)$, and the epithermal component $N_{epi}(\lambda_n)$ as [2]

$$N_{tot}(\lambda_n) = J(\lambda_n) N_{MB}(\lambda_n) + [1 - J(\lambda_n)] N_{epi}(\lambda_n) \qquad (2.26)$$

where a "joining function,"

$$J(\lambda_n) = \frac{1}{1 + \exp[-(\lambda_n - \lambda_E)/\lambda_W]} \qquad (2.27)$$

takes the form of the Fermi distribution, and λ_E and λ_W are moderator-dependent parameters specifying a value for a joining wavelength and the width of such a distribution. The relevant definitions for all the involved quantities can be found in Appendix A.

2.3.1.1 Reactors

The coolant employed to remove the fission heat at reactor sources constitutes there the main moderator. Light or heavy water used as coolants usually attains temperatures at the core outlet of 50–80 °C. Water within the reactor pool surrounding the neutron beam tubes which is usually about 30–50 °C gives as a result neutron spectra peaked at about 1–2 Å. Such neutron wavelengths are optimal to perform many kinds of experiments but are inadequate for applications that may require high fluxes of say 5 or 0.35 Å neutrons. To significantly improve fluxes of cold or hot neutrons as required, the distribution given by Equation (2.22) is shifted to lower or higher temperatures than that of the reactor pool. To such an avail, structures containing moderation materials are usually immersed within the pools from where beam extraction of hot or cold neutrons by beam tubes or neutron guides is achieved.

The practice of extracting cold neutrons from a reactor dates back to the mid-1950s and activities carried out at the Harwell reactors. Different moderating materials such as liquid hydrogen or deuterium kept at 20 K, methane (CH_4), or propane (C_3H_8) have been extensively used for the purpose. For accelerator-driven sources, methane is perhaps the best option due to the appropriate inelastic scattering channels into which the neutron energy is transferred so that moderation occurs within a small volume and within times comparable with those characteristic of short-pulse spallation sources [2]. As far as reactors, where moderation time is not an important parameter and the volume of the neutron source may be large, the main option to generate cold neutrons is liquid D_2 moderators, where moderation takes place at a slower pace than methane but its low absorption cross section allows large-volume moderators with an optimum performance. As a matter of fact, those liquid D_2 moderators in operation at the ILL give neutron distributions peaked at 4 Å and fluxes up to 5×10^{12} n cm^{-2} Å^{-1} sr^{-1}. The moderator vessel of some $\approx$40cm diameter allows up to five neutron guide tubes plus an inclined beam. The ultimate limits on the performance of reactor cold sources mostly arise from the difficulties in removing radiation-induced heat from the moderator walls.

The production of hot neutrons is in a way less involved than its cold counterpart since what usually requires is to raise the moderating material, typically a graphite block, at temperatures of the order of 2000 °C or somewhat above by means of nuclear heating. Such devices are therefore of passive nature where the temperature is controlled by means of the interplay of heat shields and exchange gases. On the other hand, such moderators are best located as close as possible to the reactor core to get an enhanced contribution from the epithermal spectrum.

2.3.1.2 Accelerator-Driven Sources

The optimal moderators for an accelerator-driven source should be located as close as possible to the source generating fast neutrons. In addition, pulsed sources require the spreading in time of the neutron pulses coming out from the moderator to be as short as achievable. This usually involves the addition of neutron absorbers or *poisons* which are added deliberately to the moderating media. In fact, in terms of the source performance, one has to bear in mind that the achievable resolution attainable at the sample position of a given neutron-scattering instrument basically depends upon the spread in wavelengths of the incoming beam,

$$\Delta/E = \frac{\delta\lambda_n}{\lambda_n} = \frac{\delta_{\text{eff}}}{L_0} \qquad (2.28)$$

where δ_{eff} is an effective moderator thickness which includes effects due to poisoning and L_0 stands for the neutron primary flight path (see Appendix A).

The full treatment of the pulse coming out from a moderator at an accelerator source requires the evaluation of the distribution of neutrons emitted from the viewed surface of the moderator per unit time, per unit energy, per unit solid angle, all taken around a time t after a source pulse, around a given energy E, around the beam direction, and per neutron generated by the source. Such a quantity $I(E, \Omega_0, t)$ is related to the neutron density at the point r on the moderator surface S_M, per unit energy, per unit solid angle, at a time t after the source pulse, per neutron generated by the source by [1]

$$I(E,\Omega_0,t) = \int_S da v_0 N(r,E,\Omega_0,t) \qquad (2.29)$$

where v_0 stands for the component of the neutron velocity in the direction of Ω_0 and r denotes a position on the viewed moderator surface. The explicit evaluation of such quantity gets pretty involved. Some simplifying assumptions can be applied if one only requires evaluation of the pulse shape at high (i.e., $E > 5\ kT$) energies and leads to time-averaged spectra such as [1]

$$I(\overline{E}) = EI(E)|_{E_0} \frac{1}{E}\left(\frac{E}{E_0}\right)^\alpha \qquad (2.30)$$

where α is a constant related to the neutron leakage probability with typical values $\alpha \approx 0.2$ and E_0 is a reference energy, usually taken as 1 eV. Closed-form evaluation of the relevant parameters can be achieved by recourse to neutron diffusion theory (see Appendix B) and yields,

$$\alpha = \frac{DB^2}{U_1 \Sigma_s} \qquad (2.31)$$

which is given in terms of quantities such as the neutron diffusion constant D, a "geometric buckling" B^2 which explicitly depends upon the shape and size of the moderator of dimensions $a \times b \times c$,

$$B^2 = \pi^2\left(1/a^2 + 1/b^2 + 1/c^2\right) \qquad (2.32)$$

the lethargy U_1 and the macroscopic scattering cross section Σ_s (see Appendix A). The term "buckling" has been coined within neutron slowing-down theory and refers to the fact that such quantity provides a measure of the neutron flux profile at a given point in space and time. Under some limiting assumptions, that is for small bucklings, the smallest eigenvalue [36] for a moderator mode, thus defining the decay constant comes to be,

$$\gamma_{n0} = \langle \Sigma_{abs} \cdot v \rangle + \langle v \cdot D(v) \rangle \tag{2.33}$$

where $D(v)$ is the diffusion coefficient and the averages are taken over the energy distribution of such a mode.

Calculation of the time- and energy-dependent distributions at low energies represents, however, a more involved problem. Even approximate solutions to calculate the relevant quantities require the use of time-, space-, and energy-dependent neutron transport theory. On semiempirical grounds, Carpenter and Yelon [1] proposed a form to represent the spectrum as a sum of Maxwellian and slowing-down components as

$$I(E) = I_{MB}(E) + \Delta E I_{SD}(E)$$
$$= \left[\frac{I_{Th}}{I_{epi}} \frac{E}{E_T^2} \exp\left(\frac{-E}{E_T}\right) + \frac{J(E)}{E} \left(\frac{E}{E_0}\right)^{\alpha} \right] I_{epi} \tag{2.34}$$

$$I_{MB}(E) = \frac{E}{E_T^2} \exp\left(\frac{-E}{E_T}\right) I_{Th} \tag{2.35}$$

where the joining function $J(E)$ can be taken as equivalent to that given above by Equation (2.27) in the wavelength domain, although simpler parametric formulae such as that due to Taylor [1],

$$J(E) = \left[1 + \exp\left(\frac{A}{\sqrt{E-B}}\right) \right]^{-1} \tag{2.36}$$

where A and B are empirical constants are often used. Also, in Equation (2.34) $E_T = k_B T_{eff}$ stands for some effective temperature somewhat larger than that of the moderator, $I_{epi} = EI(E)|_{E_0}$ is the current per unit lethargy evaluated at E_0, and I_{Th} is the total thermal neutron current per source neutron. For energies above E_T the spectrum smoothly joins the $1/E$ slowing-down component which is cutoff for lower energies for $E \simeq 5E_T$.

For energies below 1 eV, the time distribution and shape of a neutron pulse coming out of a moderator can be evaluated by means of time-dependent diffusion theory of the kind sketched in Appendix B. The calculations involve the solution of an eigenvalue problem plus a continuum term which dominates at higher energies. The discrete eigenvalues lead to exponentially decaying terms of the form $\exp(-\lambda_j t)$ with all λ_j being positive. Explicit expressions for such discrete eigenvalues can be derived only under strong simplifying assumptions (i.e., within limits of small bucklings or infinite media such as done in Equation 2.33). At any rate, the main results tell [35] that at long times we can expect an exponential decays of the pulse which becomes faster as the moderator size diminishes as well as the presence of several superposed exponential decays corresponding to different bucklings, that is,

$$I(t) = I_0(t)\exp(-\gamma_0 \cdot t)$$
$$= \Sigma_{abs} \cdot v + DB^2 - CB^4 \ldots \quad (2.37)$$

where γ_0 is the inverse of the fundamental mode decay time, D is the diffusion coefficient, and C is a cooling coefficient. The main significance of Equation (2.37) concerns the relative weights of the quadratic and quartic terms. The term in B^2 is basically governed by the moderator thickness, and for cryogenic hydrogenous moderators such as those installed in most large spallation sources, it represents the dominant term. The full width at half height of the neutron distribution emerging from such moderators is approximately given by $\Delta t_{1/2} = \ln 2/\gamma_0$. For proton pulses of duration about 1 μs such quantity amounts to some 100–300 μs. For not-too-thin moderators, the energy distribution approaches a Maxwellian at long times and a second mode which has a faster decay than the fundamental mode may become apparent in the neutron spectrum.

Finally, the time distribution of neutrons emitted from the moderator surface shows a pretty complex dependence on the neutron energy as well as on the moderator thickness. An accurate account of this quantity needs to be taken for wavelength versus time calibration. The theory of neutron moderation can be used to derive closed-form expressions for some limiting cases only. In fact, within the limit of an infinitely thin moderator, the mean moderation time and the associated RMS deviation can be written down as [2]

$$\bar{t} = \frac{(1 + 2/\zeta)\zeta m_n \lambda_n}{\varepsilon N \Sigma_s h} \quad (2.38)$$

$$\delta \bar{t} = \frac{\left(\sqrt{1 + 2/\zeta}\right)\zeta m_n \lambda_n}{\varepsilon N \Sigma_s h} \quad (2.39)$$

where ε stands for the mean increment in lethargy, N is the number of moderator atoms, and ζ is a complicated function of the mass of the moderator particles which approaches unity for $M = 1$ (protons) and can be approximated as a power law as $\zeta = 1.042 M^{-0.925}$. For finite width moderators, Equations (2.38) and (2.39) can only be used as a guide. As a compromise, most calculations are cast in terms of the product of the neutron velocity times the mean emission times (a quantity with units of distance). The time distribution of neutrons leaking a moderator as a function of $v_n \bar{t}(E_n)$ shows in fact a noticeable spatial dependence, that is, the neutron brightness varies significantly over the moderator surface. The effect is enhanced for wing moderators compared to those with slab geometry, and this substantially hinders attempts to derive closed-form expressions to estimate the relevant parameters of the emission time distributions. Actual practice however shows that product $v_n \bar{t}(E_n)$ displays a significant structure but approaches finite limits of some 1.8 cm in the epithermal region for thin moderators and about 15 cm for thick devices [1]. On such grounds, an approximate representation of the dependence of the mean emission time has been derived from fits to experimental data and yields [37],

$$\bar{t} = \sqrt{3}/v\Sigma_s + P/\gamma_0 \qquad (2.40)$$

where $P = \exp(-E_n/E_0)$ and E_0 is a constant.

The advent of large computing capabilities has enabled in the last decades to address the problems related to moderator design in a more quantitative fashion than that allowed by existing theory. For such an avail, use is made of the current knowledge of the details of neutron interactions with candidate moderating materials which is encompassed within the material and thermodynamic-state-dependent $S(Q,\omega)$ dynamic structure factor which turns out to be the double Fourier transform of the time and space van Hove correlation function (see Chapter 1). On the other hand, the nature of the problem does not require a detailed knowledge of $S(Q,\omega)$ over large momentum-$\hbar Q$ and energy transfer $\hbar\omega$ ranges and, in fact, nuclear data libraries are usually built upon available information on the main dynamical processes of the atoms or molecular units constituting a given moderator, usually cast in form of simplified dynamical models. Some of these approaches employ a *synthetic scattering function* (SSF) [38] which allows a truly quantitative prediction of the moderating properties of hydrogenous materials. The output from such calculations are usually interfaced with standard computational tools such as the NJOY code [39] and lead to the computation of total scattering cross sections which may be scrutinized against experiment. Figure 2.6 displays a comparison of the calculated cross section for *para*-H_2 and normal (mostly *ortho*) hydrogen at 20 K using the SSF.

2.3.1.3 The Peak Shapes at Reactor and Accelerator-Driven Sources

Most of the experimental data analysis carried out at steady-state sources can be performed using well-established tools. This is best exemplified for the case of crystal analyzer spectrometers where the instrument response functions can be calculated analytically from knowledge of the basic beam, monochromator, and crystal analyzer parameters, under the assumption, usually justified, of Gaussian beams. The case of accelerator-driven sources significantly departs from this since the time and energy dependence of the beam coming out from most moderators is too complex to be used directly. Under such circumstances, the usual practice has been to model the resolution of time-of-flight peaks on heuristic grounds using the basic physics underlying the moderation processes as a guide.

Figure 2.7 displays the pulse shapes for a wide energy range as coming out from a cryogenic coupled moderator at LANSCE. A glance to the figure reveals that the pulse is markedly asymmetric, showing a trailing edge which becomes more pronounced at low energies.

The asymmetry shown by the data displayed in Figure 2.7 is what one should expect since the time-of-flight peaks comprise both slowing-down and storage terms. The reference [37] gives a compilation of several reported efforts to model such response functions. The one due to Ikeda and Carpenter

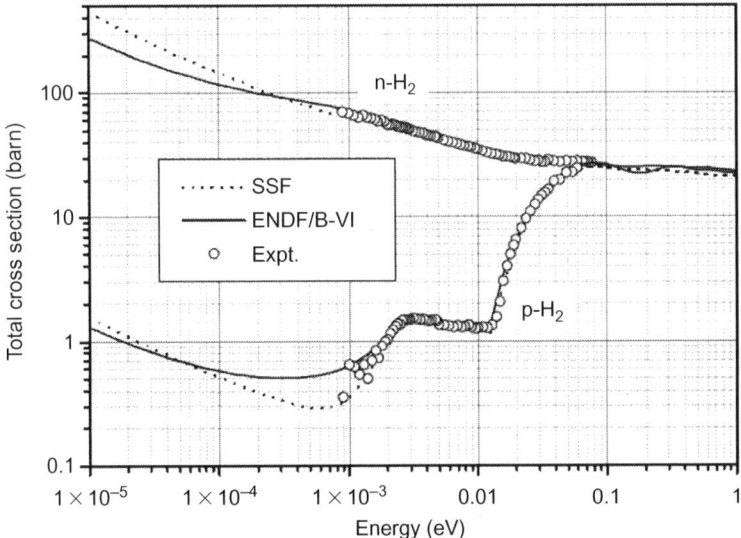

FIGURE 2.6 A comparison of the experimental total cross section for liquid hydrogen (open symbols) and that predicted using the synthetic scattering function (dots) and data within the ENDF nuclear library (solid line). The dynamical model used to represent the dynamics of H_2 uses the Young–Koppel formalism to describe the rotational elastic and one-phonon inelastic cross sections, together with three additional Einstein oscillators with energies of 0.005, 0.0147, and 0.546 eV introduced to represent the low-energy collective excitations, which are not adequately treated otherwise. See Ref. [39]. Courtesy of Dr. R. Granada (CNEA). *Reprinted with permission.*

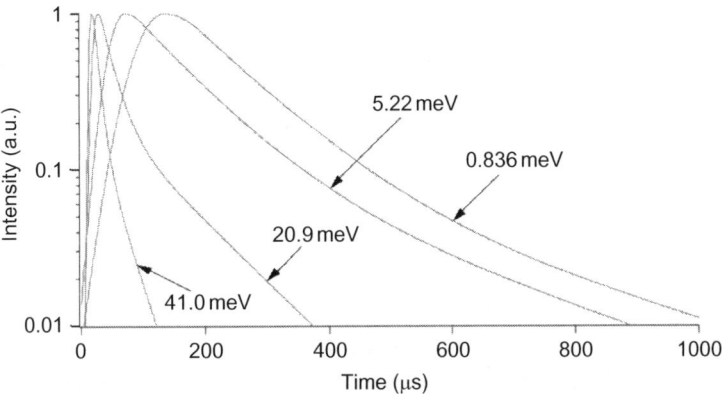

FIGURE 2.7 The shapes of pulses of 0.836–41.0 meV neutrons from the partially coupled liquidhydrogen moderator at Lujan Center [40]. *Reprinted with permission.* (For color version of this figure, the reader is referred to the online version of this chapter.)

consists on a sum of a delta function and a decaying exponential with arbitrary relative areas convoluted with a slowing-down function and reads [37]

$$R(\lambda_n,t) = \frac{2}{a}\left[(1-P)(at)^2\exp(-at) + \frac{2a^2\gamma_0 P}{(a-\gamma_0^3)}\right.$$
$$\left. \times\left\{\exp(-\gamma_0 t) - \exp(-at)\left[1 + (a-\gamma_0 t) + \frac{1}{2}(a-\gamma_0 t)^2 t^2\right]\right\}\right] \quad (2.41)$$

where the first term stands for the slowing-down function and the second for the "storage term." The parameters are assumed to show some wavelength dependence, the coefficient $a \simeq \Sigma_s \cdot v$ at high energies P is defined in Equation (2.40) and γ_0 is the decay constant of the fundamental moderator mode.

For an in-depth discussion on the peak shapes to be accounted for when fitting experimental diffraction patterns in a neutron-scattering time-of-flight measurement, the interested reader may consult Ref. [41].

2.3.1.4 Production of Very Cold and Ultracold Neutrons

The main niche of interest for neutron scattering in the use of neutrons having wavelengths larger than 4 Å or so mostly focuses on the promising applications of ultra-small-angle scattering [42] as shown by current instruments, the capability of which could be largely expanded if use is made of longer wavelengths. The quest for further efforts into this technique stands from the fact that measurement of scattering-length density fluctuations within the micrometer range cannot be carried out using the techniques of choice such as electron microscopy, light scattering, and atomic force microscopy if materials under scrutiny show low contrast or are opaque to light probes, or the property of interest is related to some magnetic structure.

A strong interest in further developing techniques for the production of ultracold neutrons (UCNs) focuses in their potential for basic fundamental particles, hadron, and nuclear and quantum-physics research using such neutron sources. In fact, cutting-edge experiments have been carried out at neutron centers such as the ILL, which demonstrate the accomplishment of world-class research into fundamental topics such as the quantum states of neutrons in the gravity potential of the earth, enabling to test Newton's Gravity Law at small distances, a number of fundamental tests of quantum mechanics performed with neutron interferometry, and novel experiments on neutron β-decay trying to scrutinize some scalar or tensor interactions which are predicted by supersymmetric field theories. Also, important tests for predictions made from the Standard Model, concerning the search for a finite value of the neutron electric dipole moment, the A coefficient in the correlation for the neutron β-decay, or the measurement of the neutron–antineutron oscillation times, are currently being carried out at several neutron sources.

To the authors knowledge, the main routes for UCN production use cold neutrons as an input onto a cryogenic converter (superfluid ^{4}He, solid $ortho$-D$_2$, solid O$_2$, and solid CD$_4$) [59]. Such UCN converters show a large cross section able to "scatter-down" neutrons to lower energies due to the available density of roto-vibrational states, while keeping "up-scattering" and absorption to a minimum. Typically, UCNs are labeled all those having velocities $v_n < 10$ m s^{-1} (energies of the order of a few hundreds of nanoelectron-volts), figure to be compared with that of about 10^3 m s^{-1} of those neutrons coming out from the moderator. Once such velocities are reached, the UCNs are transported into storage devices, usually carefully shielded storage bottles. Such bottles are lined with materials able to reflect with a positive neutron optical potential. The height of such potential varies from 335 neV for ^{58}Ni which defines the upper limit of the kinetic energy range of UCN. The most widely used materials for wall coatings also include beryllium, beryllium oxide, and more recently also diamond-like carbon. Production densities of about 10 UCN cm^{-3} are achieved at the and expected to rise up to 55 UCN per cubic centimeter, and an increase of such figure up to 6500 UCN cm^{-3} is expected to be achieved in future at the MLSF [43], where there are hopes of attaining significant gains from novel transportation systems (a "rebuncher" system rather than the simpler transport system employed before). Also, the TRIGA [44] reactor facility at Mainz (Germany) has reported on the production of 2×10^5 UCN cm^{-3} when operated in pulsed mode.

A new proposal due to Rivlin [43] consists in slowing-down thermal neutrons by repeated collisions with nuclei precooled down to a temperature of ≈ 1 μK by known methods of laser manipulation of neutral atoms. The feasibility condition for this UCN production process is its short duration, compared with the neutron lifetime. Estimates for a simplified model of the process indicate the possibility of producing UCN concentrations several orders of magnitude higher than those obtained with the present method of UCN extraction from the low-energy wing of the Maxwellian energy distribution.

2.4 BASIC BUILDING BLOCKS OF ACCELERATORS TO DRIVE NEUTRON SOURCES

The first generation of accelerators ever built used for the purpose electrostatic fields and were developed around 1919 following different concepts due to J. Cockcroft and E. T. S. Walton at Cambridge, UK using a voltage multiplier, by M. Tuve at the Carnegie Institution of Washington employing an air transformer due to Nikola Tesla, by R.J. Van de Graaff, who designed his well-known electrostatic generator and by C. Lauritsen who exploited the high-voltage facilities available at the time at the California Institute of Technology. A relatively simple electrostatic accelerator was reported in 1933 to produce neutron yields which could be as high as 10^{10} n s^{-1} [45], from accelerated deuterons which is regarded as the pioneer work on neutron production driven by energized particles leaving an accelerator.

A first proposal for a linear accelerator was due to Rolf Wideroe, who did indeed build a machine which in 1928, and as a proof of principle (POP), accelerated alkali ions by means of an accelerating potential corresponding to only half the energy. The technique banked on the fact that the accelerating potential is used twice so that the ions were pulled into one end of a tubular electrode and pushed from the other by an electric field that had meanwhile reversed direction. Further on, pondering on the Wideroe proposal led to E.O. Lawrence to develop a device able to recycle the particles by bending their paths in a magnetic field perpendicular to the plane of their orbit and to accelerate them twice a turn. The particle tracks move into a wider orbit as they gain energy, increasing their velocity and thus enlarging their path to make the interval between successive accelerations constant. The concept led to the construction of the first operating cyclotron, built by Lawrence, Livingston, and Sloan in 1930, which demonstrated acceleration of a tenuous beam of hydrogen-molecule ions up to an energy of 80 keV. Since each ion received an accelerating kick twice in a circuit as it entered through the center of the machine and left the single flat semicircular electrode or *dee*, those that managed to reach full energy and fall into the collecting cup 4.50 cm from the center of the instrument had made no fewer than 40 turns.

Most particle accelerators used for neutron production comprise machines which operate under different principles, and in general, they combine segments which make recourse to disparate technologies. In what follows, we give an overview of the basic components needed for the operation of existing installations.

2.4.1 Beam Injectors

The simplest way of producing usable fluxes of neutrons geared for experimental purposes by means of an accelerator is to generate photoneutrons by means of (γ,n) reactions for which relatively simple electron linear accelerators, synchrotrons, or betatrons can be employed. Since a variety of machines, from electron linacs to fully fledged proton or H^- ion accelerators, are able to produce intense neutron beams, it is worth delving some time to explore the diverse kind of accelerating devices built so far. The machines can be classified by means of somewhat simplified criteria as shown in the accompanying Figure 2.8.

Most of the structures employed nowadays within proton accelerator-driven neutron sources make use of electromagnetic, resonant structures as well as combinations of linear (linacs) and circular (cyclotron or synchrotron) machines, the latter used for further acceleration up to higher energies or for pulse compression purposes. Hybrid structures are widely used in electron machines of several kinds. Nonresonant machines such as induction accelerators have been significantly developed and are nowadays able to provide high-current electron beams well within the kiloampere range and energies of

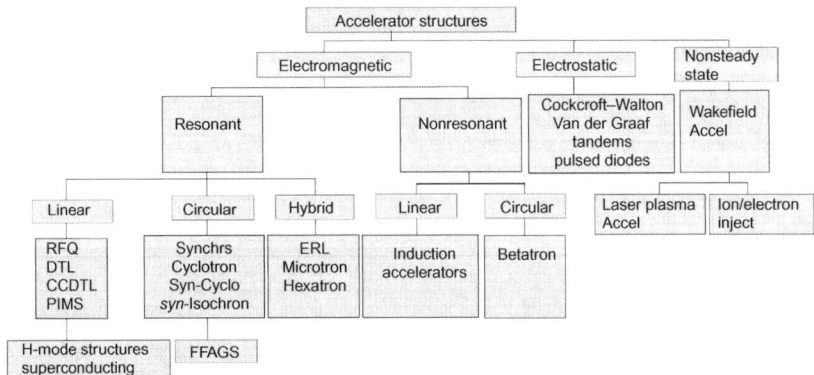

FIGURE 2.8 A simple, conceptual classification of accelerator structures. (For color version of this figure, the reader is referred to the online version of this chapter.)

several tens of megaelectron-volts. Their main use focused onto military and heavy-ion fusion applications. The technology is now mature, and therefore, their use as drivers for neutron sources is briefly addressed in Section 2.8.1. Similar qualifications apply to novel techniques such as those based upon the excitation of shock waves within confined plasmas, usually referred to as wakefield accelerators. Finally, purely electrostatic accelerators, widely used in the past and still employed as first acceleration steps, can be used to provide fast-neutron production rates of $\approx 10^{10}$ n s^{-1} arising from d–d or d–t reactions. The Cockcroft–Walton voltage multiplier was a glaring example of this, or they were also used as the only accelerating structure for low-power, low-energy machines such Van der Graaf, pelletrons, tandems, etc.

Most of the machines used as drivers for neutron production by means of either electrons or protons use the electric fields provided by radio frequency (RF) transmitters as the source for the accelerating electric fields. To such an avail, the particles are first bunched into packets of typically 10^{10} particles and then accelerated by alternating fields with phases which are made to have a fixed relationship with the multiple accelerator gaps. Such gaps, which are all at the same phase than the RF wave, are located at distances d that are proportional to the particle velocity. Because of their low rest mass, electrons attain relativistic kinematics as measured by

$$\beta^2 = \frac{v^2}{c^2} = 1 - \frac{1}{\sqrt{1 + E_{\text{kin}}/mc^2}} \qquad (2.42)$$

in terms of its kinetic energy E_{kin} and particle mass, at energies within the kiloelectron-volt range reaching velocities $v \approx 0.95c$ at energies as low as 1.1 MeV. This makes electron linacs to be composed by an injector plus a series of identical accelerating structures, with all cells having a length proportional to $\beta\lambda$, where λ stands for the RF radiation wavelength. In contrast, protons behave as classical particles up to some tens of MeV and reach a comparable velocity $v \approx 0.95c$ at about 2 GeV. This forces proton and ion linacs to be made of a sequence of different accelerating structures which change cell length, frequency, operating mode, etc. so that they are optimally matched to the ion velocity. This makes the distance between gaps variable and follows $d = \beta\lambda/2$, that is, the distance between gaps increases proportionally to the particle velocity, to keep synchronicity between the particle and the electromagnetic wave.

Further differences between electron and proton/ion accelerators concern the precise characteristics of RF wave propagation inside accelerator structures. The RF power injected into such structures is decomposed into electromagnetic field distributions or *modes*, which arise as a consequence of the superposition of waves reflected by the metallic walls of the accelerating structure. To accelerate particles, we need to select a mode with longitudinal field components on the beam axis. The precise selection of an accelerating mode depends upon the purpose for which the cavity has been built for. For electrons, because $v \approx c$ even for low energies, relatively simple structures supporting a *traveling wave* can be used for acceleration purposes. Recourse is often made to short, high-frequency (GHz) RF pulses allowing to reach large accelerating gradients, typically 30 MV m^{-1} or above. Proton and ion linacs are however built making use of *standing waves* generated by the sum of two waves traveling in opposite directions, adding up in the different accelerating cells. Such machines are usually run using long (hundreds of µs) pulses and attain moderate gradients of the order of 2–5 MV m^{-1}. Such values can only be surpassed by the use of superconducting resonators as we refer to below.

At low proton/ion energies, the equations of motion for a particle of charge q and mass m with position **x** and velocity **v** under the action on electric **E** and magnetic **B** fields are given by

$$\frac{d\mathbf{x}}{dt} = \mathbf{v} \tag{2.43}$$

$$\gamma \frac{d\mathbf{v}}{dt} = \frac{q}{m}(\mathbf{E} + \mathbf{v} \times \mathbf{B}) \tag{2.44}$$

with $\gamma = 1/\sqrt{1 - v^2/c^2}$ and c is the velocity of light. In other words, the particle acquires kinetic energy only from the electric field, whereas the magnetic force remains perpendicular to the beam axis of motion. Also, when velocities approach the relativistic limit, the radial forces from the RF field cancel with their magnetic component. At low energies, however, strong electrical repulsion is to be taken care of, specially for high-intensity beams.

A relatively novel [46] structure known as the *radio frequency quadrupole, RFQ*, deals with this problem in most modern accelerators. In itself, it constitutes a versatile section for an accelerator able to *bunch, focus,* and *accelerate* a proton beam from a few kiloelectron-volts up to a few megaelectron-volts. The idea of such an instrument was first proposed by Kapchinskii and Teplyakov [46]. However, it was not until 1980 that the first

RFQ named POP [47] was actually built and run at LANL by Stovall, Crandall, and Hamm. Since then, the RFQs have dominated the low-energy range of the linear accelerators. All such structures built so far comprise four electrodes, either in a rod or in a vane configuration. Such four electrodes generate a quadrupole field once fed by the RF injected in the cavity by means of a coupler. For a 4-vane configuration, the electrodes are accurately machined with a sinusoidal modulation that varies longitudinally with a separation of $\beta\lambda/2$, each cell having a minimum and a maximum distance from the axis to the vane. The minimum radius is called the aperture a, and the ratio between the maximum and the minimum is the modulation factor, m. The field at the center axis oscillates between positive and negative values. Negative field values decelerate a positive ion, while positive values do the opposite. The action of such oscillatory fields forces particles to execute longitudinal motions about a tagged particle synchronous with the phases of RF field. The phase between the synchronous particle at the beginning of a cell and the RF is called the *synchronous phase*. The RFQ is then designed for the synchronous particle to feel accelerating fields every half period. A schematic view of the motion of a bunch of particles through a period is shown in Figure 2.9.

Typically, the first part of the structure comprising somewhat less than 10 $\beta\lambda/2$ cells starts with a *Radial Matcher* section that has the longitudinal shape of a quarter of circle, starting with large aperture which gradually closes down. Its aim is to funnel the particles, which come with very crossed paths, into the RFQ. The next section named the *Shaper* has as its main purpose to focus the beam although some acceleration is also provided. A referred to as the *Gentle Buncher* then follows. This is precisely where the input beam is separated into smaller bunches by the action of a much stronger field. Finally, the *Accelerator* section provides most of the energy gain. Substantial acceleration is thus achieved at the final part of this structure. A view of an accelerator structure of this kind is shown in the accompanying Figure 2.10.

For high-power accelerators, two basic schemes come to mind: either the acceleration and extraction of the beam are quasicontinuous, or they are separated by some intermediate storage device such as the stretcher ring in the case of electrons or a compression and/or further acceleration ring when proton machines are involved. In the first case, one needs a continuous wave (CW) linear accelerator (linac) which can either be a straight multisection machine through which the beam passes only once, or it can be a one-, two-, or three-section machine which allows many recirculating orbits (this is the microtron), or it can be a multisection linac of intermediate length through which the beam is recirculated a few times by means of independent magnetic channels.

Let us now delve into some considerations concerning the efficiency of existing accelerators, which can be understood in terms simplified power-balance considerations. A prime quantity is the *shunt impedance* which

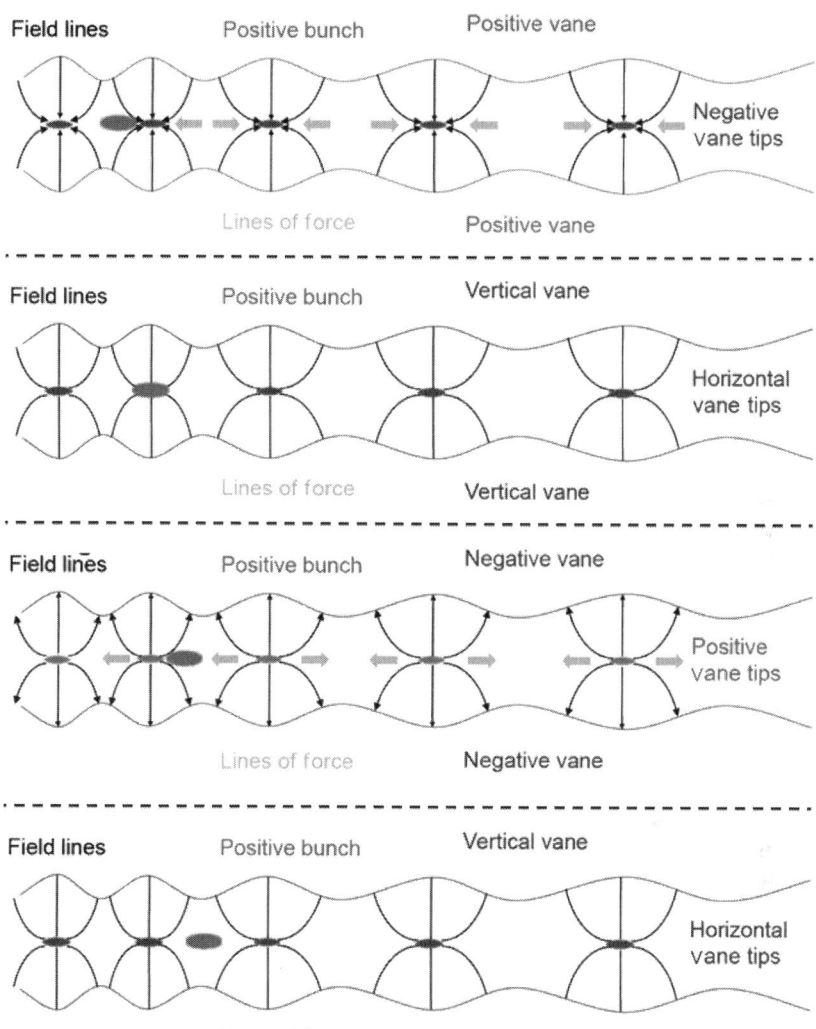

FIGURE 2.9 Sequence to show how the vane modulations accelerate a particle bunch. Only vertical vanes are shown. From top to bottom: (1) Particle bunch feels an accelerating force. (2) 1/4RF period later, the field drops to zero and the bunch feels no accelerating force. (3) Another 1/4RF period later, the field is at a maximum again but the sign is reversed, the bunch feels an accelerating force. (4) Another 1/4RF period later, the field drops to zero and the bunch feels no accelerating force once more. For details on how to control the RF intensity and phase, see Ref. [48]. Courtesy of Dr. S. Jolly (UCL). *Reprinted with permission from the author.* (For color version of this figure, the reader is referred to the online version of this chapter.)

FIGURE 2.10 A view of several RFQ accelerator structures employed in different applications. (a) The picture shows the URAL30 RFQ at IHEP-Protvino, Moscow region. The structure accelerates from 100 keV up to 2 MeV a current of 200 mA protons. The structure corresponds to an H-mode cavity operating at a resonance frequency of 148.5 MHz. (b) Front view of a section of the RFQ installed at the Spallation Neutron Source at ORNL. The structure is able to accelerate 60 mA of H$^-$ ions from 65 keV up to 2.5 MeV. It operates for an RF bunch frequency of 402.5 MHz and duty factors D up to 5%. The figure displays the front view showing the four vane electrodes together with mode-stabilizer rods. (c) The figure displays one of the segments as actually mounted showing the vacuum, RF, and tuning ports. (d) The ISIS RFQ as actually installed at the pulsed neutron source. This four-rod structure (see left frame) accelerates up to 30 mA of H$^-$ ions from 35 keV up to 665 keV, with a bunch frequency of 202.5 MHz and duty factor of 2.5%. The frame to the right shows the structure as mounted for operation. (e) A structure made of trapezoidal-shaped electrodes built at IAP-Frankfurt, able to accelerate protons from 15 up to 500 keV. The actual dimensions and shape of the electrode modulation can be seen in the frame at the right. (f) Experimental RFQ built for the Beam Experiment Aboard a Rocket, built at LANL and declassified in 1989. The structure which was meant to be installed within a vehicle for suborbital flights was able to accelerate 20 mA of H$^-$ beams from 30 keV up to 1 MeV using a bunch frequency of 425 MHz and a duty factor of 1%. *Photographs courtesy of Dr. J.E. Stovall (LANL). Reprinted with permission from the author.* (For color version of this figure, the reader is referred to the online version of this chapter.)

provides us with a hint of the accelerator efficiency. Let us name this as rL for a, say a proton linac of length L and voltage gain V. Its value is given by

$$rL = \frac{V^2}{P_{\text{losses}}} \qquad (2.45)$$

where P_{losses} mostly comprises the RF power lost in the walls of the structure. Then, for a given duty factor D, the time-averaged RF power that needs to be supplied is as follows:

$$P_{RF} = \frac{DV^2}{rL} + Vc_{av} \qquad (2.46)$$

where c_{av} stands for the average beam current. The product Vc_{av} is thus the total RF power actually transferred to the beam. Put into real numbers, this means that if for simplicity we set $D=1$ (a CW accelerator) and consider, say $V=1$ GV (i.e., to deliver a 1-GeV beam) and a time average current of some 10 mA, this makes $Vc_{av}=10$ MW. If we now plug in a reasonable value for the shunt impedance of present-day room-temperature proton accelerator structures, $rL \approx 50 \times 10^6 \, \Omega \, m^{-1}$ [49], we get a value for the lost power per unit length $P_{losses} = 4 \times 10^4 \, m^{-1}/L$ MW, which put into different words means that an accelerator of 1 km in length will lose some 40 MW of power which mostly transforms into heat by the accelerator structures. Room-temperature structures are able to provide accelerating gradients of the order of 4–5 MV m^{-1} at best. Superconducting structures, which are adequate for energies above some 50 MeV, are nowadays able to increase this figure up to about 8–10 MV m^{-1}, and last-generation superconducting resonators adequate for larger particle energies should be able to reach 25–30 MV m^{-1}. Operation below some 2 K would make superconducting resonators highly efficient as regards RF power consumption, reducing the losses by a factor of 5–6. In contrast, most of the reduction in cost arising from a more efficient use of RF power is offset by the large increase in construction costs of these accelerating structures. A way out frequently followed in the past to increase the efficiency of low-energy machines consists in recirculating the beam through the accelerating structures some N_t times, making the energy gain to be of the order of V/N_t, which reduces the required RF power by N_t^2 since the required power is now, in contrast with Equation (2.46),

$$P_{RF} = \frac{DV^2}{rLN_t^2} + Vc_{av} \qquad (2.47)$$

a fact that motivated the construction of simple recirculating machines such as microtrons and other recirculating structures.

Cyclotron machines merit separate consideration. Lawrence's invention constituted the pivotal point for a whole variety of recirculating accelerators, from the smallest microtron to the largest synchrotron. It had an enormous impact in almost every branch of science with more than 50 medium or large cyclotrons around the world devoted to research and in several areas of medicine and industry. In fact, at the time of writing, there are about 800 or so small and medium cyclotrons used to produce radioisotopes for medical and other purposes, and a rapidly growing number of 230/250-MeV proton cyclotrons have being commercially built for cancer therapy. Although the cyclotrons of several varieties are resonant RF machines, their operation principles differ from those described above. In a nutshell, a cyclotron consists of two D-shaped regions known as *dees* sandwiched between two electromagnets, providing an external magnetic field perpendicular to them. A uniform electric field is established in the gap separating the *dees* pointing from one dee to the other. When a charge is released from rest in the gap from an ion source, it is accelerated by the electric field and carried into one of the *dees*. The magnetic field in the *dee* causes the charge to follow a half-circle that carries it back to the gap. While the charge is in the *dee*, the electric field in the gap is reversed, so the charge is once again accelerated across the gap. The cycle continues with the magnetic field continually bringing the charge back to the gap. Every time the charge crosses the gap it picks up speed. This causes the half-circles in the *dees* to increase in radius, and eventually, the charge emerges from the cyclotron at the required energy.

The basic parameters of cyclotron components for say, a proton machine, consist of particle energy E_p, magnetic field density B, usually limited by magnetic saturation of iron below 1.7 T, RF frequency typically in the tens of MHz although a higher harmonic is customarily used, and the radii of the *dees* r and are linked through the equations

$$\omega_p = \frac{B}{2\pi m_p} 1.53B, \quad E_p = \frac{B^2 r^2}{2m_p} = 0.48 B^2 r^2 \tag{2.48}$$

where numerical values give the frequency in MHz, field in kG, energy in MeV, and radius in meters. A charge once enters the cyclotron through its center will move in a circular path under the influence of a constant magnetic field with period $T = 1/\omega_p$. If an RF wave is applied at angular frequency B/m, between the two sides of the magnetic poles, the charge will spiral outward, increasing in speed. The equations given above are valid for nonrelativistic energies, which for protons come to be below 10 MeV or so. Relativistic corrections involve normalizing the cyclotron frequency by the Lorentz factor, that is, $\omega_p = 1.53B/\gamma$ and the radius of particles moving in a magnetic field now becomes, $r_p = \gamma \beta m_p c/B$. Such effects are explicitly accounted for in variants of such machines as the synchrocyclotrons where the frequency of the driving RF electric field is varied to compensate for relativistic effects as the particles' velocity begins to approach the speed of light, which basically means adjusting the cyclotron frequency to its relativistic value as energy gain proceeds. As an alternative to adjust the frequency with time, isochronous cyclotrons have magnetic fields that increase with radius. Such machines are capable of dealing with substantial beam currents, but require azimuthal variations in the field strength to provide strong focusing conditions and keep the particles captured in their spiral trajectory. To such an avail, the applied field is changed as $B = \gamma B_0$. Such a radial field gradient leads to transverse defocusing effects that need to be compensated by ridges on the magnet faces which vary the field azimuthally as well and result in strong focusing conditions.

To summarize the section, the basic ingredients of a proton/electron accelerator to be used as drivers for a neutron target are thus as follows:

- *The ion/electron source.* Such a component appears at any conceivable accelerator chain. The choice of the source depends on the particle to be accelerated. Electrons are generated by a cold cathode, a hot cathode, a photocathode, or RF ion sources. Protons are generated within an ion source, which can have many different designs [50] depending upon the kind of ions to be produced, that is bare protons or negatively charged hydrogen ions. The H^- ions are generated at facilities that need to inject pulses into a synchrotron, accumulator, or pulse compression ring which is filled with bare protons resulting from complete electron stripping of the incoming H^-. Such a procedure is employed to avoid the consequences of the Liouville theorem which states that the local particle density in phase space in a collision-free regime must remain constant. Since such a constraint only holds for identical particles, injecting in a different charge simply lifts it. Active research into the generation of negative ion beams is also driven by activities within nuclear fusion devices [51].

The mechanism of ion production involves the formation of a hydrogen plasma which has then to be extracted and transported out to the accelerator structure. Such plasma is usually composed by H^+, H_2^+, and H_3^+ together with free electrons which execute orbital or spiral motions within the electron cyclotron resonance (ECR) field, Penning, or some other field. Such closed electron trajectories increase their path length and hence the number of further ionizations they can induce. The production of singly charged H^+ or H^- ions involves quite some number of processes within the atomic physics realm. Bare protons are generated by successive electron impact leading to molecular dissociation, by photons or by hot surfaces. In contrast, H^- ions require, in addition, to attach or charge exchange an electron to a positive ion resulting from molecular fragmentation, on a hot surface or in metal vapor (mainly Cs), which is added to substantially lower the work function so that a neutral H^0 may pick an additional electron. Most source developments up to the early 1980s used duoplasmatrons able to produce currents as high 5 A of protons for low duty operation ($\approx 10^{-3}$) and duopigatrons able to deliver pulsed hydrogen ion beams of 10–15 A beam current in the 20–40 keV energy range for a duration of a few tenths of a second [52]. Present-day hydrogen ion sources mostly rely on the ECR technique for proton production or surface Penning, volume, and a variety of RF-driven sources for negative ion production. Proton sources based upon the ECR technique generate the plasma via sequential electron impact ionization where the electrons orbit at the cyclotron resonance frequency. Plasmas are confined within the ECR chamber for times of the order of 10 ms under an applied field the magnitude of which is roughly given in tesla units by $B=f/28$, where f stands for the RF frequency. The RF source is usually provided by a conventional $f=2.45$ GHz magnetron or a S-band klystron with $f=2.7$ GHz such as that shown in Figure 2.11 which is able to deliver up to 1.2 kW of RF power. Production efficiency of such sources is better than 100 mA kW^{-1}, being the long-term reliability of such sources the remaining issue to address. In contrast, production efficiencies of H$^-$ sources vary within 8–12 mA kW^{-1} for the surface Penning source at ISIS, 36 mA kW^{-1} for the BNL magnetron source, 1–2 mA kW^{-1} for filament-multicusp devices, and within 1.4–4.2 mA kW^{-1} for three H$^-$ ECR sources tested at ANL, TRIUMF, and CEA-Saclay. However, contrary to the case of proton sources, H$^-$ production devices have not yet achieved operation with high currents and high duty factors at the same time.

- A *high-voltage platform* for initial injection of particles, in the case of protons or ions.
- A *low-energy beam transport (LEBT)* section is required to transport and condition the beam coming out of the electrostatic accelerator column attached to the ion source into the first accelerator structure. There are two basic approaches which employ either solenoids or Einzel lenses as focusing/defocusing elements.

FIGURE 2.11 The left-hand side shows a cutaway view of the proton ECR source operating at the ESS-Bilbao facility. From left to right, the graph shows the microwave waveguide ending at a vacuum window. The step structure shows the coupler to deliver the electromagnetic field into the cylindrical ECR plasma generating cavity. The cyclotron resonance field is that corresponds to $f=2.7$ GHz is controlled by means of two solenoids that can be accurately positioned with respect to the plasma cavity. The beam extraction system is formed by a set of three movable electrodes drawn in green. The accelerator column (light blue) connects the ECR cell that is mounted on a high-voltage (75 kV) platform with the first diagnostics chamber set to ground. The figure at the right shows a view of the source as actually built. The rightmost part of the figure shows the S-band klystron driver as well as the molecular H$_2$ bottle which feeds the ECR cell. The central part displays the automatic tuning system together with the control mechanism. The leftmost part partially shows the ECR chamber as well as the accelerator column. (For interpretation of the references to color in this figure legend, the reader is referred to the online version of this chapter.)

- *A buncher/RFQ* which divides an initially DC beam into small bunches and may provide initial acceleration.
- *For linear accelerator, a medium energy beam transport (MEBT)* section is required to match the beam coming out of the RFQ column into the next accelerator structure. Contrary to the case of the LEBT, which is mostly composed by optical elements, the MEBTs usually include active elements such as rebunching cavities to control the longitudinal beam emittance as well as pulse-shaping devices such as electrostatic choppers. These are both required for injection into a ring whereby chopping the beam at the ring revolution frequency is essential for low-loss acceleration and also for proton linacs to get accurately profiled pulse shapes.
- *Accelerator structures.* The nature and length of those will vary with the application, typically some tens of meters for electron-driven sources up to thousand meters long in last-generation proton linacs. In the latter case, room-temperature structures are used to prepare and accelerate the beam up to some tens of hundreds of MeV. Typical room-temperature structures comprise the drift-tube linac (DTL) accelerator built upon the principles that guided the Alvarez design which led to the first DTL built at Berkeley in 1947, the coupled cavity linac (CCL) composed by active cavities together with coupling field-free cells, side-coupled DTL (SCDTL) which consists of short DTL tanks, rather than single cavities, coupled together by a side cavity, as well as several other designs, are still used in most proton injectors. Typical energies achievable using these structures may easily reach 800 MeV as done at LANSCE at LANL although the current tendency is to transit into superconducting structures at lower energies. Present-day projects consider the transition into a superconducting section for beam energies as low as 50–80 MeV. Such superconducting resonators are built within cryomodule structures which typically hold from two to six of such resonating cavities. For relatively low energies (50 MeV or so), spoke resonators, or half- or quarter-wave resonators are today the structures of choice. For $\beta \geq 0.5$ or so standard, multicell elliptical cavities such as those used in electron accelerators are the obvious choice. It goes without saying that circular machines such cyclotrons may provide the accelerated beam from a single structure or, if needed, from a chain of such devices.
- *RF transmitters*, used to provide the RF fields. Electron accelerators may use solid-state microwave amplifiers as such sources. High-power proton accelerators still have to rely on the RF technology of vacuum tubes (tetrodes, klystrons, etc.) since the power requirements cannot be easily matched by recourse to solid-state amplifiers. The sources must operate at precise power, frequency, and phase appropriate to the particle type to be accelerated to obtain maximum device power. As regards to optimal operation frequencies for such transmitters, one has to bear in mind that although the use of higher frequencies is economically convenient since it leads to shorter accelerators (efficiency scales as $\sqrt{f}$) and less RF power

consumption, inherent limitations come in due to the mechanical precision in construction required (tight tolerances) as well as to beam dynamics for ion linacs at low energy. Electron linacs usually operate at frequencies within the 500 MHz to 12 GHz interval, while proton linacs restrict their operation to the 100–1300 MHz range.

- *Optical focusing/defocusing elements* such as quadrupole magnets which are used to help focus and transport the beams.
- *An intermediate structure*, such as a synchrotron or a compressor ring able to accumulate the beam injected by the linac into a single pulse, compress the pulse length of typically several hundreds of microseconds or a few milliseconds down to a few microseconds and also, to increase the beam energy by means of multiple recirculation of particle bunches over as many turns as required. Such a structure is required for short-pulse spallation machines but is not required in continuous wave (CW) or long-pulse proton machines. The rapid cycling synchrotron (RCS) provides capabilities for these tasks. Injection of beams delivered by the linac into a synchrotron is usually accomplished by means of multi-turn H^- charge where the beam from the linear accelerator is stripped of two electrons, leaving only the protons, by a thin stripping foil, the efficiency of which is required to be very high, in order to minimize beam loss caused by unwanted unstripped or partially stripped (H^0) species. At low energy and low powers, facilities such as ISIS use 0.3 μm alumina foil, which is around 97% efficient. Higher powers require different materials such as amorphous or hybrid, boron-doped carbon thin films. For *quasi*-CW proton accelerators such as SINQ or long-pulse machines, pulse compression is not required. Beams are finally extracted from the synchrotrons by means of fast-switching magnets (kickers).
- A *high-energy beam transport (HEBT)* line which leads the beam to the target. Its main purpose is to provide the requested beam footprint and profile to the target entrance, which is accomplished by means of a set of defocusing quadrupoles and octupoles. The HEBT ends at the target entrance by a beam window which isolates the relatively high vacuum of the accelerator from the target.
- *An appropriate target*. If electrons are accelerated to produce γ-rays, water-cooled heavy-metal targets are usually employed. Various target materials are used when protons are accelerated, depending upon the specific investigation, as described below.

2.4.2 Targets

As referred to above, the main bottleneck limiting the progress of electron-driven sources as power neutron sources arises from the small volumes in which a significant amount of heat needs to be dissipated. The crucial issue regards the small dimensions of the target in the direction normal to the incoming beam which are to be kept within stringent limits in order to provide efficient coupling to the moderators. The magnitude of such constraints is easily assessed remembering that a typical 50-kW electron target has a power density of ≈ 5 MW l^{-1}, similar to that of an experimental fission reactor.

Most of the targets designed for electron accelerators have been made of series of tantalum or clad uranium plates having thicknesses within the 1–20 mm range which linearly arranged keep an inverse relation with the expected heat deposition, leaving narrow gaps between such plates to allow the coolant (usually pressurized water) to take off the heat. Most of the electron linacs listed in Table 2.2 use depleted U or Ta as target materials and water of Hg as the coolant.

Proton beam-driven machines do have a record of fairly continuous development since 1977, when the first meeting of the International Collaboration of Advanced Neutron Sources took place [53]. Facilities at the time were basically the ZING-P, the forerunner of IPNS, the neutron facility at the Los Alamos Weapons Neutron Research (WNR) facility, the SNS project at the Rutherford Appleton Laboratory in Britain, and the KENS facility at KEK in Tsukuba, Japan. Some details pertaining the basic parameters of these sources are given in Table 2.6.

While the choice of target materials was at first restricted to a few (U, Ta, or W), cooling of fixed targets with rod or plate geometries was there identified as a very important issue. In addition, exploitation of the existing facilities evidenced the need to operate with very reduced backgrounds between subsequent pulses in order to get significant gains in signal to noise ratios.

From the above paragraphs, it should become clear that efficiency of neutron production is just one of the issues to be addressed in the construction of efficient, accelerator-driven neutron targets. In fact, operational experience with MW range sources reveals that problems related to radiation damage, materials stress and radiation-induced fatigue, and management of the radioactive inventory, together with heat removal issues just referred to above are as important to merit due consideration as much as the neutronic performance. In what follows, we will briefly review such issues.

2.4.2.1 Target Materials for Proton Drivers

From parametric equations reproduced above in Equation (2.8), it becomes evident that materials with a high atomic number are to be used as prime candidates for neutron-producing reactions. Use of uranium, either depleted or somewhat enriched, significantly increases the neutron yield, but its use has been discontinued because of swelling problems that take place above a structural phase transition at $T_c \approx 933$ K. Although a U–Mo (9%) alloy can be used to circumvent the problem, it is not commonly considered as a candidate material due to problems that arise regarding its radiotoxic inventory. Other issues apart from the high-A value concern the material densities which need

TABLE 2.6 High-Power Accelerator-Driven Sources

Facility	Status	Target material	Beam pulse duration (μs)	Repetition rate (Hz)	Proton energy (GeV)	Time average beam power (MW)	Peak time average power density (GW m⁻³)	Peak energy density (MJ m⁻³)
ISIS (UK)	Operating	W	0.4	50	0.8	0.16+0.04	0.25	5
LANSCE-Lujan (LANL)	Operating	W	0.3	20	0.8	0.16	0.5	25
NuMi (FNAL)	Operating	C	8.6	0.53	120	0.4	0.32	600
SINQ/Solid Target	Operating	Pb, Zr Clad	CW	–	0.57	1	1	–
SINQ/MEGAPIE (PSI)	Completed	Pb-Bi	CW	–	0.57	1	1	–
MLSF/J-PARC (Japan)	Operating	Hg	–	25	3	1	0.63	25
SNS (ORNL)	Operating	Hg	0.7	60	1.3[a]	2	0.8	13
n-TOF (CERN)	Operating	Pb	6×10⁻³	1/2.4	20	0.04	0.1	
TRIUMF (Canada)	Operating	Pb	CW	–	0.5	0.125		
AGS (BNL)	Operating	Hg/Pb	1	0.5	24[a]	0.14	0.02	2
ESS-LP (Sweden) (ASTE Coll.)	Proposed	W	2.857	14	2.5	5	2.5	150

MYRRAH-SCK (Belgium)	Under design	PbBi	CW	—	0.6	2.1	250	—
ENNG-ITEP (Russia)	Under construction	Be	220	25	0.04	0.02	4.6	—
EURISOL	Proposed	Hg	3	50	2.2	4	100	2.000
IFMIF (Japan)	Proposed	Li	CW	—	0.04 (D$_2$)	10	100	—
LANSCE-MTS (LANL)	Proposed	Pb-Bi/W	1.000	120	0.8	0.8	2.4	20
Project-X/Stage II (FNAL)	Proposed	Injector	CW	—	3	3	—	—

Figures quoted for ISIS comprise the beams serving the two target stations. Data concerning LANL are given for both the neutron-scattering center and the materials irradiation facility. Data for nonneutron-scattering facilities concern the NuMi and Project-X neutrino facilities at FNAL, the n-TOF high-energy spallation facility at CERN, the accelerator-driven-reactor systems MYRRHA at the SCK-CEN and ENNG at ITEP, the proposed IFMIF fusion materials accelerator and the EURISOL radioactive beam facility and their inclusion on the table is done for comparison purposes. The NuMi injector at Fermilab is used to generate intense π and kaon beams which then decay after some distance into muons and muon neutrinos (i.e. $\pi^+ \rightarrow \mu^+ + \nu_\mu$). The quoted beam energy for SNS refers to that mentioned within its specification. Actual value of operation still is 1 GeV. Figures quoted for the multistaged Project-X refers to stage II. The reference design envisages the construction of a pulsed linac for acceleration of a beam from 3 to 8 GeV. This beam is delivered to the recycler/main injector complex in support of the long baseline neutrino program. TRIUMF figures correspond to the NIF irradiation facility. Data for the Alternating Gradient Synchrotron at Brookhaven Natl. Lab. mostly refer to experiments carried out by the AGS-ASTE collaboration.

[a]Experiments have been performed with final proton energies of 1.94, 12, and 24 GeV.

to be high enough to increase the collision probability as given by Equation (2.11), its stability against radiation which is dictated by durability reasons, a low neutron capture cross section and finally to yield a small quantity of after heat as possible. From such considerations, tungsten and lead seem to be the best bets while mercury still is the material used in the two highest-power pulsed sources in operation (SNS and J-PARC) and the PbBi eutectic was used in the continuous source at SINQ within the MEGAPIE (MEGAwatt PIlot Experiment) project. Its use in CW sources is motivated by the need of reducing neutron absorption and thus gets a high time-averaged neutron flux.

The use of a liquid Hg target was motivated by concerns about radiation damage of materials as a consequence of the impact of GeV protons. The material has a relatively large thermal neutron cross section ($\sigma_{abs} \approx 6$ barns) which makes it inadequate for its use for steady-state sources. The intensity distribution of the leakage neutron current from thick targets somewhat favors Hg as the material of choice, especially along the target surface [54], although the measured differences with corresponding results pertaining to W especially for target depths up to 20 cm are rather modest. Additional advantages concern the absence of radiation damage since the target is kept as a room-temperature liquid, the absence of coolants such as water in the beam (leads to ^{3}H and ^{7}Be production) and the manageable specific activity within the irradiated material due to its large and circulating volume. The main concerns regarding the use of such a material arise from operational experience at SNS and J-PARC as well as for management problems regarding the processing of irradiated target materials. Specifically, the remaining issues regard the mitigation of pressure waves inside the liquid metal which develop as a consequence of the formation of cavitation bubbles which arise as the result of a tensile stress that the liquid cannot sustain. Most attempts to inject within the metal bubbles of noncondensable gases such as He aiming to scatter the pressure wave or isolating the vessel wall by means of a gas layer between it and the liquid metal did not give the expected results [55]. In turn, detailed investigations carried out within the ESS project lead to the identification of additional problems such as radiation damage for the window and liquid target container leading to the proposal of thick (3 mm) windows and target containers.

While the PbBi eutectic was considered to be a good candidate from the neutronics point of view since 1966 from work carried out within the ING project [25], it took several decades to scrutinize its performance. This was demonstrated by the MEGAPIE project carried out at the Paul Scherrer Institut, and the material still is in serious consideration for applications such as accelerator-driven transmutation devices [56]. There are however serious concerns such as the need to keep the target within 240–380 °C to avoid solidification which leads to a remarkable increase of the target volume or the problems which arise as a result of the α-activity of the inventory mostly resulting from α-active ^{209}Po and ^{210}Po generated by neutron capture from Bi. The latter fact requires to pay special attention to safe enclosure of the target as well as of the gases and other volatile components produced during irradiation. In fact, data at hand show that as much as 10 TBq of the isotopes ^{209}Po–^{210}Po were produced within the MEGAPIE target by spallation reactions and an additional 100 TBq of ^{210}Po were found resulting from thermal neutron capture of Bi. In addition, sizeable amounts of volatile vapors of Hg and some smaller quantities of thallium were also generated.

An assembly of surface cooled Pb-filled zircalloy tubes in a cross-flow configuration has been used at SINQ. The rods are hosted by an Al–Mg alloy block having an hexagonal lattice where some 350 Pb rods of some 20.7 mm in diameter are inserted in a direction perpendicular to the proton beam. The target has been shown to be able to take a time-averaged current of 1.4 mA protons of 570 MeV for long periods leading to a energy deposition of 20 dpa (displacements per atom, a measure of radiation damage). The limitation in the use of such targets mostly stems from melting of the target material. In fact, full power operation of SINQ most likely leads to partial melting of the Pb rods.

For reasons given above, solid targets for pulsed MW range sources require to be of rotating type in order to distribute the thermal and mechanical loads. Taking account of the considerations given above, tungsten appears as the best choice for a solid rotating target. The concept developed time ago within the context of the German SNQ project [57] has been revisited as an option for a second target station at SNS [58] and constitutes the prime candidate for the target of ESS [59]. On the other hand, recent studies show that such targets when used with large coupled hydrogen moderators have neutronic performances equal to or better than that with a mercury target, together with a greatly increased lifetime [58]. The concept has proven to be able to take up to 3 MW of beam power and can be modified to cope with larger powers taking special precautions to keep the thickness of the target elements limited by heat-flux density on the surface to avoid nucleate boiling and parallel flow instabilities, and at the same time, the coolant volume has to be kept within reasonable limits in order to avoid unwanted moderation.

Future candidates for target material may benefit from results obtained at experimental reactors in their quest for low-enrichment-fuels (see Section 2.7.1). In particular, the alloy U-10% Mo is being considered in the Reduced Enrichment Research and Test Reactor RERTR program [60] and it is found that, up to 70% burn-up, the material remains stable. Its expected yield comes to about 1.7 more neutrons per proton as a spallation target than best lower-Z materials. The material could then be considered as a candidate in high-power applications in connection with a heavy liquid metal coolant to avoid neutron moderation within the target volume.

2.4.2.2 Target Cooling

As referred to above, the performance of neutron sources, either steady or pulsed, is ultimately limited by the ability to remove heat generated by nuclear processes. Heat-transfer mechanisms are fundamentally different for liquid metal and solid targets. In the former case, the circulating liquid (i.e., Hg, Li, or PbBi) transports the heat within the bulk material to regions away from the primary source where it can be removed using established heat-exchange technologies. Cooling targets, whether static or rotating, pose additional constraints to target design since neutronic performance purposes dictate that the coolant volume should be minimized.

Efficient heat transport in a solid target requires specific temperature gradients to be kept. On the other hand, the heat produced within the target volume must be conducted to the surface, which leads to large axial thermal gradients and stresses. Usual dense metallic target materials show thermal conductivities at the operation temperatures that are large enough to suit such constraints. The quantity to be maximized is thus the rate of heat flux per unit area q. To grasp the adequate parameters characterizing an efficient coolant, we consider a plate of thickness a uniformly heated with power density p. The heat flow is governed by the difference in temperatures of the plate surface T_s and bulk coolant T_f scaled by a constant h called the film, convective or heat-transfer coefficient:

$$q = h(T_s - T_f) = -\kappa \frac{dT}{dx}\big|_{a/2} = \frac{pa}{2} \qquad (2.49)$$

where k stands for the thermal conductivity. From the above equation, we see that an efficient coolant must have large values of h which in general involve large flow velocities and the corresponding pressure gradients. The latter also set practical limits which arise from the need to prevent mechanical instabilities or high pumping power requirements. To get an estimation of the heat-transfer coefficient useful for many applications, recourse can be made to some empirical correlations such as that of Ditius–Boelter which for the case of forced convection in the external plane surface of a solid gives

$$h = \frac{k_f}{D_H} Nu \qquad (2.50)$$

$$Nu = \frac{hL}{k_f} = 0.036\, Re^{4/5} Pr^{1/3} \qquad (2.51)$$

in terms of the characteristic length L, the thermal conductivity of the fluid k_f, the hydraulic diameter D_H, as well as the Nusselt Nu, Reynolds Re, and Prandtl Pr numbers. The approximation is valid within the limits $0.6 \leq Pr \leq 160$, $Re \sim > 10.000$, and $L/D_H \sim > 10$ [61].

At power levels of a few kilowatts, gaseous cooling becomes practicable. At higher powers, liquid coolants are sought. These may be single or multiple phase. The latter can boil when in contact with the hot surface, while the temperature of the bulk liquid remains below boiling conditions, thus providing larger h values. For a single-phase coolant such as water flowing past a target plate at some 35 °C below saturation (i.e., the temperature for a corresponding saturation pressure at which a liquid boils into its vapor phase), the linear regime predicted by Equation (2.49) where heat transfer is proportional to the difference in plate and bulk liquid temperatures is achieved remains valid up to differences in temperatures of about 35–40 °C (depending upon the fluid flow velocity). Beyond them, a *nucleate boiling regime, NBR*, is attained where bubbles grow creating a turbulence on the plate surface which further improves heat transfer. Further increase of the heat flux some tens of degrees beyond the limit of the NBR drives the system into *critical heat flux, CHF* conditions where the bulk fluid, or in cases, regions of the bulk fluid may boil yielding large bubbles which, sometimes, may block the fluid passage. Under such conditions, the rate of heat transfer actually drops and may thus lead to burn out of the target elements. In consequence to profit from NBR conditions require stringent controls of the fluid parameters such as the pressure and flow velocity. If such requirements were met, it would then be possible to reach high heat-flux values (500 W cm^{-2}) provided that one can cope with larger pressure drops.

The use of pressurized He is being at present under consideration as a candidate coolant for the ESS target. Its heat-transfer properties of the fluid are however modest. For temperatures about 90 °C and pressures of 10 atm. (flow velocity of 500 m s^{-1}), one gets a modest 1.1 W cm^{-2} K^{-1} for its film coefficient and a relatively large pressure gradient of 0.170 atm.cm^{-1}. Such figures are remarkably worse than those for water circulating at some 15 m s^{-1} where the value for the film coefficient raises up to 9.8 W cm^{-2} K^{-1} and the pressure gradient remains at 0.102 atm.cm^{-1}. In view of such differences, it seems that the choice of He as a coolant is only justified if it leads to a significantly better neutronic performance.

Molten metals such Na or NaK provide film coefficients of ≈ 50 W cm^{-2} K^{-1} [1] for flow through millimeter-thick channels while keeping the pressure gradient at a reasonably low value of 0.116 atm.cm^{-1}. The technology for managing such refrigeration system is already known from work carried on fast reactors, and in fact, a design along these lines was completed for the IPNS-II target project. The use of such coolants seems, however, restricted to cases where the spallation targets contain a significant amount of fissile materials.

Most solid targets built so far contain many cooling channels, and particular care has to be exercised to prevent the development of well-known flow instabilities that develop within the confined fluid flow.

For megawatt-range applications, efficient enough cooling can hardly be met by a water-cooled assembly of static plate targets such those used in 100 kW power sources [1]. The concept of a rotating high-power target was developed in the 1980s within the German SNQ effort. The idea, akin to that exploited for building X-ray rotating anodes, leads to the development of a concept able to take 5.5 MW of beam power deposited by 3×10^{14} protons per pulse. The heat dissipation characteristics of such a device which basically consists on a disk target mounted on a rotating drive turning at angular frequencies of 1 Hz or less had to be able to cope with thermal loads of about

3 MW. A far more developed concept has evolved as an option for the SNS second target station. The cooling systems were required to be both simple and robust. Two configurations were considered, cooling on the top, bottom, and outer face of the segments, while the other included a horizontally split segment with cooling in the center channel as well as the top, bottom, and outer face. Flow requirements for both configurations assumed a rate of 25 l s^{-1} with 1.5 mm high ducts which yielded heat-transfer coefficients of approximately 2×10^4 W m^{-2} K^{-1} while keeping a modest flow velocity of 4.4 m s^{-1} that leads to coolant temperature rise below 20 °C. The initial design was based upon a disc with an outer diameter of 1.2 m, expected to last up to 6 years based on a limit of 10 dpa for the target shroud at 3 MW. The target temperature was to be kept below 700 °C for removal of decay heat by conduction through air and thermal radiation to the surrounding structure during a loss of coolant scenario. A target height of 70 mm was used to allow some tolerance for an off-center beam. A tungsten zone depth of 0.25 m was used for adequate stopping of the proton beam, resulting in an inner diameter of 0.7 m. Rotation rates near 30 and 60 rpm were considered to avoid overlapping of consecutive beam pulses into the same region of the target.

A rotating target test stand has been built within the SNS/ESS-Bilbao collaboration to investigate in detail the fluid flow properties inside a full-scale mockup of the rotating disk. Figure 2.12 shows the installation built at Bilbao to perform fluid flow velocimetry studies, and a map of the obtained velocity fields is also displayed in the figure.

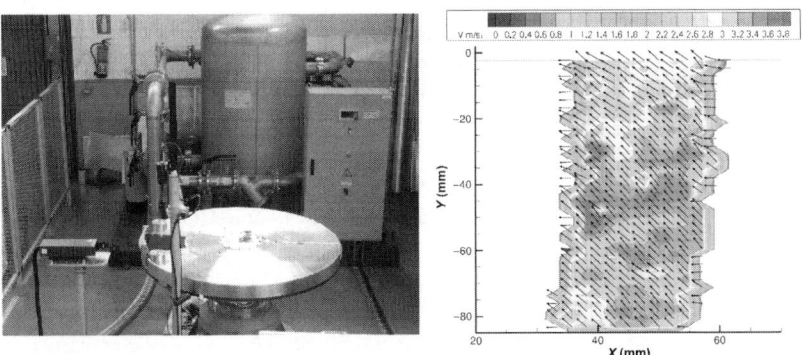

FIGURE 2.12 The left-hand side shows a view of the rotating target flow test stand installed at the ESS-Bilbao facility and developed in collaboration with SNS-ORNL. The device has been built to carry out characterization tests on 1:1 models for rotating solid targets for multi-MW sources. It is equipped with a drive to rotate the target up to 60 rpm and a water loop to provide coolant flowrates up to 30 l s^{-1}. The fluid flow is visualized by means of a particle image velocimetry (PIV) system which relies upon a frequency-doubled pulsed laser delivering up to 150 mJ pulses at 10–15 Hz and a CCD camera. The right-hand side shows a fluid velocity map as actually measured using the device. (For color version of this figure, the reader is referred to the online version of this chapter.)

A gallium-cooled system able to keep temperatures $\approx 100\ °C$ above ambient with coolant velocities of $16\ l\ s^{-1}$, achievable with moderate pumping ≈ 1 atm. pressure drop, and target coolant pressure about 0.3 atm. below driving pressure, has been demonstrated at the POP level at the ABCNT source at MIT. The excellent heat-transfer properties of liquid Ga are up to some extent compensated by difficulties in the containment of this highly reactive semimetal. Cooling for powers above a few megawatts still constitutes an open issue where research into feasible strategies to remove heat (about 2.4 MW estimated for a 5-MW target) safely and efficiently is being actively pursued.

In summary, issues pertaining heat deposition and the associated mechanical stresses are the main fetters when aiming the design of high-power, multimegawatt-range targets. If the source is to operate in a pulsed regime, the complications of having to deal with very large instantaneous energy deposition (of the order of $150\ MJ\ m^{-3}$ for the ESS project) add to the heat removal problem.

On conceptual grounds, windowless liquid metal targets appeared as the most promising configuration, being able to take some $6\ mA\ cm^{-2}$ of proton current up to a few gigaelectron-volts energies on CW mode [26]. However, the actual practice of ongoing projects such as MYRRHA based upon such a concept disregards the lack of physical isolation of the target from the accelerator vacuum and has brought back a beam window which needs to stand high stresses induced by direct heat loads.

At the time of writing, and account made of the several years of experience gained at SNS and MLSF on the use of liquid metal targets for pulsed operation, nonstationary solid targets such as rotating disks or toroids appear as the best bet to cope with the thermal deformation induced by the beam footprint. Cooling can surely be achieved by means of adequate liquid coolants. There is however a strong proviso to consider these in much detail since large amounts of coolant will surely have a deleterious effect on the target neutronic performance.

2.4.2.3 The Target–Moderator–Reflector Complex

As mentioned in Section 2.2.1, accelerator-driven sources yield neutron distributions peaked at about 1 MeV together with an *evaporation tail* which extends up to energies matching that of the incident particles. To provide neutron beams of high intensities and energies within the ranges useful for neutron-scattering applications, much effort has been devoted within the past 30 years to a better understanding of issues concerning the behavior of the target–moderator–reflector (TMR) complex. The purpose of such studies is to optimize the neutron fluxes leaking from the target leading to intense beams at the sample position of a given neutron-scattering instrument, usually located several tens of meters away from the source. For such an avail, reflectors surrounding the target active zone reduce the leakage of neutrons undergoing slowing-down processes by scattering them back to the moderator. In other words, neutrons which may undergo an initial collision within the moderator may leak out before reaching the sought energies. To get an estimate of such an effect, let us consider the value for the probability of a 1-MeV neutron to thermalize down to 1 eV without leaking from a moderator with buckling $B^2 = 0.25$ cm^{-2}, which becomes as low as 0.1, meaning that 90% of the fast neutrons intercepted by the moderator are lost before being moderated down to 1 eV. Enclosure of the moderator within reflecting materials can reduce such large losses in much the same way as done in experimental fission reactors. As an additional bonus, some reflector materials such as Be can also contribute significantly to neutron production by means of (n,2n) reactions. A schematic view of such an arrangement is shown in Figure 2.13.

2.4.2.4 Reflector

From a neutronics point of view, optimal reflector materials are those having large scattering cross sections and low atomic masses. For most existing accelerator-driven sources, Be has been the material of first choice, although lower power or CW machines such as SINQ have used and are using water as an option. Indications however abound about the gains offered by the use of composite reflectors made for instance by surrounding the Be blocks with a heavy material such as Ni in regions near the moderator. In fact, the present evidence tells [63] that reflectors made from heavy metals such as lead can

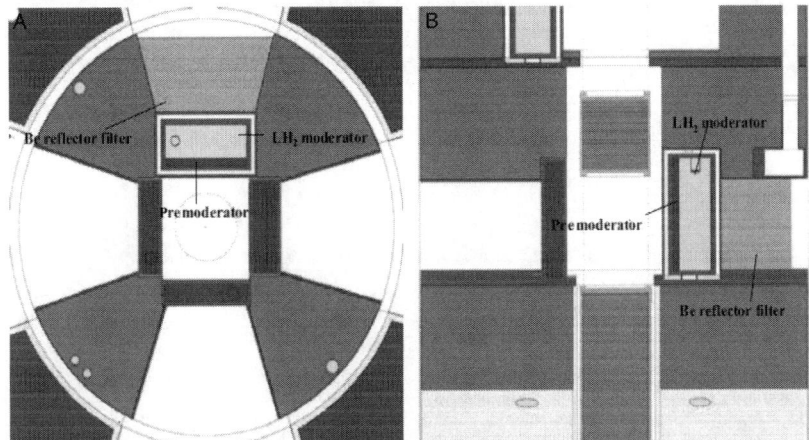

FIGURE 2.13 A layout of a target–moderator–reflector complex [62], showing a split-target and a flux-trap moderator together with a premoderator structure and a cryogenic Be filter. The figure displays a schematic cross section through the lower tier hydrogen moderator with a light-water premoderator and a beryllium reflector filter in horizontal (a) and vertical (b) direction. (For color version of this figure, the reader is referred to the online version of this chapter.)

also be used, although showing a somewhat poorer performance. In a nutshell, the pros and cons of usual reflector materials can be summarized as follows:

- Apart from its good reflector properties, ^{9}Be exhibits significant moderation properties (see Table 2.5). Its low threshold for the (n,xn) reactions and the low capture cross section result in large neutron production which manifests in a long tail in the trailing edge of the resulting pulse.
- ^{209}Pb has a very high density, poor moderating properties (see Table 2.5), and a modest threshold for (n,xn) reactions. The net result of this is much lower neutron production within the reflector and a rather short tail in the time distribution of the pulse.
- Light-water reflectors yield results between those of ^{209}Pb and ^{9}Be. This comes as a consequence of its modest moderating power and relatively high absorption cross section.
- Heavy water has large moderating power and very low absorption resulting in rather large tails in the time distribution of the pulse associated with high time-averaged values.

A further point to consider is the cooling needed to remove nuclear heat from the reflector structure, which for megawatt-range sources is obviously large. Projects such as the Second Target Station at SNS are now considering to use heavy water as a coolant which will take some 15% of the reflector volume. The resulting performance of such a reflector would then result from the interplay of neutronic processes taking place in both materials, beryllium and heavy water.

As just referred to, most reflector materials do also behave as partial moderators, leading to buildup of thermal and epithermal fluxes. Such partially moderated neutrons can reach the moderator having as a consequence a further broadening of the pulse, a fact that may be an important issue for short-pulse target stations. To prevent thermal and epithermal neutrons from entering the moderator while allowing fast neutrons to pass, a *decoupling layer* is frequently installed to cut the moderator surface away from the direct view of the reflector. Its purpose is to stop neutrons below a certain cutoff or decoupling energy, say E_{dec}. Typical values for such an energy are of the order of a few electron-volts. Absorbing materials such as boron carbide ($E_{dec} = 2.5$ eV), Cd ($E_{dec} = 0.4$ eV), or Ag–In–Cd (AIC) alloy ($E_{dec} \approx 1$ eV) are usually employed to build the decoupling layer. The net effect of turning back into the reflector neutrons with energies below E_{dec} is to somewhat extend the moderation time [2] (Figure 2.14):

$$\Delta t(\lambda) = \sqrt{\Delta t^2 + \bar{t}_{refl}(E_{dec})^2} \quad (2.52)$$

so that the resulting time can be cast as a convolution of the moderator and $\bar{t}_{refl}(E_{dec})$ reflector times. This translates for long wavelengths (i.e., for $\Delta t \gg \bar{t}_{refl}$) into a flux gain from the reflector, whereas at shorter wavelengths

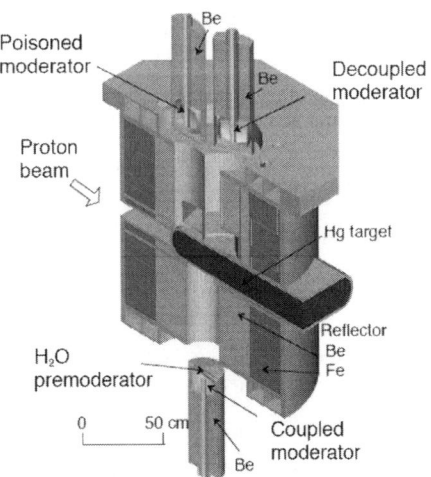

FIGURE 2.14 A view of the TMR arrangement as installed at the MLSF at J-PARC. The figure displays the geometry of the Be/Fe reflector arrangement, also showing some of its cooling channels. Note that all the three moderators are displaced out from their operating positions. *Taken from Ref. [64]. Reprinted with permission.* (For color version of this figure, the reader is referred to the online version of this chapter.)

the effect will depend upon our particular choice of E_{dec} so that a trade-off between intensity gain and lengthening the moderation time needs to be made.

The need for a better understanding of the TMR system as a whole leading to an optimal use of the neutrons produced within the target is best assessed by the data taken at SNS, which shows that only a fraction as small as 15% efficiently leaks out of the moderator toward the beam tubes [65] (Figure 2.15).

2.4.2.5 Moderator Practice

Although the detailed physics behind neutron slowing-down processes may seem too involved, the practical design of a moderator showing optimal performance can be accomplished with relative ease. In fact, only the final brightness is explicitly dependent upon its particular geometry since it is proportional to direct and reflected neutron fluxes entering the moderator. Several candidate materials together with some of their basic properties are listed in Table 2.5.

The main parameters that need to be set are the optimal moderator temperature, its thickness as well as whether to add a poison is advisable. In what follows, we briefly consider such issues.

2.4.2.5.1 Moderator Temperature

The issues here are rather different if the moderator is to operate at ambient or low temperatures. For ambient or high-temperature moderators, a decrease in

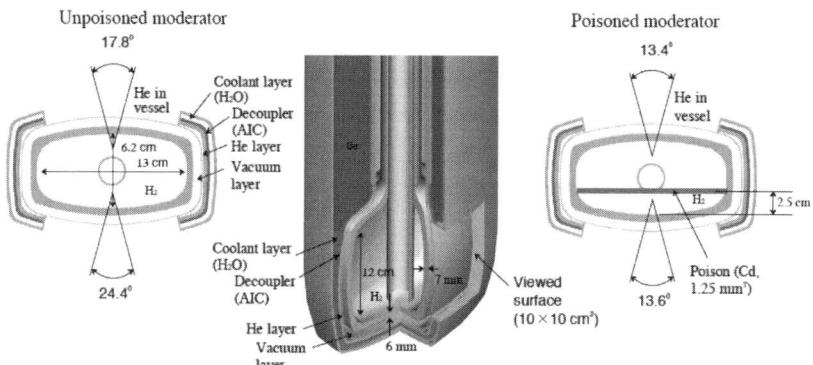

FIGURE 2.15 A schematic view of one of the decoupled H_2 cryogenic moderator as installed at the MLSF at J-PARC. The drawing shows the details of the inner moderator structures including the location of the light-water premoderator and part of the Be reflector. The moderator shown here corresponds to one of those depicted in the TMR arrangement shown in Figure 2.14. The location of a decoupling AIC layer within the structure is also visible. At both sides of the figure are shown the horizontal cross sections for unpoisoned and poisoned moderators, also depicting the location of the decoupling and cooling layers. The figure also shows the main dimensions of the moderating volume as well as the position of the Cd poison plate.*Taken from Ref. [64]. Reprinted with permission.* (For color version of this figure, the reader is referred to the online version of this chapter.)

temperature usually leads to sharpening of the neutron pulse and the flux is enhanced for wavelengths corresponding to that lower temperature. On such grounds, even those moderators used to produce thermal neutrons such as methane are usually kept at relatively low temperatures to improve performance. For hydrogenous moderators such as those composed by supercritical H_2 used in last-generation facilities, the issue of *ortho*→*para* conversion needs to be taken care of [66]. Experiments carried out by Ooi *et al.* [67] show that the performance of a liquid hydrogen moderator significantly improves if the *para*-H_2 concentration is kept close to its equilibrium value at 20 K, namely, 99.8%. The figure goes down to ≈25% at room temperature. On the other hand, the conversion between both rotational states is too low unless an activated paramagnetic catalyst is brought into contact with the fluid. Experiments carried out at LANL as well as at the Hokkaido electron linac have shown that keeping large *para*-H_2 concentrations not only improves the moderator peak intensity but also allows the pulse trailing edge to decay much faster, especially for the decoupled moderators. A comparison of moderator performance at 20 K for solid methane, mesitylene, and hydrogen was reported in Ref. [68]. While mesitylene showed a poorer performance than hydrogen or methane, the differences between the latter two were not large. In fact, experience accumulated over the past few years shows that combination of a cryogenic hydrogen moderator and mesitylene as a premoderator provides excellent moderator performance [69]. A drawback of methane compared to hydrogen concerns the radiochemical processes which are common to organic materials which are not well understood. Evidence at hand from close to three decades of operation at ISIS show that liquid methane forms radiochemical deposits after several hours of irradiation, whereas solid methane needs to be cycled daily above its melting point to release accumulated gases (known as the "burp" phenomena) [70]. Other materials such as hydrogen hydrides have been explored as potential moderators since long. Their properties may appear optimal as reflector or premoderator materials but behave poorly as candidate materials for moderating purposes due to a significant loss in transmission.

2.4.2.5.2 Optimal Thickness

For a simple case, say a slab moderator, the sharpness of the neutron pulse will vary inversely with the moderator thickness. In turn, a too thin moderator will leave too many fast neutrons to pass through. The dependence of the pulse width with moderator thickness for most materials is usually a smooth function, at least within the ranges of practical interest (1–7 cm of thickness or so). An effect that needs to be taken care of concerns the total cross section of the moderating material which will lead to a reduction of moderator brightness beyond a certain thickness. A figure of merit is usually cast in terms of the ratio between the neutron flux and the pulse width squared. Most of the room-temperature moderators are optimal for thicknesses of about 5 cm.

Cryogenic moderators are usually somewhat thicker, usually of the order of 5.5–14 cm, depending upon the density of the chosen moderator material. Real dimensions of actually installed moderators are shown in Figures 2.16 and 2.17. A quantity also dependent on moderator thickness concerns the thermal to epithermal flux ratio which for most materials shows a severalfold increase with thickness following a quasilinear behavior.

2.4.2.5.3 Moderator Geometries

A sketch of practical moderator geometries as well as their relative position with respect to the target is shown in Figure 2.16.

The basic arrangements of the target–moderator configurations shown in Figure 2.16 concern three configurations usually referred to as *slab*, *wing*, and *flux trap*. These represent extreme geometries so that combination of features pertaining to both of them can also be of interest. The first two are widely used at existing sources, and in fact, the *slab* geometry provides the higher neutron fluxes [71]. Present-day target systems such as the ones installed at SNS and MLSF have opted for wing configurations and have located their moderators both above and below the target. An example of such moderators as built at MLSF is shown in Figures 2.15 and 2.17. The chosen option in turn allows the installation of further moderators of lower intensity at positions behind those where the more intense units are located.

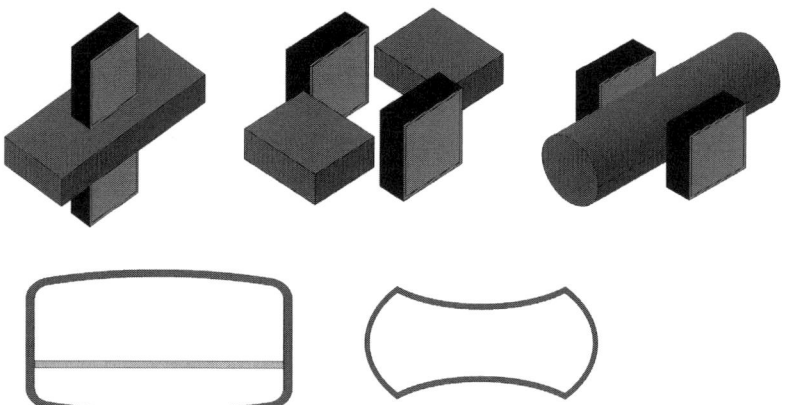

FIGURE 2.16 Moderator geometries. The upper left frame shows two moderators in wing geometry both above and below a rectangular-section target. The upper middle frame shows the moderators within a flux-trap arrangement. The upper right frame shows the moderators in slab configuration located at both sides of a cylindrical thick target. The lower left frame displays the horizontal cross section of a cryogenic moderator similar to those implemented at the MLSF. The thin plate crossing the moderator volume displays a Cd plate asymmetrically located with respect to the viewed surfaces. The geometry corresponds to the moderator arrangement shown in Figure 2.15. The rightmost frame shows the horizontal cross section of a candidate shape which was also evaluated as an alternative. (For color version of this figure, the reader is referred to the online version of this chapter.)

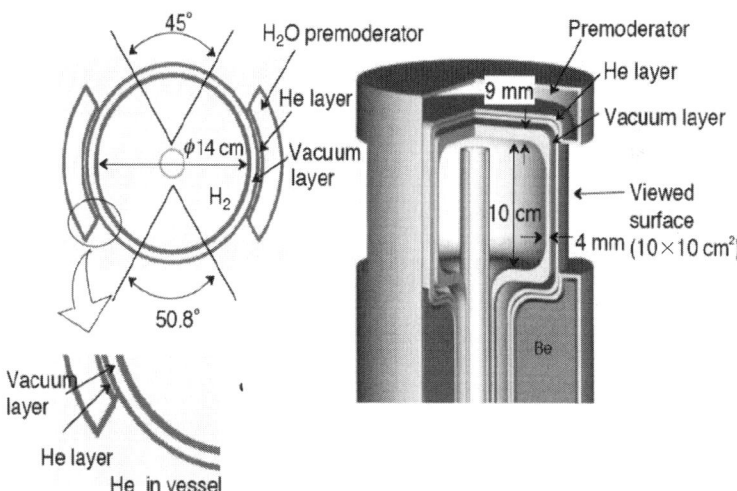

FIGURE 2.17 A schematic view of one of the coupled H_2 cryogenic moderator as installed at the MLSF at J-PARC. The drawing shows the details of the inner moderator structures including the location of the light-water premoderator and part of the Be reflector. The moderator shown here corresponds to one of those depicted in the TMR arrangement shown in Figure 2.14. *Taken from Ref. [64]. Reprinted with permission.* (For color version of this figure, the reader is referred to the online version of this chapter.)

A double *slab* moderator thus provides the most intense beams. It should be located as close as possible to the target aiming at distances d as close as 10–15 cm. Its brightness is proportional to the ratio cos $(\phi)/d^2$, where ϕ stands for the angle subtended between the normal to the moderator surface and the beam line. The main drawback of such a configuration is the potential contamination of the outgoing neutron beam with fast neutrons and γ-rays since the beam tube sees, at least in part, the target through the moderator. Such a disadvantage is in part compensated by the large solid angle subtended at the target per unit area. As an alternative, the *wing* configuration ensures that fast neutrons and γ-rays produced within the target do not go beyond the biological shield. However, such a configuration subtends a low solid angle at the target resulting in a lower efficiency than its slab counterpart. The *flux-trap* moderator configuration involves a split-target and has been used at facilities such as the Manuel Lujan Center at LANL [62] in conjunction with the cold beryllium reflector filter concept which was tested in an experiment at the WNR facility at LANSCE in January 2003 [72] as well as with a light-water premoderator.

A still further option to be considered when intensity is of prime concern is given by *grooved* moderators formed by grooved surfaces. Intensity gain factors of 1.1 or 2.0 over others with flat surfaces have been reported, depending upon the reflector. Apart from such a possible advantage, the pulse structure coming out of a grooved moderator usually shows a small subsidiary peak in the leading edge due to neutrons arising from the tips of the fins and bottoms of the grooves. A somewhat related development concerns moderating materials which are interleaved with single crystals of silicon. This extension of the grooved moderator idea provides greater flexibility and reduces the moderator volume lost, due to the absence of a need for construction materials to define the groove. Some results already obtained [73] showed that such a design provides a greater flux without a significant degradation of the pulse shape compared to a monolithic moderators. In addition, such vaned design also generates spectra that depend sensitively on the angle with respect to the orientation of the vanes themselves at which an instrument views the moderator, which is also a feature of interest. Finally, there is also interest in the performance of the *bispectral* moderator [74] that bases itself on a beam extraction system where the instrument ends up viewing the cold part of a side-by-side cold-thermal moderator for the long wavelength neutrons and the thermal part for short wavelength. The concept utilizes the fact that a Si wafer is almost transparent to thermal neutrons and the cutoff angle of its supermirror coating is proportional to the neutron wavelength.

The design of a moderator assembly for a multi-instrument source is, however, a difficult task that involves trade-offs of intensity versus background and the number of moderators and beam holes versus flux. A further compromise concerns the intensity per beam and the number of beams. Large incident proton energies of about 2.5 GeV lead to the neutron production inside the target taking place over some 45 cm, whereas such a range is restricted below 20 cm or so in kilowatt-range sources. Profit can be made from such a wider spatial distribution of the neutron production volume which thus allows placing additional moderators besides those located where neutron leakage is most intense.

2.4.2.6 Poisoning

As stated above, neutron absorbing materials or poisons are sometimes added to the moderator materials with the aim of exercising some control on the width of the moderator pulse. Current practice has shown that heterogeneous poisoning, that is, adding a layer of absorbing material beneath the moderator surface viewed by the beam tubes is the option of choice. The net effect of poisoning is thus to narrow the pulse at low energies since the most commonly used materials such as Gd or Cd capture only neutrons from the Maxwellian part of the spectrum. The price to pay is some loss of intensity, and it also affects the moderator mean emission time.

The conceptually simplest form of a heterogeneously poisoned moderator consists on a sandwich slab of width W where a sheet of absorbing material, typically about 1 mm is inserted parallel to the viewed surface at a given depth. Absorbers like Cd have absorption values of $1/e$ at 0.31 eV corresponding to a sheet of 0.25 mm. Ideally, the absorber should be transparent to neutrons with energies above the ambient Maxwellian while absorbing those at equilibrium. Those with energies within the epithermal range remain and are further slowed down in the region facing the beam tubes. The optimal thickness of the absorbing sheet and its position will depend upon the sought figure of merit. If we take the ratio of flux over pulse width, we found that the position does not seem to be a critical parameter [2]. Last-generation poisoned, decoupled cryogenic moderators [75] use 1.25-mm Cd plates asymmetrically located at 2.5 and 4.0 cm beneath the viewed surfaces, as schematically shown in Figure 2.16. With the advent of megawatt-range sources, it is expected that the lifetime of the poisoning element will be largely reduced due to the significant number of neutron captures occurring within such elements.

2.4.2.7 Premoderator Stage

The effect of adding a premoderator structure to cryogenic moderators of methane (ISIS) and a decoupled hydrogen (JSNS) moderator has been investigated in some detail. The main results show that the premoderator increases the pulse intensity without affecting the pulse shape. Experiments carried out at Hokkaido [76], Japan, show that the intensity gains depend upon the chosen reflector material. The intensity gain recorded using a graphite reflector increased a meager 7%, whereas remarkable intensity gains were achieved using a Pb reflector. As far as the premoderator material is concerned, H_2O appears as the best bet, getting maximal intensity increases using thicknesses of 1–2 cm.

The premoderator structures built so far consist on light-water-cooled thin-walled structures made of Al alloy that are installed around a liquid hydrogen or methane moderators. Because of the proximity to the target, a significant amount of deposited nuclear heat needs to be removed by means of smooth water flow without flow stagnation. Also, the structural integrity of the thin-walled structure should be kept against the water pressure. Experience so far tells that addition of a premoderator structure will also help in removing up to 33% of the heat load generated in the adjacent moderator.

2.4.2.8 Beryllium Filter

Studies at Los Alamos National Laboratory [77] showed a significant increase of brightness in flux-trap moderators introducing a beryllium filter very close to the moderator surface. This beryllium at low temperature (below 80 K) has a very low cross section for neutrons below 3.8 Å and regular beryllium cross section above that limit. This means that neutrons below 5 meV escape from the system and neutrons above that energy are reflected back to the moderator so they have a second chance to be moderated. In the case of the flux-trap configuration of the Lujan Center, an increase of flux close to a factor 2 was confirmed in the experimental measurements. The drawback of the configuration with beryllium filter is that the neutrons above 5 meV are significantly reduced due to absorption in the filter and a second cryogenic loop is required for filter cooling.

2.4.2.9 Shielding Issues

The twofold aims of an efficient shielding strategy are to reduce the background experienced by the neutron instruments as well as to reduce radiation doses. While a given value to minimize radiation doses is easily achieved, optimization of the amount of shielding becomes quite a more involved matter. An excess of shielding in places can do more harm than good and can also be spoiled if holes are drilled in places far from optimal. While Monte Carlo tools to carry out detailed evaluations are widely available, order of magnitude estimates can be guessed from results from not very refined theories. As an example, a simple formula for the attenuation tells that the dose as a function of the distance L should go as

$$D(L) = \frac{\phi \exp(-\mu w)}{L^2} \quad (2.53)$$

where ϕ is the dose at 1 m, and μ the attenuation coefficient of a shield of thickness w. The attenuation coefficient is usually energy dependent, and for some materials such as concrete, its value decreases with increasing energy. Reported values for μ on concrete go from 0.018 cm^{-1} as measured at ISIS for a 0.8-GeV beam and 0.016 cm^{-1} for a 12-GeV beam as measured at KEK. Equivalent figures concerning cast iron are 0.048 and 0.037 cm^{-1}, respectively.

Proton accelerator-driven neutron sources or those relying on pulsed reactors have to cope with large doses of fast neutrons and γ-rays in addition to that of slower neutrons characteristic or steady reactors. Typical numbers for doses at 1 m from the source are up to 1 kSv h^{-1} kW^{-1} (ISIS) and 0.4 kSv h^{-1} kW^{-1} (IBR-II), respectively.

Actual measurements of energy spectra of leaked radiation above the target carried out at ISIS [78] show the neutron spectra having two well-defined peaks, one at energies about 100 MeV directly coming from the target, a flat distribution with a broad maximum at about 1 MeV caused by scattering from light elements embedded within concrete and 10–500 keV arising from neutrons which are slowed down by inelastic scattering processes.

For spallation targets, the needs of shielding are dominated from those required to stop the small fraction of neutrons produced at energies up to the initial proton energy. This not only dictates a large portion of biological shielding around the target containing several meters of steel but also imposes to the designer to take special precautions when shielding the outgoing beam tubes since they will receive a flux of fast neutrons and become secondary sources. A simple order of magnitude estimate of the brightness of a conical beam tube [2] tells that at some 10 m from the moderator surface, a beam tube still has a brightness due to fast neutrons of 10^{-7} times that of the source. Partial remedies to such leaked radiation components have been sought by inserting collimation devices such as irises and spaces within the tubes.

Shielding problems for electron linacs also comprise the strong γ-ray doses which for 100 MeV electrons come up to some 3×10^{14} photons s^{-1} kW^{-1} which translates into a dose of 10 Sv h^{-1} kW^{-1}. A Pb shield of several centimeters around the target may reduce the dose by a few orders of magnitude, but the epithermal neutron flux will be significantly depleted.

2.5 ACCELERATOR-DRIVEN SOURCES: SOME PREDECESSORS

The story of accelerator-driven high-flux neutron production begins right after the end of the Second World War. Driven by military purposes, namely, the generation of ^{239}Pu, ^{3}H, and ^{233}U, Lawrence proposed to build a linear accelerator able to handle large (320 mA) deuteron currents and energies up to 500 MeV which after hitting a target had to be able to yield "a gram of neutrons per day." Although uranium was first considered as target, the chosen system finally consisted of a NaK-cooled beryllium primary and depleted uranium secondary target for neutron multiplication, within a water-cooled depleted uranium lattice for breeding plutonium. Lawrence proposed a prototype working at 25 MeV to the U.S. Atomic Energy Commission in 1950 [79]. The Commission approved the construction of the prototype accelerator, Mark I at a site then called the Livermore Naval Air Station as well as the design studies. The machine was codenamed as the Materials Testing Accelerator (MTA), and in March 1951, the Commission deemed Mark I sufficiently promising to justify siting studies for Mark II. Only the front end of Mark I was ever built. The estimated cost for 7 million dollars proved to be overoptimistic (it had consumed 21 million by the end of 1951). The prototype however demonstrated the possibility of achieving relatively high vacuum on an evacuated space of about 20 m long and 20 m in diameter, of maintaining a large accelerating gradient, and of introducing focusing magnets in the drift tubes. In achieving these ends, the MTA team had improved vacuum technology and diminished the lightning in the gaps. They built an ion source that emitted an ampere of deuterons or protons collimated to within 4°. The RF power was generated by 18 oscillators totalling 18 MW of power.

As an alternative, Lawrence formally proposed in 1951 to make a prototype production cyclotron giving 15 mA of 300 MeV deuterons at a cost of 20 million dollars. The fate of the project was fatally linked to the reasons for its very inception. Discovery of new uranium ore deposits in the Colorado plateau in 1952 as well as the offer of bonuses for new domestic sources of uranium killed the urgency for building such a machine which was reduced to a small research program by August 1952.

The Chalk River Laboratory of Atomic Energy of Canada Limited launched the Intense Neutron Generator (ING) Project in 1964. The goal was to build a "versatile machine" providing a high neutron flux for isotope production and neutron beam experiments. The machine was conceived as a proton linac able to accelerate protons up to 1 GeV under CW conditions and currents of 65 mA, thus delivering 65 MW of proton beam power. Detailed study for its target portrayed it as a flowing, liquid PbBi eutectic, fed by a vertical proton beam. Together with a Be multiplier and a D_2O moderator, the target assembly was specified to be able to generate a neutron production rate of 10^{19} n s^{-1} and thermal neutron fluxes reaching 10^{16} n cm^{-2} s^{-1}.

Work on ING developed apace and continued until 1968 when the project was finally canceled due to the perceived high costs and insufficient political support in the Canadian scientific community. However, even if the project did not finally lead to the construction of a facility, the entire present-day development of spallation sources owes a great debt to such groundbreaking project.

The facility at Los Alamos National Laboratory (LANL), hosted by the Los Alamos Meson Physics Facility (LAMPF), used the 800-MeV beam to produce cold, thermal, and epithermal neutrons for research into the condensed matter sciences.

It was the most powerful linear accelerator in the world when it was opened in June 1972. In 1977, a pulsed spallation neutron source was commissioned to supply moderated and fast-neutron experiments carried out at the WNR Center. Neutron-scattering experiments were started in 1983 under the auspices of the Department of Energy's Office of Basic Energy Sciences. Beginning in 1985, with the completion of the Proton Storage Ring (PSR) that compresses proton pulses from 750 μs to a quarter of a microsecond, the Los Alamos Neutron-Scattering Center (LANSCE), now known as the Lujan Center, was established, while WNR was expanded to other spallation sources on the accelerator beam. The facility operates concurrently with WNR, the Lujan Center, an isotope production unit, a facility for proton radiography, and Ultracold Neutron Facility. The facility is unique in the sense that the main accelerator is able to transport both protons and H$^-$ ions at the same time. Also, the facility has embarked on the construction of what may well be the most intense UCN source available.

The development of techniques related to spallation neutron facilities owes a great debt to predecessors such as ZING-P and ZING-P′ built at Argonne, forerunners of IPNS and to KENS, built at KEK. Both facilities showed that cutting-edge science could be made with modest intensities and moderate beam power (12 μA, 0.5 GeV protons yielding 1.5×10^{15} n s^{-1} at IPNS and 0.5 GeV protons yielding 3×10^{14} n s^{-1} at KENS).

Table 2.6 lists some of the most intense proton accelerators. Data for nonneutron-scattering facilities are also listed in order to provide a comparison of machine characteristics either existing or under development.

2.6 STATE-OF-THE-ART ACCELERATOR DRIVERS FOR NEUTRON SOURCES

2.6.1 Last-Generation Megawatt-Range Sources

2.6.1.1 The Oak Ridge Spallation Neutron Source

A sketch of the SNS facility which at the time of writing comprises a 1-GeV linear accelerator, a high-energy beam transport line into an accumulator ring, a ring-to-target beam transport line, and a liquid Hg target is shown in Figure 2.18. An RF ion source generates a current up to ≈50 mA of H$^-$ ions which are transported into the RFQ shown in Figure 2.10 through an electrostatic LEBT system. The ion source delivers 1 ms pulses with a repetition rate up to 60 Hz. The RFQ accelerates beams up to 2.5 MeV and delivers some

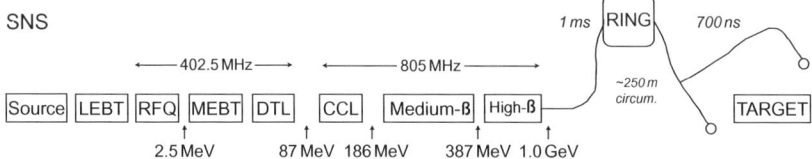

FIGURE 2.18 Schematics of current layout for the SNS source at ORNL. The room-temperature accelerating structures raise the H$^-$ beam up to 186 MeV having a jump in frequency from 402.5 up to 805 MHz at 87 MeV. The transition to superconducting sections takes place at the latter frequency. Two families of elliptical superconducting resonators are used to raise the beam to 1 GeV.

38 mA peak current. Further acceleration within room-temperature structures is achieved up to 86 MeV by a DTL as well as a CCL which raises the beam energy up to 185 MeV. A superconducting linac based upon two families of elliptical cavities with $\beta=0.61$ hosted within 11 cryomodules and a further set of 12 high $\beta=0.81$ modules rises the beam energy up to specification. The total length of the linear accelerator comes to be ≈ 340 m, at the end of which the beam is transported by an ≈ 170 m beam transport line into an accumulator ring with ≈ 250 m of perimeter. Such a circular structure compresses the ≈ 1 ms macropulses down to 0.695 µs which is then transported to the target delivering some 1.45×10^{14} protons per pulse.

The energy deposited per pulse reached 24 kJ out of which some 60% correspond to heat and a maximum power density of 9.2 J cm^{-3}. A significant amount of induced radiotoxicity was expected to be at the level of 3.4×10^4 TBq.

The TMR system of SNS comprises an ambient H$_2$O moderator together with three cryogenic H$_2$ supercritical moderators, located above and below the target volume. These are surrounded by a water-cooled Be reflector. The temperatures of the four moderating elements are kept at 300 and 20 K, respectively. The water and hydrogen moderators located upstream of the proton beam are poisoned/decoupled using Gd and Cd as absorbing materials. The geometry of the TMR complex is shown in Figure 2.19.

A power upgrade to get the beam energy up to 1.3 GeV and the average current up to 2.3 mA to reach a final power of 3 MW is envisaged. Such an upgrade foresees the development of an additional target, for the production of intense beams of cold neutrons at a second station [81]. Within the concept for the facility upgrade, it is envisaged that the SNS accelerator will operate in a *pulse-stealing* mode at 60 Hz with every third pulse going to the second target station (20 Hz second target station operation) and the remainder of the pulses going to the first target (40 pulses per second in a *pseudo-60 Hz mode*). For the 20 pulses per second going to the second target station, the long proton pulse (~ 1 ms long) from the linac is sent directly to the second target station with no accumulation in the ring, thus giving 50% more power to the second target station with no increase in peak linac current. This *long-pulse* operating mode enables the delivery of more power to the second target station for the same pulse duration because the chopping of the proton beam necessary for storage and extraction in the ring can be eliminated.

2.6.1.2 The Materials and Life Science Facility at the Japan-Proton Accelerator Research Complex

The building of the Japan-Proton Accelerator Complex has largely benefited from the long-standing experience in accelerator science and technology accumulated by years of successful operation of installations such as the Tohoku and Hokkaido electron linacs as well as the KEK complex at Tsukuba.

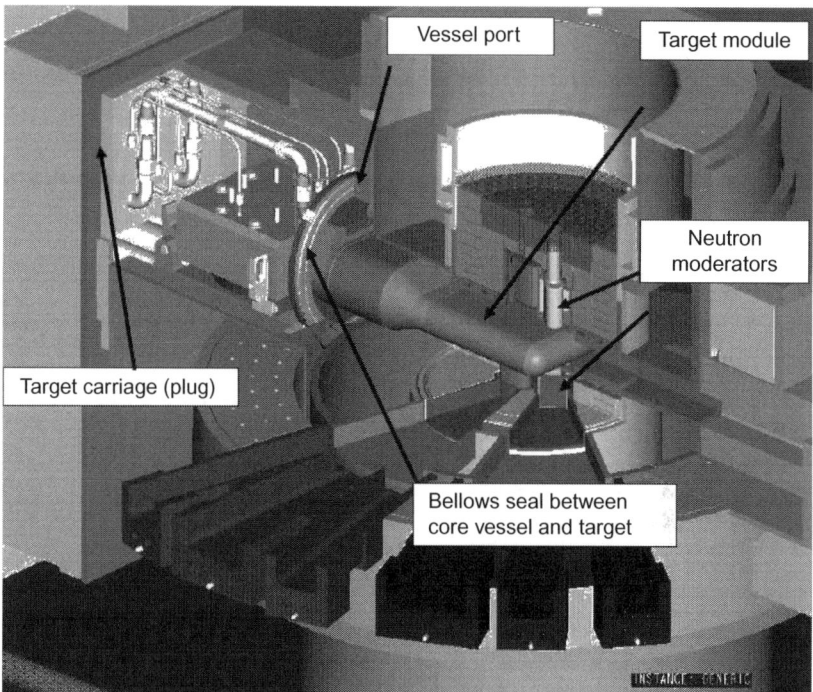

FIGURE 2.19 A cutaway view of the SNS Hg target. The target carriage is shown in production position. The picture also shows the relative position of two moderators above and below the target hull, as well as a portion of the Be reflector. *Figure taken from Ref. [80]. Reprinted with permission.* (For color version of this figure, the reader is referred to the online version of this chapter.)

The operation of the neutron-scattering KENS facility at KEK (a modest 3 kW source serving 17 instruments!) served to nucleate a user base which constituted an important part of what was originally named as the Japan Spallation Neutron Source. The facility now built at the Tokai Research Establishment of the JAERI delivers a 3-GeV beam to a spallation target also based upon Hg technology which has been renamed as the MLSF.

Contrary to the case of devoted installations such as SNS or ISIS, the MLSF is hosted within an accelerator complex which apart from neutron scattering serves other purposes such as waste transmutation, provision of a neutrino beam to Superkamiokande, and 50 GeV for use within hadron and particle physics studies. Most of the expected uses for this high-energy beam concern experiments using kaon beams to get insight into rare decays ($K^0 \to \pi^0$ or $\nu \to \bar{\nu}$) as well as to progress in the understanding of neutrino oscillations.

The facility is sketched in Figure 2.20 [82]. The linear accelerator comprises a linac that accelerates H^- ions up to 0.4 GeV delivering a peak current of 50 mA within 0.5 ms pulses at a repetition rate of 50 Hz. The beam is injected into an RCS of some 348 m of circumference at a repetition rate of

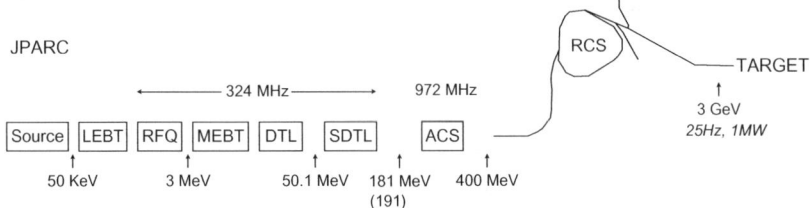

FIGURE 2.20 Schematics of current layout for the MLSF within the J-PARC accelerator complex. The block diagram shows the room-temperature accelerating structures which extend up to beam energies of 400 MeV. A frequency jump from 324 MHz up to a third harmonic of 972 MHz takes place at 191 MeV. Injection into the RCS compresses and accelerates the proton pulse up to 3 GeV.

25 Hz, reaching a final energy of ≈ 3 GeV and 0.33 mA pulses lasting about 1 μs containing 8.3×10^{13} protons. As such, the linac comprises a first 400 MeV normal conducting section of about 250 m in length followed by a superconducting section that delivers 400 MeV to the RCS or 600 MeV to an adjacent waste transmutation facility. In turn, part of the beam produced within the 3-GeV ring is directed to a hadron physics/astroparticle physics facility.

The basic accelerating structures of the linac comprise, apart from the ion source, an RFQ admitting a beam of 50 keV and delivering a bunched beam of 3 MeV and a DTL of some 27 m followed by a separated DTL (SDTL) of 91 m. An increase in energy up to 400 MeV is achieved by an annular-ring coupled structure. The superconducting section is here restricted to increase the energy up to 600 MeV previous to delivering it to the transmutation facility.

The target details do not largely differ from the design implemented at SNS. Target maintenance is done using horizontal access ports and a retractable target trolley which supports the target vessel, the heat exchanger, the surge tank, and the required piping elements.

The TMR complex shown in Figure 2.14 includes three supercritical H_2 moderators kept at 20 K as well as a Be reflector. Two of the moderators are decoupled and yield time-integrated neutron fluxes of the order of 10^8 n s^{-1} cm^{-2} and pulses of about 33 μs of full width at half height for an outgoing neutron energy of 10 meV. Decoupling is achieved by surrounding part of the moderator by an AIC layer. A coupled moderation is also placed at a position below the Hg target and also includes a water premoderator. Compared to the SNS design, the MLSF facility has been built with the aim of optimizing servicing to the moderating components. For such a purpose, each moderator and the associated parts of the reflector have been made easily replaceable by remote control procedures.

2.6.1.3 The European Spallation Source Project

After a series of delays and reconsiderations of recommendations, a working group was set up in 1990 by the Commission of the European Communities

[83], which delivered a first report recommending the construction of a proton accelerator with 5 MW of beam power having two target stations and a pulse length of 1 μs. A revised concept was later developed which incorporated part of the experience gained during the initial stages of the construction on SNS [84]. The new baseline envisaged an installation again serving two target stations for short pulses of 1.4 μs and a repetition rate of 50 Hz and a long-pulse target taking 2 ms pulses and $16\frac{2}{3}$ Hz of repetition rate, both having to cope with 5 MW of beam power. This concept from 2003 was endorsed by the user community in Europe and was then the subject of the ESS. After a further delay, an ESS initiative was established which considered a stepwise development, starting with a first, long-pulse target. A decision regarding the sitting, where Lund (Sweden) was favored, was taken in May 2009 a date in which a further revision of the baseline design was launched, which still is ongoing at the time of writing (Figure 2.21).

In the present moment, the baseline design for the accelerator consists on an ECR proton source similar to those already developed at CEA-Saclay and LNS-Catania able to provide a current of 50 mA feeding sections of a normal conducting accelerator able to reach 80 MeV. These consist on an RFQ not too different from those already in operation and a DTL together with matching sections at low (LEBT) and medium (MEBT) energies. Rising the power to the sought 2.5 GeV will probably involve the use of medium-$\beta=0.5$ superconducting spoke resonators as well as higher β cavities from the usual elliptical families. Transport to the target is done through a HEBT section.

Table 2.7 lists the main parameters of the ESS.

As far as the target is concerned, the ESS management has selected a rotating tungsten wheel as the baseline option for the target, although a metallic liquid lead bismuth eutectic was briefly considered as a competitive option. A rotating helium-gas-cooled tungsten target of 2.5 m diameter is, at the time of writing, the favored option. A water-cooled tungsten target of the same dimensions is also being evaluated. An equivalent liquid metal target can be made safe with passive cooling by choosing a large enough volume circulating through a large enough reservoir tank. The after-heat production is then uniformly distributed in the liquid metal volume. The safety concerns regarding the accidental release of volatile Po compounds need, however, to be taken seriously into consideration.

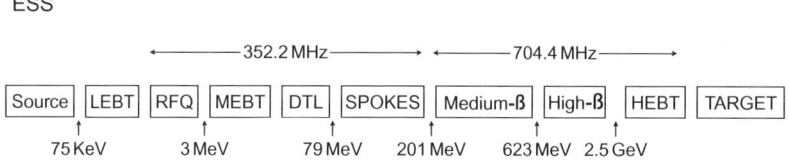

FIGURE 2.21 The proposed current layout for the ESS project as it stands at the time of writing.

TABLE 2.7 ESS Baseline Parameters as Stated at the Time of Writing

Parameter	Unit	Value
Average beam power	MW	5
Number of target stations		1
Number of beam ports		50
Number of moderators		2
Separation of ports in degrees		5
Proton kinetic energy	GeV	2.5
Average macropulse current	mA	50
Macropulse length	ms	2.86
Pulse repetition rate	Hz	14
Maximum gradient for superconducting linac	MV m^{-1}	40
Maximum linac length	m	482.5
Operating hours/year	h	5200

Two liquid hydrogen moderators of about a liter volume each, partially surrounded by water premoderators of comparable volume, are envisaged. The moderators are placed inside an inner reflector of about 1 m^3 of Be. These components will be kept at their desired operational temperature by dedicated cooling systems.

2.6.1.4 The SINQ Neutron Source at PSI

The Paul Scherrer Institut at Willigen, Switzerland, operates a cyclotron-based high-intensity proton accelerator routinely run at an average beam power of 1.3 MW. The accelerator facility relies upon an 870-keV Cockcroft–Walton electrostatic preaccelerator which injections into a four-sector cyclotron producing a 72-MeV proton beam. Further acceleration takes place at the 590-MeV ring cyclotron composed by a 0.9-T magnet, eightsectors, four resonators plus a flat-top resonator which routinely delivers DC proton beams with 2.3 mA currents to meson and neutron production targets. The beam power reaching the SINQ comes to be about 0.75 MW as a consequence of passing through the pion production target. Although the beam may look like continuous to the end user, the use of RF acceleration techniques requires the beam to have a definite time structure formed by bunches with a frequency of 50.63 MHz.

The PSI proton target is accessed from below by the proton beam which is guided vertically upward by bending magnets. The target is positioned in the center of the heavy water moderator tank which allows for a maximum number of horizontal beam tubes and neutron guides feeding the instruments. The usual neutron production target consists in an array of Pb rods enclosed in zircalloy tubes and cooled in cross-flow by heavy water coolant. To minimize neutron capture, the part of the target unit extending into the moderator tank is filled with heavy water, except for the rod bundle.

Within recent developments at the facility, the new source dedicated to the production of ultracold neutrons which has successfully started operation in August 2011 merits to be mentioned. The UCN source shares the proton beam (2.2 mA at 590 MeV) in a pulsed mode with SINQ and the meson production targets at a duty cycle of 1%. More into details, the source produces spallation neutrons from a Zr-canned Pb referred to as the *cannelloni* array, which are moderated within a D_2O tank and then converted to ultracold neutrons within a large solid *ortho*-deuterium crystal.

A fraction of the proton beam, typically about 2 mA, also strucks two pion-generating graphite targets feeding muons widely distributed in energy (from 30 keV up to 280 MeV) to several experimental stations.

Further improvements of the facility have been already planned, and in fact, an ongoing program to raise the beam power to 1.8 MW is expected to bear its fruits by 2013. Further increase in beam power is limited by losses at extraction and the resulting activation of accelerator components.

2.6.2 Medium-Power (100 kW) Sources

2.6.2.1 The ISIS Second Target Station

The past two decades have witnessed how the ISIS neutron-producing target (TS-1) has achieved a position as the world's most productive spallation neutron source. The source is driven by a 50-Hz, 800-MeV, 200-µA proton beam from a RCS, which is fed by a 70-MeV H^- DTL which in turn accepts beam from an H^- 665 keV RFQ (see the layout in Figure 2.22). Motivated by an increasing international demand, a second target station has been built (TS-2), started operation in 2008, and sustains a fully fledged program of new neutron instruments. A major upgrade has been accomplished for the accelerator system to underpin operation for at least another 15 years and also allows consideration of future upgrades. The most innovative aspects of such an upgrade concern the design and construction of a dual harmonic RF system (DHRFS) within the ISIS synchrotron as well as the implementation of a low output impedance (LOI) acceleration system. The former device works keeping the fundamental RF (1RF) cavities, running at twice the ring revolution frequency (1.3–3.1 MHz), providing up to 140 kV/turn for trapping and acceleration of the two nanosecond-scale bunches of protons and getting to work additional four (2RF) cavities, running at four times the ring revolution frequency (2.6–6.2 MHz) and at voltages of up to 80 kV/turn. Optimization of relative phases and voltages allowed to increase the phase stable

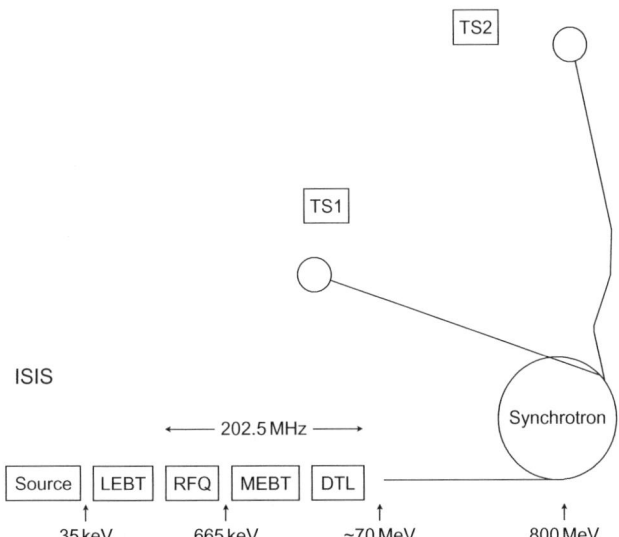

FIGURE 2.22 Schematic layout of the accelerator and targets complex for the ISIS source. The existing injector delivers a H⁻ beam of 70 MeV coming out from a DTL which is injected into the synchrotron after crossing a stripping foil. The ISIS operating frequency of 202.5 MHz significantly differs from those used in latter generation machines. Such a low-frequency field is mostly generated by tetrode vacuum tubes. At the time of writing, both target stations serve a total of 35 instruments. A secondary graphite target partially intercepts the proton beam transported from the synchrotron to TS-1. The collision of 800 MeV protons with a graphite target generates sizeable amounts of $\pi^{\pm}$ mesons which rapidly decay into $\mu^{\pm}$ muons which are transported to seven experimental stations where muon rotation, relaxation, and resonance experiments as well as muon-catalyzed-fusion activities are carried out.

regions and enhanced bunching factors which, as a net effect, leads to smaller beam loss. The installed DHRFS is able to produce currents of 187 μA to TS-1 at 50 Hz and 60 μA to TS-2 at 10 Hz keeping beam intensities approaching the theoretical maximum of 3.75×10^{13} protons per pulse. On the other hand, the LOI system has undergone initial tests that show that beam loading is now about 40 times lower than that in the existing acceleration systems, which in practical terms imply that if finally implemented for the whole ring, the LOI system could possibly be used to accelerate up to 40 times more current than that presently in operation. Practical operation of the ISIS machine at the time of writing keep the beam currents at values totaling some 210–220 μA, in order to guarantee reliable neutron production schedules.

A portion of the 800-MeV proton beam, usually below 10% of the outgoing current, is also made to collide with a carbon target devoted to pion production. The target made of graphite with dimensions $50 \times 50 \times 7$ mm is oriented at 45° to the incoming proton beam and yields about 1.6×10^4 surface muons per double proton pulse which mostly result from the decay of charged pions following:

$$\pi^+ \to \mu^+ + \mu_\nu \tag{2.54}$$

$$\pi^- \to \mu^- + \overline{\mu}_\nu \tag{2.55}$$

The current TS-2 consists of a tantalum-coated solid tungsten cylinder 68 mm in diameter by 307 mm long, which operates at a repetition rate of 10 Hz and is optimized for the production of cold neutrons. The target material is cooled from the surface by D_2O, and its design allows for two moderators in wing geometry, a 35-K decoupled solid methane and a 20-K coupled liquid hydrogen unit. Both are surrounded by a reflector built from Be rods which is contained within a vessel also cooled by D_2O. The core target components are contained within a cylindrical void vessel surrounded by shielding. The void vessel provides a controlled environment for the target and acts as a containment vessel in the case of a target component failure or water leak. The helium-filled vessel is embedded within a 12 m diameter and 7 m high steel and concrete shield which allows vertical access to the proton beam window, the neutron beam windows, the upper reflector assembly, and the shutter assemblies. The target and lower reflector assembly may be removed horizontally. The shielding is made of cast and machined steel blocks with a poured concrete outer layer. Finally, the moderators have been designed to be placed very close to the primary target, leading to significant gains in fluxes of moderated neutrons.

Several important upgrade plans are already being considered. A first option envisages to build a new 180 MeV linac and/or 800 MeV synchrotron which will increase power up to the 0.4–0.9 MW range. A second option considers adding a new ≈ 3 GeV synchrotron, thus increasing the beam power to about 1 MW by taking the output of the existing facility. This could be built with minimal interruptions to ISIS operations, gives predictable increases in power at reasonable estimated costs, has well-defined upgrade routes, and is the favored upgrade path. Other ideas are also being studied such as upgrading the 800 MeV machine plus a 3 or 6 GeV ring or building anew a 1.3 GeV linac plus an accumulator ring much in the same way as the short-pulse target station of the 2003 ESS project envisaged.

2.6.2.2 The China Spallation Neutron Source (CSNS)

The facility is being built by the Institute of High-Energy Physics from the Chinese Academy of Sciences at Dongghuan in the Guandong province. A first phase under construction comprises a warm linac able to accelerate a 62.5 μA H^- current up to 80 MeV and then inject into a rapid cycling synchrotron to achieve a final energy of 1.6 GeV. The accelerator which is built employs the usual technologies for its accelerating structures (RFQ, DTL, and RCS). The duty factor is kept down to manageably figure keeping the repetition rate to 25 Hz. Descriptions of the target and instrument suite are available from Ref. [85]. The target within the first phase will take 100 kW of beam power onto a tantalum-cladded tungsten assembly. A set of three moderators, one thermal (H_2O) and two cryogenic (H_2 poisoned, coupled, and decoupled, respectively) and a Be reflector will supply up to 20 neutron beam lines. Neutron count rates for some of the wideband instruments such as diffractometers reach figures of 10^7 n cm^{-2} s^{-1} at the sample position which compares favorably with those available at operating installations such as ISIS. A second phase was envisaged where the ion current will ramp up to 125 µA, and the proton linac will increase its final energy up to 130 MeV, leading to a beam power of 200 kW (Figure 2.23).

Figure 2.24 displays a view of the target station as well as some details on the retractable target trolley. The proton beam enters horizontally at 1.8 m above the floor level. The target station is primarily located along the center of the target building with the solid tungsten target positioned approximately 24 m from the point of beam entry. The target is surrounded by 1 m in diameter heavy-water-cooled beryllium and iron reflector, with two moderators (one decoupled ambient water moderator plus one decoupled and poisoned supercritical hydrogen cryogenic moderator) above the target and one coupled supercritical hydrogen cryogenic moderator below the target. The helium vessel system encloses the TMR complex to limit the production of activated gas and to provide hydrogen gas containment in case of a cryogenic moderator failure. Outside the helium vessel, the surrounding radiation bulk shielding (including the neutron beam shutters) extends to a radius of approximately 6 m with 4.8 m steel plus 1.2 m heavy concrete and thus keeps radiation dose levels below 2.5 µSv h^{-1}.

The target as such is mounted on a rectangular flange. The center the target station is located 103 mm away from the tip of the target when it is in the operational position. The positioning of the whole target should be within a fairly narrow tolerance of ±0.5 mm in the vertical and horizontal directions to avoid any misalignment with the proton beam and interference with other parts. The total heat load on the target is about 60 kW at 100 kW beam power. The estimated lifetime for the target should be 3–5 years at the end of which it will be replaced via remote handling.

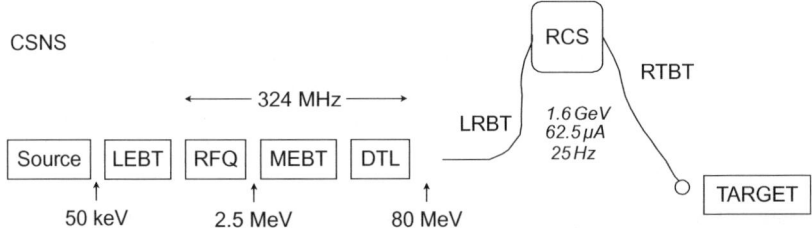

FIGURE 2.23 Schematics of current layout for the Chinese spallation source. The block diagram shows the room-temperature accelerating structures which extend up to beam energies of 80 MeV. Injection into the RCS compresses and accelerates the proton pulse up to 1.6 GeV.

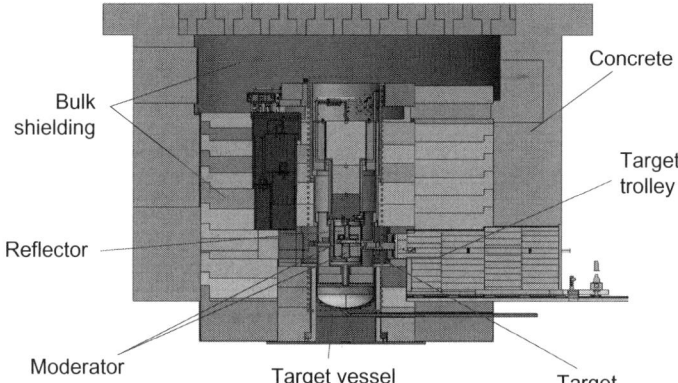

FIGURE 2.24 Details of the monolith hosting the solid target built at CSNS. The removable target trolley with a total length of 7 m and weighting 120 tons is shown in the production position at the lower right-hand corner. When not in operation the target is transferred to a hot cell area for servicing purposes. The target has a rectangular section with dimensions 130×50 mm and a thickness of 570 mm. The central part of the figure shows the helium vessel into which the target is hosted. The moderator/reflector comprises three different sections corresponding to an upper plug of 1.3 m diameter and 1.4 m length, an intermediate section of 1.150 m diameter and 1.8 m length, and a lower plug with 1m diameter and 1.3 m length which consists of an aluminum shell that holds the Be reflector. The dimensions of the reflector are 0.7 m diameter and 0.8 m length. Each moderator and its transferring pipeline within the vessel can be fully detached. *Courtesy of Dr. Xuejun Jia. Reprinted with permission from the author.* (For color version of this figure, the reader is referred to the online version of this chapter.)

2.6.3 Compact, Accelerator-Driven Sources

The development work that led to the design and construction of the existing large-scale facilities was mostly carried out and supported by networks of smaller research laboratories that had previously agreed to collaborate in the field and to provide input to each other's projects. Recent advances in accelerator technology and neutronic design have made possible the construction of small-scale accelerator-driven neutron facilities that will be able to play a significant role in future advancements in neutron technology and science. This opened up new opportunities for smaller-scale organizations to enter the field of neutron physics with modest investments and without the proliferation and safety concerns associated with building any large facility. Such low-to-medium scale facilities can be used in fields as diverse as materials science, nuclear physics, medical physics, engineering, and cultural heritage. Their value and scientific and technical relevance were emphasized in 2004 within a document originated by the IAEA [4], where it is envisaged that,

There is a clear synergy between the operations of high and low to medium power sources, which is beneficial to both types. Due to their cost, the high-power sources tend to be unique to the continent on which they reside, and their operation is geared

to reliably providing intense neutron beams to a wide user community. Such a mode of operation has a number of consequences:

- Overload of available beams with respect to both, the range of instruments that can be accommodated and the beam time available.
- Experimental time is restricted, and usually very little flexibility in schedule can be tolerated to accommodate unknowns.
- Little opportunities to perform tests of more speculative nature.
- Experiments not requiring the high intensity of the source are relegated to other, lower power sources; hence these sources must be available.
- Justification of the full experimental program is necessary. This can lead to situations in which measurements are favored over experiments.
- The pressure to make most efficient use of beam time to produce results, limits the possibility of carrying out training of students or new users.

Small to medium power sources provide a network of facilities that provide an invaluable experimental resource, which also serves for the development of the technique and for training of the community. Not all measurements or experiments require the beam intensity offered by the high-power sources, and excellent science programs can be carried out at smaller facilities. The science programs carried out at smaller sources may be adapted to a specialized community, which may better reflect the regional requirements. The use of the facility will certainly not be restricted to neutron-scattering experiments. Scientific and technological experience and know-how developed at such sources can be shared effectively with the larger facilities. Hence, covering all the needs of the user community, low-, medium-, and high-power neutron sources should be considered as complementary, each playing an important role in the application of neutron techniques to science and technology.

A number of small-to-medium range installations worldwide are now in operation, under construction or planned. Electron accelerators have been already referred to in Section 2.2.3 and continue to play a very valuable role in the development of large-scale facilities. Such clear synergies have been glaringly shown by work performed on moderator and instrument development carried out at laboratories such as the Low-Energy Neutron Facility at the University of Indiana [86] in relation to the Oak Ridge project or the activities developed within the Hokkaido University electron linac regarding the MLSF-J-PARC facility [87]. In addition, a number of neutron facilities are now under construction mostly relying on proton accelerators as drivers. Most of them cover a wide spectrum of activities which apart from neutron scattering purposes also envisage activities in other areas such as accelerator-driven systems for waste transmutation, hadron therapy, neutron imaging, etc. This merits a brief description as follows:

The Compact Pulsed Hadron Source (CPHS). The project of CPHS in Tsinghua University, Beijing, consists of an accelerator front-end, a high-intensity ion source, a 3 MeV RFQ linac and a 13 MeV DTL, a neutron target station, a beryllium target with solid methane, and room-temperature water moderators/reflector and experimental stations for neutron imaging/radiography, small-angle scattering and proton irradiation. The proton platform will in the future serve multiple stations for bioapplications, fuel cell and nanoapplications, and space irradiation and detection. CPHS may also serve as an injector to a ring accelerator that subsequently accelerates the beam to up to 300 MeV for proton therapy and radiography. Schematics of the facility as being built together with the envisaged expansion are shown in Figure 2.25 [88].

The Peking University Neutron Imaging Facility (PKUNIFTY). A neutron imaging facility, PKUNIFTY, based on an RFQ accelerator-driven compact neutron source, is at present under commissioning at the Beijing University. The machine consists of a deuteron linear accelerator, a neutron target–moderator–reflector assembly, and a thermal neutron imaging system. Neutrons are generated via the deuteron–beryllium reaction with an expected yield of 3×10^{12} n s^{-1} and an estimated thermal flux of 5×10^5 n cm^{-2} s^{-1}. The injector system provides 50 keV and 50 mA of D$^+$ beams to the entrance of the 2 MeV RFQ with 10% duty factor (1 ms, 100 Hz) [89].

The Proton Engineering Frontier Project (PEFP). The Project (PEFP) under development at KAERI, Daejeon, Korea has built a 100 MeV proton linac to provide 20 and 100 MeV proton beams. The linac consists of a 50 keV proton injector, a 3 MeV RFQ, a 20 MeV DTL, an MEBT, and the higher energy part (20 MeV ≈ 100 MeV) of the 100 MeV DTL. The MEBT is located after the 20 MeV DTL in order to extract 20 MeV proton beams as well as to match the proton beam into the higher energy part of the linac. The 20 MeV part of the linac was completed and is now under beam test. The higher energy part of the PEFP linac was designed to operate with 8% beam duty and is now under construction. To extend the output beam energy up to 1 GeV, superconducting technology is under consideration. In addition, a conceptual design of a rapid cycling synchrotron as an extension of the PEFP linac has also been accomplished [90]. In the initial stage, the target beam power with 15 Hz repetition rate is about 58 kW at 1 GeV extraction. The characteristic feature of the PEFP RCS is that it is applicable to the

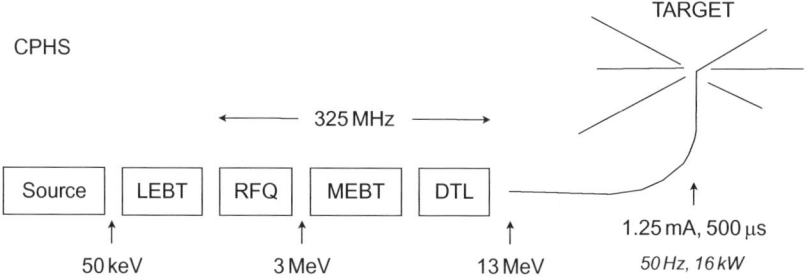

FIGURE 2.25 The layout of the first phase of the accelerator being built at Tsinghua University.

spallation neutron source facility via fast extraction and the medical research as well as the radioisotope production facility via slow extraction.

The Low-Energy Neutron Source (LENS) at the Indiana University Center for Exploration of Energy and Matter (CEEM). The source utilizes a low-energy direct reaction in Be coupled with a high-current, variable-pulse-width proton accelerator to produce either short or long neutron pulses. A highly optimized moderator produces cold and very cold neutrons for use by a suite of neutron-scattering instruments and development facilities. The accelerator delivers 13 MeV protons after acceleration through a DTL section and transports about 23 mA of peak current to the LENS target. The source is usually run at variable repetition rates up to 20 Hz, delivering pulses of 150 or 600 μs that feed the neutron production target on which two instruments are operational.

The Bilbao Neutron Facility (ESS-B) at present under development by the ESS-Bilbao Consortium. The project envisages the construction of a 50 MeV/75 mA (peak) proton source that feeds a rotating Be target for neutron production as well as allows the use of proton beams by several other irradiation facilities. The design at the time of writing foresees the construction of a target station able to take 112 kW of beam power hosting two optimized moderators which should be easily changeable to carry out studies on their performance. The estimated performance of the source yields a value for moderated neutrons with energies below 0.4 eV of 1.3×10^{11} n cm^{-2} s^{-1} sr^{-1} [91] which compares to values of 10^{11} n cm^{-2} s^{-1} sr^{-1} for ISIS-TS2 or 2.1×10^{12} n cm^{-2} s^{-1} sr^{-1} for the hydrogen moderator installed at SNS. Typical applications envisaged for a first construction phase comprise material irradiation facilities by proton beams, small-angle neutron scattering, and high-resolution spectroscopy.

The target station has been designed to be able to cope with a power upgrade which may then rise the neutron production by an estimated factor of three.

2.7 RESEARCH REACTORS

For more than five decades, research reactors have constituted the main workhorses for the production of neutron beams. According to the IAEA inventory, 651 research reactors have been built [92]. These devices are more simple and compact than the power reactors, and for many decades, their fuels were based on highly enriched $^{235}_{92}$U (HEU). The situation however changed during the 1990s because of proliferation concerns, and this forced some of the existing reactors to seek transformation of their cores to adapt them for the provision of beams generated from low-enrichment fuel elements (LEU).

Although most of the existing reactors were built along the 1960s and 1970s, new additions to the list such as the OPAL reactor operated by the Bragg Institute, the PIK reactor at Gatchina (St. Petersburg) still under construction and significant refurbishment and modernization programs carried out at existing sources show that the technique is alive and well and will continue at the forefront of neutron science for years to come.

The research reactor technology has not experienced large changes, and the current-day designs such as that implemented at FRM-II are largely optimized with respect to those built in previous decades.

Table 2.8 displays a list of the most productive research reactors over the world.

In what follows, we will briefly review some of the developments.

2.7.1 Core Designs

The design of research reactors geared for high thermal and cold neutron flux production significantly differs from that of power reactors where the goal is to have a flux profile as flat as possible. In fact, experiment-driven research reactors operate under substantial undermoderation conditions in order to provide a strong fast-neutron source at the core edge so that full thermalization is achieved at the noses of the beam tubes. Reactors which use light water as a cooling element need to be more compact than those using D_2O as a coolant and operate at higher power density to provide such fast-neutron fluxes. The former reactors generally use HEU as fuel, and in some cases, conversion into LEU sources may represent an important engineering problem.

There is a marked difficulty if one attempts to describe the core of existing reactors in a unified fashion since an exception can be found for every rule. The first generation of powerful research reactors at U.S. universities such as the Massachusetts Institute of Technology (MIT) reactor (1958) and the Missouri University Research Reactor (1966) as well as developments at the British Atomic Energy Research Establishment at Harwell shared some basic design concepts. Second-generation reactors such as the HFBR at BNL were designed to maximize the neutron flux in the reflector, which, as a result, yielded high fluxes to small neutron beam tubes in the reflector. The reactor built at the National Bureau of Standards incorporated several innovative elements in its design. The uranium fuel elements were "split" with a central gap so that nine very large thermal beam tubes (four 16 cm diameter and five 13 cm diameter) could be inserted radially to provide high-intensity beams with low fast-neutron and γ-ray backgrounds to samples closer to the reactor core than in most reactors. Also and being aware of the potential of cold neutrons for neutron-scattering applications, the NBS engineers installed a very large (56 cm diameter) "thimble" near the high-flux region in the reactor to enable the easy installation, servicing, and operation of a cold source.

Last-generation facilities such as the Heinz Maier-Leibnitz reactor (FRM-II) at the Technical University of Munich use a single fuel element as shown in Figure 2.26 to provide a power density of 1 MW per liter. Its inner diameter is 118 mm, its outer diameter 243 mm, and its height is ≈ 700 mm. The fuel element with its packaging is ≈ 1.3 m high. Inside the fuel element, there is

TABLE 2.8 A List of the Most Significant Research Reactors

Name	Power (MW)	Coolant	Reflector	Thermal flux (n cm^{-2}s^{-1})	Fuel
HFR (Grenoble)	58.3	D$_2$O	D$_2$O	1.5×10^{15}	UAlx–HEU*
FRM-2 (Munich)	20	H$_2$O	D$_2$O	8×10^{14}	U$_3$Si$_2$–HEU
HZB (Berlin)	10	H$_2$O	Beryllium	$\sim 10^{14}$	HEU
BNC (Budapest)	10	H$_2$O	Beryllium	$\sim 2.2 \times 10^{14}$	VVR-SM 36%^{235}U*
OPAL (Sydney)	20	H$_2$O	D$_2$O	$\sim 4 \times 10^{14}$	UAl$_x$–LEU
HFIR (Oak Ridge)	85	H$_2$O	Beryllium	$2-6 \times 10^{15}$	U$_3$O$_8$–Al–HEU*
NIST (Gaithersburg)	20	D$_2$O	D$_2$O	1.5×10^{14}	U$_3$O$_8$–Al–HEU
CARR (China)	65	D2O	Graphite	8×10^{14}	HEU
JRR-3M (Japan)	20	H$_2$O	D$_2$O	2×10^{14}	U$_3$O$_8$–U$_3$Si$_2$Al$_x$ LEU
Orphee (Saclay)	14	H$_2$O	D$_2$O	3×10^{14}	UAl$_x$–HEU
RA-10 (Argentina)	30	H$_2$O	D$_2$O	$\approx 10^{14}$	U$_3$Si$_2$–LEU
TRIGA-class MarkII	0.14	H$_2$O	Graphite	4×10^{12}	UZrH–LEU
–Pulsed mode–	250			10^{15}	
IBR-II/JINR (Dubna)	1850	Na	Ni/steel	10^{16}	PuO$_2$–HEU
PIK/PNPI (Gatchina)	100	H$_2$	D$_2$O	5×10^{15}	UO$_2$/Cu/Be–HEU

The asterisk marks those reactors where well-advanced studies of conversion to LEU fuels have been undertaken. Data quoted for the TRIGA-class reactor at the Johannes Guttenberg Universitat–Mainz comprise steady and pulsed operation. In the latter case, the quoted figures are per-pulse quantities. Data quoted for IBR-II refer to quantities per pulse. In terms of time-averaged quantities, the neutron flux at the surface of the moderatos comes to be $\approx 10^{13}$ n cm^{-2} s^{-1}.

Chapter | 2 Neutron Sources

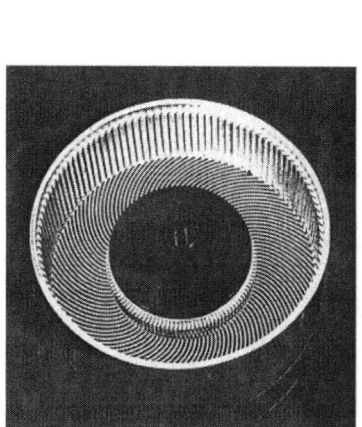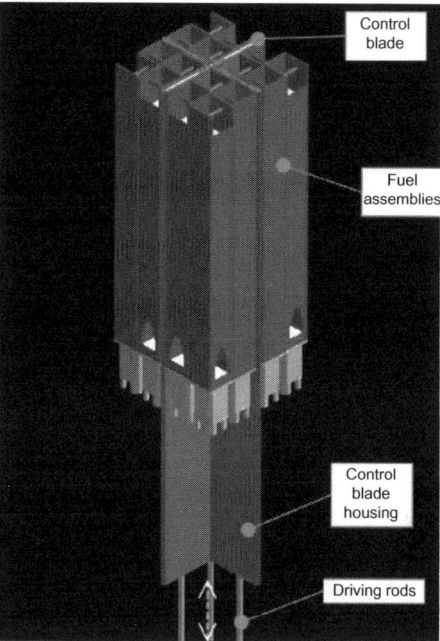

FIGURE 2.26 Core of the research reactor FRM-II (left).Fission neutrons are provided by uranium plates (93% enrichment) close to the nose of a beam tube. This geometry avoids neutron moderation in water to a high extent and delivers a fast-neutron flux of 2×10^9 cm^{-2} s^{-1} at the exit of this beam tube. The core also provides neutrons to be thermalized by a hot and a cold source. The hot neutrons with energies within 100 meV to 1 eV emerge from a radiation heated graphite block that reaches temperatures up to 2900 K. The cold neutron spectrum peaked around 5 meV is supplied by a liquid D_2 moderator located at a distance of 40 cm from the core axis and kept at 25 K. A future source of ultracold neutrons with energies below 10 μeV will be provided by a solid D_2 or a superfluid He source. The OPAL core (right).A cold source is installed in the reflector vessel at some 50 cm from the reactor core. The source is made by liquid D_2 and is cooled by helium vapor circulating through a heat exchanger and in an outer cooling jacket. Installation of a hot neutron source is also envisioned. *Left: Taken from Nucl Eng Des 178 (1997) 127-133. Figure 1. ELSEVIER. Right: Courtesy of ANSTO. Reprinted with permission.*

the central control rod. The fuel element is made up of 113 involuted, curved fuel plates. The fuel plates consist of three layers. The fuel U_3Si_2 composed of a mixture of HEU uranium silicide and aluminum powder is embedded in an aluminum matrix. Each fuel plate consists of a fuel zone sandwiched between Al-cladding layers. The fuel plates of involute shape are arranged in a circle to provide cooling channels of constant width of 2.2 mm. The core diameter of 24 cm is exceptionally small, leading to a very high thermal peak flux of 8×10^{14} n cm^{-2} s^{-1} in the moderator, given a fairly small reactor power of 20 MW. The compact reactor core is in the center of a D_2O moderator tank with 2.5 m in diameter and 2.5 m high. Twelve beam tubes (10 horizontal, 2 inclined) are tangentially arranged around the core with their noses in close vicinity to the maximum flux of thermal neutrons or to one of the secondary sources. The spectrum of neutrons delivered by these tubes is determined by the characteristic source temperature in front of the nose, which will be thermal to those viewing D_2O (300 K), cold for those tubes facing the cold D_2 source, or hot to those close to the graphite source (2900 K). The beam tubes are made out of Al alloys to minimize neutron losses from passage through the walls.

The new Australian research reactor, the Open Pool Australian Light-water Reactor (OPAL), has a core composed by 16 square fuel assemblies which configure an square core as shown in Figure 2.26 and is located under 10 m of water in an open pool. The assemblies are located in a four by four square matrix, and each of these assemblies is 8 cm^2 and holds 21 fuel plates. The fuel plates contain LEU $^{235}_{92}$U (<20%) and 80% $^{238}_{92}$U. The uranium is sandwiched between Al alloy plates. Light water is used as the coolant and moderator, while heavy water is used as the neutron reflector.

The reactivity of the research reactors, that is, the quantity $\rho = (k-1)/k$ with k defined in Appendix B, has to be kept at values about zero by means of two basic mechanisms such as the introduction of absorption elements, usually a single control rod such does HFR or FRM-II, or reducing the efficiency of the reflector such does a pulsed reactor as IBR-2. If $\rho > 0$, the reactor becomes supercritical and the neutron population increases at each successive generation. In the opposite case, $\rho < 0$, the neutron population is successively attenuated. The aim of this is to keep conditions at criticality under control so that excursions on ρ above and below zero can be compensated automatically within a time lapse significantly shorter than the reactor time constant. This is achieved by insertion of control rods which are made from materials which show a large absorption cross section for thermal neutrons. Such control rods are introduced in the center of the cylinder in single element reactors (HFIR), or as plates in between the fuel elements (OPAL). In the latter case, the control mechanism relies on five neutron absorbing elements made of Hafnium within the core located between the fuel assemblies. One of these control rods is cruciform in shape and centrally located within the core, while the other four are plates that divide the core into four quadrants. The rods also provide a means by which the reactor can be rapidly shutdown or tripped if required.

Because of the realization of potential dangers of proliferation as stated by the Global Threat Reduction Initiative, the use of HEU is severely compromised. There are at present some 18 research reactors using an amount of ≈ 500 kg of HEU per year that require the development of adequate LEU fuels [93]. For such a purpose, six nations actively pursuing LEU fuel development formed the RERTR International Fuel Development Working Group in 2004 to pursue research into new very high density fuels which are required for conversion of the world's high-performance research reactors. In fact, existing reactors such as HFIR at ORNL are at present carrying out activities geared at converting the HFIR from HEU fuel to LEU fuel. Conversion from

HEU to LEU will require a change in fuel form from uranium oxides to an U–10% Mo alloy. With axial and radial grading of the fuel foil and an increase in reactor power to 100 MW, calculations indicate that the HFIR can be operated with LEU fuel with no degradation in reactor performance from the current level.

A short list of the most relevant fuels, both in use and as potential candidates, comprise

- uranium oxides (U_3O_8 or UO_2) [94]
- plutonium dioxide (PuO_2) [95]
- U–Mo alloy (U–10% Mo) [96]
- uranium silicide (U_3Si_2) [97,98]
- uranium aluminum powder (UAl_x) [99]
- low-enriched uranium aluminum powder (UAl_x, 40%)
- low-enriched uranium silicide (U_3Si_2, 55%)
- uranium-zirconium hydride alloys ($UZrH_x$) [100]

The fuel employed in research reactors of the TRIGA-class, that is, $UZrH_{1.7}$, has as its main advantage its chemical stability. It can be quenched at 1200 °C with no interactions in water. The high-temperature strength and ductility of the stainless steel or Alloy 800 fuel cladding provide total clad integrity at fuel temperatures as high as 1150 °C in an operating reactor. It also offers far superior retention of radioactive fission products compared with aluminum-clad plate-type fuel since it could retain more than 99% of its volatile fission product inventory even if all the cladding were removed.

Most of the cladding materials are made of Al or Zr alloys due to the low absorption cross section of both materials. In order to increase the neutron flux, the core is surrounded by a reflector material: heavy water or beryllium. The moderators for cold and hot neutron production are to be placed in the reflector, as close as possible to the core. The reactor core is usually surrounded by a heavy water vessel as reflector, and reducing the water level is possible to change the reactivity of the core.

The cooling system depends on each reactor design. In the single element reactor, a forced water cooling system introduces the water in between the involute fuel plates; nevertheless, in the low power density reactors, natural convection may be used. Regarding the shutdown and the residual heat, as the research reactors have relative low thermal power compared to power reactors, the emergency cooling can be achieved solely by natural convection.

2.7.2 Reactor Vessel

Around the core, there is an area where the moderators and other experimental devices are placed. This area is usually filled with heavy water to increase the neutron flux and reduce the neutron capture. As an example, Figure 2.27 shows the vessel of the OPAL reactor. Here, the vessel has the hot and cold

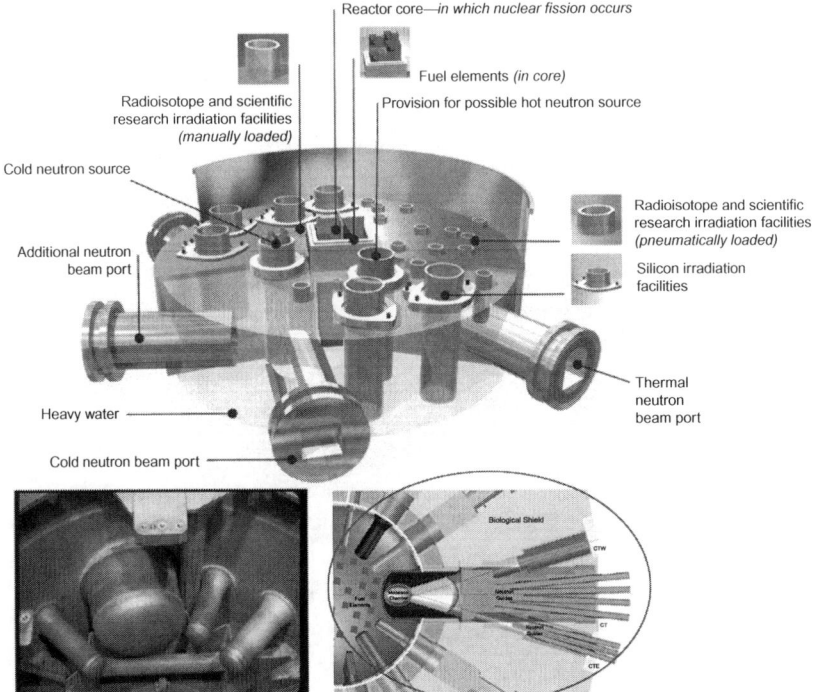

FIGURE 2.27 The upper frame shows the reactor vessel of the OPAL reactor. The lower frame shows the vessel of the NCNR (previously NBS) reactor displaying the large "thimble" outside which the cold source is located [101], as well as the location of the cold source and the groups of neutron guides. *Upper panel: Courtesy of ANSTO. Reprinted with permission. Lower panel: Courtesy of Dr. R. Capelletti. Reprinted with permission.* (For color version of this figure, the reader is referred to the online version of this chapter.)

moderators for neutron-scattering purposes, several areas for materials irradiation, and others for isotope production. The neutron guides directly view the reflector within which the neutrons have thermal energy spectra. The hot neutrons are usually produced in a graphite block at very high temperature (~2000 K) and for the production of cold neutrons large deuterium volumes at 20 K are used. As it can be shown in the figure, the hot and cold moderators are placed as close as possible to the core in order to increase the neutron flux.

Also in Figure 2.27, the vessel structure of the NCNR reactor is shown, indicating the position of the cold source as well as the insertion of the outgoing neutron guides.

Most experimental reactors are steady neutron sources, where the nuclear chain reaction is controlled by rods made of a neutron absorber. Exceptions to this are the IBR-2 pulsed reactor operated at the Frank Laboratory of Neutron Physics at the Joint Institute of Nuclear Research (Dubna, Russia) which first went critical in 1979 as well as the TRIGA-class reactors such as that in operation at Munich which could work on both steady and pulsed modes. The former facility consists [95] on a pulsed reactor of 2 MW of average thermal power delivering neutron pulses of 10^{16} resulting from a peak power of 1.85 GW. The device consists of a plutonium dioxide core, forming a compact lattice having a multiplication factor just below unity. Reactivity is thus controlled by means of two mobile external reflectors which rotate in opposite directions with different velocities (600 and 300 rpm, respectively). When both reflectors coincide near the reactor core, a power pulse is generated yielding pulses of 340 μs of thermal neutrons. The facility can operate at 5–10 Hz of repetition rate yielding neutron fluxes as high as 2.6×10^{17} of fast neutrons per pulse and 5×10^{15} thermal neutrons at the surface of the grooved water or cold moderators.

In October 2007, the French Nuclear Safety Authorities (Autorité de Sûreté Nucléaire, ASN) officially signed off the Refit Programme, a major initiative by the ILL to ensure the seismic stability of its high-flux reactor. This signature confirmed the successful conclusion of almost 6 years of intensive renovation work. Under such a program, several key reactor components were subjected to nondestructive testing and these tests have confirmed the complete integrity of the reactor system after 35 years of operation. This, together with the ongoing modernization of ILL's instrument suite—the Millennium Programme—has enabled the institute to provide the user community with 14 new or refurbished instruments that, together with marked improvements in infrastructure, have hugely increased the average detection rate across the instrument suite. Such remarkable efforts will help secure the Institute's position at the forefront of neutron science well beyond the coming decade as has been aptly envisaged in a recent document [102].

As regards to the largest power experimental reactor, that is the PIK facility at Gatchina (see Table 2.8), it was expected to start operation within 2012 after a long series of delays in its construction which extends over a period of about 30 years. At the time of writing, licensing problems due to the use of HEU fuel seem to have resulted in yet another delay and startup has again been delayed to 2014.

2.8 FUTURE PROSPECTS

2.8.1 Accelerators

Several developments are taking place at the time of writing concerning accelerator science and technology which will probably have a significant impact on future accelerator-driven projects. Between those, here we select some significant advances within fields not usually visited in reviews of neutron science such as those regarding induction accelerators, fairly recent breakthroughs registered in wakefield accelerators, which use traveling plasma waves as an accelerating structure [103] or developments concerning fixed-field alternating gradient synchrotrons (FFAG) [104]. On a more futuristic vein, recent achievements on laser-plasma acceleration where extremely large acceleration is achieved on mesoscopic length scales as well as the onset of extremely cost efficient methods for the production of usable neutron fluxes will certainly have an impact on the field and thus deserve to delve some time to consider their future potential.

2.8.1.1 Induction Machines

The drive registered in the early 1960s for the development of kiloampere electron beams, that is, well above ranges accessible to RF accelerators, was mainly originated by developments within controlled fusion projects. The main aim was to get an electron ring intense enough to develop a self-magnetic field well in excess of that generated by the applied external fields. Such an avail required beam currents well above the few hundred amperes then available from RF accelerators. Most developments ushered in national laboratories such as Lawrence Livermore National Laboratory, LLNL (Livermore, USA) and Lawrence Berkeley National Laboratory, LBNL (Berkeley, USA) and mostly followed the pioneering ideas of Christofilos and others [105]. The basic concept on which such machines rely on consists in an *induction cell* which is basically a low impedance structure that provides the direct coupling to the beam and becomes the functional analogue to resonant cavities employed within RF accelerators. Such a cell contains a magnetic core that acts as a transformer which, in turn, is driven by a pulse power system. If compared to conventional RF accelerating cavities, induction cells are nonresonant structures which serve as autotransformers allowing the electromagnetic energy from the power sources to flow directly into the beam (i.e., without having to fill the cavities as RF structures do).

Typical performances of present-day induction machines [106] comprise final energies up to 50 MeV and currents up to 50–100 kA, for low duty machines as LIU-30 at VNIEF (Russian Federal Nuclear Center at Sarov, USSR), and 20 MeV and 3 kA for high repetition-rate accelerators such as DRAGON-I at the Chinese Institute of Fluid Physics.

The possibility of employing induction accelerators as injectors to drive a neutron source was pointed out by Keefe and Hoyer [107] back in 1981, although the technology was at the time not mature enough to merit serious consideration. More recent developments within the field of heavy-ion fusion [108] provide, however, hints pertaining the expected performance of such a driver if used for neutron production. For a 5 MW spallation target driven by a 1 GeV proton beam containing a charge of 100 µC per pulse one may dimension [109] an injector delivering a 1.6 µs proton pulses with 60 A current feeding an accelerator using 3 T solenoids and a transport system with a charge density of 2.6 µC m^{-1}. The current delivered at the target would reach some 600 A with pulse lengths of 160 ns. Compared to conventional

RF machines of similar characteristics, an induction driver will surely be a cost-effective solution.

2.8.1.2 Wakefield Accelerators

In plasma-based accelerators, a laser or a high-energy charged particle beam excites a plasma wave that is generated within a confined material and such a wave is then used as the accelerating structure. The physical basis for this phenomenon is grounded upon the longitudinal charge-separation trailing wave generated by the laser or charged particles beam which shows a phase velocity matching the velocity of the laser or charged particle pulses. While the research field has been active for some decades now [110], maximum energy gains achieved using the technique were moderate [111]. However, a significant breakthrough was reported a few years ago [112] when energy-doubling of 42 GeV electrons in a meter-scale plasma was reported.

Up to the present date, plasmas excited by either lasers or bunches of relativistic electrons have been shown to be able to generate acceleration gradients within the range of 10–100 GV m^{-1}. However, the possibility of proton-bunch-driven plasma-wakefield acceleration has been explored by means of numerical simulations that show that a teravolt energy regime could be reached in a single accelerating stage [113]. In fact, after 450 m of acceleration, the electron bunch reaches a mean energy of 0.62 TeV per electron showing an energy spread of about 1% at the highest energies.

Ongoing experiments use the long proton bunches available from the Super Proton Synchrotron ring at CERN and use the self-modulation of the proton bunch in the plasma to excite the wakefields. These experiments will demonstrate the acceleration of externally injected electrons to more than 1 GeV in a plasma 5–10 m long.

The results obtained so far will provide experience and guidance toward the accelerating a high-quality electron bunch to high energy. Plasma-based particle accelerators could be the future technology to make linear particle colliders shorter and therefore more affordable.

2.8.1.3 Fixed-Field Alternating Gradient Synchrotrons

The original ideas to develop a resonant circular accelerator combining constant magnetic fields as cyclotrons and the strong focusing conditions characteristic of synchrotrons ushered in basically at the same time (circa 1952) from Okawa in Japan, Simon in the USA, and Kolomna in the USSR, where the thing was baptized as a phasotron ring [114,115]. A first prototype built in 1956 at the Midwestern Universities Research Association (MURA, Wisconsin) consisted on a machine of 54 cm orbit radius able to accelerate a beam from 20 to 400 keV [116]. The machine was built using eight sectors or superperiods, each consisting of one large and one small magnet. The magnetic field increases with the radius as $B(r) = B_0 \left(\frac{r}{r_0}\right)^k$, $k \equiv \frac{r}{B_a v} \frac{dB_a v}{dr}$ is the angle-averaged field index, and B_0, r_0 are fixed machine parameters. Such a relationship or scaling law forces the orbit shapes to stay the same as particles spiral out. One important consequence of such a scaling law is to avoid beam destabilization to betatron resonances. The field in the smaller magnet is reversed, which then provides the alternating gradient, the advantage of which is to make the orbit circumference about five times larger than it would be for an uniform magnetic field. A spiral sector machine where the magnet edges are twisted which generates a strong edge focus was proposed the next year by D. Kerst. Progress continued leading to the operation at MURA of an electron machine able to deliver a 50 MeV beam in 1961. Although a design of a 12.5 GeV proton FFAG was proposed at MURA, only a design for a scaled down version 720 MeV proton accelerator [117] was actually funded. The machine which contained 16 spirals and a maximum orbit radius of 9 m needed to operate 4 MW of electric power to feed its 72 magnets.

From about the mid-1970s, the field went into a dormant period until the early 1980s. The main drawbacks to continue research on such sort of machines stem from the requirement of rather large, very complex magnets which drove the cost up to unacceptable levels. The activity within the field awoke up in the 1980s when proposals to build next-generation spallation sources gave a new appraisal for such machines as possible proton drivers [118]. A POP machine was built at KEK in Tsukuba, able to deliver a proton beam of 500 keV [119]. In turn, FFAG accelerators have been the focus of recent interest due to their potential medical applications in proton therapy, noninvasive security inspections of cargo containers, acceleration of muon beams, or drivers for accelerator-driven subcritical reactors. On a different vein, FFAGs were also examined as candidates for the acceleration of muons for Neutrino Factories and for Muon Colliders. Within such projects, the task for FFAGs was to accelerate muons to an energy of about 20 GeV, and it had to be done very fast because of the very short lifetime of the particles (2.2 μs).

The early FFAG machines used bending fields which, as stated above, scale as a high power of the radius. The advantage of such a design, called *scaling* FFAGs, is basically the avoidance of betatron oscillations, which are resonances in transverse beam stability leading to poor performance of cyclic accelerators rendering substantial beam losses. Such machines were, however, limited to low energies and modest luminosity until the concept of *nonscaling* FFAGs was introduced. The rationale behind a nonscaling FFAG lies in the fact that one can avoid the deleterious effects of betatron resonances if particles are fast enough so that they have no time to build up to a damaging amplitude. The radial dependence of the magnetic field now becomes somewhat simpler $B(r) = B_0(1 + \frac{k}{r_0}(r-r_0)r)$ which provides greater flexibility with respect to the scaling case and eliminates higher field orders such sextupole and octupole fields. The remarkable features of these machines are that they generate a much smaller momentum compaction than scaling FFAG and hence lead to a smaller orbit excursion. In turn, such reduced orbit excursions can be achieved with linear magnetic fields, which open up the possibility of both the use of large dynamic apertures and the ability to use higher RF frequencies than scaling FFAGs. The latter fact then enables the acceleration up to relativistic energies without having to recourse to a change of the RF frequency.

Getting such ideas down to the ground took, however, quite some time, and to the authors knowledge, the first nonscaling FFAG ever built codenamed EMMA—the Electron Model of Many Applications—[104] has been built at the STFC Daresbury Laboratory in the UK and achieved acceleration in March 2011, with the purpose of serving as an electron model for a muon accelerator which is planned for a neutrino factory. The machine is built using a double lattice from 42 cells of about 40 cm length yielding a fairly compact ring structure of ≈ 16.6 m of circumference and should be able to raise an electron beam up to 20 MeV. The acceleration is provided by 1.3 GHz RF fields injected from inductive output tube into 19 cavities.

The future use of nonscaling FFAGs as proton drivers for intense neutron sources will be surely dependent upon the outcome of some ongoing projects such as PAMELA (Particle Accelerator for MEdicaL Applications) or CONFORM (COnstruction of a Nonscaling FFAG for Oncology, Research and Medicine) aiming to build a proton/carbon accelerator for medical purposes. A recent reappraisal of the use of nonscaling FFAGs as proton drivers is given in Ref. [120]. As in the case on wakefield accelerators, a successful nonscaling FFAG will help to significantly bring down the costs of construction of future accelerator-driven neutron sources.

2.8.1.4 New Cyclotron Concepts

A new concept for a multicavity proton cyclotron accelerator has been proposed [121]. It consists of a cascade of RF cavities embedded in a nearly uniform axial magnetic field. The resonant frequencies differ in adjacent cavities by a fixed amount $\Delta\omega$ so that protons injected into the first cavity at intervals of $\Delta T = 1/\Delta\omega$ or integer multiples thereof can experience continuous energy gain throughout the structure provided the relative phases of the fields in each cavity are properly adjusted. As protons can make a substantial number of orbital turns in each cavity, the voltage gain can be much larger than the product of the peak electric field and the cavity length. Also since the protons are heavily constrained to follow spiral orbits in the strong magnetic field, no further focusing magnets are needed. Put into real numbers, the calculation shows that for an average current of 122 mA, using a frequency separation $\Delta\omega = 8$ MHz, a final energy of 952.7 MeV can be reached employing for the purpose an average gradient of 39.7 MV m^{-1}. The design uses just eight cavities to reach such a final energy and the total length of the device reaches just 25.2 m. It goes without saying that if a doable prototype can be built out of such a concept, it will surely revolutionize the field.

2.8.1.5 Neutron Production by Laser-Driven Plasmas

Recent achievements in ultraintense laser technology ushered in a new generation of particle accelerators. The fully relativistic plasmas generated during the interaction of multi-TW laser pulses are now seriously considered as a brilliant source of GeV electrons, MeV ions, as well as versatile photon beams. The resulting ion beams show several remarkable characteristics such as very low emittance and ultra-high peak currents, while requiring acceleration lengths of just several tens of micrometers. A very relevant breakthrough for the potential of such systems has recently been demonstrated with the production of electron and ion beams showing a narrow distribution of energies.

A proposal for proton acceleration by the use of circularly polarized laser pulses with powers of 10^{21}–10^{23} W cm^{-2} has been considered [122]. The laser beam is then directed to a target composed of a thin dense proton-rich foil located at the front followed by an underdense gas region behind with an effective Z/A ratio of the order of 1/3. As the narrowly collimated laser pulse passes through the thin foil and propagates in the underdense gas region, it excites high-amplitude electrostatic fields moving at a high speed, which appear like a moving double layer. The preaccelerated protons can get trapped and accelerated in the moving double layer and quasi-monoenergetic proton beams at tens of gigaelectron-volts result, provided that laser intensity and plasma density are properly chosen. Another proton production based upon a two-stage acceleration scheme has also been reported by Pfotenhauer and others [122]. The proposal is grounded on a cascaded acceleration concept, leading to the generation of spectrally controlled ion beams. A megaelectron-volt proton beam produced during a relativistic laser–plasma interaction on a thin foil target is spectrally shaped by a secondary laser–plasma interaction on a separate foil, reliably creating well-separated quasi-monoenergetic features in the energy spectrum. The observed modulations are fully explained by a one-dimensional (1D) model supported by numerical simulations. The findings demonstrate that laser acceleration can, in principle, be applied in an additive manner. On a more futuristic vein, a recent report [123] ponders on the potential to develop sources for neutron-scattering science leaning on a merger with the rapidly developing field of inertial fusion energy. Their results suggest that such a route could in future provide a major stepchange in performance.

2.8.1.6 Tabletop Production of Neutrons

Within the past few years, several reports have appeared [124] where neutron production making use of the electric fields generated by pyroelectric crystals has been proven. Their principle of operation relies upon the use of a pyroelectric crystal (i.e., LiTaO$_3$), the polarization of which is controlled by means of changes in the crystal temperature. The change in polarization creates free charges on the crystal faces which, because of the small capacitance, create a high electrical potential between such a crystal face and another face placed at ground potential. If such a crystal, or in cases, a pair of such crystals are placed within low-pressure deuterium gas, the net result consists in gas ionization leading to ion formation and subsequent acceleration of such ions inside the electric field. If such deuterium ions are then directed to struck a deuterated or tritiated target, production of measurable fluxes of neutrons arising from d–d and d–t reactions is produced. The setup using two crystals can double the acceleration potential and thus increase neutron production. Such a device, which operates in the absence of any power supply to generate the accelerating field, has been shown to generate X-rays of some 200 keV as well as 10^5 neutrons at 2.5 MeV per heating cycle. Results reported from LLNL by Tang and others [124] show that by incorporating a pulsed, spark-plasma ion source, delivering pulses of a few hundreds of nanoseconds the system increases its neutron production up to 10^4 per pulse thus reaching the 10^{10} neutrons per second. Such a yield compares to that obtained using small, laboratory-based electrostatic accelerators quite some time ago [45], which was based upon the acceleration of deuterons up to energies within the 100–300 keV range. Impact into solid deuterium or tritium targets showed that monoenergetic neutron beams were produced by means of ^{2}H(d,n)^{3}He and ^{3}H(d,n)^{4}He at right angles from the deuteron beam. The schemes here referred to are, however, able to reach similar neutron fluxes without employing any accelerating structure beyond the ion source.

2.8.2 Hybrid Systems

Significant efforts have been devoted since the 1990s to develop high-power accelerator sources coupled to an array of subcritical fissile or fertile materials with the purpose of either transmutation of nuclear waste (ATW), production of tritium (APT) or even, accelerator-driven energy production (ADEP). Common to all of these is the requirement of an electron or a high-power proton or deuteron accelerator operating in continuous wave mode with optimal energies within 0.8–1 GeV (minimum cost–performance optimized) and currents from 10 mA needed for ATW or ADEP to 100 mA for APT and up to 300 mA to breed fissile materials. Spallation targets will either contain a heavy metal (W, Hg, PbBi) or depleted uranium and will be surrounded by an array of fissile or fertile fuel assembly. Operation with a subcritical regime is envisaged for multiplication constants with typical values of $k \simeq 0.9$–0.955.

Practical experience with these kind of systems mostly rely on work carried out at demonstrators such as the YALINA facility at the Radiation Physics and Chemistry Problems Institute in Minsk-Sosny, Belarus as well as experiments carried out at operating fast reactor facilities such as MASURCA at Cadarache, France. The former consists on a zero-power, subcritical assembly which may operate in a *thermal* mode driven by a conventional neutron generator a static ^{252}Cf source, or using small deuteron accelerators that yield neutrons generated from reactions, d(d,n)^{3}He, d(t,n)^{4}He or in a *booster* configuration where the external sources can be located in different places relative to the core center, which allows to study criticality conditions up to $k \leq 0.98$. The facility can also be operated in pulsed mode delivering pulses within 0.5–100 µs and repetition frequencies within the interval 1 Hz to 10 kHz. When operated in the booster configuration, a large gain in neutron production from the flux values injected by the external source (10^{10} n s^{-1} for the d–d reaction and 1.5×10^{12} n s^{-1} for the d–t reaction) is achieved [125]. Also, experiments carried out within the MUSE program (multiplication with an external source) have been carried out at MASURCA using the GENEPI neutron generator (Générateur de Neutrons Pulsés Intenses) which also consists on a small deuteron accelerator able to deliver a 50 mA beam onto a deuterium or tritium titanium target (TiD or TiT, respectively), in pulses shorter than 1 µs and repetition rates that can vary from a few hertz up to 5 kHz [126].

The MYRRHA project now under development at SCK-CEN in Belgium aims to build a flexible fast spectrum research reactor (50–100 MW$_t$h), able to operate in subcritical and critical modes. It contains a proton accelerator of 600 MeV, a spallation target, and a multiplying core with mixed oxides (Pu-U, MOX) fuel, cooled by liquid lead–bismuth (Pb–Bi). Expected neutron production under subcritical operation reaches 2×10^{17} n s^{-1} which compares to that expected for ESS project but comes at a far reduced cost.

A few proposals for a facility of this kind to be used as a neutron source have already appeared [127,128]. These refer to either a facility based upon a proton accelerator or another fed by an electron accelerator for the production of photoneutrons. Furthermore, design studies carried out at the Research Reactor Institute of the Kyoto University, Japan, led to the conception of an accelerator-driven subcritical reactor as a future neutron source [128]. The results, on the basis of a proton accelerator with energies from 100 MeV up to 1 GeV and a current of 1 mA impinging on a target made from W, Pb, or depleted uranium, show that a thermal neutron flux in the core of approximately 10^{14} cm^2 s^{-1} under the 5 MW thermal output can be obtained with multiplication constants $k \leq 0.98$.

Our current appraisal of the advantages of such a concept to be used as a neutron source for condensed matter research if compared with last-generation spallation facilities portrays them as cost-effective solutions for either continuous or long-pulse (a few milliseconds) sources. In contrast, short- (i.e., less than a few hundred microseconds) neutron pulses cannot be provided by such facilities due to the long characteristic times involved in the neutron multiplication processes of these kinds of sources.

2.8.3 Reactors

Reactor technology has reached full maturity [129]. Prospects of reaching fluxes beyond 10^{16} n cm^{-2} require power densities of the order of 10 MW l^{-1}

and are well beyond the cooling capabilities of light or heavy water. Innovative reactor concepts have been advanced in the past as well as within current efforts to develop accelerator-driven subcritical assemblies. Many of these concepts seem to be feasible, and its operational cost would surely come well below that of multi-megawatt-range spallation sources.

While there are not at present ongoing projects to develop high-flux research reactors, the experience gained with the remarkable exploitation of the NIST reactor as well as the coming online of facilities like OPAL suggests that a global strategy for the fruitful development of the technique ought to consider the deployment of medium-sized installations within regions devoid of neutron-scattering facilities.

At this stage, it is worth mentioning the capabilities of research reactors which started in the 1960s with developments headed by General Atomics leading to a series of turn-key small-scale reactor facilities. The TRIGA-class MarkII (Training, Research, Isotopes, General Atomics) reactors are now installed at some 24 countries with power levels up to 16 MW in static operation and up to 22 GW in pulsed mode. Such reactors aim to production of radioisotopes for medical, industrial, and environmental uses, metallurgy and chemistry, implementation of nuclear analytical techniques such as neutron activation analysis and nondestructive examination techniques, as well as carrying out basic research programs in solid-state physics, including basic neutron-scattering instrumentation. The potential of such sources is best exemplified by pioneering results from the facility at Mainz, referred to in the previous sections. Also attention should be paid to the INVAP reactor project RA-10 (see Table 2.8) now under development, which constitutes a recent important addition to these capabilities. The design profits from experience gained during the design and construction of OPAL and aims to the development of a multipurpose facility suitable for radioactive isotope production, material and fuel irradiation, neutron techniques, and silicon doping. The flux characteristics of $\approx 10^{10}$ n cm^{-2} s^{-1} at the reactor face make it a rather interesting option as a source for neutron-scattering purposes. In fact, thermal or cold neutron fluxes of the order of 10^9 n cm^{-2} s^{-1} are estimated at 50 m from the reactor core, which, if confirmed, would make such facilities entirely competitive and would then fit within the global two-tiered structure of neutron facilities proposed by relevant international organizations.

2.9 NONNEUTRON-SCATTERING USES OF NEUTRON SOURCES

Although most of the neutron sources here reviewed have as their main mission neutron-scattering research, a number of additional capabilities have been successfully exploited within some of the facilities. Here, we will briefly mention some of those taken as the most relevant.

2.9.1 Isotope Production, In-Vessel Irradiation, γ-Radiation, and Neutron Activation Analysis

The main purpose of reactors built back in the 1960s such as HFIR was the production of transuranium isotopes such as plutonium and curium. To such an avail, its design included numerous vertical irradiation facilities of various sizes located within the center of the reactor core or fluxtrap, throughout the beryllium reflector. Such facilities are built using pneumatic tubes which enables insertion and removal of samples while the reactor is operating.

Isotope production of the heavy elements such as ^{252}Cf as well as other α-emitting radionuclides such as ^{225}Ac or ^{211}At or active research into increased production of ^{3}He constitutes some of the stated priorities of the Isotope Development and Production for Research and Applications subprogram of the Office of Nuclear Physics in the U.S. Department of Energy Office of Science [130].

Irradiation facilities for isotope production have also been installed within the OPAL reactor core (see Figure 2.26), with the stated aim of achieving commercial production of radionuclides employed in nuclear medicine such as ^{99}Mo and ^{131}I.

Several facilities designed to carry out research activities into severe neutron damage to materials as well as irradiation of large samples capable to accommodate silicon ingots for doping for the semiconductor industry have been installed at OPAL and are available at reactor and quasicontinuous sources such as SINQ. Such facilities are built from tubes within the reflector vessel that access neutron fluxes of varying wavelengths as required for the material being irradiated and its end purpose. Also, high γ doses for studies of the effects of radiation on materials are available at installations such as HFIR where a γ Irradiation Facility has been built which employs intense γ-radiation from spent fuel elements.

Neutron activation and delayed neutron activation facilities, used for chemical analysis purposes and the amount of uranium in samples, respectively, are also available at most reactor-based sources.

2.9.2 Nuclear Physics and Engineering: Astroparticle Physics, Nuclear Structure and Reactions, and Transmutation of Nuclear Waste

A number of Nuclear Physics applications have been developed at both reactor- and accelerator-based facilities and are at present being exploited. Among them, studies on stellar nucleosynthesis are particularly appealing. Some answers to fundamental questions such as why is iron the limit for stellar nucleosynthesis in red giants? and where did heavier elements than iron come from? can be derived from experiments being carried out at neutron production sources. In fact, our knowledge on the processes leading to the stellar synthesis on nuclei beyond $Z=50$ (Fe) portrays them in terms of neutron capture and disintegration steps within stellar media where available neutrons exhibit a Maxwellian spectrum with energies below 100 keV generated from He burning steps, either along the valley of stability (known as slow-process) or far from stability as may happen within supernovae (referred to as the fast process). Experiments being carried out at reactor-based sources as well as at accelerator facilities where a low-energy (2–5 MeV) proton beam impinges on a low-energy neutron generation target such as ^{7}Li are expected to show light on some of these questions.

The excited states of nuclei far from stability in nuclei with even numbers of protons and neutrons show a transition from a spherical to deformed shape which occurs when the number of neutrons increases from 58 to 60. This shape change concerns the whole shape of the nucleus and is not simply due to the extra volume added by the two neutrons. This onset of deformation is particularly sudden in the Sr and Zr nuclei. Experiments using γ-ray spectroscopy tools such as those carried out at ILL have significantly contributed to our understanding of the structure of such nuclei.

Studies on the basic laws and physical phenomena associated with nuclear reactions continue to attract considerable attention from both basic science and possible applications. On the one hand, studies of fission processes have now reached a pretty refined level. In fact, studies on dynamic phenomena such as collective nuclear motions of the fissioning system upon neutron capture as well as subsequent processes that take place within timescales of the order of 10^{-21} s are currently being carried out by means of detailed analysis of the spectra of fission products measured at reactor facilities. On the other hand, studies pertaining the details of intra- and internuclear processes usually referred to as spallation reactions continue to be carried out at some of the smaller facilities.

The idea of transmutation of the radioactive waste produced in nuclear reactors into stable isotopes or less hazardous and shorter-lived radionuclides has been discussed in the open literature since 1958 [131]. Since that date, several transmutation concepts have been discussed and include the use of thermal and fast-fission reactors and high-intensity particle accelerators. A recent review of the latter class of facilities is given in Ref. [132].

Some of the new accelerator facilities such as J-PARC, host of the MLSF source, also include a facility dedicated to such endeavor. In addition, purpose built facilities such as MYRRHA (see Table 2.6) are at present at project development steps.

2.9.3 Hadron Physics: Neutrino-Related Phenomena

Nuclear reactors generate electron antineutrinos from β-decays of fission products (mostly from ^{253}U, ^{238}U, ^{239}Pu, and ^{241}Pu) in numbers of about $10^{20}\bar{\nu}_e s^{-1}$ per nuclear core [133]. The number of decays is given by the reactor power so that if its thermal power is measured, and account made that six electron antineutrinos result per fission event, the total neutrino flux reaching a given detector can be estimated. The energy distribution of a neutrino beam leaving a reactor core can also be calculated and it has been confirmed that such beams of $\bar{\nu}_e$ particles have energies below some 12 MeV. The emitted neutrinos can be detected by the inverse-β-decay process, that is being captured by the free protons within some target following:

$$\bar{\nu}_e + p \rightarrow e^+ + n \qquad (2.56)$$

Pioneering experiments were carried out in the 1980s and 1990s at a few tens of meters from reactor cores such as that from the ILL Grenoble. The main interest of such measurements was to put some bounds on the neutrino oscillation parameters. This refers to the convincing evidence for flavor conversion of atmospheric, solar, reactor, and accelerator neutrinos, which put in other words means that electron ν_e, muon ν_μ, and tau ν_τ neutrinos interconvert while traveling on free space. Reactor neutrino experiments measure the survival probability $P(\bar{\nu}_e \rightarrow \bar{\nu}_e)$ of the antineutrinos emitted by the reactor at a given distance. Because of the usually short distances considered, forward scattering from matter electrons can be disregarded and therefore any significant loss could be attributed to oscillatory phenomena. Also, the quest to determine some of the oscillation parameters led to the setup of the Karlsruhe–Rutherford Medium Energy Neutrino using for the purpose the intense $\bar{\nu}_\mu$ beams generated by pion and muon decays. Both sets of experiments did not yield convincing evidence of or $\bar{\nu}_\mu \rightarrow \bar{\nu}_e$ oscillations, but served to clear the ground for new experiments now being run at some other facilities such as the Daya Bay (China) or Chooz (France) reactor experiments.

2.9.4 Fundamental Physics: Foundations of Quantum Mechanics, Effects of Gravity on Isolated Particles, Search for Dark Matter Using Ultracold Neutrons, Tests, and Validations of the Standard Model of Particle Physics

The development of neutron interferometry techniques has enabled fundamental tests on quantum mechanics to be carried out experimentally at some neutron sources. In particular, developments within a field now referred to as "neutron quantum optics" have now opened a new field of research into fairly basic issues regarding nonlocality in quantum mechanics which may have some practical relevance for future quantum cryptographic systems [134].

The availability of intense ultracold neutron beams as well as improvements in UCN storage devices at several neutron centers has ushered in a new era in which tests pertaining fundamental issues such as the identification of feasible mechanisms giving rise to dark matter or detailed studies on the quantum states of neutrons in the Earth's gravitational field.

There are a number of stringent tests on predictions made on the ground of the Standard Model of Particle Physics that have been made at some neutron facilities and, in fact, the interest in pursuing such studies is alive and well at the time of writing. Between those, the measurement of accurate bounds for the neutron electric dipole moment carried at the ILL and precision experiments to accurately measure the asymmetry parameters describing the correlation between the spin of the decaying neutron and the momentum of the resulting electron antineutrino. Intense cold neutron beams with a polarization up to 99.7%, which yield a lower bound for the mass of the W_R boson of 290.7 GeV c^{-2}, merit to be mentioned.

2.9.5 Use of Muon Beams for Condensed Matter and Fusion Research

Positive and negative muons are leptons with spin 1/2 carrying one elementary electric charge. Such particles may be considered as heavy electrons since their rest mass comes to be some 207 times that of the electron or light protons because it is about one-ninth of the proton rest mass. Because of its sizable magnetic moment, that for a positive muon is about three times larger than that of the proton, muons can be used as very sensitive microscopic magnetometers.

Both positive and negative muons can be produced in relatively large quantities from the decay of pions and kaons resulting from proton interactions with targets at energies beyond several hundreds of megaelectron-volts. Once implanted into the sample under study, positive (negative) muons decay yielding a positron (electron) and two neutrinos, after an average lifetime of 2.2 μs. The emitted positrons (or electrons) are preferentially directed following the muon spin, which in turn is highly polarized. Muon polarization can thus be followed from measurements of the decay of positrons or electrons.

Positive muons can also yield to the formation of exotic atoms such as muonium (Mu) by picking an electron from the sample material. Furthermore, the hyperfine interaction between muon and electron spin makes of Mu an extremely sensitive magnetic probe, which can be used as a substitute for hydrogen in molecular materials, giving much detailed information on the structure, dynamics, and reactions of the samples under study.

Negative muons are attracted by the atomic nuclei, form muonic atoms (i.e., atoms in which one electron is replaced by the negative muon), a fact that has been used to demonstrate the muon-catalyzed fusion process.

Because pion production targets only take a small percent of the proton beam, facilities to carry out muon spectroscopy have been installed at major facilities such as ISIS, PSI, TRIUMF, and some others.

ACKNOWLEDGMENTS

The support of the ESS-Bilbao team is warmly acknowledged. Dr. J.E. Stovall from LANL and Dr. S.J. Jolly from UCL are kindly acknowledged for allowing us to reproduce some of their unpublished material.

APPENDIX A   SOME BASIC RELATIONSHIPS

Most measurements performed at accelerator-driven sources rely on the time-of-flight technique where the time-averaged neutron spectrum $N(t)$ is measured within discrete time channels δt as a function of the time t_0 taken to travel a total flight path $L_0 = L_i + L_f$ between the moderator and the detector. The time of flight t_0 is thus proportional to the neutron wavelength

$$t_0 = m_n L_0 \lambda_n / h \tag{A.1}$$

where m_n stands for the neutron mass. The spectrum is measured at discrete time bins of width δt or wavelength $\delta \lambda_n = h \delta t / m_n L_0$. The neutron counts recorded by each time bin can also be written in terms of time or wavelength distributions as

$$N(t)\delta t = N(\lambda_n)\delta \lambda_n \tag{A.2}$$

so that the neutron spectrum per unit wavelength is given by

$$N(\lambda_n) = N(t)\delta t / \delta \lambda_n = m_n L_0 N(t) / h \tag{A.3}$$

If we now allow for a finite moderator area S_M as well as for a finite solid angle $1/4\pi L_0^2$ subtended by the fact that the detector is located at a position L_0 at the end of a neutron guide, we arrive at an expression giving the time-averaged counts from all directions crossing a unit area of the moderator surface per unit wavelength and unit time as

$$N_{L_0}(\lambda_n) = \frac{S_M N_0(\lambda_n)}{4\pi L_0^2}, \tag{A.4}$$

where $N_0(\lambda_n)$ refers to a time average over the number of neutrons crossing a unit area at the moderator surface and is given per unit wavelength and per unit time. The result can also be recast in terms of a flux $\Phi_0(E_n)$ of neutrons leaking the whole moderator surface per unit solid angle as $\Phi_0(E_n) = S_M N_0(E_n)/4\pi$.

A useful quantity named as *lethargy* is defined in terms of a relative fast-neutron energy, typically $E_{fast} = 1$ MeV, as

$$U_l = \ln(E_{fast}/E_n) \tag{A.5}$$

which goes from zero at the beginning of the thermalization process and increases as the neutron slows down. An increase in lethargy of one unit thus corresponds to a decrease of neutron energy by $1/e$. The lethargy increment δU_l is given by

$$|\delta U_l| = \left|\frac{\delta E}{E}\right| \tag{A.6}$$

and is thus related to the fractional resolution in the incident neutron energy. The total number of collisions of a neutron with the moderating material to achieve a final energy E_i is given by

Chapter 2 Neutron Sources

$$nU_1 = \ln\left(\frac{E_f}{E_i}\right) \quad (A.7)$$

For slowing down in moderators composed by atoms heavier than the neutron mass, the time dependence of lethargy growth can be written in approximate form as

$$\frac{dU_1(t)}{dt} = \varepsilon v \Sigma_s \quad (A.8)$$

$$U_1(t) = 2\ln\left[1 + \frac{1}{2}\varepsilon t v_f \Sigma_s\right] \quad (A.9)$$

where ε stands for the mean increment of lethargy per collision and Σ_s is the total (macroscopic) scattering cross section. The next step would be to estimate the distance a neutron with lethargy U_1 has traveled from its point of production. From basic kinetic theory, we know that the mean square distance traveled between collisions is

$$\langle r_1^2 \rangle = \int_0^\infty dr p(r) r^2 = 2l^2 \quad (A.10)$$

where l stands for the collision mean free path. The mean square distance traveled after n collisions becomes $\langle r_n^2 \rangle 2 = 2nl^2$, while the mean number of collisions in a time t is vt/l. For times that are large enough, the mean square distance may be written in terms of the lethargy as

$$\frac{1}{6}\langle r^2(U_1) \rangle \simeq \frac{1}{3}\frac{lU_1}{\varepsilon \Sigma_s} \quad (A.11)$$

The equation written above formally obeys a diffusion equation. If we now consider a slowing-down density $\phi(r, U_1)$ rather than single neutrons, the time evolution of such density assuming weak loses satisfies the equation

$$\nabla^2 q(r,\tau) = \frac{\partial q(r,\tau)}{\partial \tau} \quad (A.12)$$

usually referred to in nuclear engineering literature as the "age equation," where the neutron or "Fermi age" is given by

$$\tau(U_1) = \int_0^{U_1} \frac{D(U_1')dU_1'}{\varepsilon \Sigma_s(U_1')} \quad (A.13)$$

where $D(U_1')$ stands for a diffusion coefficient at lethargy U_1'. The function $\tau(U_1)$ has dimensions of surface and serves to define a slowing-down length $L_1^2 = \langle r^2(\tau) \rangle / 6$. The slowing-down time is given by

$$t_s(\tau) = \tau/\langle Dv \rangle \quad (A.14)$$

where $\langle Dv \rangle$ stands for the weighted average of $D(u)v$ over the lethargy range. The latter quantity is usually referred to as the conventional diffusion coefficient on the grounds of the Fick's diffusion law derived within the theory of Brownian motion.

Other relevant moderator parameters are the slowing-down power and the moderating ratio defined as

$$S_p = U_1 \Sigma_s, \quad N_{MB}(\lambda)/N_{epi}(\lambda) = U_1 \Sigma_s / \Sigma_{abs} \tag{A.15}$$

in terms of the total (macroscopic) scattering cross section at epithermal energies and the spectrum-averaged macroscopic absorption cross section of thermal neutrons.

Treatment of the details of the time dependence of the slowing-down processes requires the specific evaluation of neutron velocities in all directions for which recourse is made to neutron transport theory, a sketch of which is given in Appendix B.

APPENDIX B THE TRANSPORT EQUATION FOR NEUTRONS

For Appendix B.[1] Here we sketch some of the steps followed in the derivation of a somewhat simplified form of the neutron Boltzmann equation, also called the neutron transport equation. It has been derived to allow the characterization of a relatively small number of neutrons colliding within a large number of nuclei. For such an avail, the neutron is considered as a point particle, experiencing deflection from or capture by a given nucleus.

In general terms, the classical description of the neutron motion requires the use of six independent variables which are the three components of its position vector r, and those of its velocity v. We also consider an additional vector Ω that specifies a point on the surface of a fictitious sphere of unit radius surrounding the neutron and pointing in its direction of travel.

To start with, we define several magnitudes. The neutron flux at time t is defined as

$$\Psi(r,v,t) = vN(r,v,t) \tag{B.1}$$

where t is time as measured from some reference and $N(r,v,t)$ is neutron density.

The density of neutrons for each energy range within a certain volume follows

$$\frac{\partial N}{\partial t} + \frac{\partial N}{\partial r}\frac{\partial r}{\partial t} = \text{Source} - \text{losses} \tag{B.2}$$

The source term could comprise neutrons which lose energy after a collision. Its form for fission reactions could be written as

$$\frac{\chi(v)}{4\pi} \int_0^{4\pi} \int_0^{v_0} v \Sigma_f(r,v') \Psi(r,v',t) dv' \tag{B.3}$$

---

1. The reader should notice that because of the rather specific topics covered within this appendix, adherence to the symbols convention employed for the rest of the book is here relaxed.

where $\chi(v)$ is the probability of generating a neutron with velocity v from a fission induced by a neutron with velocity v'. Σ_f is the fission cross section for energy E and v is the fission performance.

A collision leads to a loss of a neutron within a given velocity and a gain of another with a different velocity. If $\Sigma_s(r,v')$ is the scattering cross section for v' and $p(v' \to v)$ is the probability of generating a neutron with velocity v after scattering of a neutron with v', the scattering source term will be

$$\int_0^{v_o} \int_0^{4\pi} \Sigma_s(r,v')p(v' \to v)\Psi(r,v',t)\mathrm{d}v' \quad \text{(B.4)}$$

The total cross section comprises the absorption and scattering cross sections and reads

$$\Sigma_t(r,v)\Psi(r,v,t) = \Sigma_s(r,v)\Psi(r,v,t) + \Sigma_a(r,v)\Psi(r,v,t). \quad \text{(B.5)}$$

Inserting all terms into Equation (B.2), we obtain the differential form of the transport equation which reads as

$$\frac{1}{v}\frac{\partial \Psi(r,v,t)}{\partial t} + \Omega \nabla \Psi(r,v,t) = \frac{\chi(v)}{4\pi}\int_0^{4\pi}\int_0^{v_0} v\Sigma_f(r,v')\Psi(r,v',t)\mathrm{d}v' \\ + \int_0^{v_o}\int_0^{4\pi}\Sigma_s(r,v')p(v' \to v)\Psi(r,v',t)\mathrm{d}v' + S(r,v,t) - \Sigma_t(r,v)\Psi(r,v,t) \quad \text{(B.6)}$$

Such an equation has no known closed-form solution so that some simplifying assumptions are often required to find a solution. The usual way to proceed is to drop the velocity direction Ω as an independent variable. This means that neutrons can travel with the same probability in each direction. Once such assumption is taken, the previous expression can be written as

$$\phi(r,v) = \int_{4\pi} \Psi(r,\Omega)\mathrm{d}\Omega \quad \text{(B.7)}$$

The neutron current is defined as

$$J(r) = \int_{4\pi} \Omega\Psi(r,\Omega)\mathrm{d}\Omega \quad \text{(B.8)}$$

And the leak rate can be described by

$$\text{Leak rate} = \int_s J(r)\mathrm{d}S = \int_v \nabla \cdot J(r)\mathrm{d}V \quad \text{(B.9)}$$

Introducing these new variables in the expression (B.6)

$$\frac{1}{v_t}\frac{\partial \phi(r,t)}{\partial t} = S(r,t) - \sum_a(r)\phi(r,t) - \nabla \cdot J(r,t) \quad \text{(B.10)}$$

If the current term can be described by Fick's law, we have the following equations:

$$J = -\frac{1}{3\Sigma_s}\nabla\phi \qquad (B.11)$$

$$\nabla \cdot J = \nabla \cdot \frac{-1}{3\Sigma_s}\nabla\phi = \frac{-1}{3\Sigma_s}\Delta\phi \qquad (B.12)$$

Substitution into Equation (B.10) gives the complete expression of the monoenergetic neutron diffusion equation:

$$\frac{1}{v_t}\frac{\partial\phi(r,t)}{\partial t} = \chi(v)\int_0^{v_0} v\Sigma_f(r,v')\phi(r,v',t)dv'dr \\ + S(r,t) - \Sigma_a(r)\phi(r,t) + \frac{1}{3\Sigma_s}\Delta^2\phi(r,t) \qquad (B.13)$$

This expression is only valid under conditions stated at the beginning of the appendix which still are relevant for reactor physics as well as for most neutron sources.

If we apply this equation to a reactor in steady-state conditions with "no fission sources" ($S(r,t)=0$), the sum of absorptions and losses must equal the fission production rate. The neutron economy of the reactor is specified starting from the multiplication constant k defined as

$$k = \frac{\int_r \int_0^{v_0} v\Sigma_f(r,v')\phi(r,v',t)dv'dr}{\int_r \Sigma_a(r)\phi(r)dr + \int_r \frac{1}{3\Sigma_s}\Delta^2\phi(r)dr} \qquad (B.14)$$

If criticality is achieved, and thus k becomes slightly larger than unity then the neutron flux will grow at a rate

$$\varphi(r,t) = \varphi(r,0)\exp[t/T] \qquad (B.15)$$

where the reactor period is given by

$$T = l_\infty/\rho k \qquad (B.16)$$

where ρ stands for the reactivity and $l_\infty = 1/(\Sigma_a v_0)$ is the mean lifetime of a thermal neutron, a quantity typically of the order of 10^{-3}–10^{-4} s. In a real situation, the exponential growth predicted by Equation (B.15) does not take place because a relatively small fraction (some 0.65% for ^{235}U) of the fission neutrons are delayed, which then makes the multiplication constant k to comprise both prompt and delayed neutrons simultaneously ($k = k_p - k_d$). This makes the reactivity ρ to be dependent upon several groups of precursors of delayed neutrons and transforms the problem into one specified by a time-dependent set of diffusion equations (i.e., as many as the number of precursors). The specific values of their roots explain how reactor control is easily achievable by mechanical means. A detailed discussion on these issues is given in chapter 5 of Ref. [6].

REFERENCES

[1] Carpenter JM, Yelon WB. Methods of experimental physics, vol. 23A. Orlando, FLA, USA: Academic Press; 1986. p. 99. [chapter 2].
[2] Windsor CG. Pulsed neutron scattering. London: Taylor and Francis Ltd.; 1981.
[3] Data mostly taken from publicly accessible web-pages of the relevant neutron centers as listed in Ref. [5]. A subset of the data here displayed has been taken from tables and graphs prepared by the directorate of the Institut Laue Langevin, Grenoble (France).
[4] See *Development opportunities for small and medium scale accelerator driven neutron sources*, IAEA-TECDOC-1439, Vienna 2005 and *Working paper on: Improved production and utilization of short pulsed, cold neutrons at low-medium energy spallation neutron sources*, Report of the 2nd Research Coordination Meeting, IAEA, Vienna 2009.
[5] See http://www.ncnr.nist.gov/nsources.html [last accessed on March 11, 2013].
[6] Byrne J. Neutrons, nuclei and matter: an exploration of the physics of slow neutrons. London: IOP Publishing Ltd.; 1994.
[7] Butler ST. Phys Rev 1957;106:272–86.
[8] Tilquin I, Froment P, Cogneau M, Delbar T, Vervier J, Ryckewaert G. Nucl Instrum Methods Phys ResA 2005;545:339–43.
[9] Lavelle CM, Baxter DV, Bogdanov A, Dercnchuk VP, Kaiser H, Leuschner MB, et al. Nucl Instrum Methods Phys ResA 2008;587:324–41.
[10] The IFMIF Engineering Validation and Engineering Design Activities (EVEDA) is now building a test accelerator at Rokkasho/Aomori, Japan aiming at the specification of a detailed, complete and fully integrated engineering design of IFMIF and at validating continuous and stable operation of prototypes of each IFMIF subsystem.
[11] IFMIF International Team. IFMIF Comprehensive Design Report, International Energy Agency; 2004.
[12] Sato T, Shin K, Yuasa R, Ban S, Lee H. Nucl Instrum Methods Phys Res A 2001;463:299–308.
[13] Lynn JW. Contemp Phys 1980;21:483–500.
[14] Swanson WP. Radiological safety aspects of the operation of electron linear accelerators: IAEA Technical Report Series 188. Vienna: IAEA; 1979.
[15] Guber KH, Bigelow TS, Ausmus C, Brashear DR, Harvey JA, Koehler PE, et al. In: Bersillon O, Gunsing F, Bauge E, Jacqmin R, Leray S, editors. Proceedings of international conference on nuclear data for science technology, ND 2007. EDP Sciences; 2008. p. 441–3.
[16] Ene D, Borcea C, Kopecky S, Mondelaers W, Negret A, Plompen AJM. Nucl Instrum Methods Phys Res A 2010;618:54–68.
[17] Bedogni R, Quintieri L, Buonomo B, Esposito A, Mazzitelli G, Foggetta L, et al. Nucl Instrum Methods Phys Res A 2011;659:373–7.
[18] Bohr N. Nature 1936;137:344–8. Weisskopf V. Phys Rev 1937;52:295–303.
[19] O' Connor PR, Seaborg GT. Phys Rev 1948;74:1189–90.
[20] Serber R. Phys Rev 1947;72:1114–5. Cuningham BB, Hopkins HH, Lindner M, Miller DR, O'Connor PR, Perlman I, et al. Phys Rev 1947;72:739.
[21] Metropolis N, Bivins R, Storm M, Turquevich A, Miller JM, Friendlander G. Phys Rev 1958;110:185–203. Dostrovsky I, Rabinowitz P, Bivins R. Phys Rev 1958;111:1658–76.
[22] Mancusi D, Charity RJ, Cugnon J. Phys Rev C 2008;82:044610. See also Kowalczyk A. Proton induced spallation reactions in the energy range 0.1–10 GeV. Ph.D. Dissertation, Cracow: Jagiellonian University, Institute of Physics; 2008, Available at http://arxiv.org/abs/0801.0700 [last accessed on March 11, 2013].

[23] Filges D, Goldenbaum F. Handbook of spallation research. Weinheim: Wiley-VCH; 2009.
[24] Russell GJ. Proceedings of ICANS-XI; 1990. Misawa M, Furusaka M, Ikeda H, Watanabe N, editors. KEK Report 90-25; 1991. p. 291–9.
[25] Fraser JS, Green RE, Hilborn JW, Milton JCD, Gibson WA, Gros EE, et al. Phys Can 1965;21(2):17. See also, The AECL study for an intense-generator (ING), Bartholomew GA and Tunnicliffe PR, editors, Atomic Energy of Canada Limited., Report AECL-2600; 1966. Intense Neutron Sources. In: Motz HT, Keepin GR, editors, U. S. Atomic Energy Commission, CONF-660925; 1966. p. 637.
[26] Sordo F. Desarrollo de criterios nucleares de diseño de fuentes de neutrones de espalación. Ph.D. Dissertation, Spain: Universidad Politécnica de Madrid, Madrid; 2008, Available at http://oa.upm.es/1069/1/Fernando_Sordo_Balbin.pdf [last accessed on September 24, 2012].
[27] Vassilkov RG, Yurevich VI. Proceedings of ICANS-XI; 1990. Misawa M, Furusaka M, Ikeda H, Watanabe N, editors. KEK Report 90-25; 1991. p. 340–53.
[28] Barashenov VS, Toneev VD, Chigrinov SE. At Energ 1974;37:480. Bauer GS et al. Proceedings of ICANS-VI; 1982.Carpenter JM, editor. Argonne National Laboratory Report, ANL-82-80; 1983.p. 620–7.
[29] SeeFilges D, Goldenbaum F. Handbook of spallation technology. Weinheim: Wiley-VCH; 2009, p. 37 and p. 260.
[30] Antinucci M, Bertin A, Capiluppi P, D'Agostino-Bruno M, Rossi AM, Vannini G. Lett Nuovo Cimento 1973;6:121–8. See also Catanesi MG, Radicioni E, Edgecock R, Ellis M, Soler FJP, Gößling C, et al. Phys Rev C 2008;77:055207.
[31] Meigo S, Ikeda Y, Ino T, Kasugai Y, Maekawa F, Nakashima H, et al. Proceedings of ICANS-XV; 2000. Itoh S, Suziki J, editors. KEK Proceedings 2000–22; 2001. p. 941–54.
[32] Enqvist T, Wlazlo W, Armbruster P, Benlliure J, Bernas M, Boudard A, et al. Nucl Phys A 2001;686:481–524. For information on the radiotoxicity of PbBi targets see Atchison F. Nuclide production in the SINQ target: SIN report SINQ/816/AFN-702; 1997.
[33] Reeves H, Fowler WA, Hoyle F. Nature 1970;226:727–9.
[34] Hurwitz H, Zweifel PF. J Appl Phys 1955;26:923–31. Marshak RE, Brooks H, Hurwitz H. Nucleonics 1949;5:59–68.
[35] Williams MMR. The slowing down and thermalization of neutrons. Amsterdam: North Holland; 1966.
[36] This refers to eigenvalues of the operator ∇^2 in a scalar Helmholtz equation for the scalar flux $\phi(r, t).\nabla^2 \varphi(\mathbf{r},t) + B^2 \varphi(\mathbf{r},t) = 0$ (2.88).
[37] Kropf F, Granada JR, Mayer RE. Nucl Instrum Methods 1982;198:515–25. Ikeda S, Carpenter JM. Nucl Instrum Methods Phys Res A 1985;239:536–44. Carpenter JM, Taylor AD, Robinson RA, Picton DJ. Nucl Instrum Methods Phys Res A 1985;234:542–51.
[38] Granada JR, Gillette VH, Mayer RE. Phys Rev A 1987;36:5585–93. Granada JR, Gillette VH. Phys B (Amsterdam, Neth.) 1995;213–214:821–3. Morishima N, Nishikawa Y, Mitsuyasu T. Phys B (Amsterdam, Neth.) 2004;350:e679–e681. Keinert J, Mattes M, Bernnat W. Thermal neutron cross-section data for light-water ice, liquid hydrogen and solid methane for the temperature range from 14 K up to 273 K in MCNPX format: University of Stuttgart, Institute of Nuclear Technology and Energy Systems, Report IKE 6-198; 2002.
[39] Chadwick MB, Herman M, Oblozinsky P, Dunn ME, Danon Y, Kahler AC, et al. Nucl Data Sheets 2011;112:2887–996. MacFarlane RE. RSICC peripheral shielding routine collection: NJOY99.0, code system for producing pointwise and multigroup neutron and photon cross sections from ENDF/B Data: Oak Ridge National Laboratory Report, PSR-480/NJOY99.0; 2000. Young JA, Koppel JU. Phys Rev 1964;135:A603–A611.

[40] Ino T, Ooi M, Kiyanagi Y, Kasugai Y, Maekawa F, Takada H, et al. Nucl Instrum Methods Phys Res A 2004;525:496–510.
[41] Izumi F, Asano H, Murata H, Watanabe N. J Appl Crystallogr 1987;20:411–8, See also http://www.ccp14.ac.uk/ccp/web-mirrors/plotr/Tutorials&Documents/TOF_FullProf.pdf[last accessed on March 11, 2013].
[42] Most of the existing USANS spectrometers use an optical scheme of the conventional Bonse-Hart Double-Crystal-Diffractometer.The system is basically a double crystal (or triple axis) spectrometer with two perfect Silicon channel cut crystals as monochromator and analyzer mounted in parallel geometry on a vibration isolated optical bench. Typical areas of application are the study of supramolecular structure of polymers, the macroscale self-similarity of rocks, the nanostructure of colloidal crystals and alloys or the hydration of cement pastes.
[43] Atchinson F, Blau B, Bodek K, van den Brandt B, Bryś T, Daum M, et al. Europhys Lett 2011;95:12001. Rivlin LA. Quantum Electron 2011;41:659–62, Yoshioka T. Available at http://fnp.kek.jp/workshop/20100408/proceedings/28-tyoshioka.pdf [last accessed on March 11, 2013].
[44] See http://www.uni-mainz.de/eng/14192.php [last accessed on March 11, 2013]. See also Altarev I, Daum M, Frei A, Gutsmiedl E, Hampel G, Hartmann FJ, et al. Eur Phys J A 2008;37:9–14.
[45] Crane HR, Lauritsen CC, Soltan A. Phys Rev 1933;44:692–3, ibid. 514.
[46] Kapchinskii IM, Tepliakov VA. Prib Tekh Eks 1970;119:19–22.
[47] Stovall JE, Crandall KR, Hamm RW. IEEE Trans Nucl Sci 1981;28:1508–10. Hamm RW, Crandall KR, Fuller CW, Hansborough LD, Potter JM, Rodenz GW, et al. Proceedings of the International Conference on Low Energy Ion Beams, vol. 2; 1980. Also as Los Alamos Scientific Laboratory Report, LA-UR-80-1091.
[48] Hassanzadegan H, Garmendia N, Etxebarria V, Bermejo FJ. Phys Rev ST Accel Beams 2011;14:052803.
[49] Plostinar DC, Letchford AP. In: Proceedings of the XXIV linear accelerator conference; 2008, Available at http://accelconf.web.cern.ch/accelconf/LINAC08/papers/thp065.pdf [last accessed on March 11, 2013].
[50] Brown IG, editor. The physics and technology of ion sources. 2nd ed. Weinheim: Wiley-VCH; 2004.
[51] Takieri Y, Tsumori K, editors. Second international symposium on negative ions, beams and sources (NIBS-2010). AIP conference proceedings 1390; 2012.
[52] Stirling WL, Tsai CC, Ryan PM. Rev Sci Instrum 1977;48:533–6.
[53] Notes on the First meeting of the international collaboration on advanced neutron sources, ICANS-I, Argonne National Laboratory; 1977.
[54] Filges D, Neef RD, Schaal H. Nucl Technol 2000;132:30–48.
[55] Riemer B, Haines J, Wendel M, Bauer G, Futakawa M, Kogawa H. J Nucl Mater 2008;377:162–73. Thomas K. J Nucl Mater 2006;256:321. Futakawa M, Kogawa H, Hasegawa S, Ikeda Y, Riemer B, Wendel M, et al. J Nucl Mater 2008;377:182–8.
[56] Bauer GS, Salvatores M, Heusener G. J Nucl Mater 2001;296:17–33.
[57] Bauer GS, Sebening H, Vetter JE, Willax H. Report Jül-Spez 114 and KfK 3171; 1981.
[58] MacManamy T, Rennich M, Gallmeier F, Ferguson P, Janney J. J Nucl Mater 2010;398:35–42.
[59] Kharoua C, Ene D, Mezei F, Noah E, Plewinski F, Sabbagh P, et al. In: Proceedings of 4th high power targetry workshop; 2011, Available at http://hep.princeton.edu/mumu/target/Kharoua/Kharoua_050211.pdf [last accessed on March 11, 2013].

[60] Wachs DM. RERTR fuel development and qualification plan: Idaho National Laboratory Report, INL/EXT-05-01017; 2007, Available at http://www.inl.gov/technicalpublications/Documents/3610331.pdf [last accessed on March 11, 2013].
[61] Incropera FP, DeWitt DP, Bergman TL, Lavine S. Fundamentals of heat and mass transfer. 6th ed. New York: Wiley; 2007, p. 514.
[62] Muhrer G, Pitcher EJ, Russell GJ. Proceedings of ICANS-XVII: Also as Los Alamos National Laboratory Report, LA-UR-05-3168; 2005.
[63] Taylor AD, Johnson MW. Juelich KFZ Report, ISSN 0244-05789;1980.
[64] Teshigawara M, Harada M, Saito S, Kikuchi K, Kogawa H, Ikeda Y, et al. J Nucl Mater 2005;343:154–62.
[65] Riemer B. Proceedings of UCANS-III, 2012. To appear in *Physics procedia*.
[66] Magán M, Terrón S, Sordo F, Bermejo FJ, Perlado JM. In: Baxter DV, editor. Proceedings of UCANS I& II. Physics procedia, vol. 26; 2012. p. 211–8.
[67] Ooi M, Ino T, Muhrer G, Pitcher EJ, Rusell GJ, Ferguson PD, et al. Nucl Instrum Methods Phys Res A 2006;566:699–705. Kai T, Harada M, Teshigawara M, Watanabe N, Ikeda Y. Nucl Instrum Methods Phys Res A 2004;523:398–414. Oi M, Konno M, Iwasa H, Kamiyama T, Furusawa M, Kiyanagi Y. J Nucl Sci Technol 2004;41:1138–44.
[68] Kiyanagi Y, Carpenter JM, Kosugi N, Iwasa H, Hiraga F, Watanabe N. Phys B (Amsterdam, Neth.) 1995;213:854–6. Kiyanagi Y. Nucl Instrum Methods Phys Res A 2006;562:561–4.
[69] Granada JR, Cantargi F, Márquez Damián JI. In: Proceedings of ICANS-XX; 2012, Available at http://www.icansxx.com.ar/proceedings.php, report 358 [last accessed on March 11, 2013].
[70] BroomeTA, Hogston JR, Holding M, Howells WS. Proceedings of ICANS-XII; 1993. Steigenberger U, Broome T, Rees G, Soper A, editors. Rutherford-Appleton laboratory report, RAL 94-025; 1994. p. T/156-T/163.
[71] Watanabe N. Rep Prog Phys 2003;66:339–81.
[72] Pitcher EJ, Russell GJ, Muhrer G, Jarmer JJ, Corzine RK. Proceedings of ICANS-XVI; 2003. Mank G, Conrad H, editors. Report ESS 03-136-M1; 2003. p. 849.
[73] Baxter DV, Ansell S, Ferguson PD, Gallmeier FX, Iverson EB, Kaiser H, et al. In: Baxter DV, editor. Proceedings of UCANS I& II. Physics procedia, vol. 26. Elsevier; 2012. p. 153–60.
[74] Zannini L, Batkov K, Mezei F, Takibayev A, Bermejo FJ, Magán M, et al. Proceedings of UCANS-III, 2012. To appear in *Physics procedia*.
[75] Arai M et al. Proceedings of ICANS-XVI; 2003. Mank G, Conrad H, editors. Report ESS 03-136-M1, vol. I; 2003. p. 1.
[76] Ooi M, Konno M, Iwasa H, Kamiyana T, Furusaka M, Kiyagani Y. J Nucl Sci Technol 2004;41:1138–44.
[77] Muhrer G, Pitcher EJ, Russell GJ. Nucl Instrum Methods A 2005;536:154–64.
[78] Nakamura T, Nunomiya T, Yashima H, Yonai S. Prog Nucl Energy 2004;44:85–187.
[79] See Lawrence and his laboratory. Lawrence Berkeley National Laboratory Magazine; 1981 [chapter 5], Available at http://www.lbl.gov/Science-Articles/Research-Review/Magazine/1981/81fchp5.html [last accessed on March 11, 2013].
[80] Mansur LK. J Nucl Mater 2003;318:14–25.
[81] A second target station for the spallation neutron source. White Paper, SNS 100000000-TR0029-R00, Oak Ridge National Laboratory; 2007.
[82] Yamazaki Y. In: Proceedings of the 8th European particle accelerator conference; 2002. p. 169–73. Ikeda Y. Proceedings of ICANS-XVI; 2003. Mank G, Conrad J, editors. Report ESS 03-136-M1, vol. I; 2003. p. 13–24.

[83] Bauer G, Broome T, Filges D, Jones H, Lengeler H, Letchford A, et al. The European spallation source study: Report ESS 96-53-M, ESS-Council; 1996, vol. III.
[84] The ESS Project, Vol. III, Technical Report, ISBN 3-89336-303-3: 2002 and The ESS Project, Vol. III Update, Technical Report, ISBN 3-89336-345-9; 2003.
[85] See http://csns.ihep.ac.cn/english/index.htm [last accessed on September 26, 2012].
[86] See http://www.indiana.edu/~lens/ [last accessed on March 12, 2013].
[87] As a representative example see Ref. [76].
[88] Wei J, Chen HB, Huang WH, Tang CX, Xing QZ, Fu SN, et al. In: Proceedings of the 23rd particle accelerator conference; 2009. p. 1360–2, Available at http://accelconf.web.cern.ch/accelconf/PAC2009/papers/tu6pfp035.pdf [last accessed on March 12, 2013].
[89] Ren HT, Peng SX, Lu PN, Zhou QF, Yuan ZX, Zhao J, et al. Rev Sci Instrum 2012;83:02B711.
[90] Chung B, Cho YS, Lee YY. In: Proceedings of the 22nd particle accelerator conference; 2007. p. 2817–9, Available at http://accelconf.web.cern.ch/AccelConf/p07/PAPERS/THPMN046.PDF [last accessed on March 12, 2013].
[91] Sordo F, Terrón S, Magán M, Muhrer G, Ghiglino A, Martínez F, et al. Nucl Instrum Methods Phys Res A 2013;707:1–8.
[92] Safety considerations for research reactors in extended shutdown. IAEA-TECDOC-1387, IAEA, Vienna; 2004.
[93] The global stockpile of HEU fuels at the time of writing is estimated to be of 1440 tones. See http://fissilematerials.org/ [last accessed on March 11, 2013].
[94] Xoubi N, Primm III RT. Modeling of the high flux isotope reactor cycle 400. ORNL/RM-2004/251.
[95] See http://flnp.jinr.ru/563/ [last accessed on March 12, 2013].
[96] Primm III RT, Chandler D, Ilas G, Jolly BC, Miller JH, Sease JD. Design study for a low-enriched uranium core for the high flux isotope reactor. Oak Ridge: Office of Scientific Technical Information; 2009. ORNL/TM-2009/87.
[97] Kennedy SJ. Phys B (Amsterdam, Neth) 2006;385:949–54.
[98] Axmann A, Böning K, Rottmann M. Nucl Eng Des 1997;178:127–33.
[99] Campioni G, Calzavara Y, Mollier M, Atkins D, Veler R, Sebie P, et al. Ann Nucl Energy 2009;36:1319–24.
[100] Simnad MT. Nucl Eng Des 1981;64:403–22.
[101] Taken from Rush JJ, Cappelletti RL. The NIST Center for Neutron Research: over 40 years serving NIST/NBS and the Nation, NIST Special Publication 1120; 2011.
[102] See Report from the ILL Associates' working group on neutrons in Europe for 2025, 2011, Institut Laue-Langevin, Available at http://goo.gl/VDgU3 [last accessed on March 12, 2013].
[103] Tajima T, Dawson JM. Phys Rev Lett 1979;43:267–70.
[104] Edgecock TR, Kelliher D, Beard CD, Bliss N, Clarke J, Hill C, et al. In: Proceedings of the 11th European particle accelerator conference; 2008. p. 3380–2, Available at http://accelconf.web.cern.ch/accelconf/e08/papers/thpp004.pdf [last accessed on March 11, 2013].
[105] Christofilos N, Hester R, Lamb W, Reagan D, Sherwood W, Wright R. Rev Sci Instrum 1964;35:886–9.
[106] Takayama K, Briggs RJ, editors. Induction Accelerators, Springer Series on Particle Acceleration and Detection. Berlin: Springer-Verlag; 2011.
[107] Keefe D, Hoyer E. Induction Linacs as high intensity neutron sources: LBL-12855, Technical Report, CONF-8106120-4; 1981.
[108] See for instance Logan B, Perkins L, Barnard J. Phys Plasmas 2008;15:072701.

[109] Barnard JJ, Briggs RJ. Induction accelerators. In: Takayama K, Briggs RJ, editors. Springer series on particle acceleration and detection. Berlin: Springer-Verlag; 2011 [chapter 10].
[110] Rosenzweig JB, Cline DB, Cole B, Figueroa H, Gai W, Konecny R, et al. Phys Rev Lett 1988;61:98–101.
[111] Esarey E, Sprangle P, Krall J, Ting A. IEEE Trans Plasma Sci 1996;24:252–88.
[112] Blumennfeld I, Clayton CE, Decker F-J, Hogan MJU, Huang C, Ischebeck R, et al. Nature 2007;445:741–4.
[113] Caldwell A, Lotov K, Pukhov A, Simon F. Nat Phys 2009;5:363–7.
[114] Ruggiero AG. Brief history of FFAG accelerators. Brookhaven National Laboratory Report BNL-75635-2006-CP; 2006, Available at http://www.bnl.gov/isd/documents/31130.pdf [last accessed on March 12, 2013].
[115] Jones LW, Sessler AM, Symon KR. Science 2007;316:1567.
[116] Symon KR. In: Proceedings of the particle accelerator conference; 2003. p. 452–6, Available at http://accelconf.web.cern.ch/accelconf/p03/PAPERS/WOPA003.pdf [last accessed on October17, 2012].
[117] Cole FT, Parzen G, Rowe EM, Snowdon SC. In: Howard FT, Vogt-Nilsen N, editors. Proceedings of the international conference on sector-focused cyclotron and meson factories; 1963. p. 351–2. CERN 63-19, Available at http://accelconf.web.cern.ch/AccelConf/c63/papers/cyc63g05.pdf [last accessed on March 12, 2013].
[118] Martin SA, et al. In: Steigenberger U, Broome T, Rees G, Soper A, editors. Proceedings of ICANS-XII; 1993, RAL 94-025.
[119] Aiba M, Koba K, Machida S, Mori Y, Muramatsu R, Ohmori C, et al. In: Proceedings of the European particle accelerator conference; 2000. p. 581–3, Available at http://accelconf.web.cern.ch/AccelConf/e00/PAPERS/MOP1B21.pdf [last accessed on March 12, 2013].
[120] Johnstone C, Berz M, Makino K, Snopok P. arXiv:1208.5798. FERMILAB-CONF-11-170-AD; 2012.Sheehy SL, Johnstone C. In: Proceedings of the international partticle accelerator conference; 2012. p. 3234–6, Available at http://accelconf.web.cern.ch/AccelConf/IPAC2012/papers/theppb003.pdf [last accessed on March 21, 2013]
[121] LaPointe MA, Yakovlev VP, Kazakov SYu, Hirshfield JL. In: Proceedings of the 23rd particle accelerator conference; 2009. p. 3045–7, Available at http://accelconf.web.cern.ch/accelconf/PAC2009/papers/we6rfp105.pdf [last accessed on March 11, 2013].
[122] Pfotenhauer SM, Jäckel O, Polz J, Steinke S, Schlenvoight H-P, Heymann J, et al. New J Phys 2010;12:103009. Yu L, Xu H, Wang W, Sheng Z, Shen B, Yu W, et al. New J Phys 2010;12:045021.
[123] Taylor AD, Dunne M, Bennington S, Ansell S, Gardner I, Norreys P, et al. Science 2007;315:1092–5.
[124] Geuther J, Danon Y, Saglime F. Phys Rev Lett 2006;96:054803. Naranjo B, Gimzewski JK, Putterman S. Nature 2005;434:1115–7. Danon Y. J Instrum 2012;7:C04002. Tang V, Meyer G, Falabella S, Guethlein G, Sampayan S, Kerr P, et al. J Appl Phys 2009;105:026103.
[125] Berglöf C. On measurement and monitoring of reactivity in subcritical reactor systems. Ph.D. Dissertation. KTH, School of Engineering Sciences, Department of Physics, Division of Reactor Physics; 2010, Available at http://kth.diva-portal.org/smash/get/diva2:315354/FULLTEXT01.pdf [last accessed on June 28, 2013].
[126] Soule R, Assal W, Chaussonnet P, Destouches C, Domergue C, Jammes C, et al. Nucl Sci Eng 2004;148:124–52.
[127] Gohar Y. In: Raja R, Mishra S, editors. Applications of high intensity proton accelerators. Proceedings of the workshop held 19–21 October 2009 in Fermilab, Chicago. Singapore: World Scientific Publishing; 2010.

[128] Shiroya S, Unesaki H, Kawase Y, Moriyama H, Inoue M. Prog Nucl Energy 2000;37:357–62. Shiroya S, Yamamoto A, Shin K, Ikeda T, Nakano S, Unesaki H. Prog Nucl Energy 2002;40:489–96. Yamamoto A, Shiroya S. Ann Nucl Energy 2003;30:1425.

[129] The Advanced Neutron Source project developed at Oak Ridge constituted the last attempt by the western countries to overcome such flux limit. A brief description of the project is given by Hayter JB, West CD. Neutron News 1992;3:16–9. A technical design report is available as Peretz FJ *et al.* Conceptual design summary. ORNL/TM-12184, DE92 041230; 1992, Available at http://www.osti.gov/bridge/servlets/purl/7282459-8hvDfS/7282459.pdf [last accessed on March 12, 2013].

[130] See http://science.energy.gov/np/research/idpra/ [last accessed on March 24, 2013].

[131] Steinberg MV, Wotzak G, Manowitz B. Neutron burning of long-lived fission products for waste disposal: Report BNL-8558, Brookhaven Nat. Lab., Upton NY; 1958.

[132] Gokhale P, Deokattei S, Kumar V. Prog Nucl Energy 2006;48:91–102.

[133] Laserre T, Sobel HW. C R Phys 2005;6:749–57.

[134] Bartosik H, Klepp J, Schmitzer C, Sponar S, Cabello A, Rauch H, et al. Phys Rev Lett 2009;103:040403.

Chapter 3

Experimental Techniques

Masatoshi Arai
Materials and Life Science Division, J-PARC Center, Japan Atomic Energy Agency, Tokai, Ibaraki, 319-1195, Japan

Chapter Outline

- 3.1. Introduction — 246
- 3.2. Scattering Measurements — 247
  - 3.2.1. Cross Section — 247
  - 3.2.2. Integrated Intensity and the Lorentz Factor — 249
- 3.3. Useful Neutrons for Condensed Matter Science — 254
  - 3.3.1. Neutron Flux from Moderators — 254
  - 3.3.2. Pulse Peak Structure — 255
  - 3.3.3. Pulse Peak Width — 257
  - 3.3.4. Choice of Parameters in Spallation Sources — 257
  - 3.3.5. High-Energy Neutron Background — 259
- 3.4. Diffraction Techniques — 261
  - 3.4.1. Powder Diffraction — 262
  - 3.4.2. Single-Crystal Diffractometers — 271
- 3.5. Inelastic Scattering Techniques — 275
- 3.5.1. Triple-Axis Spectrometer — 275
- 3.5.2. Chopper Instruments — 277
- 3.5.3. Inverted-Geometry Instrument — 288
- 3.6. Instruments for Semi-Macroscopic Structures — 293
  - 3.6.1. Small-Angle Neutron Scattering Instruments — 293
  - 3.6.2. Neutron Spin-Echo Spectrometers — 296
- 3.7. Neutron Detectors — 300
  - 3.7.1. ^{3}He-Gas Detectors — 301
  - 3.7.2. Scintillation Detectors — 301
- 3.8. Beam Transport and Tailoring — 305
  - 3.8.1. Neutron Optics — 305
  - 3.8.2. Choppers — 313
- References — 318

3.1 INTRODUCTION

The neutron is a neutral particle that easily penetrates most materials. It is scattered by nuclei and magnetic fields and behaves as a particle wave with a mass about 2000 times that of the electron. These physical properties give the neutron a very unique character as a probe for studying condensed matter at the atomic scale with scattering techniques. The relations between neutron's energy E, wavelength λ and wave vector k are

$$E_n = k_B T = \frac{1}{2} m_n v_n^2 = \frac{h^2}{2 m_n \lambda^2} = \frac{\hbar^2 k_n^2}{2 m_n} \tag{3.1}$$

Therefore, if we want to study the atomic structure of materials with neutrons of a wavelength 1 Å, for example, they have an energy of about 80 meV, which is comparable to the energies of fundamental excitations in condensed matter such as phonons, magnons, and other basic excitations. Because of this important feature, neutrons are simultaneously sensitive to both atomic structure and dynamics and thus probe the space–time correlation function of the system under study.

Atomic structure, given by the spatial correlation function, is generally measured by diffraction techniques. These are either crystal-monochromator diffraction or time-of-flight (TOF) diffraction, depending on the character of neutron source either a reactor or a pulsed source (cf. Figure 3.1). These techniques are discussed in Section 3.3.4.

The dynamics of atoms and spins are studied by inelastic neutron scattering techniques. Several general concepts are illustrated in Figure 3.2, including the use of crystal monochromators to select the incident neutron wavelength, TOF techniques, or a combination of the two. The triple-axis technique is commonly used in steady-state sources, while the direct-geometry and inverted-geometry TOF techniques are mainly used at pulsed sources. A hybrid crystal-TOF technique has recently been adopted in a high-intensity spectrometer, HYSPEC, at SNS [1]. These techniques are discussed in Section 3.5.

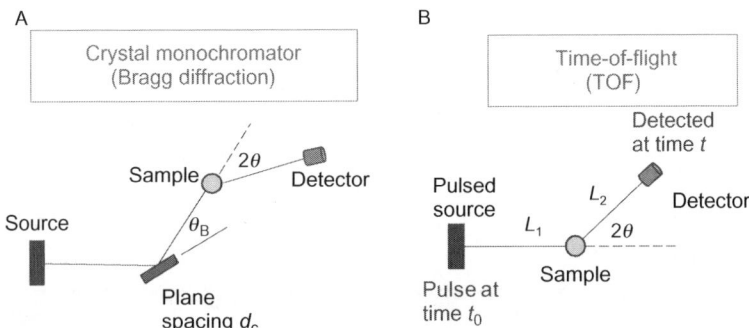

FIGURE 3.1 Diffraction technique: (a) crystal-monochromator diffraction and (b) time-of-flight diffraction. (For color version of this figure, the reader is referred to the online version of this chapter.)

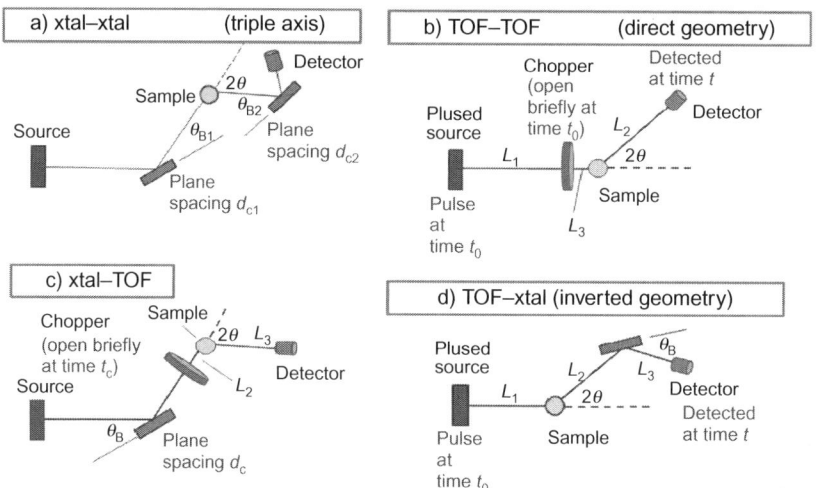

FIGURE 3.2 Typical inelastic scattering techniques. (For color version of this figure, the reader is referred to the online version of this chapter.)

Each technique has of course a merit/demerit and an advantage/disadvantage, depending on the system under study and the purpose of the experiment.

In this chapter, we study general aspects of these techniques with more emphasis on those developed for pulsed sources, which are becoming increasingly prevalent but have fewer references in the literature for scientists trying to learn about the techniques. Polarized neutron techniques are not described here except for the neutron spin-echo technique, as these will be treated in greater depth in later volumes of this series.

3.2 SCATTERING MEASUREMENTS

3.2.1 Cross Section

Neutron scattering is described by momentum and energy conservation between the incident and scattered neutrons and the scattering object

$$\mathbf{Q} = \mathbf{k}_i - \mathbf{k}_f \tag{3.2}$$

$$E_n = \hbar\omega = E_{\eta_f} - E_{\eta_i} = \frac{\hbar^2 k_i^2}{2m_n} - \frac{\hbar^2 k_f^2}{2m_n} = E_i - E_f \tag{3.3}$$

where $\mathbf{Q}$ is the momentum transfer and E the energy transfer (cf. Figure 3.3); η_i and η_f stand for the initial and final states of the scatterer.

In case of elastic scattering, $k_i = k_f$, the momentum transfer is given by

$$Q = 4\pi \sin(\theta)/\lambda \tag{3.4}$$

When we study the atomic structure of a material, we choose the wavelength as, say, 1 Å, to be of the same order of magnitude as lattice constants.

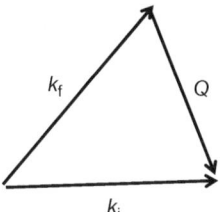

FIGURE 3.3 Scattering triangle showing the relationship between momentum transfer and initial/final wave vectors.

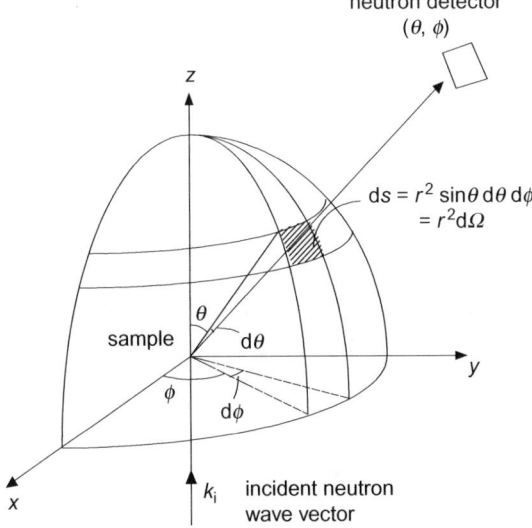

FIGURE 3.4 Incident neutron along the z-direction, specified by k_i, is scattered into a scattering angle ϑ with azimuthal angle φ in a solid angle $d\Omega$, defined by an area ds on a detector.

Scattered neutrons are registered by a detector and we can describe the scattering probability in terms of a cross section, which gives information about the material's atomic structure and dynamics.

A scattering diagram in three-dimensional space is illustrated in Figure 3.4, in which an incident neutron comes along the z-direction and scattered into the ϑ, φ direction spanned by a solid angle $d\Omega$ of a detector in an energy range between E_f and $E_f + dE_f$.

The double differential cross section is defined by the following formula,

$$\left(\frac{d^2\sigma}{d\Omega dE_f}\right) = \frac{\begin{pmatrix}\text{Count rate in a solid angle } d\Omega \\ \text{in a scattering angle } \theta, \phi \\ \text{in the energy range } E_f \sim E_f + dE_f \text{ (n/s)}\end{pmatrix}}{(\Phi d\Omega dE_f (\text{n/cm}^2/\text{s})(\text{meV}))} (\text{cm}^2/\text{meV}) \quad (3.5a)$$

where Φ is the incident flux in units of $(\text{n/cm}^2/\text{s})$.

When the energy analysis is not performed, then the cross section becomes a single differential cross section

$$\left(\frac{d\sigma}{d\Omega}\right) = \frac{\left(\begin{array}{c}\text{Count rate in a solid angle } d\Omega \\ \text{in a scattering angle } \theta, \phi \text{ (n/s)}\end{array}\right)}{(\Phi d\Omega(\text{n/cm}^2/\text{s}))} (\text{cm}^2) \qquad (3.5b)$$

Neutrons scattered into all directions define a scattering cross section,

$$\sigma_S = (\text{total scattered neutron count rate})/\Phi \, (\text{cm}^2) \qquad (3.6)$$

The relation between the above three cross sections is

$$\frac{d\sigma}{d\Omega} = \int_0^\infty \left(\frac{d^2\sigma}{d\Omega dE_f}\right) dE_f \qquad (3.7)$$

$$\sigma_S = \int \left(\frac{d\sigma}{d\Omega}\right) d\Omega = \int_0^\pi \int_0^{2\pi} \left(\frac{d\sigma}{d\Omega}\right) \sin\theta d\theta d\phi \qquad (3.8)$$

3.2.2 Integrated Intensity and the Lorentz Factor

We now estimate the observed scattering intensity in a specific solid angle of a detector. This quantity depends on the method used for the measurement with respect to angle and wavelength. We examine three kinds of measurement: angle scanning, powder diffraction, and TOF diffraction.

3.2.2.1 Angle-Scanning Diffraction (Rotating Crystal)

Angle-scanning diffraction is the method used for single-crystal diffraction with a monochromatic beam. The scattering angle and crystal are rotated to scan along the wave vector **Q** and neutrons are detected within a given solid angle. Rotating the scattering angle φ covers a volume in Q-space through the solid angle $d\Omega$, which defines a cylindrical volume with an area dS projected on the Ewald sphere (cf. Figure 3.5).

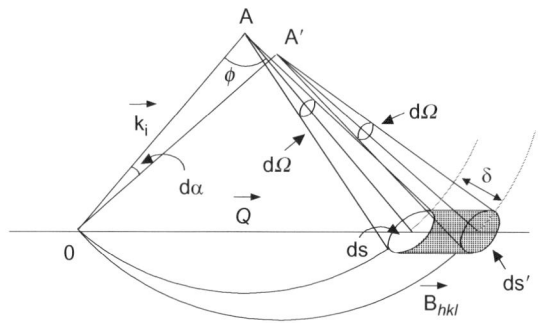

FIGURE 3.5 Angle scanning of a volume in **Q**-space. Angle-scanning diffraction.

The volume dV_q is given by

$$dV_q = ds\delta = k_i^2 d\Omega d\alpha k_i \sin\phi = \frac{(2\pi)^3}{\lambda^3} d\Omega d\alpha \cdot \sin\phi \tag{3.9}$$

The integrated intensity is obtained by rotating the crystal by $d\alpha$:

$$I = \int \left(\frac{d\sigma}{d\Omega}\right) d\Omega d\alpha \tag{3.10}$$

If we explicitly describe dV_q as a volume in the Q-space, $dq_1 dq_2 dq_3$, then

$$d\Omega d\alpha = \frac{\lambda^3}{(2\pi)^3 \sin\phi} dq_1 dq_2 dq_3 \tag{3.11}$$

The structure factor $F(Q)$ is a smooth function in Q. Therefore, it can be taken out from the integration and it becomes,

$$I = |F(Q)|^2 \frac{\lambda^3}{(2\pi)^3 \sin\phi} \int G(Q) dq_1 dq_2 dq_3 \tag{3.12}$$

here $G(Q)$ is Laue function. Its integration is,

$$\int G(Q) dq_1 dq_2 dq_3 = \frac{N_C (2\pi)^3}{v_a} \tag{3.13}$$

Here, v_a and N_C are the unit-cell volume and the number of unit cells in the sample.

When the Bragg condition is satisfied, $\phi = 2\theta_B$,

$$I = |F(Q)|^2 \frac{\lambda^3}{\sin 2\theta_B} \frac{N_C}{v_a} = |F(Q)|^2 \frac{\lambda^3}{\sin 2\theta_B} \frac{V_S}{v_a^2} \tag{3.14}$$

where $V_S = N_C \times v_a$ is the sample volume.

The Lorentz factor, Λ, is defined as

$$I = \Lambda V_S \tag{3.15}$$

Therefore, the Lorentz factor for the angle-scanning diffraction is

$$\Lambda = \frac{|F(Q)|^2 \lambda^3}{\sin 2\theta_B} \frac{1}{v_a^2} \tag{3.16}$$

that modifies the observed intensity.

3.2.2.2 Powder Diffraction (Debye–Scherrer Method)

To measure diffraction from a powder sample, we utilize the same kind of instrument as for single-crystal diffraction. In this case, we just rotate the detector around the sample. Scattering takes place into any direction within

Chapter | 3 Experimental Techniques   251

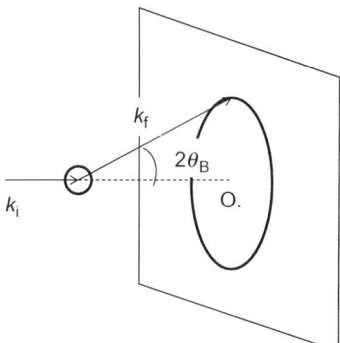

FIGURE 3.6 Debye–Scherrer diffraction ring for a powder sample.

a cone around the direction of the incident beam (Figure 3.6). The shape of the cross section of the cone vertical to the incident direction defines a ring (Debye ring). A detector counts neutrons in the solid angle spanned by the detector in the ring.

The integrated intensity for this method is calculated by taking the solid angle, $d\alpha\Delta\beta$, along the scattering vector **Q**. Here, $d\alpha$ is the angle along the scanning direction, and $\Delta\beta$ is one vertical to it, which corresponds to the detector width.

If we make the scattering vector **Q** equal to the Bragg vector $\vec{B}_{hkl}$, the intensity from the Bragg diffraction is distributed on the surface of a sphere of radius $\vec{B}_{hkl}$. Now we consider the angular spread $(\chi+d\chi, \varphi+d\varphi)$ along **Q**, into which some crystallites in the sample with an appropriate orientation contribute to the observable intensity. Here, χ is the polar angle of **Q** for the polar axis $\mathbf{k}_i$, and φ is the radial angle of the projection of **Q** on the plane perpendicular to $\mathbf{k}_i$, as shown in Figure 3.7.

Now we count the number of such crystallites as a fraction of the solid angle spanned by $d\chi$ and $d\varphi$:

$$W(x,\varphi)dxd\varphi = N_t \frac{dx}{4\pi}\sin x d\varphi \qquad (3.17)$$

where N_t is the total number of the crystallites in the sample.

Therefore, the observed intensity δI in the solid angle $d\alpha\Delta\beta$ is

$$\delta I d\alpha\Delta\beta = N_t \int \left(\frac{d\sigma}{d\Omega}\right) \frac{1}{4\pi}\sin\chi d\chi d\varphi d\alpha\Delta\beta$$

Since the structure factor $F(Q)$ displays a smooth variation with Q, it can be taken out from the integration,

$$\delta I d\alpha\Delta\beta = N_t |F(Q)|^2 \frac{1}{4\pi}\sin\chi \int G(Q,\chi,\varphi)d\chi d\varphi d\alpha\Delta\beta \qquad (3.18)$$

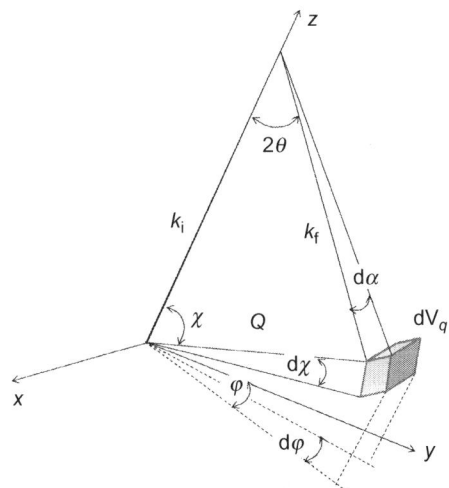

FIGURE 3.7 Scattering diagram for the Debye–Scherrer method. (For color version of this figure, the reader is referred to the online version of this chapter.)

where $G(Q)$ is the Laue function, characterized by a very sharp peak structure as a function of Q.

Therefore, the integrated intensity along α is

$$I = \int \delta I d\alpha \Delta\beta = N_t |F(Q)|^2 \frac{1}{4\pi} \sin\chi \iint_{\varphi,\chi,\alpha} G(Q,\chi,\varphi) d\chi d\varphi d\alpha \Delta\beta \quad (3.19)$$

The volume in reciprocal space δV_q spanned by $d\chi$, $d\varphi$, $d\alpha$ is (see Figure 3.7)

$$\begin{aligned}\delta V_q &= Q d\chi k d\alpha \sin\chi Q \sin\chi d\varphi = dq_1 dq_2 dq_3 \\ &= 4\sin^2\theta_B \cos^2\theta_B k^3 d\chi d\varphi d\alpha \\ &= 2k\sin 2\theta_B \sin\theta_B \sin\chi d\chi d\varphi d\alpha\end{aligned} \quad (3.20)$$

$$x = \frac{\pi}{2} - \theta_B$$

From Equation (3.20) it follows that

$$\sin x dx d\varphi d\sigma = \frac{dq_1 dq_2 dq_3}{k^3 2\sin 2\theta_B \sin\theta_B}$$

From Equation (3.19) it follows that

$$\begin{aligned}I &= N_t |F(Q)|^2 \frac{1}{8\pi}\left(\frac{\lambda}{2\pi}\right) \int G(\vec{Q}) dq_1 dq_2 dq_3 \Delta\beta \frac{1}{\sin 2\theta_B \sin\theta_B} \\ &= N_t V_c |F(Q)|^2 \frac{\lambda^3}{8\pi \sin 2\theta_B \sin\theta_B} \frac{\Delta\beta}{v_a} \frac{N_c}{} = \Lambda V_c \Delta\beta N_t\end{aligned}$$

where N_c is the number of crystallite, V_c the volume of crystallite, and v_a is the cell volume.

Therefore, the Lorentz factor for powder diffraction becomes

$$\Lambda = \frac{|F(Q)|^2 \lambda^3}{8\pi \sin\theta_B \sin 2\theta_B v_a^2} \quad (3.21)$$

We note that the observed intensity is proportional to $\Delta\beta$, which corresponds to the detector length parallel to the Debye ring.

If we wish to measure the entire intensity around the Debye ring, then since $\Delta\beta = 2\pi \sin 2\theta_B$, the total intensity for sample volume V_s becomes

$$I_t = \frac{|F(Q)|^2 \lambda^3}{4\sin\theta_B} \frac{1}{v_a^2} V_s \quad (3.22)$$

3.2.2.3 Time-of-Flight Diffraction (Laue Method)

In a pulsed neutron source, it is essential to use a white beam and a TOF method to analyze neutron energies. A diffraction intensity satisfying the Bragg condition is recorded at the detector:

$$2d_{hkl}\sin\theta_B = \lambda \quad (3.23)$$

A measurement is made by keeping the scattering angle constant, which gives a diffraction signal along the crystal axis as depicted in Figure 3.8. The Q-space volume for this measurement is given by

$$dV_q = dq_1 dq_2 dq_3 = k_i^2 d\Omega dk(1-\cos\phi)$$
$$= 2k_i^2 d\Omega dk \sin^2(\phi/2) \quad (3.24)$$
$$dk d\Omega = dq_1 dq_2 dq_3 / 2k_i^2(\theta)$$

here we used the relation $\phi = 2\theta$.

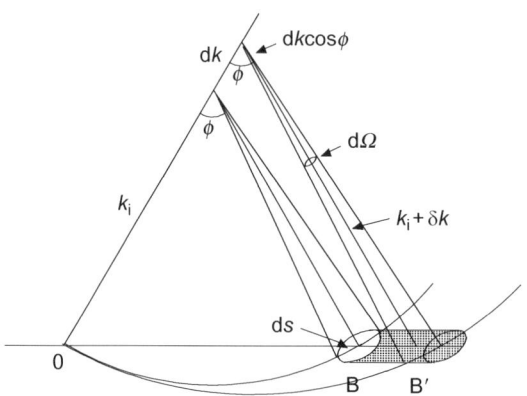

FIGURE 3.8 Scattering diagram for time-of-flight method.

$$t = \frac{L}{v} = \frac{m_n L}{\hbar k_i} \tag{3.25}$$

Its derivative is

$$dt = \frac{-mL}{\hbar} \frac{dk}{k_i^2}$$

$$dt d\Omega = \frac{-m_n L}{\hbar} \frac{dq_1 dq_2 dq_3}{2k_i^4 \sin^2 \theta_B} \tag{3.26}$$

$$I = \int \left(\frac{d\sigma}{d\Omega}\right) d\Omega dt = |F(Q)|^2 \frac{m_n L}{\hbar} \frac{\lambda^4}{2(2\pi)^4 \sin^2 \theta_B} \int G(Q) dq_1 dq_2 dq_3 \tag{3.27}$$

Using the relation of Equation (3.13), therefore the Lorentz factor for the TOF method becomes,

$$\Lambda = |F(Q)|^2 \frac{m_n L}{4\pi \hbar} \frac{\lambda^4}{\sin^2 \theta_B} \frac{1}{v_a^2} \sim \frac{\lambda^2}{Q^2} \tag{3.28}$$

This result has a very important implication for diffraction using the TOF method. If we consider a Bragg peak at a particular value of Q, it is good option to observe it with a longer wavelength at a higher scattering angle if the flux at the longer wavelength is sufficient.

3.3 USEFUL NEUTRONS FOR CONDENSED MATTER SCIENCE

3.3.1 Neutron Flux from Moderators

Neutron production in a reactor and in a target of a spallation source is described in Chapter 2.

For neutron scattering experiments for condensed matter research, it is necessary to reduce the neutron energy in moderators to less than 100 meV, corresponding to a wavelength >1 Å, comparable to atomic spacings. Moderators contain hydrogenous materials such as water, heavy water, methane, and hydrogen. Hydrogen nuclei have a large cross section and a mass almost the same as the neutron, so that neutrons are strongly scattered and lose their energies precipitously. Neutrons in moderators reach thermal equilibrium after a large number of scattering events and have an energy spectrum known as a Maxwell distribution, centered at an energy close to the temperature of the moderator, in case of sufficient thick moderators.

In terms of neutron wavelength λ, this distribution is given by

$$n_f(\lambda) d\lambda = 2n_F \frac{\lambda_T^4 d\lambda}{\lambda^5} e^{-\lambda_T^2/\lambda^2} \tag{3.29}$$

where $\lambda_T = \frac{h}{m_n v_T}$, $v_T = \left(\frac{2k_B T}{m_n}\right)^{1/2}$, $n_F = \frac{2n_D}{\sqrt{\pi}} v_T$, and n_D is the number density of neutrons.

This spectrum has a maximum at $\lambda_{\max} = \frac{h}{(5k_B T m_n)^{1/2}}$, a factor of about 1.5 smaller than the value estimated from simple thermal statistics, $\lambda_{\text{Therm}} = \frac{h}{(2k_B T m_n)^{1/2}}$.

It is also useful to have a description in energy,

$$n_f(E_n)dE_n = 2n_F \frac{E_n}{(k_B T)^2} e^{-E_n/k_B T} dE_n \tag{3.30}$$

which gives a maximum at $E_n \sim k_B T$.

On the other hand, short-pulse spallation sources need to have a time distribution of sharp pulses of thermalized neutrons. Therefore, the moderator dimensions need to be relatively small, typically around $10 \times 10 \times 10$ cm^3, as shown in Chapter 2. Such moderators allow neutrons to escape before reaching thermal equilibrium, which adds a flux component, proportional to $1/E$, in an energy region above the thermal equilibrium region. The total distribution is then described by [2]

$$n_f(E_n)dE_n = A\left(\frac{E_n}{(k_B T)^2} e^{-E_n/k_B T} + \eta \frac{1}{E_n}\right) dE_n \tag{3.31}$$

or in terms of wave length,

$$n_f(\lambda)d\lambda = B\left(\frac{\lambda_T^4 d\lambda}{\lambda^5} e^{-\lambda_T^2/\lambda^2} + \xi \frac{1}{\lambda}\right) d\lambda \tag{3.32}$$

Figure 3.9 shows a typical spectrum at the Institute Laue Langevin (ILL) in France and J-PARC in Japan for 1 MW proton power. The spectrum at ILL is described by Equations (3.29) and (3.30), but that of J-PARC has a $1/E$ tail at higher energy region above the Maxwellian peak as described by Equations (3.31) and (3.32). Instantaneous proton bombardment creates a very high peak flux in short-pulse spallation sources. It can be about 100 times higher in peak intensity of the coupled moderator (CM) than the cold source at ILL, the world's highest flux research reactor, although the time-averaged flux is about 1/4 of ILLs.

3.3.2 Pulse Peak Structure

Steady-state sources create a neutron flux continuous in time, including SINQ, PSI, which is an accelerator-based spallation source [4], while a pulsed source creates flux pulses with the period of the accelerator frequency or reflector rotation at a pulsed reactor at Dubna [5].

The neutron–time distribution from an infinite medium is analytically described at relatively high energies (slowing-down region) by [6],

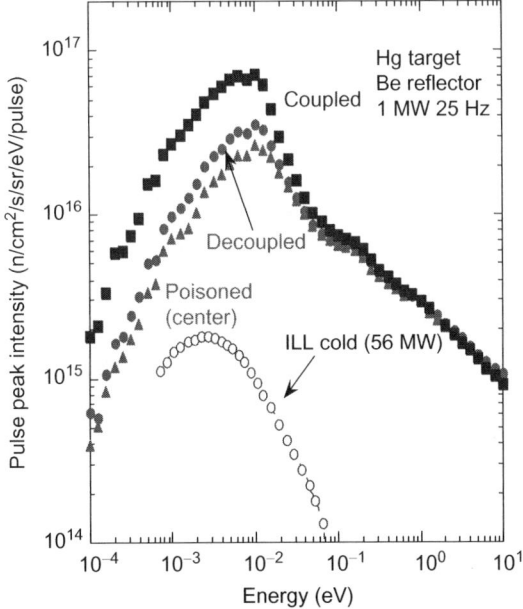

FIGURE 3.9 Calculated pulse peak intensities of CM, DM, PM (cf. Section 3.3.1) moderators at JSNS compared to the cold source at ILL (cf. Ref. [3]). (For color version of this figure, the reader is referred to the online version of this chapter.)

$$\phi(v,t) = \frac{\Sigma v}{2}(\Sigma vt)^2 \exp(-\Sigma vt) \qquad (3.33)$$

where Σ is the inverse mean-free-path, assumed constant for all energies, v is the neutron velocity, and t the time. Ikeda and Carpenter have developed a semiempirical formula to describe the pulse shape function over a wide energy range. They assumed two types of energy region: a slowing-down region, where neutrons come out after a small number of collisions, and a storage-region, where they reach thermal equilibrium. The former has a narrow distribution and dominates at high energies, while the latter has a broad distribution and dominates at low energies. The pulse shape is described by the so-called Ikeda–Carpenter function [7]:

$$\phi(v,t) = \frac{\alpha}{2}\left\{(1-R)(\alpha t)^2 e^{-\alpha t} + 2R\frac{\alpha^2\beta}{(\alpha-\beta)^3} \right. \\ \left. \times \left[e^{-\beta t} - e^{-\alpha t}\left(1+(\alpha-\beta)t+\frac{1}{2}(\alpha-\beta)^2 t^2\right)\right]\right\} \qquad (3.34)$$

for $t>0$ and is otherwise $=0$, where $\alpha=\Sigma v$ and β is an inverse storage time independent of neutron velocity. R is the ratio of the two terms. This function, shown in Figure 3.10, describes the observed pulse structure rather well.

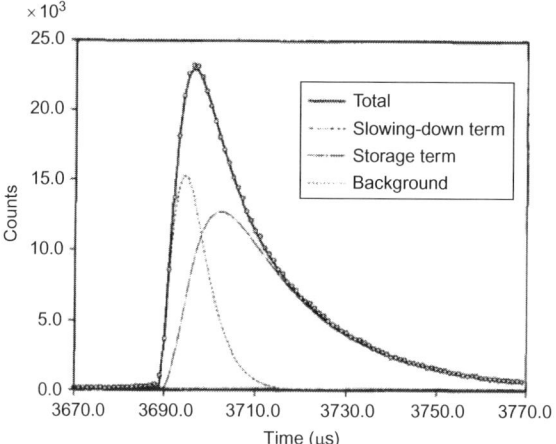

FIGURE 3.10 Pulse peak shape described by the Ikeda–Carpenter function.

3.3.3 Pulse Peak Width

The pulse width is of key importance for short-pulse spallation sources, since it determines the performance of the instruments, and the pulse peak intensity is another important measure. The width is proportional to $1/E^{1/2}$ or to the neutron wavelength λ in the $1/E$ slowing-down region, where neutrons escape from the moderator before equilibrium. It broadens in the thermal equilibrium region, then saturates at the low energies due to the storage process. The broadening starts to occur at about 300 meV (0.5 Å wavelength) for ambient-temperature moderators and about 15 meV (2.5 Å wavelength) for a 20 K methane moderator, as clearly seen in Figure 3.11 [8].

The pulse time width in the slowing-down region has a simple character that is proportional to wavelength or root-mean-square neutron energy:

$$\Delta t_m (\mu s) = 7\lambda \left(\text{Å} \right) = 2/\sqrt{E_n(\text{eV})} \qquad (3.35)$$

The proportionality between the pulse width and wavelength is of great importance, especially for high-resolution instruments such as powder diffractometers, as we will see in the later section.

3.3.4 Choice of Parameters in Spallation Sources

The design of instruments is tailored to specific scientific objectives and users' demands. This is also true in designing source characteristics. In particular, the characteristics of the moderators govern the performance of the instruments: resolution, intensity, dynamical range, etc. Therefore, conceptual design of moderators is a very important process, especially for in spallation sources.

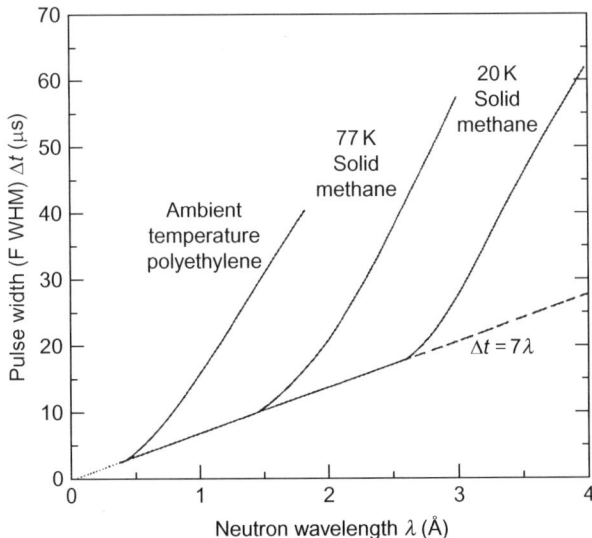

FIGURE 3.11 Pulse widths for typical moderators. In the $1/E$ region for each moderator, the pulse width is proportional to $\Delta t\ (\mu s) \sim \frac{2}{\sqrt{E_n\ (eV)}} \sim 7\lambda\ (\text{Å})$.

(1) Temperature, (2) neutronic structure, and (3) materials are the key parameters for characterizing moderators. Low temperature can naturally enhance the cold-neutron flux at the expense of the thermal neutron flux. More importantly, this concept can extend the slowing-down region to the lower energy and make sharp pulse structures in the extended energy range.

In recent advanced neutronics technology, there are three kinds of moderator concept: (1) the CM, for high flux while sacrificing peak width; (2) the decoupled moderator (DM), offering good resolution; and (3) the poisoned decoupled moderator (PM), which gives very high resolution with very sharp peak structure but with substantial sacrifices in intensity. Figure 3.12 shows characteristic of peak width for these moderators.

The CM is a hybrid device, in which a cold moderator is surrounded by an ambient-temperature water moderator layer. This concept, invented by Watanabe and Kiyanagi, considerably enhances the neutron flux from a cold moderator without serious heat deposition in it [9]. Recently constructed spallation sources are deriving great benefit from this moderator by obtaining high fluxes (Figure 3.9). The DM has a neutron decoupling layer around the moderator, neutronically separating it from the reflector and other components of the target-reflector assembly. A decoupler (absorber) with a decoupling energy of typically 1 eV, cuts off neutrons from the reflector after full thermalization and thus sharpens the time structure of the neutron pulses. To obtain further sharpness in the peak structure, an absorbing plate is

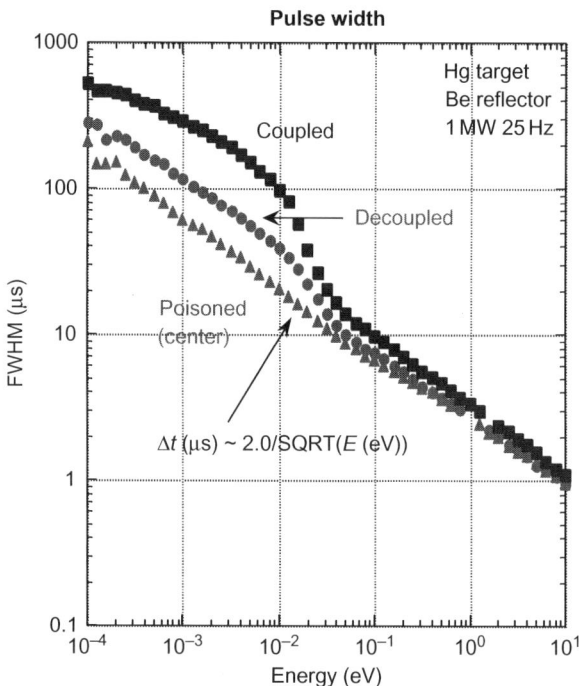

FIGURE 3.12 Pulse time width for the coupled, decoupled, and poisoned moderators at J-PARC. (For color version of this figure, the reader is referred to the online version of this chapter.)

inserted in the middle of the moderator, leading to the poisoned moderator. This type of moderator is used for extremely high-resolution instruments, such as the high-resolution powder diffractometer.

In Figure 3.13, the pulse peak structure of the three moderators is compared. The choice of moderator depends on the requirement or design concept of the instruments viewing it [10].

All three are hydrogen moderators at 20 K but the neutronics structure is quite different, in order to be optimized for different instruments.

3.3.5 High-Energy Neutron Background

When creating neutrons by spallation reactions, nuclear evaporation is the major process producing neutrons, giving a maximum flux at around 2 MeV. After thermalization in the moderators, part of the neutrons reduce their energies, but the majority of neutrons coming from the moderators are energetic neutrons above 1 eV, as shown in Figure 3.14. Neutrons above the keV energy range can easily go through steel shields and produce

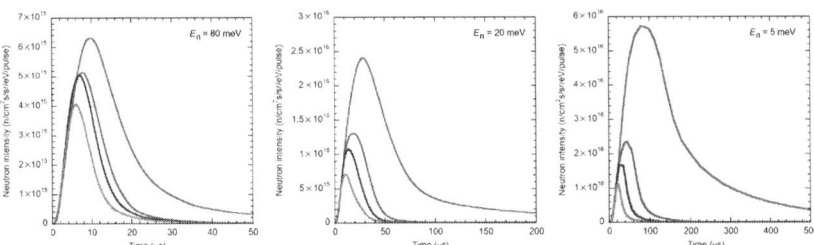

FIGURE 3.13 Pulse structures of neutrons from CM (blue), DM (green), PM (black); poisoning depth 3.7 cm, red; poisoning depth 2.5 cm at 5, 20, and 80 meV respectively. (For interpretation of the references to color in this figure legend, the reader is referred to the online version of this chapter.)

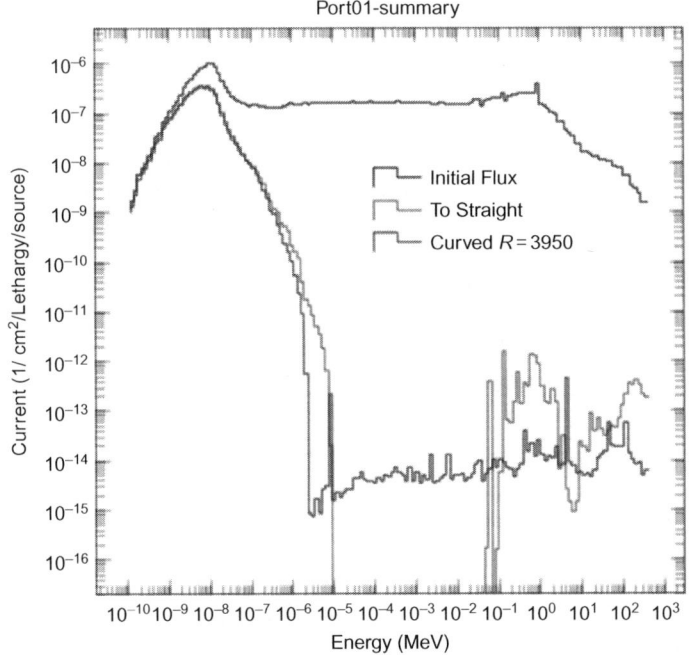

FIGURE 3.14 Neutron intensity as a function of energy at a beam port of JSNS; background suppression by either a curved guide or a T_0 chopper is clearly shown [11]. (For color version of this figure, the reader is referred to the online version of this chapter.)

backgrounds after thermalization in the instruments. These emerge sometime later around the main time for the TOF measurement, so it is very important to remove them before they pass into the instruments.

There are two methods to dramatically reduce high-energy backgrounds. If an instrument has a long flight path and it is possible to limit the energy

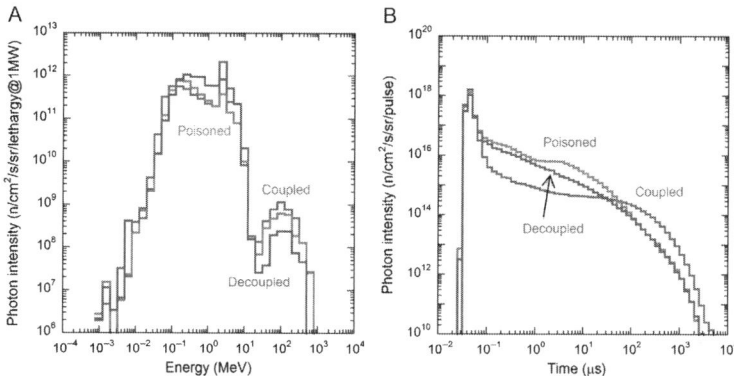

FIGURE 3.15 Gamma spectrum from JSNS moderators (a) and the time structure (b). (For color version of this figure, the reader is referred to the online version of this chapter.)

band width, then a good way to introduce a curved guide. Otherwise a background-suppression chopper (so-called T_0 chopper) can be an alternative choice. Figure 3.14 shows how the two methods can effectively suppress backgrounds [11].

Recent neutronics calculations also give detailed information on gamma rays [12]. Figure 3.15a shows a gamma spectrum at the moderator. A peak at around 2 MeV is caused by neutron-capture gamma rays from hydrogen nuclei in moderator and a peak at 100 MeV comes from π_0 decay in the target. The intensity of the gamma rays is comparable to that of the neutrons, and the photon at 2 MeV survives for 2 ms after proton bombardment. Therefore, it is also crucial to reduce the gamma flash with either of two methods described above.

3.4 DIFFRACTION TECHNIQUES

There are two kinds of diffraction technique. One is the angle-scanning method with monochromatic neutrons, commonly used in steady-state sources (cf. Figure 3.16). The scattering vector, Q, is scanned by changing the scattering angle and crystal angle in either a "$\theta/2\theta$" or a "$\omega/2\theta$" mode, with a constant wavelength as discussed in Chapter 2. Another is the wavelength-scanning method by TOF utilized in pulsed neutron sources. The scattering vector is scanned by automatically changing the wavelength while keeping the scattering angle constant. Both of them, of course, satisfy the Bragg's law:

$$Q = 4\pi \sin(\theta)/\lambda \qquad (3.36)$$

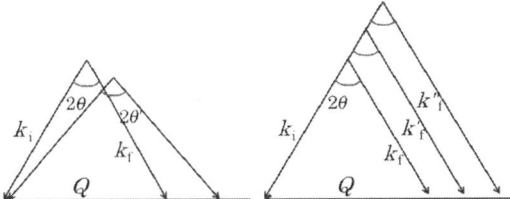

FIGURE 3.16 Q-scan methods. Angle scanning with a monochromatic beam and TOF with a white beam. (For color version of this figure, the reader is referred to the online version of this chapter.)

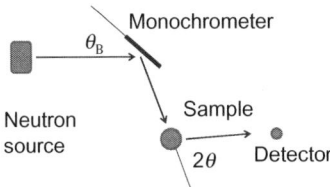

FIGURE 3.17 Angle-scanning diffraction with a monochromatic beam. (For color version of this figure, the reader is referred to the online version of this chapter.)

3.4.1 Powder Diffraction

3.4.1.1 Diffraction with a Monochromated Beam

This method was developed in the early days of neutron scattering at reactor sources. Incident neutrons are monochromated with a single crystal, and neutrons scattered by a powder sample are detected by a detector as a function of scattering angle 2θ by rotating the detector angle (cf. Figure 3.17).

This method is based on a very simple principle and has been well established in steady-state sources.

Figure 3.18 illustrates a typical powder diffractometer, the D2B diffractometer at the ILL. Neutrons monochromated at the monochromator crystal arrive at the sample and the scattered neutrons are detected in a detector array surrounding the sample. Thus, diffraction patterns are obtained as a function of scattering angle [13].

3.4.1.2 TOF Powder Diffraction

3.4.1.2.1 Principle of TOF Diffraction

The energy or wavelength of neutrons arriving at a diffractometer is analyzed by TOF recorded at detectors. Hence, all neutrons emitted from the moderator through the flight path come to the sample are scattered and then detected by the detector. Since there is no monochromator and no neutron loss there, it is a very efficient way to use neutrons. Figure 3.19 shows an illustration of principle of a TOF diffractometer. Neutrons created at the neutron source pass

Chapter | 3 Experimental Techniques

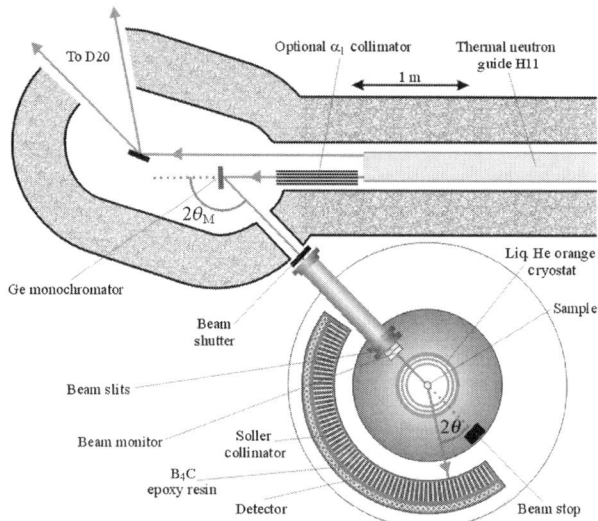

FIGURE 3.18 Illustration of D2B diffractometer at ILL. (For color version of this figure, the reader is referred to the online version of this chapter.)

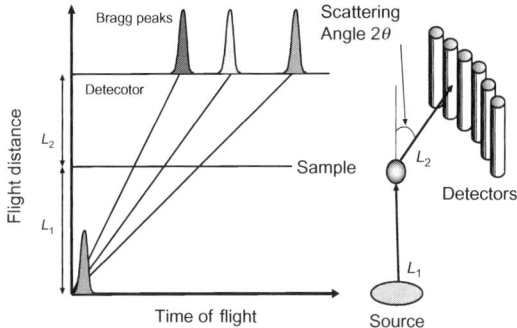

FIGURE 3.19 Layout of time-of-flight diffraction. (For color version of this figure, the reader is referred to the online version of this chapter.)

along the flight path L_1, are scattered at the sample and detected by detector at a distance L_2 from the sample. The neutron count is recorded as a function of TOF, with reference to the time of proton bombardment of the target.

The pulse structure created at the moderator is retained during the flight. Therefore, either sharp peaks or long flight paths provide high resolution for the instrument. Spallation neutron sources are very rich in epithermal neutrons, for which neutron guides are not effective to transport. Therefore, if an instrument such as a liquid-amorphous diffractometer needs to utilize epithermal neutrons, it should have a short flight path.

The resolution of a diffractometer is given by

$$\left(\frac{\Delta Q}{Q}\right)^2 = (\cot\theta\cdot\Delta\theta)^2 + \left(\frac{\Delta t}{t}\right)^2 \tag{3.37}$$

There are two important terms in Equation (3.37). One is an angular term and the other is a timing term. The first term comes from any angular uncertainty and the major contributions come from those in the incident and scattered beams, related to the finite size of the moderator, sample and detector. (Sometimes we should consider beam divergence created by a super-mirror guide as described in Section 3.8.1) The relation between TOF and wavelength is:

$$t(\mu s) = \frac{m_n \lambda L}{2\pi\hbar} = \frac{\lambda(\text{Å})\cdot L(\text{cm})}{0.3956} \tag{3.38}$$

Here, we note the importance of the relation ($\Delta t_m = 7\lambda$, Equation 3.35) in order to achieve a constant Q-resolution. This is a highly demanded feature for high-resolution powder diffractometers, whose diffraction pattern has a higher density of Bragg peaks at higher Q. Equation (3.35) makes the second term in Equation (3.37) constant. Because of this characteristic, such diffractometers are traditionally installed at a beam port viewing either DM or PM.

Assuming an instrument viewing a PM, then from Equations (3.37), (3.38) with (3.35), the resolution possible at a high-angle detector is easily estimated. For instance, for a flight path length $L_1 = 100$ m, the resolution becomes $\Delta Q/Q = 0.04\%$ with the condition of Equation (3.35).

3.4.1.2.2 Time Focusing

TOF diffraction gives a momentum transfer as a function of both of scattering angle, θ, and TOF, t, from Equations (3.4) and (3.25), as

$$Q = 4\pi\frac{\sin(\theta)}{\lambda} = \frac{2\pi m_n L \sin(\theta)}{\hbar \, t} \tag{3.39}$$

Therefore, the diffraction pattern for a different scattering angle θ has Bragg peaks at different TOFs. For powder diffraction, this is a very effective way to increase the statistics by adding up signals over an extended angle. A simple integration gives the diffraction pattern (a) in Figure 3.20. However, if we take a reduced TOF, t^*, for each scattering angle θ_i, where θ_0 is a reference angle,

$$t^* = t_i \frac{L_o \sin(\theta_o)}{L_i \sin(\theta_i)} \tag{3.40}$$

then, the reduced diffraction pattern from different scattering angles merges into a single pattern as shown in Figure 3.20b [14]. This is a so-called *in silico*

Chapter 3 Experimental Techniques

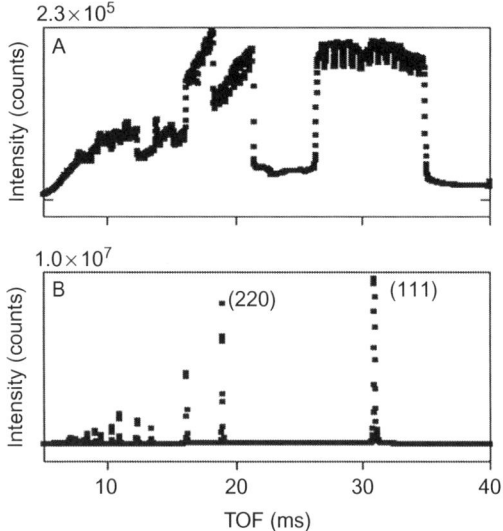

FIGURE 3.20 (a) Summed-up intensities from a range of detectors. (b) After time focusing as a function of the reduced TOF (Equation 3.40).

time focusing. In former times, a detector arrangement was designed so that the TOF from each detector satisfies Equation (3.40). Nowadays, the detector arrangement is optimized in terms of resolution, cost effectiveness, etc., and time-focusing integration is easily done on a computer.

3.4.1.2.3 Inelastic Effects with TOF Method

When a diffraction measurement is made without energy analysis of the scattered neutrons, we need to take care of inelastic effects. The effect can be negligible for powder diffraction, where the intensity is concentrated in Bragg peaks. However, diffraction from liquid and amorphous samples involves an obvious inelastic effect, which arises from the dynamics of the atoms. Figure 3.21 shows a distance–TOF diagram with three possible paths to reach at a detector at the same TOF. The paths with a kink at the sample show that inelastic scattering took place in the sample. Any path satisfying $t = t_1 + t_2$ gives an inelastic contribution to the observed intensity. Therefore, the observed diffraction intensity is an integrated intensity along all possible paths [15].

L_1 and L_2 are the distances between the source and sample, and sample and detector, respectively.

We define a fraction

$$f = L_2/(L_1 + L_2)$$

and a scattering vector for elastic scattering, $Q_e = 2k_e \sin \theta$.

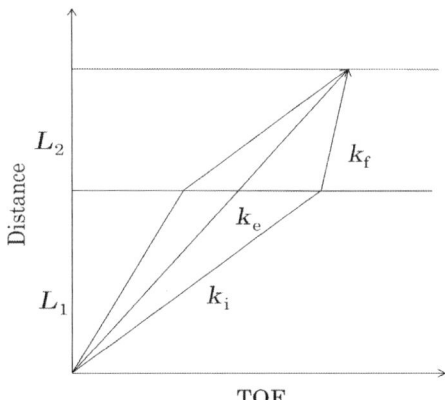

FIGURE 3.21 Schematic diagram illustrating the inelastic effects in TOF diffraction.

We further define

$$E_Q = \hbar^2 Q_e^2/2m$$
$$x = k_i/k_e - 1 \qquad (3.41)$$

We can obtain a locus, along which observed intensity is integrated in (E,Q):

$$\begin{cases} Q^2 = \left(\dfrac{Q_e}{2\sin\theta}\right)^2 (1+x)^2 \left[1 + \dfrac{f^2}{(f+x)^2} - \dfrac{2f\cos 2\theta}{(f+x)}\right] \\ \hbar\omega = E_Q \left[\dfrac{1+x}{2\sin\theta}\right]^2 \left[1 - \dfrac{f^2}{(f+x)^2}\right] \end{cases} \qquad (3.42)$$

In Figure 3.22, the loci are illustrated for several different conditions by changing the path length and the scattering angle.

In general, when the scattering angle $\varphi = 2\theta$ is small, the trajectory is a straight line along the energy axis and integration is properly performed as Equation (3.43). This condition corresponds to a static approximation, where $k_i \sim k_f$, both with large values:

$$S(Q) \sim \int \dfrac{k_f}{k_i} S(Q,E) dE \qquad (3.43)$$

Because of this relation, a liquid and amorphous diffractometer has detectors at smaller angles and uses epithermal neutrons, as we will see in Section 3.4.2.2.

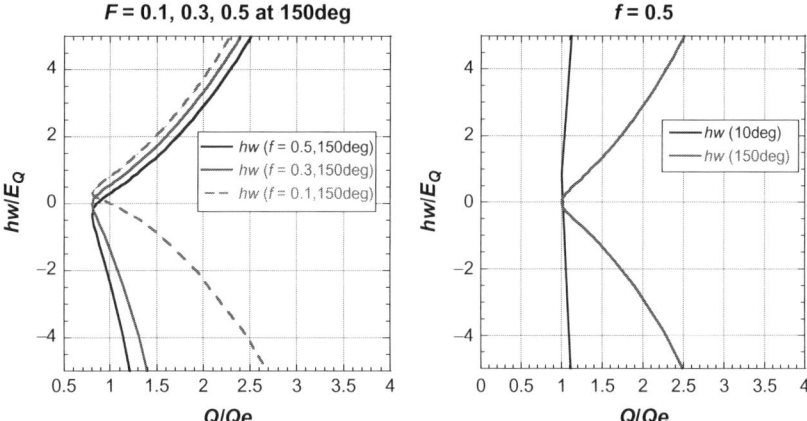

FIGURE 3.22 Scanning trajectory in the energy–momentum space for TOF for several different conditions of the path length and the scattering angle. (For color version of this figure, the reader is referred to the online version of this chapter.)

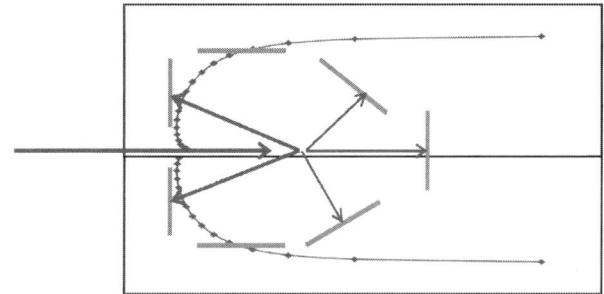

FIGURE 3.23 The detector arrangement denoted by the black lines gives a constant resolution at any scattering angle. However, it needs an infinite detector tail along the small-angle direction and is thus unrealistic. The compromise arrangement is shown in red. (For interpretation of the references to color in this figure legend, the reader is referred to the online version of this chapter.)

3.4.1.3 Detector Arrangement for a Resolution-Oriented Instrument (Powder Diffractometer)

If we want to have an instrument that gives a constant resolution in $\Delta Q/Q$ at any detector angle, a balance between the angular term and the timing term in Equation (3.37) determines the detector arrangement shown by the black line in Figure 3.23. This arrangement, however, involves an enormous cost for detectors and is unrealistic because of the infinite tail in the small angle direction.

If the source frequency is high enough, say 50 or 60 Hz, the detector should have a continuous range of scattering angles to cover a required

Q-range, as has been adopted in the POWGEN diffractometer at SNS [16] (Figure 3.24). The wavelength band is fairly narrow because of the source frequency. Therefore, the Q-range is spanned by scattering angle and not by TOF. On the other hand, we can discontinuously arrange detectors for a low frequency source, such as J-PARC, since a wide wavelength band in a TOF frame can itself give a wide Q-range for each detector, bank, as shown by the red lines in Figure 3.23, and actual instruments are shown in Figures 3.24 [16] and 3.25 [17]. In this case, the data from each detector bank

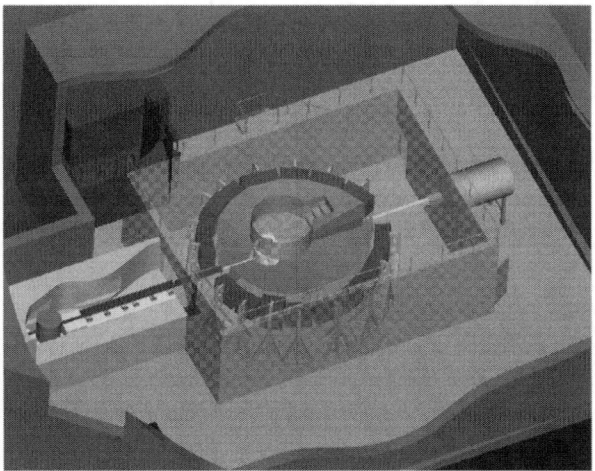

FIGURE 3.24 Layout of POWGEN at SNS. The detector arrangement covers a continuous range of scattering angle. (For color version of this figure, the reader is referred to the online version of this chapter.)

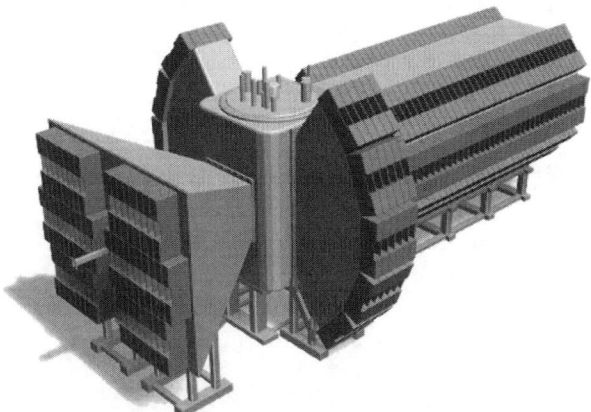

FIGURE 3.25 Detector arrangement of iMATERIA. The detector coverage is a discrete arrangement, separated into high-, medium-, and small-angle banks. (For color version of this figure, the reader is referred to the online version of this chapter.)

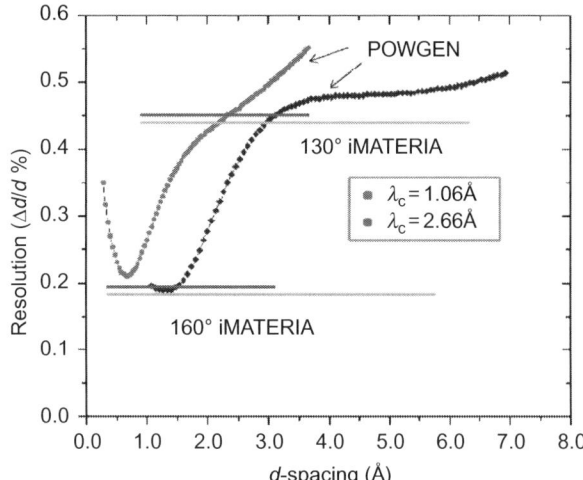

FIGURE 3.26 Resolution of POWGEN for a typical wavelength and iMATERIA for a single frame (red) and double frame mode (blue). (For interpretation of the references to color in this figure legend, the reader is referred to the online version of this chapter.)

are integrated in a time-focusing mode into a single diffraction pattern, as described in the previous section.

The resolution for the two instruments has quite different characteristics. One has a varying angular component because of the continuous angular detector coverage in order to scan a diffraction pattern by angle. The other has constant resolution because of discrete detector banks as illustrated in Figure 3.26. A diffraction pattern with constant resolution is easier to analyze, however, an instrument with a continuous detector arrangement can utilize the entire detector area to collect very high intensity.

3.4.1.4 Total Scattering Diffractometer

Another kind of diffractometer has a quite different concept for its design. This is the total scattering diffractometer, which is used to analyze the structure of liquid and amorphous systems, whose diffraction does not show any sharp Bragg peaks, but has a smooth structure consisting of inelastic intensity integrated with respect to energy through a scattering locus, giving an instantaneous atomic pair correlation. These materials do not have a regular atomic periodicity, so it is important for the instrument to have a very wide dynamical range in the momentum transfer Q, so that the Fourier transformation truncation error is minimized in the real-space pair correlation function. However, the energy integration leads to a serious difficulty in obtaining precise structural information, especially for light elements. Those have a recoil effect [18] on scattering, and have more pronounced inelasticity effects in the diffraction data [15]. Therefore, design of the detector arrangement should take into account both the resolution effect and the inelasticity effect at the same time.

An extreme case for a scattering from hydrogen (H_2) was examined with a free gas model [18], Equation (3.44), which gives the dynamical structure factor as follows, shown in Figure 3.27:

$$S(Q,\omega) = \left(\frac{\beta}{4\pi E_\gamma}\right)^{1/2} \exp\left(-\frac{\beta}{4\pi E_\gamma}(\hbar\omega - E_\gamma)^2\right), \qquad (3.44)$$

where $\beta = \frac{1}{k_B T}$, $E_\gamma = \frac{\hbar^2 Q^2}{2m_P}$

The elastic component is not elastically scattered anymore because of the recoil effect, and the dynamical structure factor has a quadratic shape in Q–E space. On the other hand, the scattering locus is described as Equation (3.42), which strongly depends on both scattering angle and neutron energy, as seen Figure 3.27 [19].

Scattering at smaller angles with higher energy makes the locus more straight along the energy axis and reduces the inelasticity effect, as discussed in the previous section. The detector arrangement is chosen by taking these effects into account, so that the difference between the integrated intensity at an edge of a detector bank from that of the next detector bank is minimized. The resulting detector arrangement is shown in Figure 3.28.

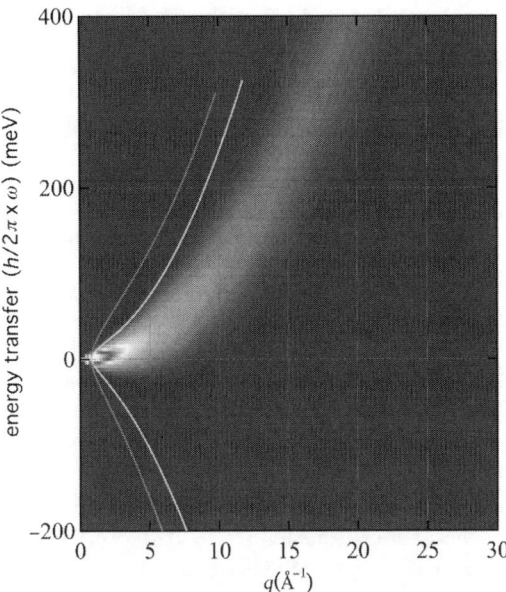

FIGURE 3.27 $S(Q,E)$ for a free gas model and scattering locus for a TOF measurement. The two lines are scanning locus at the nearest edges of the small (7deg, red) and medium angle (14deg, blue) bank for the same Q_e. (For interpretation of the references to color in this figure legend, the reader is referred to the online version of this chapter.)

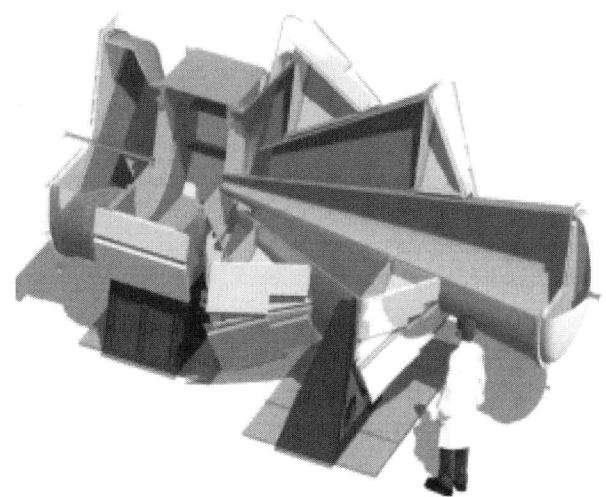

FIGURE 3.28 Detector arrangement for the total scattering instrument NOVA. The detector banks are inclined so as to optimize the resolution and inelasticity effects. (For color version of this figure, the reader is referred to the online version of this chapter.)

3.4.2 Single-Crystal Diffractometers

3.4.2.1 Four-Circle Diffractometer in a Steady-State Source

The single-crystal diffractometer is the most essential instrument for crystal structure analysis. On a steady-state source, the traditional four-circle diffractometer technique is normally used, with a monochromatic beam and a single detector. Rotations of the crystal are used to scan each reflection sequentially as discussed in Section 3.2.2.1. For each reflection, it is necessary to scan in reciprocal space in order that the reflection passes entirely through the Ewald sphere, correctly measuring its intensity. The scan is normally made in either a $\theta/2\theta$ or a $\omega/2\theta$ mode, in which the crystal and detector rotations are linked and rotated at the same time with a 1:2 ratio between the incident and scattering angles. An area detector makes it possible to increase the region of reciprocal space accessed in one measurement. However, it is still necessary to rotate the crystal to measure the intensities correctly [20].

Figure 3.29 shows the FONDER four-circle diffractometer installed at JRR3 [21], which is equipped with a wide-area detector system, covering a wide area in the reciprocal space simultaneously.

A structure factor observed with a monochromatic beam has some advantages in accurate structure analysis, since (1) the methodology is well established including X-ray diffraction measurements worldwide on four-circle diffractometers, and (2) there is no need for wavelength-dependent corrections such as absorption, extinction, and incident flux profile, simplifying error corrections of the observed data and giving reliable results in a straightforward way.

FIGURE 3.29 FONDER diffractometer at JRR3. (For color version of this figure, the reader is referred to the online version of this chapter.)

3.4.2.2 The TOF Single-Crystal Diffractometer

The time-of-flight Laue diffraction (tofLD) technique has demonstrated the efficiency of pulsed neutron source to access large volumes of reciprocal space in a single measurement. The technique was firstly developed in the 1970s at an early-pulsed source [22]. Now it is commonly used at high-flux pulsed sources worldwide.

In TOF diffraction the single-crystal sample stays at a fixed position, while the white neutron beam, with a continuum of wavelengths, scans a wide range of reciprocal space between $1/\lambda_{min}$ and $1/\lambda_{max}$, similar to quasi Laue diffraction but analyzed precisely by TOF [23]. This situation is illustrated in Figure 3.30 for a case of a linear position-sensitive detector (PSD). White-beam TOF single-crystal diffraction with an area detector is known as tofLD [24].

The unique feature of the TOF technique is that it allows access to large volumes of reciprocal space simultaneously in a single measurement. Thus, tofLD can observe many Bragg reflections during the same measurement as illustrated in Figure 3.30. These reciprocal space volumes, moreover, are completely resolved in three dimensions with an area detector, so that scattering between Bragg reflections is also measured (e.g., diffuse scattering as discussed in Chapter 1). Such complete measurements are unlike those made with a step-scan measurement, which involves a loss of an information between the steps [25].

The intensity of the Bragg reflection by the TOF method is given by the Lorentz factor, Equation (3.28).

There are two different concepts for a single-crystal diffractometer at a pulsed source. Since pulse structure is important for the TOF measurement, one instrument chooses a DM to have a sharp pulsed peak character.

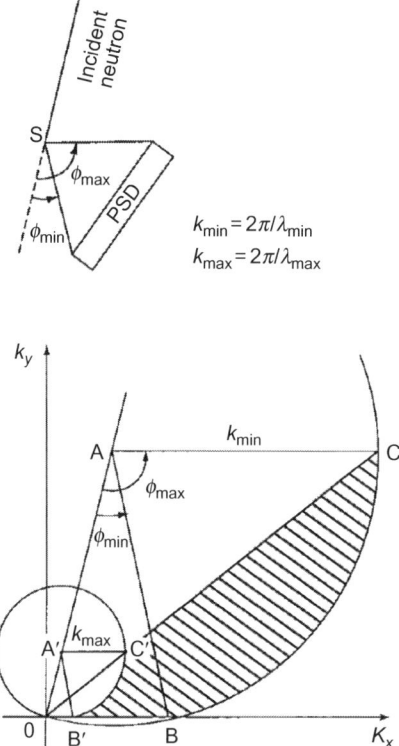

FIGURE 3.30 Top: scattering diagram for tofLD. Bottom: region of reciprocal space accessible in a single measurement.

This concept is the one so far traditionally taken on this kind of instrument, and still there are benefits with this choice, for example, the instrument can become reasonably short and can have a wide wavelength band. This kind of traditional instrument is useful and suitable for studying inorganic crystals with fairly small unit-cell size. Figure 3.31 shows the SENJU diffractometer dedicated to small crystals under extreme conditions such as high magnetic field, high pressure, etc. [26]. A counterpart is a diffractometer devoted to studies of protein crystals with a large unit cell, where intensity is a more important factor. Ibaraki Biological Crystal Diffractometer (iBIX) is such a diffractometer, installed at J-PARC, as shown in Figure 3.32, which views the CM, giving a high flux but with a broad pulse nature, with a fairly long flight path with a curved 40 m guide to eliminate direct gamma rays which can damage the sample [27]. Both these diffractometers can study a very small sample (<1 mm^3). This has been made possible not only by the source flux but also by the use of state-of-the-art scintillation detectors (described in Section 3.7) covering a very wide solid angle around the sample.

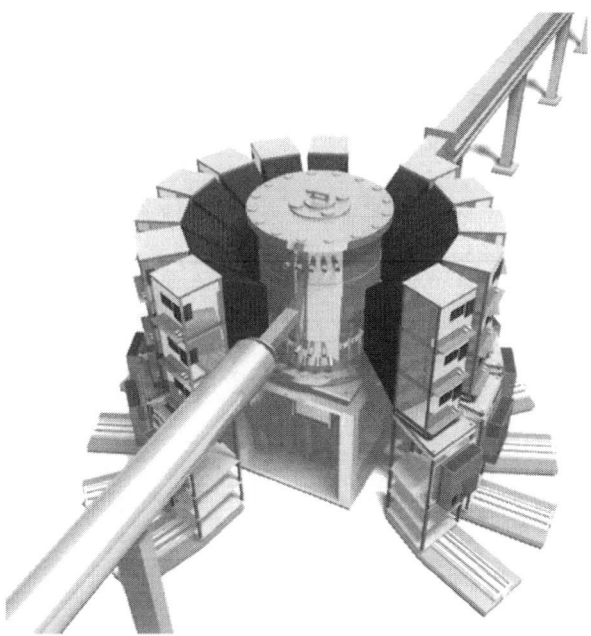

FIGURE 3.31 SENJU diffractometer for small crystal in an extreme condition. (For color version of this figure, the reader is referred to the online version of this chapter.)

FIGURE 3.32 iBIX diffractometer for protein structural analysis. (For color version of this figure, the reader is referred to the online version of this chapter.)

3.5 INELASTIC SCATTERING TECHNIQUES

3.5.1 Triple-Axis Spectrometer

One of the most typical neutron instruments that can be used to measure the double differential cross section is universally regarded to be the most versatile, the triple-axis spectrometer. This instrument was pioneered by Brockhouse [28] and the principle is illustrated in Figure 3.33. The triple-axis spectrometer has been used extensively to measure the energy dispersion of coherent excitations in single-crystal specimens.

The incident and scattered neutron wave vectors $\mathbf{k}_i$ and $\mathbf{k}_f$ are defined by Bragg reflection from the single-crystal monochromator and analyzer. A series of collimators C serve to define the resolution. At a selected scattering angle $2\theta_s$, the scattering angle defines the scattering vector

$$\mathbf{Q} = \mathbf{k}_i - \mathbf{k}_f \quad (3.45)$$

The scattering vector $\mathbf{Q}$ is of fundamental importance in defining a position in reciprocal space for an excitation such as a phonon, independent of the individual values of the wave vectors k_i and k_f and depending only on their difference.

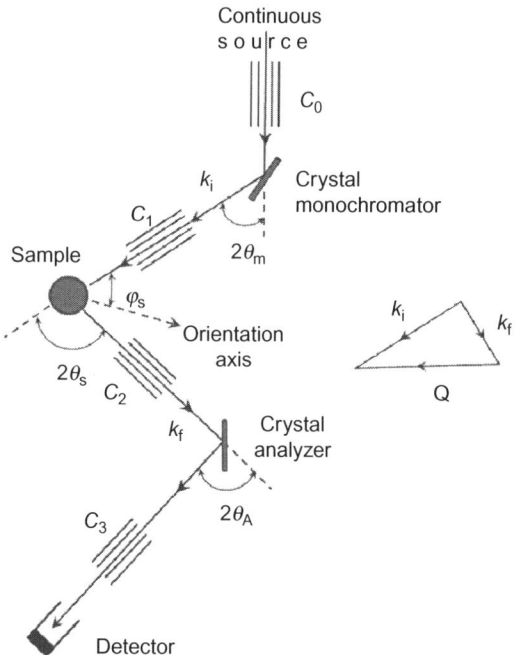

FIGURE 3.33 Illustration of a triple-axis spectrometer. Incident neutrons are monochromated with a crystal and the scattered neutron energy is analyzed by a second crystal [29].

The energy transfer $\hbar\omega$ is given by

$$\hbar\omega = \frac{\hbar^2}{2m_n}\left(k_i^2 - k_f^2\right) \tag{3.46}$$

where m_n is the neutron mass, $k_i = |\mathbf{k}_i|$, and $k_f = |\mathbf{k}_f|$. By changing the sample orientation direction φ_s, the scattering vector $\mathbf{Q}$ can be positioned anywhere in the reciprocal space.

The most frequently used scan gives a particular energy transfer at constant $\mathbf{Q}$, and involves keeping k_f constant by simultaneously changing θ_m, θ_A, θ_s, and φ_s.

3.5.1.1 An Advanced Triple-Axis Spectrometer: RITA [30]

An attempt to get beyond the point-by-point data collection is realized in the analyzer setup for RITA, a cold triple axis spectrometer developed at Risoe in Denmark (Figure 3.34). This concept is now used at several facilities. The single analyzer-detector arrangement of a conventional triple-axis spectrometer is replaced by an array of seven independently rotatable, flat vertical blades of pyrolytic graphite which scatter neutrons onto an area detector. By rotating the blades independently, as well as the entire array, it is possible to choose different modes of operation. A flat analyzer, together with Soller-slit collimators and an electronically selected window in the center of the area detector is the standard triple-axis geometry. By removing the collimators, one can also employ the standard horizontal focusing geometry onto the center of the detector. In this case, each analyzer blade is set for the same energy. All of them focus on the same point on the detector, and the analyzer array is set at an angle with respect to the scattered beam, as shown in Figure 3.34.

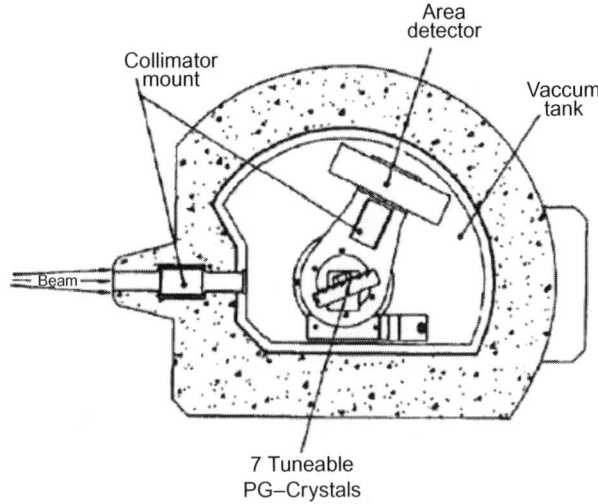

FIGURE 3.34 Schematic drawing of the RITA spectrometer.

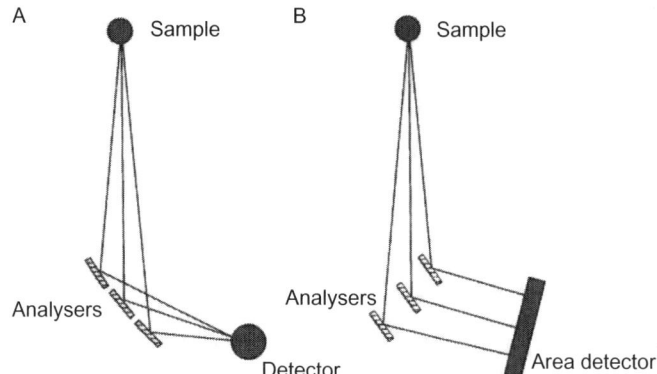

FIGURE 3.35 A multi-element analyzer operation in (a) point focus and (b) line focus mode.

Figure 3.35a illustrates the spot focusing mode where the integrated intensity in Q-space at constant energy transfer is gathered in a spot on the center of the detector, giving a fivefold increase of intensity compared to a flat analyzer.

Another arrangement is a line-focusing mode by taking an advantage of the area detector, Figure 3.35b, where the analyzer blades are displaced across the face of the detector. In this case, a constant energy scan perpendicular to $\mathbf{k}_f$ runs across the detector so that the gain in efficiency of horizontal point focusing is realized without the loss in Q-resolution.

Further developments of high-intensity spectrometers based on the RITA concept have been made in other facilities such as the National Institute of Standards and Technology (NIST) reactor. MACS is a novel cold-neutron spectrometer, having 20 independent detector channels with a double-focusing analyzer, enabling wave-vector resolved spectroscopy [31]. Because of a large focusing monochromator together with the analyzer system, the performance of this instrument exceeds by a factor 40 that of a conventional triple-axis spectrometer.

3.5.2 Chopper Instruments

The chopper inelastic scattering instrument LRMECS [32], known as a direct-geometry instrument, at a spallation source was developed at IPNS, Argonne National Laboratory, in the 1980s [33], and was subsequently improved with HET [34] and MARI [35] at a more intense sources, ISIS in the United Kingdom [36]. It has become one of the most typical and popular instruments for inelastic neutron scattering measurements at spallation neutron sources. Now MAPS [37], MERLIN [38], ARCS [39], SEQUOIA [40], and 4SEASONS [41], Figure 3.36, are equipped with wide-area detectors. These instruments monochromate neutrons ranging from the meV to the epithermal range

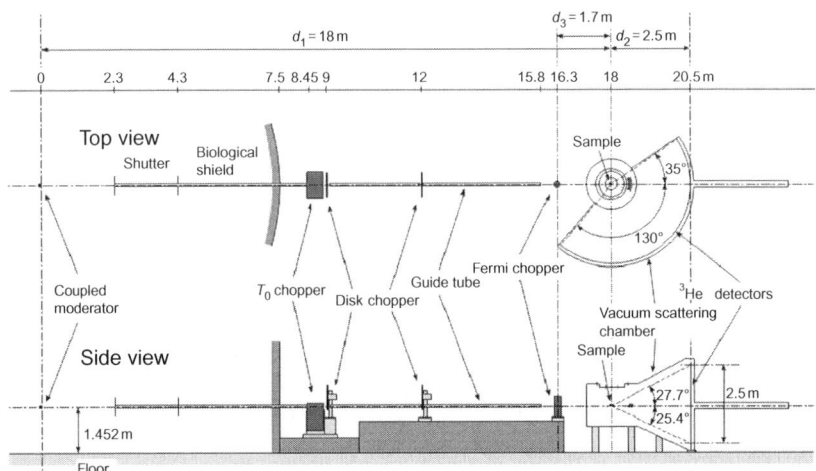

FIGURE 3.36 Layout of the 4SEASONS spectrometer at J-PARC. This is a Fermi-chopper instrument; however, it is equipped with two band-definition disk choppers as well as a T_0 chopper to eliminate unusable fast neutrons [41].

with a Fermi chopper placed in front of the sample, rotating synchronously in phase with the source.

Since chopper instruments are installed in most spallation neutron sources and frequently used for a variety of experiments, some details are described here [42]. In this section, we use the notation 1 and 2 instead of i and f, for incident and final terms, taking a traditional description for chopper instruments.

3.5.2.1 Energy Resolution of a Chopper Instrument

We can easily imagine that two dispersive terms compose the energy resolution. One is a contribution from the chopper opening time (cf. Figure 3.37). Another comes from the pulse width of moderator (cf. Figure 3.38). The energy derivative is translated into the time derivative as

$$\Delta E_1 = \frac{\partial E_1}{\partial t_1} \Delta t_1 = \frac{-2}{\left(\frac{1}{2}m_n\right)^{1/2}} (E_1)^{3/2} \frac{1}{L_1} \Delta t_1 \quad (3.47)$$

The same holds for the neutrons scattered by the sample

$$\Delta E_2 = \frac{\partial E_2}{\partial t_2} \Delta t_2 \quad (3.48)$$

The time width appearing at the detector is the sum of Δt_1 and Δt_2.

Here we note that the energy uncertainty of E_2 should be the same as E_1. Hence $\Delta E_1 = \Delta E_2$. This gives, from Equations (3.47) and (3.48),

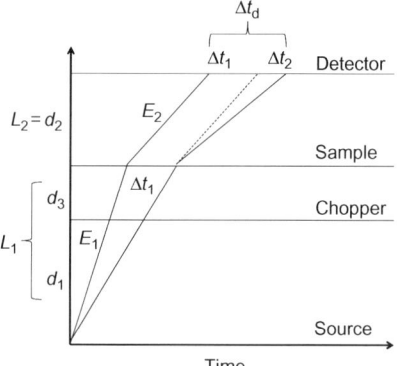

FIGURE 3.37 Ray diagram of the energy resolution starting from the chopper opening time.

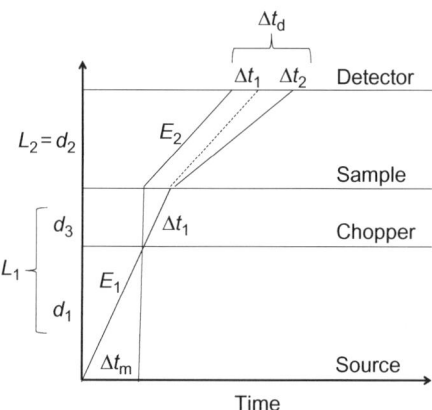

FIGURE 3.38 Ray diagram of the energy resolution due to the pulse time width in moderator.

$$\Delta t_2 = \frac{L_2}{L_1}\left(\frac{E_1}{E_2}\right)^{3/2} \Delta t_1 \qquad (3.49)$$

Hence, the time uncertainty at the detector is

$$\Delta t_d = \Delta t_1 + \Delta t_2 = \Delta t_1 \left[1 + \frac{L_2}{L_1}\left(\frac{E_1}{E_2}\right)^{3/2}\right] \qquad (3.50)$$

The energy transfer $\hbar\omega = \varepsilon = E_1 - E_2$.

Since the energy resolution at the detector is calculated with Δt_d as an uncertainty in E_2

$$\Delta\varepsilon = \frac{\partial E_2}{\partial t_2}\Delta t_d = 2\frac{E_2}{t_2}\Delta t_d \qquad (3.51)$$

with Equation (3.50) this leads to

$$\frac{\Delta \varepsilon}{E_1} = 2\frac{E_2}{E_1}\frac{\Delta t_d}{t_2} = 2\frac{\Delta t_1}{t_1}\left[1 + \frac{L_1}{L_2}\left(\frac{E_2}{E_1}\right)^{3/2}\right] = 2\frac{\Delta t_1}{t_1}\left[1 + \frac{L_1}{L_2}\left(1 - \frac{\varepsilon}{E_1}\right)^{3/2}\right] \quad (3.52)$$

In actual chopper instruments, there are three important distances, d_1, d_2, d_3 as shown in Figure 3.36, and Equation (3.52) is rewritten as

$$\frac{\Delta \varepsilon_{ch}}{E_1} = 2\frac{\Delta t_{ch}}{t_{ch}}\left[1 + \frac{d_1 + d_2}{d_2}\left(1 - \frac{\varepsilon}{E_1}\right)^{3/2}\right] \quad (3.53)$$

Another contribution to the energy resolution comes from the moderation time inside the moderator, Δt_m. In Figure 3.38, the effect of Δt_m is illustrated by assuming that the chopper opening time is decoupled and treated as a pinhole in time. Then it is easily evaluated and leads to the following contribution analogous to that shown in Equation (3.53), by just considering the process after the chopper,

$$\frac{\Delta \varepsilon_m}{E_1} = 2\frac{\Delta t_m}{t_{ch}}\left[1 + \frac{d_3}{d_2}\left(1 - \frac{\varepsilon}{E_1}\right)^{3/2}\right] \quad (3.54)$$

Therefore, the energy resolution is the root-mean-square sum of each squared term and is,

$$\frac{\Delta \varepsilon}{E_1} = \left\{\left[2\frac{\Delta t_{ch}}{t_{ch}}\left(1 + \frac{d_1 + d_3}{d_2}\left(1 - \frac{\varepsilon}{E_1}\right)^{3/2}\right)\right]^2 + \left[2\frac{\Delta t_m}{t_{ch}}\left(1 + \frac{d_3}{d_2}\left(1 - \frac{\varepsilon}{E_1}\right)^{3/2}\right)\right]^2\right\}^{1/2}$$

$$(3.55)$$

In Figure 3.39, a typical performance of Equations (3.53)–(3.55) is illustrated.

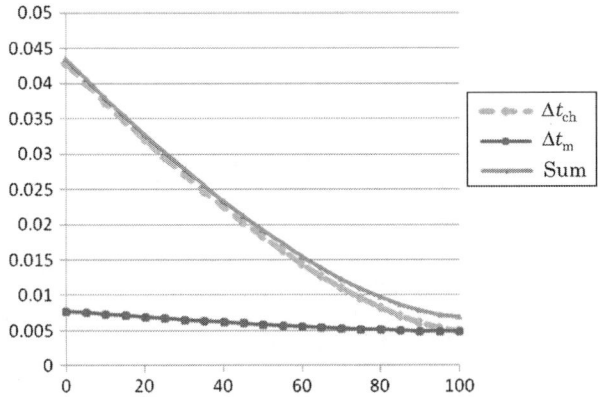

FIGURE 3.39 The individual component of the energy resolution (Equations 3.53–3.55). (For color version of this figure, the reader is referred to the online version of this chapter.)

Here, $E_1 = 100$ meV, $d_1 = 20$ m, $d_2 = 2.5$ m, $d_3 = 1.5$ m, $\Delta t_{\text{ch}} = \Delta t_{\text{m}} = 10$ μs are assumed.

When chopper instruments were developed in the 1970s they were installed at a bare beam port and aimed at utilizing epithermal neutrons, without any neutron optical equipment. In this situation, a long value of d_1 loses flux by the inverse square law, hence, a short d_1 with sharp chopper width and moderation time width were preferable together with $\Delta t_{\text{ch}} = \Delta t_{\text{m}}$, in terms of an intensity optimization. This is the reason why a DM was chosen for conventional chopper instruments in the past. However, the importance of cold neutrons has recently been recognized even for chopper instruments. When flux gain is improved by optical components such as a super-mirror guide, an innovative design concept has been proposed with a high-flux CM. This concept leads to a fairly long d_1 and tunable Δt_{ch} to match Δt_{m}. Δt_{ch} can be chosen much narrower than the broad Δt_{m} from a CM to realize reasonable resolution while keeping a high peak flux.

3.5.2.2 Q Resolution of a Chopper Instrument

Now we consider resolution in momentum space. The scattering diagram involves

$$\mathbf{Q} = \mathbf{k}_1 - \mathbf{k}_2 \tag{3.56}$$

$$\hbar\omega = \varepsilon = E_1 - E_2 = \frac{1}{2m_n}\left(\hbar^2 k_1^2 - \hbar^2 k_2^2\right) \tag{3.57}$$

$$Q^2 = k_1^2 + k_2^2 - 2k_1 k_2 \cos(\varphi) \tag{3.58}$$

$$E_Q \equiv \frac{\hbar^2 Q^2}{2m} = E_1 + E_2 - 2\sqrt{E_1 E_2}\cos(\varphi) \tag{3.59}$$

If we can assume the uncertainty in E_Q is only composed of those in E_2 and φ, then the total differential can be written as

$$\Delta E_Q = \frac{\partial E_Q}{\partial Q}\Delta Q = \frac{\hbar^2 Q}{m}\Delta Q = \frac{\partial E_Q}{\partial E_2}\Delta E_2 + \frac{\partial E_Q}{\partial \varphi}\Delta \varphi \tag{3.60}$$

This leads to the following formula by taking the error propagation into account:

$$\left(\frac{\Delta Q}{Q}\right) = \frac{1}{E_Q}\left\{\left[\frac{1}{2}\left(1 - \sqrt{\frac{E_1}{E_2}}\cos(\varphi)\right)\Delta E_2\right]^2 + \left[\sqrt{E_1 E_2}\sin(\varphi)\Delta\varphi\right]^2\right\}^{1/2} \tag{3.61}$$

We can express Equation (3.61) in a simple fashion

$$\left(\frac{\Delta Q}{Q}\right) = \left(\frac{\Delta Q(E_2)}{Q}\right) + \left(\frac{\Delta Q(\varphi)}{Q}\right) \tag{3.62}$$

The first term is caused by the uncertainty in k_2, E_2, or ε and its sine component onto Q.

$$\Delta Q(E_2) = \Delta k_2 \sin(\xi) = \frac{Q}{2E_Q}\left(1 - \sqrt{\frac{E_1}{E_2}}\cos(\varphi)\right)\Delta E_2$$

$$= \frac{Q}{2E_Q}\left(\sqrt{\frac{E_1}{E_2}}\cos(\varphi) - 1\right)\Delta\varepsilon \tag{3.63}$$

This is specified as "A" in Figure 3.40b.

The second term is the deviation caused by the uncertainty in φ and its cosine component onto Q, which is specified as "B" in Figure 3.40b. This is

$$\Delta Q(\varphi) = k_2 \Delta\varphi \cos(\xi) = \frac{Q}{E_Q}\sqrt{E_1 E_2}\sin(\varphi)\Delta\varphi \tag{3.64}$$

Figure 3.41 shows the behavior of each term, $\Delta Q(\varphi), \Delta Q(E_2), \Delta Q$, as a function of Q or E.

Most terms have a smooth behavior in the whole range except for the term related to E_2, which has a saddle point when the **Q** perpendicularly crosses $\mathbf{k}_2$.

The concept described above for the Fermi-chopper instrument is also applicable to a disk-chopper instrument such as AMATERAS, equipped with a pulse-shaping chopper viewing a CM. This correspondence can be achieved by replacing Δt_m with Δt_c, the opening time of the pulse-shaping chopper [43].

3.5.2.3 Observable Q–E Space

The accessible Q–E space can be obtained from the momentum and energy conservation conditions, $\mathbf{Q} = \mathbf{k}_1 - \mathbf{k}_2$ and $\hbar\omega = E_1 - E_2 = \frac{1}{2m}(\hbar^2 k_1^2 - \hbar^2 k_2^2)$. Figure 3.42 shows the scattering locus in $\hbar\omega$ and Qx–Qy space along with the scattering diagram of Figure 3.3, spanned by an array of detectors in the equator of the scattering plane. The scattering locus scans phase space like an umbrella. Each radial line is the locus for a detector at a certain angle. Semicircles correspond to various energy transfers.

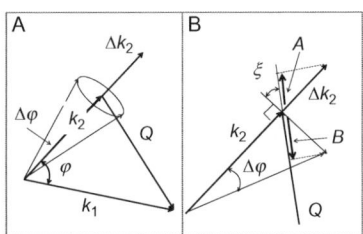

FIGURE 3.40 (a) Scattering diagram with an uncertainty in scattered momentum k_2 and scattering angle φ. (b) Enlarged diagram: A and B are the projection onto Q from the uncertainties in k_2 and φ. (For color version of this figure, the reader is referred to the online version of this chapter.)

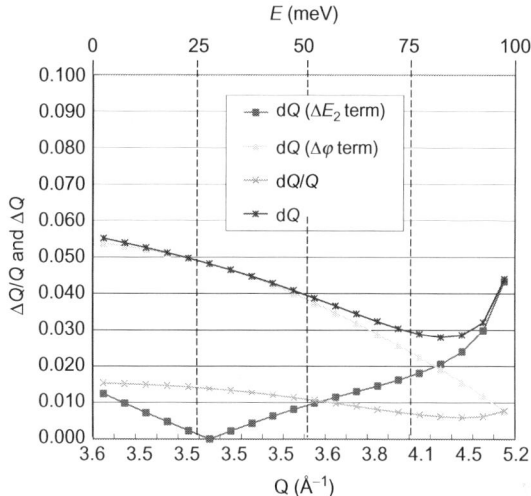

FIGURE 3.41 The behavior of $\Delta Q(\varphi), \Delta Q(E_2)$ as a function of Q or E at $\varphi=30°$. Here the calculation parameters are: $E_i=100$ meV, $d_1=18$ m, $d_2=2.5$ m, $d_3=1$ m, $\Delta t_m=\Delta t_c=6$ μs, $\Delta\varphi=0.01$ rad. The saddle point occurs when $\mathbf{Q}$ is perpendicular to $\mathbf{k}_2$, which shifts to higher energy with increasing the scattering angle in the range below 90°. (For color version of this figure, the reader is referred to the online version of this chapter.)

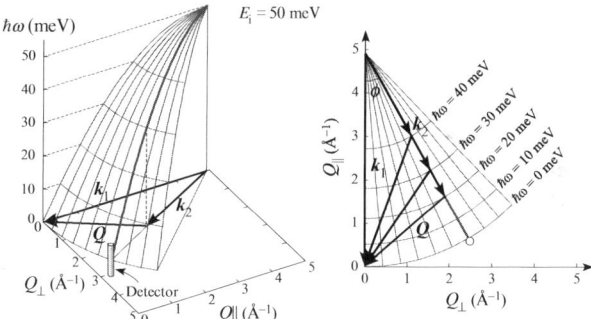

FIGURE 3.42 (a) Scattering locus in Q–E space, like an umbrella shape. (b) Projection of the umbrella onto the scattering plane. The semicircles show various energy transfers with corresponding k_2 and Q. The example shows the case for $E_1=50$ meV. (For color version of this figure, the reader is referred to the online version of this chapter.)

3.5.2.3.1 Glassy System

Only the absolute value of Q is meaningful for glass and amorphous systems and even for powder crystalline samples. The scattering locus is a projection of the umbrella as a function of the energy transfer and absolute value of Q. From Equations (3.58) and (3.59), we can obtain the relation between ε and Q as

$$\frac{\hbar^2 Q^2}{2m_n} = 2E_1 - \varepsilon - 2\sqrt{E_1(E_1 - \varepsilon)}\cos(\varphi) \qquad (3.65)$$

This equation gives scattering loci for each detector as shown in Figure 3.43.

Figure 3.44 shows the dynamical structure factor of an Al powder sample, in which phonon dispersion is clearly observed [44].

Here, we note that there is an area not easily accessed at the finite energy region at small Q. This is because of the kinematic constraint in energy–momentum space. Since the neutron is a de-Broglie wave with a mass m_n, the relation between E and Q leads to the constraints as Equations (3.56) and (3.57).

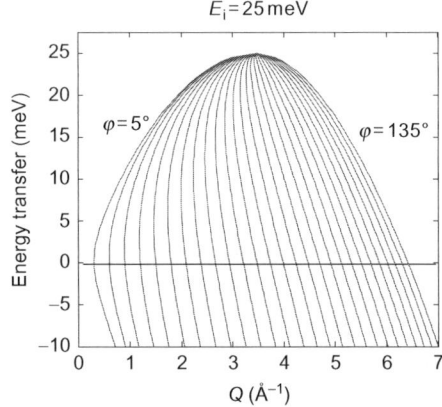

FIGURE 3.43 Scanning trajectories using Equation (3.65). (For color version of this figure, the reader is referred to the online version of this chapter.)

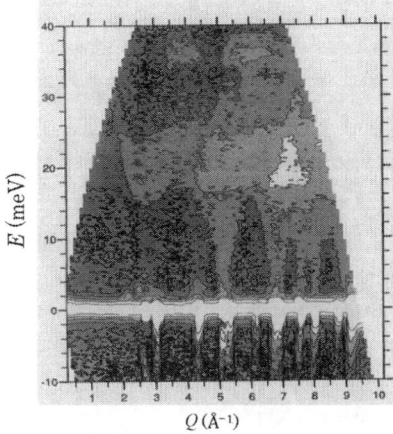

FIGURE 3.44 Dynamic structure factor $S(Q,E)$ for an Al-powder sample. (For color version of this figure, the reader is referred to the online version of this chapter.)

Higher energy can reduce this dead space; however, we need good resolution for an actual measurement. X-ray inelastic scattering (IXS) does not have this constraint and can access this area [45]. Liquids and noncrystalline system do not have reciprocal points except for the center of the reciprocal space. Acoustic phonon of those systems can be observed only in this area, and IXS and a specially designed instrument BRISP [46] are indispensable for the study.

3.5.2.3.2 One-Dimensional System

A one-dimensional system provides a typical example of a measurement with a chopper instrument [47]. The one-dimensional chain axis is perpendicular to k_1, as shown in Figure 3.45 (Figure 3.45 shows an experimental condition: b^* is parallel to k_1 and c^* is perpendicular to k_1). Then the umbrella locus cuts through the spin dispersion as shown in Figure 3.45. After projecting its cross section onto the $\hbar\omega$–c^* plane, we can get a clear picture of the spin excitation as shown in Figure 3.46, where very broad excitation due to quantum spin fluctuation is clearly detected. As demonstrated here, the chopper instrument is a very suitable and effective tool for observing broad excitations that occurs in the wide area of phase space. Because of this performance, chopper instruments are widely used in various scientific fields.

3.5.2.3.3 Two-Dimensional System

In the case of a two-dimensional (2D) system, there is a Bragg rod perpendicular to the 2D plane, along which there is no Q dependence. Therefore, if we chose a geometry such that the rod direction (c^*) is in parallel to k_1, then, we can determine an energy dispersion as in Figure 3.47. The two-dimensional detector surface at a certain energy level cuts through the dispersion and results are as shown in Figure 3.48.

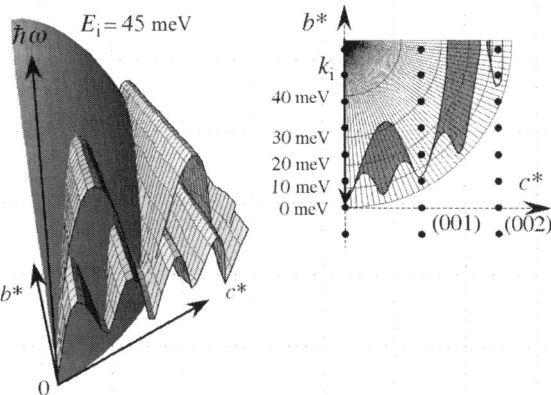

FIGURE 3.45 One-dimensional antiferromagnetic system. The chain axis is aligned perpendicular to k_1. (For color version of this figure, the reader is referred to the online version of this chapter.)

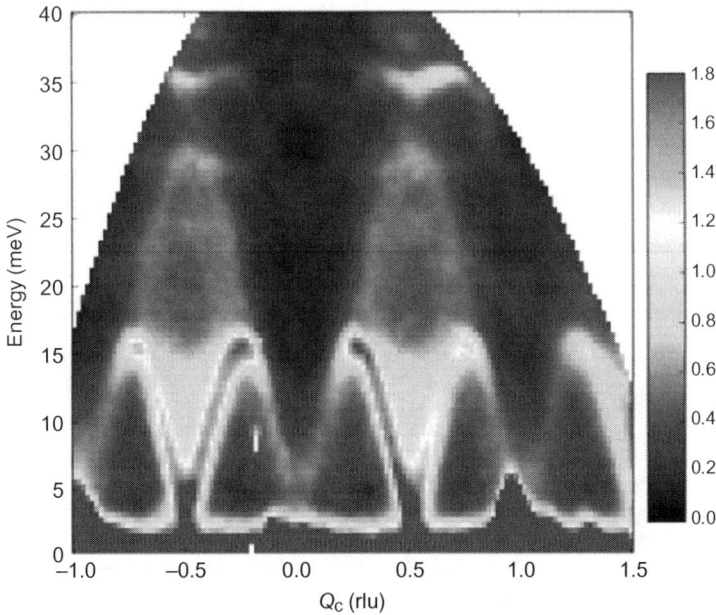

FIGURE 3.46 Spin excitation from the one-dimensional antiferromagnet $CuGeO_3$. (For color version of this figure, the reader is referred to the online version of this chapter.)

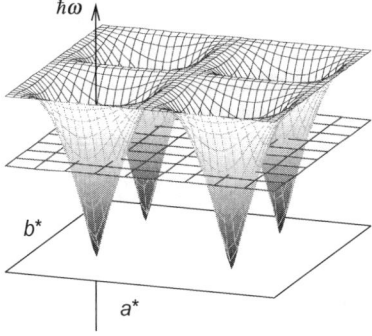

FIGURE 3.47 The rod direction c^* of a 2D system is set parallel to k_1. The a^*–b^* plane is parallel to two-dimensional detector plane. (For color version of this figure, the reader is referred to the online version of this chapter.)

3.5.2.3.4 Three-Dimensional System

Finally, if we have a three-dimensional system, the scattering locus can have an overlap in a very limited area with a spin dispersion as shown in Figure 3.49. Therefore, in order to obtain the full structure of the spin dispersion, we must either rotate the crystal or change the incident energy, so that the scattering locus scans the dispersion: Finally, we can have four-dimensional information in phase

Chapter | 3 Experimental Techniques

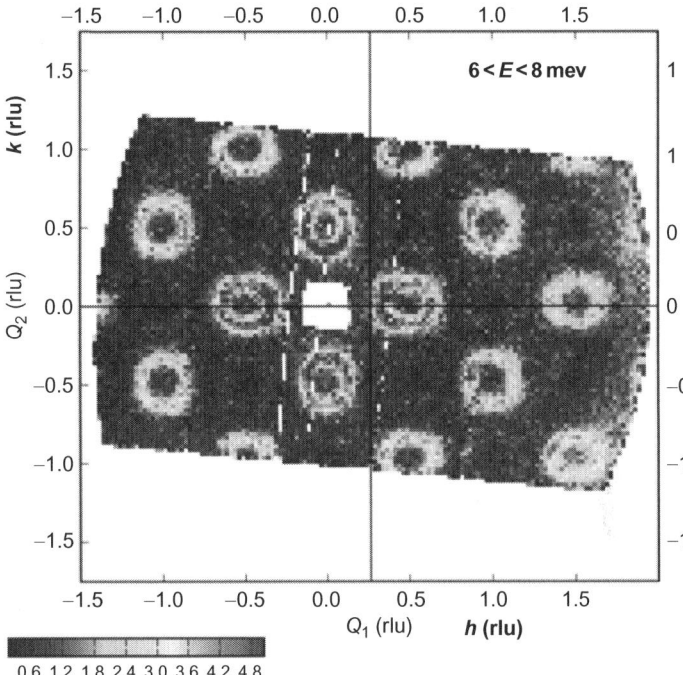

FIGURE 3.48 Energy slice of a spin-wave dispersion curves in $La_2CoO_{4.24}$ using the geometry shown in Figure 3.47. (For color version of this figure, the reader is referred to the online version of this chapter.)

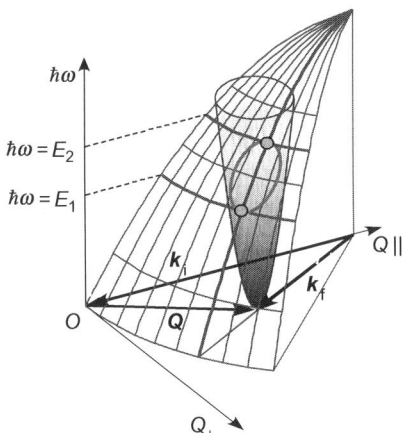

FIGURE 3.49 Spin-wave dispersion and scattering locus. Only the small overlap between them gives signal. (For color version of this figure, the reader is referred to the online version of this chapter.)

space, from which we can make a slice of the excitation at a certain energy in the Q_x–Q_y plane, etc. Figure 3.50 shows an equi-energy plane projected onto the Q_x–Q_y plane. This experiment was done by rotating a crystal 100° by 1° for each step.

3.5.3 Inverted-Geometry Instrument

There is another type of inelastic neutron scattering instrument, the inverted-geometry instrument, which analyzes scattered neutron energies with a crystal analyzer (Figure 3.51). The final energy is fixed instead of monochromating the incident energy. Thus, a wide-energy-band neutron beam reaches the sample, is scattered by it, and then scattered back to the detector by Bragg reflection at the crystal analyzer. The reflection angle at the analyzer is always the same if the analyzer is encircled on a Rowland circle. A large analyzer can collect neutrons scattered from the sample toward the detector with the same energy, known as an energy focusing analyzer. Therefore, this kind of instrument has high intensity with a relaxed Q-resolution and a wide dynamical range, but at the expense of high background caused by neutrons scattered by structures behind the analyzer. (In this section, we also use the notation

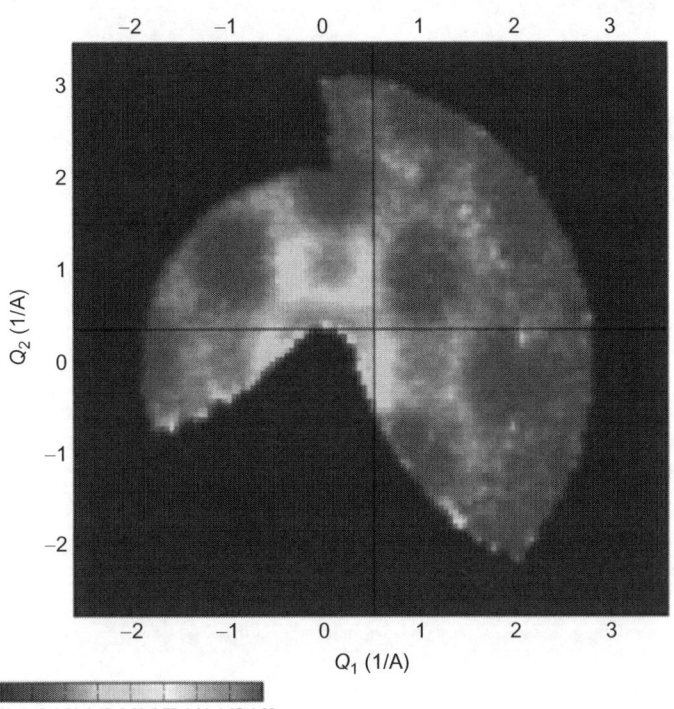

FIGURE 3.50 A $S(Q,E)$ slice on the Q_x–Q_y plane at fixed energy transfer. (For color version of this figure, the reader is referred to the online version of this chapter.)

Chapter | 3 Experimental Techniques

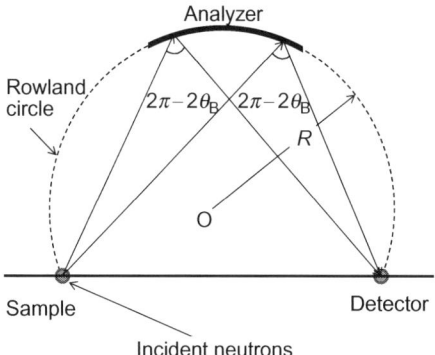

FIGURE 3.51 Schematic illustration of an inverted-geometry instrument. (For color version of this figure, the reader is referred to the online version of this chapter.)

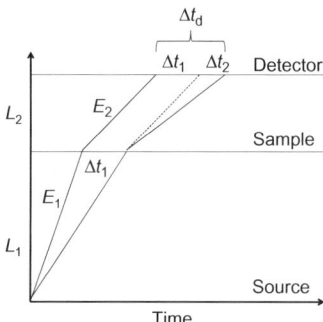

FIGURE 3.52 Contribution to the resolution function arising from the angular ambiguity of the analyzer crystal.

1 and 2 instead of i and f, for incident and final term by taking a traditional description for TOF instruments.)

3.5.3.1 Energy Resolution of an Inverted-Geometry Instrument

The energy resolution is also composed of two terms. The first term is attributed to the angular uncertainty of the analyzer crystal, whose process gives a similar diagram of the chopper instrument (Figure 3.52), but with the dispersive character created by the analyzer crystal. Therefore, the argument given in Chapter 3.5.2.1 for a chopper resolution can be taken from Equations (3.47)–(3.51). However, Equation (3.52) has to be modified to take into account the angular uncertainty,

$$\frac{\Delta \varepsilon'}{E_1} = 2\frac{E_2}{E_1}\frac{\Delta t_d}{t_2} = 2\frac{E_2}{E_1}\frac{\Delta t_1}{t_2}\left[1 + \frac{L_2}{L_1}\left(\frac{E_1}{E_2}\right)^{3/2}\right] \quad (3.66)$$

Here, Δt_1 is defined by the mosaicity of the analyzer, therefore,

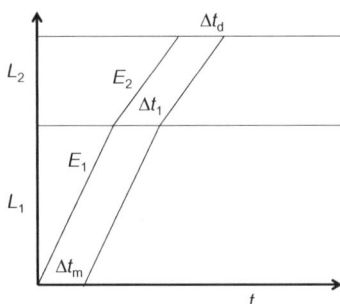

FIGURE 3.53 Contribution from the peak time width of moderator.

$$\frac{\Delta t_1}{t_2} = \frac{\Delta \lambda}{\lambda_2} = \cot\theta \cdot \Delta\theta \tag{3.67}$$

This result leads to,

$$\frac{\Delta\varepsilon'}{E_1} = \frac{E_2}{E_1}(\cot\theta \cdot \Delta\theta)\left(1 + \frac{L_2}{L_1}\left(\frac{E_1}{E_2}\right)^{2/3}\right) \tag{3.68}$$

Another contribution to the energy resolution comes from a time uncertainty of the pulse width. Here, we assume that the analyzer is a perfect crystal and the time width at moderator propagates without any change, as shown in Figure 3.53. Thus, the contribution from the pulse time width becomes,

$$\frac{\Delta\varepsilon''}{E_1} = 2\frac{\Delta t_m}{t_1} = 2\frac{\delta_m}{L_1} \tag{3.69}$$

Here, the peak time width Δt_m is converted to a length δ_m called the moderation depth. Typically, this depth is about 2.5 cm for a DM according to Equations (3.35) and (3.38), and it is largely independent of the moderating medium.

The energy resolution is a squared sum of Equations (3.68) and (3.69) and the total energy resolution can be described as

$$\frac{\Delta\varepsilon}{E_1} = 2\left[\left(\frac{\delta_m}{L_1}\right)^2 + \left\{\frac{E_2}{E_1}(\cot\theta \cdot \Delta\theta)\left(1 + \frac{L_2}{L_1}\left(\frac{E_1}{E_2}\right)^{2/3}\right)\right\}^2\right]^{1/2} \tag{3.70}$$

Here, E_2 is fixed by the analyzer. Therefore, in order to have a good energy resolution, it is important to have a small δ_m, a long L_1, a large θ, and short L_2. Because of these reasons, this kind of instrument with the high-energy resolution option was traditionally installed at a beam port viewing a DM with a long flight path L_1. There is, however, an innovative concept

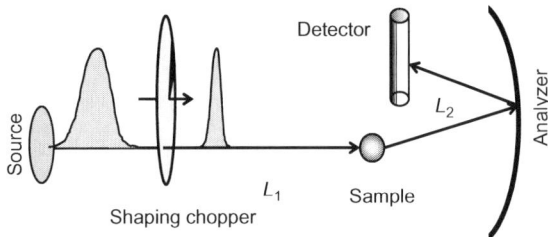

FIGURE 3.54 Illustration of an inverted-geometry instrument with a pulse-shaping chopper. (For color version of this figure, the reader is referred to the online version of this chapter.)

to realize a required energy resolution and high intensity viewing a CM by putting a pulse-shaping chopper near the source as illustrated in Figure 3.54 with a much shorter flight path L_1.

In the extreme case of perfect backscattering, Equation (3.70) has an additional term, the Darwin width $\Delta d/d$, due to a dynamical scattering process in a perfect crystal and it is described as follows at the elastic position:

$$\frac{\Delta \varepsilon}{E_1} = 2\left[\left(\frac{\delta_m}{L_1}\right)^2 + \left(\frac{\Delta d}{d}\right)^2 + (\cot\theta \cdot \Delta\theta)^2\right] = 2\left[\left(\frac{\Delta t_m}{t}\right)^2 + \left(\frac{\Delta d}{d}\right)^2 + (\cot\theta \cdot \Delta\theta)^2\right] \quad (3.71)$$

The estimated energy resolution of a typical backscattering instrument with $L_1 = 100$ m at a DM can reach 1.5 µeV as a natural consequence with a Si(111) backscattering crystal, $E_2 = 2$ meV, as shown in Figure 3.55.

3.5.3.2 DNA Instrument at J-PARC

The nearly completely near-backscattering instrument DNA has adopted a pulse-shaping chopper with a CM for the first time [48] (Figure 3.56). By shaping a broad peak of 200 µs at 2 meV to 10 µs, even with a short flight path of 40 m, three terms in Equation (3.71) become comparable to an order of magnitude 10^{-4} and can achieve a 1 µeV resolution with a Si(111) crystal.

Figure 3.57 shows a tunneling spectrum of γ-picoline [49], where the energy resolution is 3 µeV with a single-shaping chopper compared to 15 µeV for a natural pulse width without the chopper. A counter-rotating double-shaping chopper would realize a resolution of 1.5 µeV. Although the dynamical range becomes narrower because of the pulse-shaping window, and several rephasings of the chopper are necessary to scan the whole spectrum in the ±300 µeV range, very high statistics can be obtained within a reasonable measuring time—roughly half a day at 1 MW accelerator power.

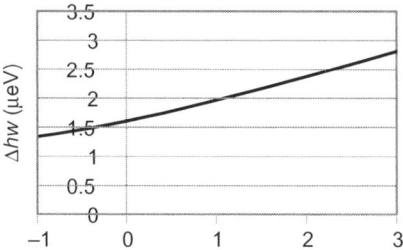

FIGURE 3.55 Energy resolution of a backscattering instrument with $L_1 = 100$ at a typical decoupled moderator, $\delta_m = 2.5$ cm. (For color version of this figure, the reader is referred to the online version of this chapter.)

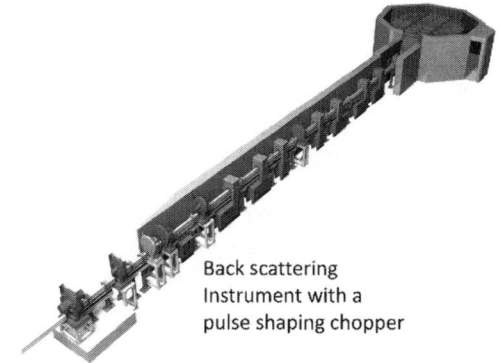

FIGURE 3.56 DNA instrument, equipped with a pulse-shaping chopper in the front. (For color version of this figure, the reader is referred to the online version of this chapter.)

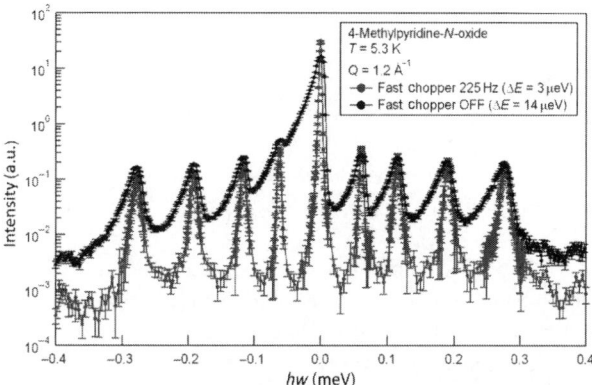

FIGURE 3.57 Tunneling spectrum from γ-picoline. The energy resolution is about 3 μeV with a single pulse-shaping chopper as designed (blue line), compared with a spectrum without operating the pulse-shaping chopper, having a resolution of 14 μeV. (For interpretation of the references to color in this figure legend, the reader is referred to the online version of this chapter.)

Asynchronous phasing of the chopper is also planned to automatically scan the entire spectrum over a wide energy band.

3.6 INSTRUMENTS FOR SEMI-MACROSCOPIC STRUCTURES

3.6.1 Small-Angle Neutron Scattering Instruments

The small-angle neutron scattering (SANS) technique was largely developed at the D11 instrument at ILL in the 1970s. Since then, this kind of instruments has received great demand for studies on large-scale structures of materials, ranging from polymer science, materials science, phase separation in alloy and glasses, biomolecule science, magnetism, superconductivity, and industrial applications. Now SANS is widely available in neutron scattering facilities worldwide.

Figure 3.58 illustrates the SANS instrument D11 at ILL [50]. It is composed of two parts. One is the collimator section including the velocity selector, 40 m long, and another is the detector tank starting right after the sample position, also 40 m long. The instrument receives neutrons from the cold source, 100 m upstream from the instrument. The polychromatic beam from the guide exit is monochromated by a helical slit velocity selector within a band of $\pm 9\%$ about a mean wavelength.

Neutrons scattered from the sample are detected on a 96×96 cm^2 wide ^{3}He multidetector in the vacuum tank. The detector can be placed at a distance between 1.2 and 39 m from the sample position, giving a total accessible momentum transfer range of 3×10^{-4} to 1 Å^{-1}.

3.6.1.1 Collimation and Resolution

The key element for SANS instrument is the collimators. Since the collimation can determine the minimum Q reachable and the background level around the small-angle signal, considerable efforts are made to have a sharp collimation.

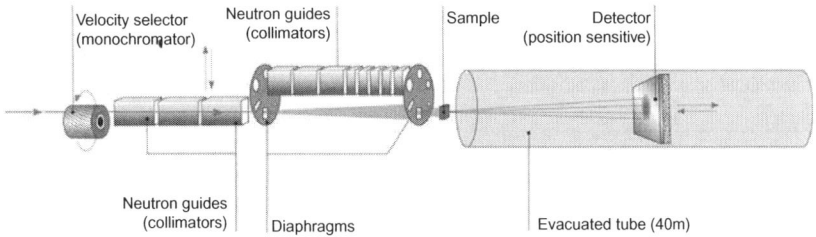

FIGURE 3.58 Illustration of D11, which is the archetype of a long, pinhole geometry SANS instrument. (For color version of this figure, the reader is referred to the online version of this chapter.)

The scattering vector for small-angle scattering can be written simply with the scattering angle φ.

$$Q = 4\pi \sin(\theta)/\lambda = \frac{2\pi\varphi}{\lambda} \tag{3.72}$$

The resolution is composed of angular term and wavelength term and can be described by

$$(\Delta Q)^2 = \left(\frac{2\pi}{\lambda}\Delta\varphi\right)^2 + \left(2\pi\varphi\frac{\Delta\lambda}{\lambda^2}\right)^2 \tag{3.73}$$

The angular term is the most crucial component for determining the performance of a SANS instrument because the wavelength term can be controlled by a velocity selector. The angular term is a geometrical contribution from collimators, sample and detector as illustrated in Figure 3.59a:

$$(\Delta\varphi)^2 = \left(\frac{d_1}{L_1}\right)^2 + \left(d_2\left(\frac{1}{L_1}+\frac{1}{L_2}\right)\right)^2 + \left(\frac{d_3}{L_2}\right)^2 \tag{3.74}$$

where d_1 is the size of collimator inlet, d_2 is the sample size, d_3 is the detector element size, L_1 is the distance between d_1 and d_2, and L_2 is that between d_2 and d_3.

Optimization of the intensity means that the four components including the wavelength term in Equation (3.73) should have approximately the same magnitude. Therefore, we can reasonably choose the dimension of each contribution as

$$L_1 = L_2 \text{ and } d_1 = 2d_2 = d_3 \tag{3.75}$$

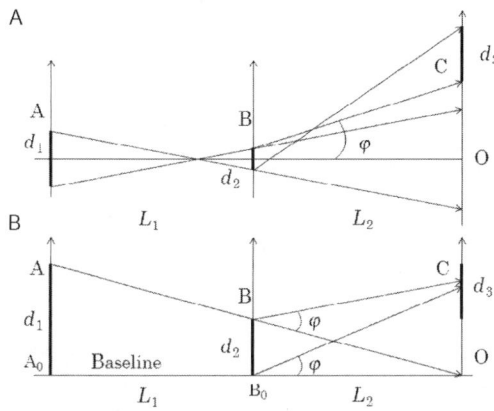

FIGURE 3.59 Ray diagram for small-angle scattering. (a) Conventional collimator, (b) converging collimator.

This result leads to the conventional configuration for instruments in steady sources [51].

3.6.1.2 Converging Collimation

A converging collimator, from which neutrons tend to go to the center of detector, is very useful for pulsed sources, where limiting the total length of instrument is important not to have frame overlap [52] and large sample can be used while maintaining resolution.

A neutron at A_0 in the baseline heading for O is scattered at B_0 on the sample and goes into position C at the detector. In the converging condition, every neutron started at any position at A heading to O scattered to C has the same scattering angle in the small-angle regime (Figure 3.59b).

Here, the converging condition satisfies Equation (3.76)

$$\frac{d_1}{L_1} = d_2 \left(\frac{1}{L_1} + \frac{1}{L_2} \right) \qquad (3.76)$$

Therefore, Equation (3.74) becomes

$$(\Delta \varphi)^2 = 2 \left(\frac{d_1}{L_1} \right)^2 + \left(\frac{d_3}{L_2} \right)^2 = \Delta \varphi_{coll}^2 + \Delta \varphi_{det}^2 \qquad (3.77)$$

Here, the angular resolution does not depend on sample size, and so a large sample does not spoil the resolution and leads to high intensity.

Since the neutron beam is convergent and make a spot at the detector, d_3 should be chosen reasonably small. Hence, the overall resolution is dominated by the performance of the collimator.

We estimate some concrete values for the parameters, starting from the conventional expression of the resolution in Equation (3.78) [53].

$$\left(\frac{\Delta Q}{Q} \right)^2 = \left(\frac{\Delta \varphi}{\varphi} \right)^2 + \left(\frac{\Delta \lambda}{\lambda} \right)^2 \qquad (3.78)$$

Consider a region at $Q = 0.01$ Å^{-1} measured with $\lambda = 6.4$ Å (2 meV). The scattering angle for this condition is $\varphi = 10$ mrad, as estimated from Equation (3.72). This is 10 cm from the center at a 10 m position of the detector. If we choose $\Delta Q/Q = 1/10$, then the detector cell size, d_3, can be 1 cm, $\Delta \varphi/\varphi$ can be 1/10 and $\Delta \lambda/\lambda$ can be also 1/10. This is a very typical example and many instruments at a steady-state source use a similar condition.

Selection of $\Delta \lambda$ directly influences the intensity. Therefore, tuning the band width in wavelength is very important in a steady-state source.

Contrary to this, $\Delta \lambda$ is not tunable and is a fixed parameter for moderators in pulsed sources, as we saw in Figure 3.12. It is about 200 μs for 6 Å neutrons from a CM. Since it takes 30 ms for 6 Å-wavelength neutrons to come to a detector at 20 m, the second term in Equation (3.78) is negligibly small,

1/150, compared to the angular term. This means that the performance of a short-pulse source is mismatched to the SANS technique and a long-pulse source can be more preferable for this purpose. However, an instrument at a pulsed source can provide a very wide dynamical range in momentum space in one measurement.

In order to increase the performance of small-angle scattering instruments, multiconverging techniques, multi-pin holes, Soller collimators, etc., have been proposed [54].

3.6.1.3 Focusing SANS

In 1998, Eskildsen *et al.* [55] demonstrated a biconcave MgF_2 neutron lens, and Choi *et al.* applied this technique to SANS to extend the smallest reachable Q-range at NIST. They succeeded to extend the small-angle region one order of magnitude and reached 0.0001 $Å^{-1}$ [56]. On the other hand, a sextupole magnet behaves as a neutron lens through the Stern–Gerlach effect. Oku *et al.* demonstrated the application of the lens to small-angle scattering at JRR3 [57]. These state-of-the-art techniques are now routinely available.

3.6.2 Neutron Spin-Echo Spectrometers

The neutron spin-echo technique, invented by Mezei in the 1970s [58], gives very high-energy resolution by measuring changes in spin polarization associated with change of velocity of the neutron after exchanging energy with the sample. Unlike the instrument concepts discussed earlier, this technique does not have any trade-offs between intensity and energy resolution because of the independence of the spin from the kinetic parameters governed by Louville's theorem. Therefore, the intensity and the resolution are decoupled.

3.6.2.1 The Principle of Neutron Spin-Echo

This technique has already been described in great detail [58]. In this section, it will be sufficient to recall that a neutron travelling through a magnetic field H_0 (assumed to be homogeneous) and polarized perpendicular to the magnetic field direction undergoes Larmor precession (Figure 3.60).

The magnetic moment μ feels a torque Γ from a magnetic field H and undergoes Larmor precession described by

$$\Gamma = \mu \times H \tag{3.79}$$

The equation of motion of the magnetic moment of neutron in the magnetic field H_0 is given by

Chapter 3 Experimental Techniques

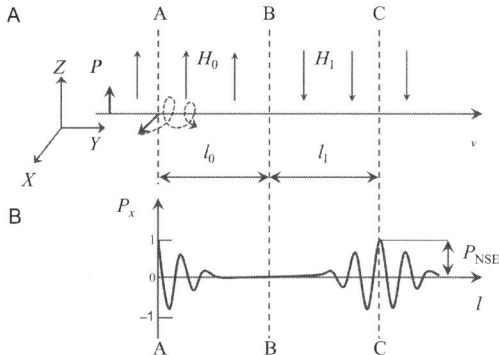

FIGURE 3.60 The principle of neutron spin-echo spectroscopy. Neutrons undergo Larmor precession in a magnetic field.

$$\frac{d\mu}{dt} = -\gamma_n \mu_n \times H_0 \qquad (3.80)$$

where $\gamma_n = 1.832 \times 10^{+8}$ s^{-1}/T is the neutron gyromagnetic ratio

Then the frequency ω is,

$$\omega = \gamma_n H_0 \qquad (3.81)$$

Therefore, the precession angle, ϕ, for a given neutron traveling a distance l in the field, H_0 with velocity, v, is given by

$$\phi = \gamma_n \frac{lH_0}{v} \qquad (3.82)$$

The polarization component, P_x, along a direction x perpendicular to H_0 is the average of neutrons with a velocity v.

$$P_x = \langle \cos\phi \rangle = \int f(v) \cos\left(\gamma_n \frac{lH_0}{v}\right) dv \qquad (3.83)$$

where $f(v)$ is the velocity distribution function. Clearly, from Equation (3.82), the Larmor-precession angles for neutrons having different velocities become more and more out of phase and the averaged polarization $P_x = \langle \cos\phi \rangle$ in the velocity distribution tends to zero along l. However, spin echo is realized by making the neutrons precess with the opposite rotation after a certain time via the use of a second field H_1 acting so as to time-reverse the initial spin trajectories defined by H_0. The total Larmor precession angle now reads:

$$\phi = \phi_0 - \phi_1 = \gamma_n \left[\frac{l_0 H_0}{v_0} - \frac{l_1 H_1}{v_1} \right] \qquad (3.84)$$

where v_0 and v_1 are the incoming and outgoing neutron velocities, l_0 and l_1 are the travel lengths in the fields H_0 and H_1, respectively, v_0 and v_1 are related with those energies E_0 and E_1 through

$$E_0 = \frac{1}{2}mv_0^2, \quad E_1 = \frac{1}{2}mv_1^2 \tag{3.85}$$

The neutron energy change $\hbar\omega$ in which we are interested is

$$\hbar\omega = E_0 - E_1 = \frac{1}{2}m(v_0^2 - v_1^2) \tag{3.86}$$

The change in the Larmor precession angle $\phi = \phi_0 - \phi_1$ and the energy transfer $\hbar\omega$ are related each other through Equations (3.84) and (3.86). We consider that there is a velocity distribution around averaged velocities $\bar{v}_0$ and $\bar{v}_1$ corresponding to $\hbar\bar{\omega}$ and $\bar{\phi}$.

Small deviations in the neutron velocities $\Delta\bar{v}_0$ and $\Delta\bar{v}_1$ about their average values will produce small changes in the energy transfer

$$\Delta\hbar\omega = m(\bar{v}_0\Delta v_0 - \bar{v}_1\Delta v_1) \tag{3.87}$$

and an uncertainty in the Larmor precession angle is also described as

$$\Delta\phi = \gamma_n \left[\frac{l_0 H_0}{\bar{v}_0^2}\Delta v_0 - \frac{l_1 H_1}{\bar{v}_1^2}\Delta v_1\right] \tag{3.88}$$

If the ratio of the magnetic fields H_0 and H_1 is chosen to be

$$\frac{l_0 H_0}{l_1 H_1} = \frac{\bar{v}_0^3}{\bar{v}_1^3} \tag{3.89}$$

then the uncertainty in the Larmor precession angle can have a very simple proportional relation to that of the energy transfer from Equations (3.87) and (3.88).

$$\Delta\phi = \left[\gamma_n \frac{l_0 H_0}{\bar{v}_0^3}\frac{\hbar}{m}\right]\Delta\omega \tag{3.90}$$

Note that the first factor on the right-hand side of Equation (3.90) has dimensions of time, what is called Fourier time. The longer the Fourier time, the better the energy resolution. The change in ϕ according to a frequency change ω is proportional to λ^3. Hence, long wavelengths help achieving a high spectral resolution.

The distribution of ϕ in the scattered beam is simply given by the scattering function $S(Q,E)$ which describes the probability that neutrons are scattered with an energy transfer $\hbar\omega$. Consequently, the polarization component P_x given from Equation (3.83)

$$P_x = \langle\cos(\phi - \bar{\phi})\rangle = \frac{\int S(Q,\omega)\cos[t(\omega - \bar{\omega})]d\omega}{\int S(Q,\omega)d\omega} = \frac{I(Q,t)}{S(Q)} \tag{3.91}$$

Therefore, the spin-echo technique can directly give the normalized intermediate scattering function $I(Q,t)$.

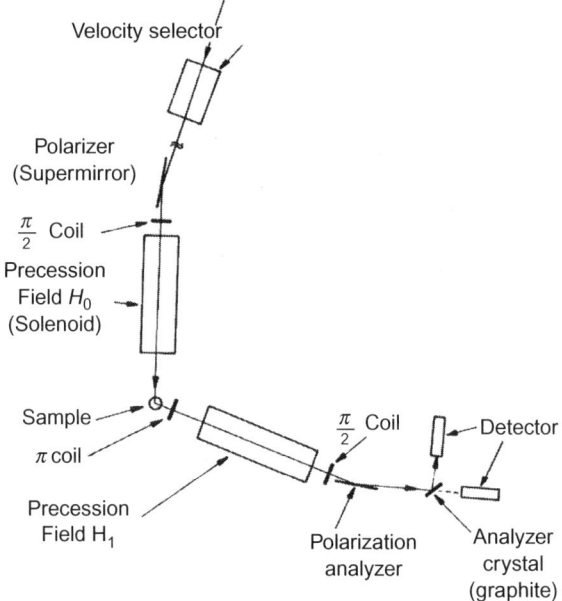

FIGURE 3.61 Schematic layout of the IN11 neutron spin-echo spectrometer.

Figure 3.61 shows an illustration of the IN11 instrument at ILL [59], which is the original form of the spin-echo instrument and it has produced a significant body of important scientific results.

3.6.2.2 Technical Explanation of the Spin-Echo Instrument

In order to understand how we can observe the spin-echo signal, we describe an instrument in more detail (Figure 3.62) [60]. Neutrons from the reactor are roughly monochromated by a velocity selector, with a typical wavelength distribution of $10\% < \Delta\lambda/\lambda < 20\%$. Neutrons are polarized by a polarizer in the z-direction, the traveling direction, then the first $\pi/2$ flipper flips spins perpendicular to it, say to the x-direction, as shown in Figure 3.62. (Here the magnetic field in the first $\pi/2$ flipper has a 45° angle to the z-direction.) Larmor precession starts to occur in the first precession field ($\mathbf{H}_0$), and the precession angle is given by Equation (3.82). Instead having an opposite magnetic field for the second precession field ($\mathbf{H}_1$) after the sample, a π-flipper (π-coil) flips the spin by 180° just before the sample position. The second precession field ($\mathbf{H}_1$), which has the same field direction as $\mathbf{H}_0$, makes the spins precess but, because of the π-flipping, the actual precession direction is opposite from that in $\mathbf{H}_0$. At the second $\pi/2$ flipper, the spins are flipped back along z-direction. Finally, the analyzer can determine the phase difference, Equation (3.84), that occurred in the sample.

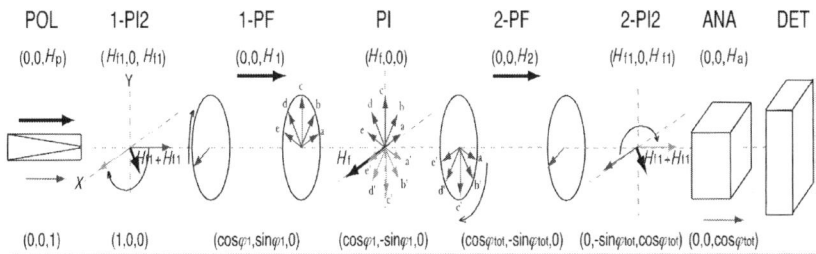

FIGURE 3.62 Illustration of a neutron spin-echo instrument. Neutrons travel from the left to the right (x-axis above) (the bold arrows indicate the magnetic field while the thin arrows show neutron spins). POL, polarizer; 1-PI2, first π/2 flipper; 1-PF, first precession field; PI, π-flipper; 2-PF, second precession field; 2-PI2, second π/2 flipper; ANA, analyzer; DET, detector.

The neutron spin at 6 Å wavelength can make one turn for $lH = 22.6$ (μTm). This is typically $l = 2$ m with $H = 10$ μT. In order to sense one-tenth of a rotation, the field sensitivity should be $\Delta(lH) = 0.3$ (μTm) for 6 Å, 0.2 for 10 Å and 0.1 for 15 Å, for instance. Typically, the total number of precessions in the field is $\phi = 10{,}000$ precessions for $lH \sim 0.5$ [T m]. Therefore, the precision of the phase sensitivity can be $\Delta(lH)/(lH) \sim 10^{-6} \sim \Delta\lambda/\lambda$. This leads to an energy resolution about 0.01 μeV or 10 ns in the time resolution for $\lambda = 10$ Å.

Another technique to realize spin-echo spectroscopy uses a resonance flipper instead of a precession field. However, here we just refer to Ref. [61].

3.7 NEUTRON DETECTORS

The detection of thermal neutrons exploits the strong absorption cross section of certain elements and energetic ionic particles emerging from the nuclear reaction to produce charge or scintillation that is converted to an electric signal to be detected. Typical and useful nuclear absorption reactions are:

^{3}He(n,p) with 5330 barns:

$$n + {}^3\text{He} \rightarrow {}^3\text{H} + p + 0.76\,\text{MeV} \tag{3.92}$$

^{10}B(n,α) with 3840 barns:

$$\begin{aligned} n + {}^{10}\text{B} &\rightarrow {}^7\text{Li} + \alpha + 2.8\,\text{MeV}\,(7\%) \\ &\rightarrow {}^7\text{Li}^* + \alpha + 2.8\,\text{MeV}\,(93\%) \\ &\rightarrow {}^7\text{Li} + \alpha + \gamma + 0.48\,\text{MeV} \end{aligned} \tag{3.93}$$

^{6}Li(n,α) with 940 barns:

$$n + {}^6\text{Li} \rightarrow {}^3\text{H} + \alpha + 4.79\,\text{MeV} \tag{3.94}$$

^{157}Gd(n,γ) with 254,100 barns:

$$n + {}^{157}\text{Gd} \rightarrow \text{Gd}^* \rightarrow \gamma(80\,\text{keV to } 6.7\,\text{MeV}) \rightarrow e^-(29 - 180\,\text{keV}) \tag{3.95}$$

^{157}Gd(n,γ) has a very large cross section; however, Gd has also a high gamma sensitivity, causing a high background in a gamma environment, not only in the beam but from the sample itself.

3.7.1 ^{3}He-Gas Detectors

The most popular material for thermal neutron detector has been ^{3}He gas, although there is a recent shortage in supply. It has reasonable absorption cross section with a very low gamma sensitivity—typically the γ/n ratio is about 10^{-7}—and has high stability and long lifetime. Figure 3.63 illustrates a typical PSD with ^{3}He gas. Neutrons absorbed by ^{3}He create an ionic cascade in the surrounding atoms by the nuclear reaction energy, as in Equation (3.92).

Because of the high voltage applied between the central wire and outer shell of the detector, electric charge accumulated/induced on the central wire comes out as a charge Q_a and Q_b at the both end of the resistive wire according to the Ohm's law. This leads to a relation;

$$Q_a/(Q_a+Q_b) = X/L \qquad (3.96)$$

giving the information of the detected position on the wire, giving a position sensitivity. Typically, the axial special resolution is about 5–10 mm for a 1/2″ to 1″ thick detector with a pressure of 10 atm. or above. The typical detecting speed is about 25 kHz. Therefore, when neutron events exceed this speed, the counter saturates detection ability. However, recent development in electric circuitry is improving the detection speed. Figure 3.64 shows an array of the position-sensitive ^{3}He detector of 4SEASONS [62].

3.7.2 Scintillation Detectors

This type of detector system is becoming increasingly important because of its higher spatial resolution and detection speed. Recent shortages in ^{3}He supply worldwide have clearly accelerated this trend.

Great efforts have already been made in the R&D activity at the ISIS facility, UK.

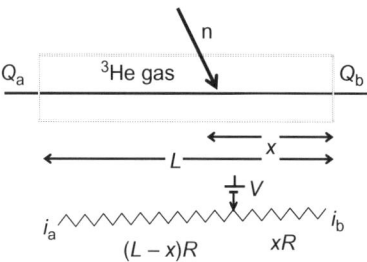

FIGURE 3.63 Illustration of a ^{3}He gas position-sensitive detector and its equivalent electric circuit. (For color version of this figure, the reader is referred to the online version of this chapter.)

FIGURE 3.64 Array of ^{3}He position-sensitive detector in a scattering vacuum tank of a chopper instrument. (For color version of this figure, the reader is referred to the online version of this chapter.)

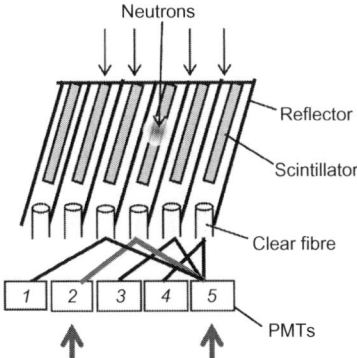

FIGURE 3.65 Venetian-type geometry for scintillator-based detection. (For color version of this figure, the reader is referred to the online version of this chapter.)

ZnS/^{6}LiF is a prototypical material for a neutron scintillation detector. According to Equation (3.94) neutrons absorbed by ^{6}Li create energetic ^{3}H and α particles, which excite the electric states and ionize the ZnS scintillator and eventually emit scintillation photons. These create electrons at the photo-multiplier tube (PMT) through the photon-electron conversion process and these are detected as a charged signal in the PMT.

One of the well established detector systems is the Venetian-type detector, Figure 3.65 [63], in which the scintillator element is inclined so as to increase detecting efficiency; photons from one cell are led to two PMTs and the signal is encoded in a $_nC_2$-fashion to determine the detecting position. Therefore, 16 PMTs are needed to encode 120 detector segments. Figure 3.66 shows the same kind of detector system but enlarged in size at J-PARC. The segment

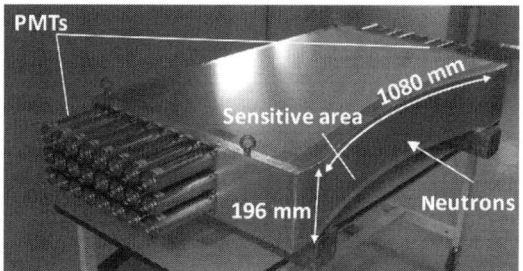

FIGURE 3.66 Scintillation detector in use at TAKUMI. (For color version of this figure, the reader is referred to the online version of this chapter.)

TABLE 3.1 Specifications of the Venetian-Type Scintillation Detector

Spatial resolution	3 mm
Detector efficiency	85% at 2.2 Å
Counting uniformity	9% (1σ)
Background counts	20 counts/h/pixel
Gamma sensitivity	5×10^{-7}
Pulse pair resolution	<5 μs
Pixel size	3 × 196 mm

size is 3 mm(horizontal) × 196 mm(vertical) and is in use at TAKUMI, an engineering powder diffractometer [64]. The specification of the detector is specified in Table 3.1 [65].

3.7.2.1 Wavelength-Shifting Fiber Detectors

This type of scintillation detector can have a two-dimensional array of light guides, which shift the wavelength of the scintillation light initially created at the neutron absorber, and the shifted light can easily transmit along the fiber to the PMT (cf. Figure 3.67). Figure 3.68 shows how the light created after neutron absorption in the scintillation layer propagates in the light guides. The spatial resolution is mainly determined by the separation of the light guide so that the resolution can be fairly tunable and can have good spatial resolution, less than 1 mm [66,67]. The event position is determined by using a direct-encoding technique with multi-anode PMTs.

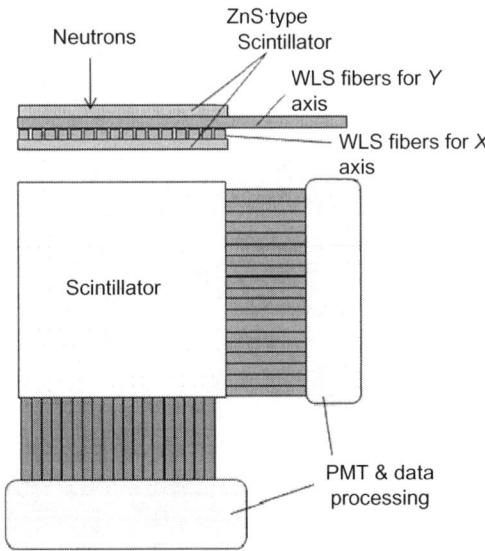

FIGURE 3.67 Illustration of the WLSF detector, 2D scintillator plate and WLS fibers in the X–Y direction. (For color version of this figure, the reader is referred to the online version of this chapter.)

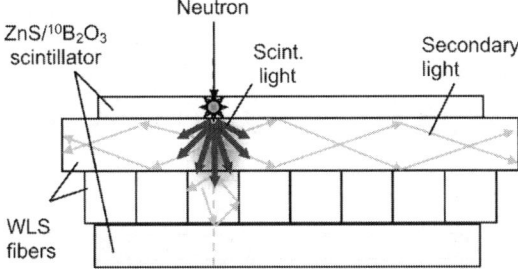

FIGURE 3.68 Cross section of the WLSF detector, which has X- and Y-axes fibers sandwiched with an upper and a lower scintillation plates to increase the detection efficiency. (For color version of this figure, the reader is referred to the online version of this chapter.)

The newly developed sintered ceramic scintillator material, $ZnS/^{10}B_2O_3$, has improved detecting efficiency and at the same time a lower gamma sensitivity. Together with the direct-encoding technique, the performance of this kind of detector is very competitive relative to other technologies as shown in Table 3.2.

Since the spatial resolution and detector area size are fairly tunable, WLSF detectors are now in use at iBIX (protein crystal diffractometer with 1.2 mm resolution) [68] (cf. Figure 3.69), SENJU (single-crystal diffractometer with 4 mm resolution) [69], and SHARAKU (reflectometer with 4 mm resolution) [70] at J-PARC.

TABLE 3.2 WLSF Detector Specification

Spatial resolution	<1.0 mm
Detector efficiency	50% at 2.2 Å
Counting uniformity	10% (1σ)
Background counts	2 counts/h/cm^2
Gamma sensitivity	10^{-6}
Pulse pair resolution	3–5 µs
Pixel size	0.52 x 0.52 mm

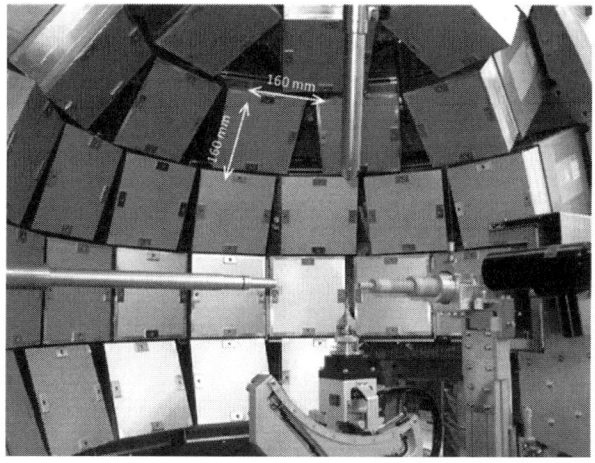

FIGURE 3.69 WLSF detectors in use at the iBIX protein diffractometer. (For color version of this figure, the reader is referred to the online version of this chapter.)

3.8 BEAM TRANSPORT AND TAILORING

3.8.1 Neutron Optics

In this section, we introduce neutron guides, used in the efficient transport of neutrons over long distances (tens of meters). Such a task requires a brief review of refractive neutron optics.

Let us consider a neutron travelling from medium 1 (assumed to be vacuum) to medium 2. The wave vector changes from $\mathbf{k}_1$ to $\mathbf{k}_2$. The refractive index n is defined as

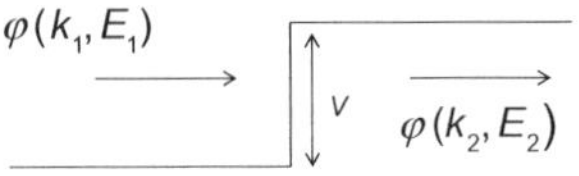

$$\varphi(k_1, E_1) \rightarrow \quad \Big\updownarrow V \quad \overrightarrow{\varphi(k_2, E_2)}$$

FIGURE 3.70 Neutron travels from medium 1 to medium 2. φ is the wave function of neutron.

$$n \equiv \frac{k_2}{k_1} \quad (3.97)$$

The kinetic energy of neutrons in medium 1 is

$$E_1 = \frac{\hbar^2 k_1^2}{2m_n} \quad (3.98)$$

The potential energy in medium 2 is V (cf. Figure 3.70). From Equations (3.97) and (3.98) the refractive index can be written as

$$n = \left(\frac{E_1 - V}{E_1}\right)^{1/2} \sim 1 - \frac{V}{2E_1} \quad (3.99)$$

where we assume that $V \ll E_1$. Actually, V is in the 100 neV range compared with E_1 in meV range.

V is the Fermi pseudo-potential introduced to describe the scattering process and define the scattering length b,

$$V = \frac{2\pi\hbar^2}{m_n} bN \quad (3.100)$$

where b is the coherent scattering length in medium 2 and N is the number density of nuclei. From Equations (3.99) and (3.100),

$$n = 1 - \frac{bN}{2\pi}\lambda^2 = 1 - \frac{2\pi bN}{k^2} \quad (3.101)$$

As shown in Table 3.3, most elements have $b > 0$, $n < 1$. Therefore, when neutrons enter these media from a vacuum, neutrons can undergo total reflection if the glancing angle is small enough.

The refractive index is also defined by reflection angles according to Snell's law for optical refraction states (cf. Figure 3.71),

$$n = \cos\theta_1/\cos\theta_2 \quad (3.102)$$

When θ_1 is lowered, θ_2 approaches to zero, then when θ_1 becomes smaller than the critical angle $\theta < \theta_c$, a total reflection suddenly appears.

The critical angle is defined as,

$$n = \cos\theta_c \sim 1 - \frac{\theta_c^2}{2} \quad (3.103)$$

Since θ_c is a small number as shown in Table 3.3, Equation (3.103) can be expanded. From Equation (3.101), therefore, the critical angle is described by

TABLE 3.3 Critical-Angle Coefficient α for a Series of Chemical Elements Typically Used in Neutron-Optical Devices

	N ($\times 10^{29}$ m^{-3})	b ($\times 10^{-14}$ m)	Nb ($\times 10^{38}$ m^{-2})	α (mrad/Å)
^{58}Ni	9	1.44	13	2.03
Be	12.3	0.77	9.5	1.73
Ni	9	1.03	9.3	1.7
Fe	8.5	0.96	8.2	1.62
C	11.1	0.66	7.3	1.61
Cu	8.5	0.79	6.7	1.39
Co	8.9	0.25	2.2	0.86
Al	6.1	0.35	2.1	0.81
Ti	5.6	−0.34	−1.9	-

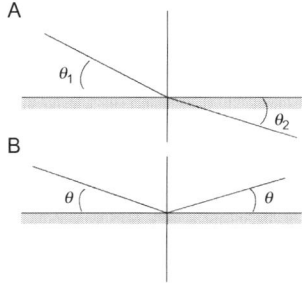

FIGURE 3.71 Refraction (a) and total reflection below the critical angle (b).

the scattering length and number density, and is proportional to the wavelength.

$$\theta_c = (2(1-n))^{1/2} = \sqrt{\frac{bN}{\pi}} \cdot \lambda = \alpha \cdot \lambda \tag{3.104}$$

Table 3.3 summarizes those values for typical elements.

3.8.1.1 Neutron Guides

The neutron guide was invented in 1960s by Maier-Leibnitz of the Munich Technical University [71]. Neutrons reflect at a shiny surface of material, similar to the way light does and can be transported to a distant site and thus reduce background.

One of the most typical elements for a neutron guide is Ni, for which the critical angle is as follows:

$$\theta_c^{Ni}(\text{mrad}) = 1.7\lambda\left(\text{Å}\right) \quad \text{or} \quad \theta_c^{Ni}(°) = 0.1\lambda\left(\text{Å}\right) \tag{3.105}$$

A critical momentum transfer for Ni can be also defined as

$$Q_c = 4\pi\frac{\theta_c^{Ni}}{\lambda} = 0.021 \text{Å}^{-1} \tag{3.106}$$

Mezei invented the super-mirror, which is a multilayer mirror with a gradually changing layer thickness that can continuously reflect a series of Bragg reflections (cf. Figures 3.72 and 3.73) [72]. The critical angle of the super-mirror exceeds that of Ni.

Therefore, the performance of a super-mirror is measured by a value $m \equiv \frac{\theta_c^{SM}}{\theta_c^{Ni}}$, and the critical momentum transfer scales as mQ_c. The m-value can in principle be increased by increasing the number of layers, N_L. However, roughly speaking $N_L \sim 4\,m^4$ and a high mQ_c mirror needs a large number of layers. For instance, an $m=3$ mirror has 500 layers; $m=4$, 1000 layers; $m=8$, 9000 layers, which is the highest m-value reached so far.

Figure 3.73 shows a Bragg reflection from typical layers in a super-mirror. Each layer causes Bragg reflection, resulting in a cumulated intensity from multi-layers. Figure 3.74 [73] shows the measured cumulative reflectivity from the entire super-mirror as a function of Q and m-value for several different mQ_c mirrors.

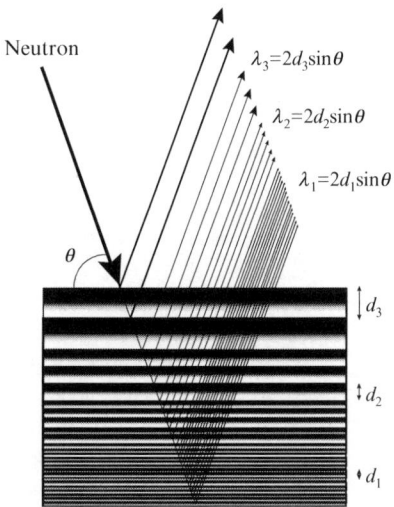

FIGURE 3.72 Illustration of a super-mirror. Reflection occurs at different depths for different wavelengths.

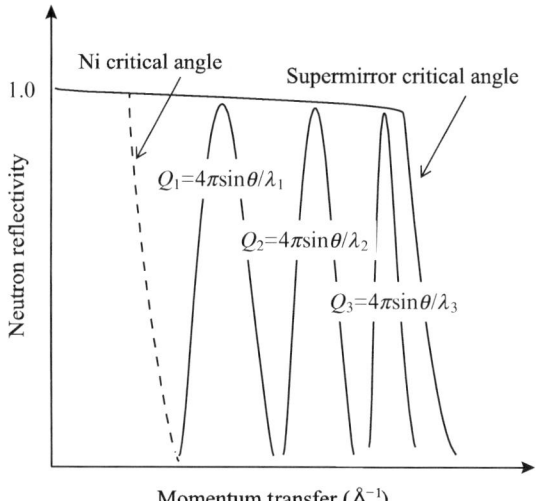

FIGURE 3.73 Neutron reflection from a super-mirror.

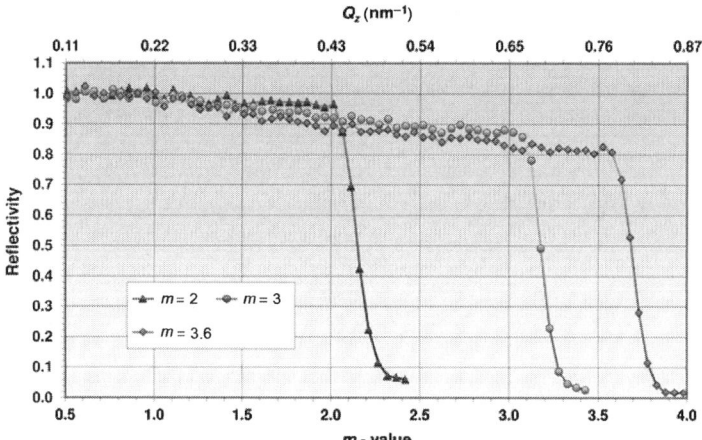

FIGURE 3.74 Reflectivity of an $m=2, 3, 3.8\ Q_c$ mirror. (For color version of this figure, the reader is referred to the online version of this chapter.)

A super-mirror is composed of pairs of layers, and this unit is repeated with a gradually changing thickness as shown in Figure 3.72. The reflectivity from the double layer, A and B, is given by

$$R = \frac{4N^2 d^4 (Nb_A - Nb_B)^2}{\pi^2 n^4} \qquad (3.107)$$

where d is the thickness of one double layer, A and B, and n is the index of the Bragg reflection [74]. Because of this, a combination of Ni and Ti, which has a negative scattering length, enhances the reflectivity and is commonly used for super-mirror material.

3.8.1.2 Straight and Curved Guides

Originally, neutron guides were designed to transfer neutrons to a distant site.

If there is no guide, the flux at a site P in Figure 3.75 is proportional to a solid angle

$$\Delta\Omega = ab/L^2 \qquad (3.108)$$

where a and b are edges of the neutron source moderator, and L is the distance between the source and the site P. On the other hand, a guide with a critical angle θ_c extends the effective solid angle to

$$\Delta\Omega_g = (2LQ_c)/L^2 = 4\theta_c \qquad (3.109)$$

and is independent on the distance L. Therefore, the gain factor between the two is,

$$\text{Gain} = 4\theta_c L^2/ab \qquad (3.110)$$

for a big enough source size.

In order to eliminate high-energy neutrons and gammas and have a low background, a curved guide is generally a good choice, since the neutron must make at least one reflection to get through the guide, while fast neutrons have extremely small reflection angles and are unlikely to be transmitted.

Neutrons reflect inside the guide in two distinct ways (cf. Figure 3.76). Reflection at only the outer surface is called Garland reflection, and at both sinner and outer surfaces is a zigzag reflection. Since the Garland reflection has a smaller number of reflections, it has better transmission through the guide. The largest reflection angle of the Garland reflection is called the characteristic reflection angle, θ^*, and is equal to the critical angle

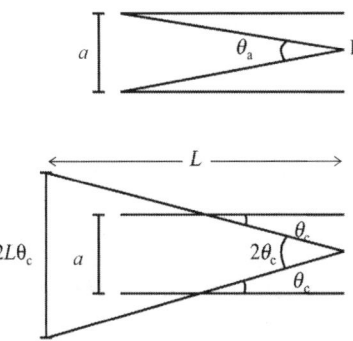

FIGURE 3.75 Effective solid angle subtended by a neutron source at the site P for a bare beam line and one with a mirror guide.

Chapter | 3 Experimental Techniques

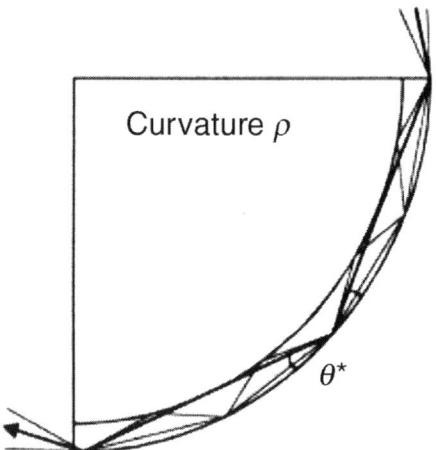

FIGURE 3.76 Curved guide section. The lines show garland-type and zigzag reflections.

$$\theta^* = \sqrt{\frac{2d}{R_g}} = \theta_c = \alpha\lambda^* \qquad (3.111)$$

where d is the width of the guide and R_g is the curvature radius. λ^* is the characteristic wavelength, the longest wavelength, for the Garland reflection, and has the highest transmission.

The line-of-sight length, the longest distance of one reflection in the guide, is

$$L^* = \sqrt{8dR_g} \qquad (3.112)$$

A curved guide section for eliminating the fast neutron background should exceed this distance. When designing a curved guide, this estimates how long should be the curved section, taking the most desired neutron wavelength λ^* into consideration at the same time [75]. For example, for $\lambda^* = 2.5$ Å, $d = 2$ cm, $m = 3$, then $R_g = 2500$ m, $L^* = 6.3$ m.

3.8.1.3 Focusing and Ballistic Guides

When the length of a guide exceeds about 40 m, reflection losses become large. Mezei has introduced a concept of ballistic neutron guide [76], that is, the initial section is a linearly tapered, thus the cross section of guide increases, followed by a regular straight guide as shown in Figure 3.77. The width of the guide is much thicker than viewing source and the exit of the guide, leading much smaller number of reflection with a high reflectivity of a low m-mirror, such as $m = 2$ or 3. Therefore, neutron can be transported over a long distance [77]. A recent precise grazing technique makes it possible to have a sophisticated shape of mirror surface, and parabolic or elliptical mirror

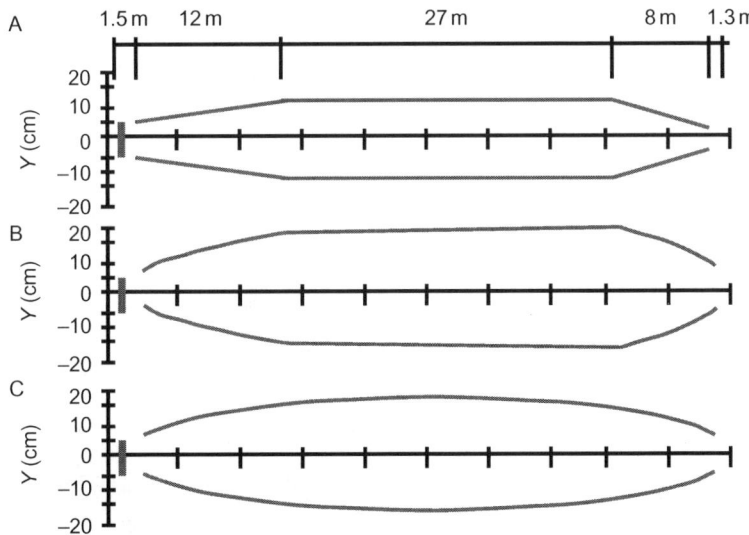

FIGURE 3.77 Ballistic guide with (a) linearly tapered sections at the entrance and the exit, (b) parabolic sections at the entrance and the exit, and (c) elliptical section for the entire guide. (For color version of this figure, the reader is referred to the online version of this chapter.)

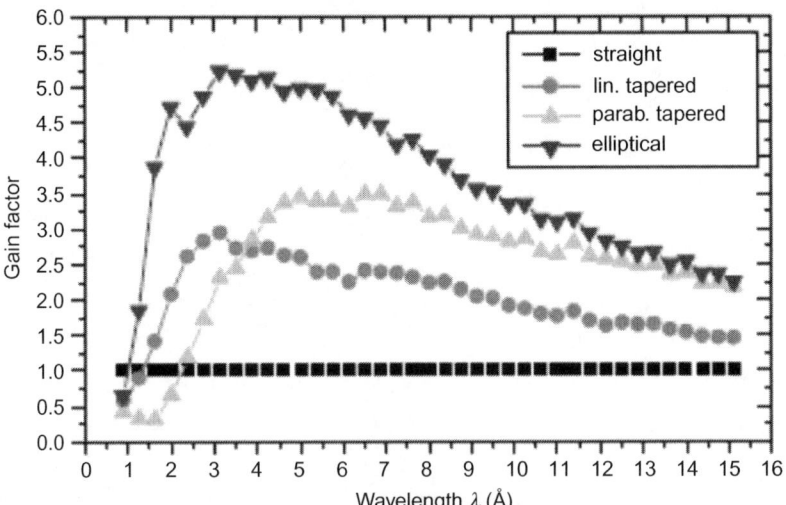

FIGURE 3.78 Estimated performance for the model guide shown in Figure 3.77. (For color version of this figure, the reader is referred to the online version of this chapter.)

is becoming available [78]. Estimation of flux with those model guides has been estimated by a computer simulation code McStas (cf. Figure 3.78) [79]. Among the models, the elliptical guide has a very good performance over a wide wavelength range. Efforts to make a focusing mirror are underway and a demonstration has shown a gain factor 50 [80].

3.8.2 Choppers

3.8.2.1 T_0 Choppers

It is an important issue for any instrument to reduce the background as low as possible. Especially this is a golden rule for inelastic scattering instrument design. Inelastic signals can be as much as five orders of magnitude smaller than the equivalent elastic signal. Low backgrounds in the epithermal neutron region is also important for a total scattering diffractometer, whose analyzed intensity at high Q region has to be unity in the absolute scale.

One of obvious "background killers" is a background-suppression chopper known as a T_0 chopper. Figure 3.79 shows such a chopper developed at J-PARC, which has an in-wheel motor to reduce the longitudinal size to fit into a neutron guide system. This apparatus has a thick blade, 30 cm long along the beam line and made of bulk Inconel alloy to reduce the radioactivity. It is placed close at the time of proton bombardment on the target in order to suppress high-energy neutrons and also gamma rays, which otherwise could easily come into the secondary spectrometer, be thermalized in the instrument shield, and come out at later time, producing a very high background. Figure 3.80 is an example of background suppression at various rotation speeds of a T_0 chopper [81]. Lower rotation speeds can suppress more background because of a longer closed time. Background can be easily suppressed by a factor of 2–4.

3.8.2.2 Bandwidth Choppers

Pulsed neutron sources produce pulsed neutrons at the accelerator frequency. Neutrons disperse along the flight path according to their different speeds. Slow neutrons reach the detector at a very late time, when faster neutrons from the

FIGURE 3.79 To chopper with an in-wheel motor, designed for a narrow gap of guide. (For color version of this figure, the reader is referred to the online version of this chapter.)

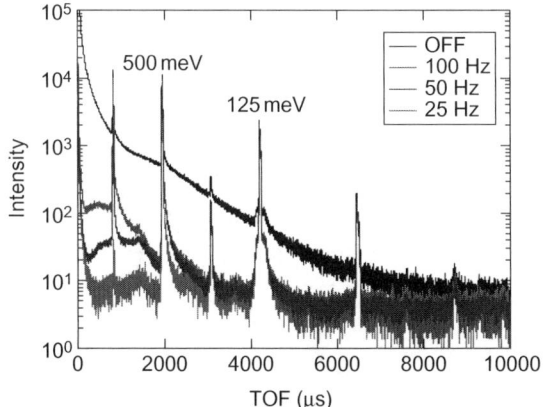

FIGURE 3.80 Background suppression by a T_0 chopper. Data were taken at a detector position on a vanadium sample with a Fermi chopper. (For color version of this figure, the reader is referred to the online version of this chapter.)

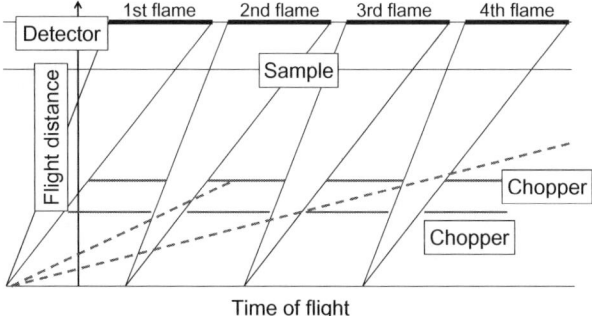

FIGURE 3.81 Illustrating how bandwidth choppers eliminate neutrons from previous frames. (For color version of this figure, the reader is referred to the online version of this chapter.)

following time frame can arrive and mix with the slow neutrons, producing a background. Therefore, it is important to eliminate unusable slow neutrons from the early frame and fast neutrons from the late frame. Figure 3.81 shows an example of frame definition by bandwidth choppers. The first chopper near the source eliminates most of unusable neutrons from the reference frame. The second chopper stops neutrons from earlier frames. However, these are not enough to eliminate neutrons from further previous frames, which may go to the sample at nine frames later. These very slow neutrons are negligible in most cases; however, they can produce a serious background for small-angle scattering or reflectivity measurements, where very slow neutrons have a very high scattered intensity or reflectivity. Therefore, in most cases three bandwidth choppers are installed for such instruments. Figure 3.82 shows a bandwidth

FIGURE 3.82 A bandwidth chopper developed at J-PARC. (For color version of this figure, the reader is referred to the online version of this chapter.)

chopper developed for J-PARC. The absorbing material is B_4C, so that it can stop even epithermal neutrons effectively.

3.8.2.3 Fermi Choppers

The application of a mechanical chopper to produce a short neutron pulse from a continuous neutron beam of a reactor for TOF measurements is a well-known technique. To attain a high chopper transmission in the slow neutron energy range ($E \sim$ meV), one has to use choppers with curved slits. If the slits are curved enough, the chopper will transmit neutrons in a certain energy range only, thus the transmitted neutron beam will be monochromated too.

3.8.2.3.1 Basic Principles of the Fermi Choppers

The speed of the rotor blade should be the same as the neutron speed v_0. Thus the curvature radius R for an angular speed $\omega = 2\pi f$ is

$$R = \frac{v_0}{4\pi f} \quad (3.113)$$

The peripheral speed of the rotor is

$$v_p = \pi D f \quad (3.114)$$

where D is the diameter of the rotor (cf. Figure 3.83).

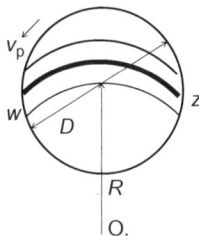

FIGURE 3.83 Cross section of a chopper rotor.

The slit width w is designed such that the chopping time is comparable to the pulse time width for a pulsed source,

$$\Delta t_m = \Delta t_p = \frac{w}{v_p} \tag{3.115}$$

The neutron transmission through the chopper is described as follows by introducing [82]

$$\beta = \frac{D}{2w} v_p \left(\frac{1}{v} - \frac{1}{v_0} \right) \tag{3.116}$$

$$\left. \begin{array}{ll} T(\beta) = 1 - \dfrac{8}{3}\beta^2 & 0 \le |\beta| < \dfrac{1}{4} \\ T(\beta) = \dfrac{16}{3}\beta^{1/2} - 8\beta + \dfrac{8}{3}\beta^2 & \dfrac{1}{4} \le |\beta| < 1 \\ T(\beta) = 0 & 1 < |\beta| \end{array} \right\} \tag{3.117}$$

For a given thickness of the absorbing plate, the chopper transmission is described by

$$T = \frac{w}{w+z} T(\beta) \tag{3.118}$$

Figure 3.84 shows $T(\beta)$ for chopper parameters optimized for 100, 200, 400, and 800 meV.

Figure 3.85 shows a chopper rotor developed at the SNS (USA).

3.8.2.3.2 Sloppy Choppers

The curved blade for a chopper rotor is needed to make a monochromatic beam at a steady source. However, in a pulsed source, the pulse width is essentially defined by the source characteristics and the neutron pulse structure after a chopper is a result of chopping timing of the chopper in phase with pulsed source, and not the result of the curved blade in the rotor. Therefore, rotors with straight blades have been used in pulsed neutron facilities since the 1980s at the IPNS facility (USA) and the ISIS facility (UK). By using this concept rotors can have a great flexibility to accommodate a wide range of

Chapter | 3 Experimental Techniques

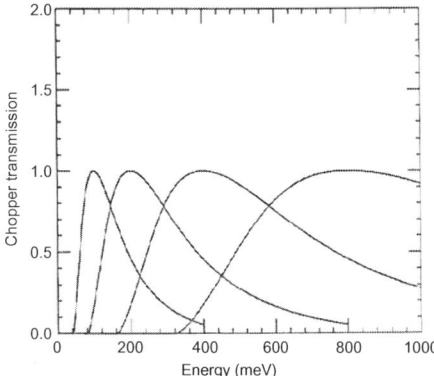

FIGURE 3.84 Transmission through a chopper rotor, $T(\beta)$, for chopper parameters optimized for 100, 200, 400, and 800 meV neutrons with a 10-cm diam. rotor at 600 Hz.

FIGURE 3.85 A chopper rotor developed at the SNS facility. (For color version of this figure, the reader is referred to the online version of this chapter.)

energies by just changing rotating speed and phase. This concept has also been developed into a multiple incident energy measurement with a Fermi chopper, whose idea was originally proposed by Mezei, known as Repetition Rate Multiplicity, with a disk chopper [76]. A chopper with a curved blade diminishes the intensity of neutrons with energies different from the optimized energy, but a chopper with a straight blade allows neutrons to pass through the chopper even at off-optimized energies [83]. Figure 3.86 is an example of the multi-incident-energy method. One experiment gives four or

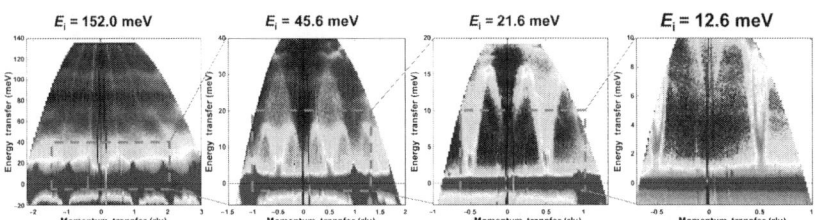

FIGURE 3.86 Spectrum observed by the multi-E_i method. Four different incident energies are used simultaneously to observe four different regions of energy–momentum space. (For color version of this figure, the reader is referred to the online version of this chapter.)

more data sets for different energy–momentum space regions and provides not only efficiency but also flexibility for observing unknown phenomena in unexpected regions of phase space.

REFERENCES

[1] http:/neutrons.ornl.gov/hyspec/ [last accessed on September 6, 2013].
[2] Carpenter JM, Yelon WB. In: Skold K, Price DL, editors. Neutron sources in methods of experimental physics, neutron scattering, vol. 23. Academic Press, Inc; 1986. p. 99, Chapter 3, Neutron Sources.
[3] http://j-parc.jp/MatLife/en/instrumentation/data/Pulse_paper.pdf [last accessed on August 7, 2013].
[4] http://www.psi.ch/sinq/ [last accessed on September 6, 2013].
[5] http:/ftp.jinr.ru/JINR-info-e [last accessed on September 6, 2013].
[6] Fluharty RG. Proposal for a pulsed-neutron facility, Los Alamos Technical Reports, LA-4138-MS.
[7] Ikeda S, Carpenter JM. Nucl Instrum Methods 1985;A239536.
[8] Mildner DFR, Sinclair RN. J Nucl Energy 1979;6225.
[9] (a) Watanabe N. Rep Prog Phys 2003;66:339–81. (b) Watanabe N, et al. In: Proceedings of 10th meeting of International Collaboration on Advanced Neutron Sources ICANS-X, October 1988, Los Alamos, USA; 1988. p. 787–97.
[10] Harada M, Maekawa F, Oikawa K, Meigo S, Takada H, Futakawa M. Prog Nucl Sci Technol 2011;2:872–8.
[11] Niita K, et al. In: Proceedings of 17th meeting of the international collaboration on advanced neutron sources, ICANS-XVII, April 25–29, Santa Fe, New Mexico, USA; 2005. p. 640–7.
[12] http://j-parc.jp/MatLife/en/instrumentation/data/Photon_paper.pdf [last accessed on August 7, 2013].
[13] http://pd.chem.ucl.ac.uk/pdnn/inst3/diff1.htm [last accessed on August 7, 2013].
[14] Ito Takayoshi, Nakatani Takeshi, Harjo Stefanus, et al. Mater Sci Forum 2010;652:238–42.
[15] Windsor CG. In: Pulsed Neutron Scattering. London: Taylor and Francis Ltd; 1981. p. 277, Chapter 7.1, Analysing glass and liquid diffraction.
[16] http:/neutrons.ornl.gov/powgen/ [last accessed on August 7, 2013].
[17] Ishigaki Toru, Harjo Stefanus, Yonemura Masao, et al. Physica B 2006;385–386:1022–4.
[18] Marshall W, Lovesey SW. Theory of thermal neutron scattering from condensed matter, vol. 1. Oxford: Clarendon Press; 1984. p. 78, Chapter 3.6 Simple model system.

[19] http:/j-parc.jp/researcher/MatLife/en/instrumentation/ns_spec.html#bl21 [last accessed on August 7, 2013].
[20] Wilson CC. Single crystal neutron diffraction from molecular materials. World Scientific; 2000. p. 86, Chapter 3, Techniques for Single Crystal Neutron Diffraction.
[21] Noda Y, Kimura H, Kiyanagi R, et al. J Phys Soc Jpn 2001;70(Suppl.):A456–A458.
[22] Niimura N, Kubota T, Sato M, Arai M, Ishikawa Y. Nucl Instrum Meth 1980;173:517.
[23] Wilkinson C, Lehmann MS. Nucl Instrum Methods A 1991;310:411–5.
[24] Niimura Nobuo, Podjarny Alberto. In: Neutron protein crystallography. Oxford Science Publications; 2011. p. 27, Chapter I-5.3 Neutron Choppers and the time-of-flight method.
[25] Takahashi Miwako, Ohshima Ken-ichi, Arai Masatoshi. J Appl Crystallogr 2007;40:799–807.
[26] Tamura I, Oikawa K, Kawasaki T, Ohhara T, Kaneko K, Kiyanagi R, et al. J Phys Conf Ser 2012;340012040. http:/j-parc.jp/researcher/MatLife/en/instrumentation/images/senju.jpg, [last accessed on August 7, 2013].
[27] Tanaka I, et al. In: Proceedings of ICANS-XVII, 17th meeting of the international collaboration on advanced neutron sources, April 25–29, 2005, Santa Fe, New Mexico, LA-UR-3904 (ICANS-XVII proc.); 2006. p. 937–45.
[28] Brockhouse BN. Inelastic scattering of neutrons in solids and liquids. Vienna: IAEA; 1961, p. 147.
[29] Gavin WW. Polarized neutrons. Oxford Science Publications; 1988. p. 14, ISBN: 0-19-851005-5, Chapter 1, Introduction.
[30] (a) Clausen KN, Lebech B, McMorrow DF, Sorensen SA, Aeppli G, Mason TE. Neutron News 1996;7(4):21–4. (b) Mason TM, Clausen KN, Appli G, McMorrow DF, Kjems JK. Can J Phys 1995;73:697–702.
[31] Rodriguez JA, Adler DM, Brand PC, Broholm C, et al. Meas Sci Technol 2008;19034023.
[32] Kolesnikov A, Zanotti J, Loong C. Neutron News 2004;15:19–22, Taylor & Francis.
[33] http:/www.aai.anl.gov/history/project_pages/ipns.html [last accessed on September 8, 2013].
[34] Taylor AD, Boland BC, Bowden ZA. In: Atchison F, Fischer W, editors. Proceedings of ICANS IX (international collaboration on advanced neutron sources). p. 349, SIN Report 40926.
[35] Arai M, Taylor AD, Bennington SM, Bowden ZA. In: Howells WS, Soper AK, editors. MARI - A new Spectrometer for liquid and amorphous materials, recent developments in the physics of fluids. Adam Hilger; 1992. p. F321–8.
[36] http://en.wikipedia.org/wiki/ISIS_neutron_source [last accessed on September 8, 2013].
[37] Perring T, Frost C. Neutron News 2004;15:30–4, Taylor & Francis.
[38] http:/www.isis.stfc.ac.uk/instruments/merlin/ [last accessed on September 8, 2013].
[39] http:/neutrons.ornl.gov/arcs/ [last accessed on September 8, 2013].
[40] http:/neutrons.ornl.gov/sequoia/ [last accessed on September 8, 2013].
[41] Kajimoto Ryoichi, Nakamura Mitsutaka, et al. J Phys Soc Jpn 2011;80(Suppl. B):SB025-1–SB025-6.
[42] Carlile CJ, Taylow AD, Williams WG. In: Neutron scattering in the 'nineties, conference proceedings, Julich, 14–18 January. Vienna: International Atomic Energy Agency; 1985. p. 421–34.
[43] Nakajima Kenji, Ohira-Kawamura Seiko, et al. J Phys Soc Jpn 2011;80(Suppl. B):SB028-1–SB028-6.
[44] Bennington SM, private communication. Data was taken on MARI, ISIS.
[45] Sette F, Ruocco G, et al. Phys Rev Lett 1995;75:850–3.
[46] Aisa D, Aisa S, et al. Phys B Condens Matter 2006;385–386:1092–4.
[47] Arai M, Fujita M, et al. Phys Rev Lett 1996;773649.
[48] Takahashi Nobuaki, Shibata Kaoru, et al. J Phys Soc Jpn 2011;80(Suppl. B):SB007-1–SB007-4.
[49] Ikeda S, et al. J Phys Soc Jpn 1991;60:3340–50.

[50] http://www.ill.eu/instruments-support/instruments-groups/instruments/d11/description/instrument-layout/ [last accessed on August 7, 2013].
[51] Schmatz W, Springer T, Shelten J, Ibel K. J Appl Cryst 1974;7:96–116.
[52] Mildner DFR, Carpenter JM. J Appl Cryst 1987;20:419–24.
[53] Ishikawa Y, Furusaka M, Niimura N, Arai M, Hasegawa K. J Appl Cyst 1986;19:229–42.
[54] Crawford RK, Carpenter JM. J Appl Cryst 1988;21, 589–601.
[55] Eskildsen MR, Gammel PL, Isaacs ED, Detlefs C, Mortensen K, Bishop DJ. Nature 1998;391:563–6.
[56] Choi S-M, Barker JG, Glinka CJ, Cheng YT, Gammel PL. J Appl Cryst 2000;33:793–6.
[57] Oku T, Iwase H, et al. J Appl Cryst 2007;40:s408–s413.
[58] Mezei F, editor. Neuron spin echo, vol. 128. Heidelberg: Springer-Verlag; 1980. p. 3, The Principles of Neutron Spin Echo.
[59] Gagleish PA, Hayter JB, Mezei F. Lecture notes in physics, vol. 128. Berlin: Springer-Verlag; 1980, p. 67.
[60] Nagao M. Hamon. Journal of the Japan Society for Neutron Science 2009;19(1):42–8.
[61] Golub R, Gahler R. Phys. Lett. 1987;A12343.
[62] Kajimoto R, et al. J Neutron Res 2007;15(1):5–12.
[63] Schooneveld EM, et al. Proceedings of ICANS-XVI (international collaboration on advanced neutron sources), vol. I; 2003. p. 455.
[64] Harjo Stefanus, et al. Mater Sci Forum 2006;524–525:199–204.
[65] Nakamura T, Sakasai K, Katagiri M, Birumachi A, Hosoya T, Soyama K, et al. Performance test of the Japanese ENGIN-X type linear scintillation neutron detector. Japan Atomic Energy Agency, JAEA-Research 2007-014.
[66] Nakamura T, Katagiri M, Hosoya T, Birumachi A, Ebine M, Soyama K, et al. Development of two-dimensional scintillator neutron detector using wavelength-shifting fibres -Development of a compact detector for iBIX instrument in J-PARC/MLF. Japan Atomic Energy Agency, JAEA-Research 2008-115.
[67] Katagiri M, et al. Nucl Instr Methods A 2004;529274.
[68] Hosoya T, et al. Nucl Instr Methods A 2009;600217.
[69] (a) Tamura I, Oikawa K, Kawasaki T, Ohhara T, et al. J Phys Conf Ser 2012;3404. (b) Nakmaura T, et al. Nucl Instr Methods A 2012;68664.
[70] Takeda Masayasu, et al. Chin J Phys 2012;50:161–70.
[71] Maier-Leibnitz H, Springer T. J Nucl Energy 1963;17:217–25.
[72] Mezei F. Commun Phys 1976;181.
[73] Hino M, Sunohara H, Yoshimura Y, Maruyama R, Tasaki S, Yoshino H, Kawabata Y. Nucl Inst Methods Phys Res A 2004;529:54–8, http://www.swissneutronics.ch/index.php?id=24.
[74] Saxena AM, Schoenborn BP. Acta Cryst 1977;A33:805–13.
[75] Kajimoto R, Nakajima K, Nakamura M, et al. J Neutron Res 2008;16:81–6.
[76] Mezei F. J Neutron Res 1997;63.
[77] Schanzer C, Boni P, Filges U, Hils T. Nucl Inst Methods Phys Res A 2004;529:63–8.
[78] Nagano M, Yamaga F, Yamazaki D, Maruyama R, Hayashida H, Soyama K, Yamamura K. J Phys Conf Ser 2012;340012016.
[79] Nielsen K, Lefmann K. Physica B 2000;283426.
[80] Nagano M, Yamaga F, Yamazaki D, Maruyama R, Hayashida H, Soyama K, Yamamura K. J Phys Conf Ser 2012;340012034.
[81] Itoh S, et al. Nucl Inst Methods Phys Res A 2012;661:86–92.
[82] Zsigmond GY, Bata L, Kisdi-Koszo E. Nucl Inst Methods 1966;45:255–60.
[83] Nakamura Mitsutaka, Kajimoto Ryoichi, Inamura Yasuhiro, et al. J Phys Soc Jpn 2009;78:093002-1–093002-4.

Chapter 4

Structure of Complex Materials

Silvia C. Capelli
Institut Laue Langevin, Grenoble, France

Chapter Outline
4.1. Introduction   321
4.2. Useful Properties
     of Neutrons   324
     4.2.1. Neutron Scattering
            Length   324
     4.2.2. A Particle with a
            Mass   324
4.3. What Can Be Learnt from
     Neutron Diffraction
     Experiments?   326
     4.3.1. Hydrogen Bonding   326
     4.3.2. Proton Migration   329
     4.3.3. Transition Metal
            Hydrides   332
     4.3.4. Porous Materials   333
     4.3.5. Diffuse Scattering 335
4.4. Outlook   339
     4.4.1. Neutron Sources   339
     4.4.2. Neutron Optics   340
     4.4.3. Detectors   341
     4.4.4. Samples and
            Sample
            Environment   342
     4.4.5. Software   347
4.5. Conclusions   348
References   349

"Why water boils at 100° and methane at -161°, why blood is red and grass is green, why diamond is hard and wax is soft, why glaciers flow and iron gets hard when you hammer it, how muscles contract, how sunlight makes plants grow and how living organisms have been able to evolve into ever more complex forms ... the answers to all these problems have come from structural analysis."

Max F. Perutz, 1996.

4.1 INTRODUCTION

Up to 1912, the world of atoms was out of reach to direct observation, but their role was already deployed in scientific theories like the kinetic theory of gases or the lattice theory of crystals. The experiment by Friedrich, Knipping, and Laue, we are celebrating the centennial of, showed the phenomenon of diffraction of X-rays by crystals [1], and it started a revolution in solid-state physics: atoms became real physical objects, experimentally "visible". A very rapid development followed this first observation, both in the direction of explaining the diffraction process and in trying to understand the physical nature of X-rays. The intuition by William Lawrence Bragg that the diffraction pattern obtained by Laue and colleagues was due to the reflection of short wavelength electromagnetic waves from planes of atoms in the crystal [2], set the first milestone in the process of understanding materials from their structure, and offered a preliminary experimental confirmation of the physical quantum theories that were being developed in those years. The basis of modern chemistry and solid-state physics were framed.

With the discovery of the neutron in 1932 by Chadwick [3], neutron diffraction became feasible and was demonstrated by Halban and Preiswerk using a radium–beryllium radioactive-decaying source, in an experiment that aimed to prove the particle-wave duality of the neutron [4]. Although the radioisotope-driven sources of neutrons allowed the first neutron measurements of Bragg intensities from inorganic crystals, see, for example, MgO [5], it was only after the remarkable work of Enrico Fermi on the characterization of neutrons [6] and with the construction of the first graphite-moderated reactor in Chicago as part of the Manhattan Project, that neutron diffraction started to taking off (for a nice historical summary, see Ref. [7]).

By 1920 most of the theory of diffraction was available, and many structures had been solved by photographic methods. However, it was only after the solution of the phase problem by Hauptman and Karle in 1956 [8], and with the increasing availability of automatic diffractometers and electronic computers, that X-rays and neutron crystallography started to be a fundamental tool in chemical, physical and material science research.

The last 30 years have seen a tremendous impact of crystallographic work on the study of materials, as can be seen, for example, in the number of organic and metalorganic structures deposited with the Cambridge Structure Database [9] (Figure 4.1). During this period, the number of X-rays structures has grown from 52,363 to 595,276, while the number of neutron structures has increased much less: only from 567 to 1534, an astonishing discrepancy. Many reasons have contributed to such a colossal difference. We have to admit it: photons have always run much faster than neutrons, but in recent years, the technical development of sources, optical components and detectors for X-rays has been remarkable and has made X-rays techniques much more accessible than neutron ones. The impressive development and availability of laboratory equipment with integrated microfocus X-ray tubes and CCD-type detectors, has transformed X-ray crystallography into a standard investigation tool in many university departments, prompting some companies to produce table-top "push-button" instruments for routine structure

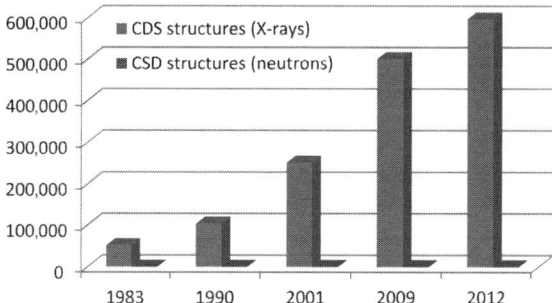

FIGURE 4.1 Increase of the total number of entries in the Cambridge Structure Database for X-rays (blue) and neutron structures (red) in the last 30 years. (For interpretation of the references to color in this figure legend, the reader is referred to the online version of this chapter.)

determination [10] for which the expertise of a crystallographer is considered almost unnecessary. The increasing interest in structural analysis of protein and nucleic acids has brought the field a lot of funding that was quickly spent in designing more and more automated crystallographic screening and data collection procedures, especially targeted at macromolecular beamlines at synchrotron sources [11, 12]. The determination in developing the field is quite visible in the new generation of liquid metal jet anode tubes [13], together with the new fast acquisition, large surface area detectors [14].

Neutron crystallography did not benefit of comparable technical progress as did X-rays. The production of neutrons is only the first in a series of important and technically challenging steps, like the need of focussing a largely divergent neutron beam onto a relatively small target, detecting a neutral particle, and growing samples with a volume that ensures a reasonable duration of the experiments. The combination of these factors have discriminated against the development of neutron techniques as compared to X-rays ones.

The "ability" of X-rays to see the electron distribution in materials has opened the way to many studies of the chemical bond, the "stabilizing interaction" that holds atoms together and accounts for the formation of molecules and solids (see the pioneering work of Debye in 1915 [15]). In this context, charge density analysis and other methods for extracting physical properties from structural data have emphasized the different types of information obtainable with X-rays and neutrons in chemical analysis and relegated neutrons to the secondary role of providing accurate nuclear positions and anisotropic displacement parameters for H atoms [16, 17].

Solid-state physicists, for their part, were aiming at understanding lattice vibrations and magnetic properties in solids; they found in neutrons "the most valuable probe of a solid we posses", as Ashcroft and Mermin put it in their milestone book [18]. Inelastic scattering of neutrons then became the field of solid-state physicist while neutron crystallography was mainly viewed as the tool to help understanding magnetic structures.

The rest of this chapter presents a brief discussion of the use of neutron diffraction in structural science from the point of view of a chemical crystallographer. The aim is not to be exhaustive about the fields in which neutron diffraction gives valuable contributions but to reinvigorate discussion about what used to be called "neutron chemical crystallography", but is no longer recognizable as a (sub-)discipline in itself. After recalling some properties of neutrons (section 2), a few selected examples will show where neutron diffraction can play a fundamental role in understanding properties and applications of various materials (section 3). The chapter concludes with an outlook (section 4). Readers interested in magnetic neutron scattering are referred to Chapter 6 of this book, completely dedicated to the subject. Here the emphasis is to chemical problems for which neutron diffraction is not the "unique" technique that can be applied but a tool that can often "make the difference".

4.2 USEFUL PROPERTIES OF NEUTRONS

4.2.1 Neutron Scattering Length

Neutrons are scattered by nuclei, the process being governed by the strong nuclear interaction. The strength of the interaction of the neutron-nucleus system is defined by the scattering length b, i.e. by a quantity that strictly depends on the details of the nuclear interactions. Therefore, in the expression of the structure factor for neutron diffraction, the X-ray atomic form factor describing the shape of the electron cloud is replaced by the parameter b.

Two properties of the scattering length are important: (i) the value of the scattering length varies irregularly for different nuclei, including isotopic variants, and it is completely independent of the atomic number Z of the element considered; (ii) the scattering power of an atom at rest does not fall off with the scattering angle as it does for X-rays.

Figure 4.2 pictures the variation of the neutron scattering length b with increasing atomic number Z for a few common elements. Note the almost identical - negative! - values of the scattering lengths for hydrogen and manganese, despite the difference of 24 electrons in Z.

The independence of the scattering length on the weight of the element makes neutrons the ideal probe for investigating materials composed of light and heavy atoms, since X-rays would struggle to see the light atoms near the heavy ones. Because the scattering power does not fall off with the increasing scattering angle, Bragg peaks at high resolution tend to be stronger on average than in the X-ray case, thus leading to very accurate atomic positions and anisotropic displacement parameters.

4.2.2 A Particle with a Mass

Neutrons are particles with a mass, bearing no electric charge but carrying a spin; therefore, they can interact with magnetic moments, including those

Chapter | 4 Structure of Complex Materials

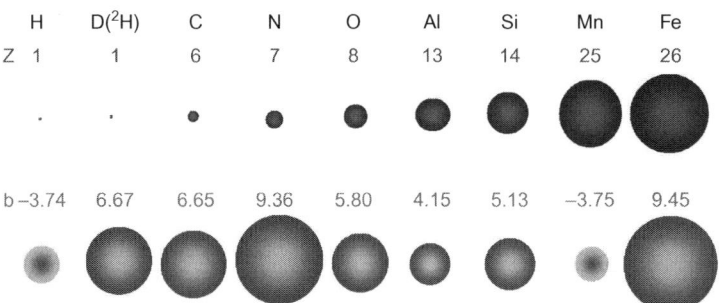

FIGURE 4.2 Pictorial representation of the variation of the scattering length b (bottom row and green dots) with increasing of the atomic number Z (top row and blue dots) for some elements. *Numerical values of the scattering lengths are taken from Ref. [19].* (For interpretation of the references to color in this figure legend, the reader is referred to the online version of this chapter.)

associated with unpaired electrons (see Chapter 6 for an accurate treatment of this kind of scattering).

The kinetic energy E_n of a particle with a mass m_n like a neutron, depends on its velocity v according to

$$E_n = \frac{1}{2} m_n v_n^2. \quad (4.1)$$

The de Broglie relation determines the wavelength λ_n for such a particle:

$$m v_n = \frac{h}{\lambda_n}, \quad (4.2)$$

where h is the Planck constant. Therefore energy and wavelength for a particle in motion are related:

$$E_n = \frac{h^2}{2m\lambda_n^2}. \quad (4.3)$$

From this relation, the energy of a neutron of 1 Å wavelength is 0.818 meV. The energy for a photon of 1 Å wavelength follows from the relation for an electromagnetic wave:

$$E = \frac{hc}{\lambda} \quad (4.4)$$

It is six orders of magnitude bigger: 12.40 keV. Considering that the energy to break a C-C bond is roughly 600 kJ/mol, that is about 6.2 eV, the comparison implies that in a neutron scattering experiment, the energy transferred onto the sample by the incoming neutron beam is too small to produce radiation damage, while that of a photon beam of the same wavelength is big enough to destroy the sample, especially a biological molecule. Therefore, a neutron scattering experiment with thermal neutrons, will not cause any beam damage to the sample.

4.3 WHAT CAN BE LEARNT FROM NEUTRON DIFFRACTION EXPERIMENTS?

In 1948, neutron diffraction was shown to be a method for locating hydrogen atoms in crystal structures or for determining the order-disorder structure of Fe-Co alloys [20]. More than 60 years later, the large majority of the experiments conducted at neutron sources aim at determining the magnetic structure and properties of materials containing elements whose electrons or nuclei interact with the spin of the neutron. The remaining experiments are aimed at locating light atoms in the presence of heavy, electron-rich elements. The search is quite often for hydrogen atoms, or for oxygen atoms in the presence of heavy metals, or for adjacent elements that would be indistinguishable by X-rays (like, e.g., carbon, nitrogen, and oxygen in disordered materials: as shown in Figure 4.2, the neutron scattering length for these second raw atoms differ more than the corresponding X-ray scattering factors).

The following examples are intended to illustrate the unique role neutron diffraction can play in solving some specific problems in chemical crystallography. Single-crystal diffraction work on molecular crystals will mainly be considered. The concepts underlying some of the examples may look outdated, e.g. "hydrogen bonding", but the specific studies reported below reveal surprising results.

4.3.1 Hydrogen Bonding

In his book "The Nature of the Chemical Bond" [21], Linus Pauling attributes the first claim of a hydrogen bond-mediated, stabilizing interaction to a study of amines in aqueous solution by Moore and Winmill in 1912 [22]. If we look deeper into the history of science, we will probably find that nearly all concepts we are using in chemistry and physics every day have been unveiled during the first two decades of the twentieth century. The concept of hydrogen bonding being born at the same time as the first Laue diffraction pattern is of significance to crystallographers who see hydrogen bonding interactions in almost every structure involving hydrogen bearing molecules.

The concept of a *bond* between an oxygen lone pair and the hydrogen nucleus was used to explain the behaviour of liquids and solvents in 1920 by Latimer and Rodebush [23]. Few years later, the X-ray structure determinations of sodium bicarbonate [24] and oxalic acid dehydrate [25, 26] set the basis for more systematic studies of donor–acceptor geometries, distances and angles.

The fundamental role of hydrogen bonding for stabilizing structures was further underpinned in 1953 through the discovery of base pairing in the double helix of DNA [27–29]. While the debate went on for a while before this hypothesis was generally accepted for DNA [30], more and more experimental results on hydrogen-mediated interactions were accumulated. Subsequently, the concept of hydrogen bonding evolved to include electropositive donors like carbon atoms [31], electronegative acceptors like fluorine atoms [32], as well as some intriguing cases of transition metal acceptors [33]. The large amount of information on hydrogen bonding reported in the literature has recently led to the formulation of a new definition of hydrogen bonding with particular emphasis on the evidence of bond formation [34].

The structure of [PtCl$_2$(NH$_3$)(N-glycine)]·H$_2$O determined by single-crystal neutron diffraction at 20 K, provides a nice example of bond formation with a metal atom as acceptor [35]. An X-ray diffraction determination had shown the axial position of the oxygen atom at a puzzlingly long distance of 3.52 Å from the Pt atom [36]. The neutron diffraction experiment confirmed the direct interaction of the square planar neutral Pt^{II} atom with a proton of the axial water molecule in the "H-ahead" or "inverse hydration" configuration (Figure 4.3). Such a configuration had been predicted by second-order perturbation Møller–Plesset calculations [37]. The importance of the neutron diffraction result is not just the accurate position of the hydrogen atom between the O and Pt atoms, but – in combination with the quantum chemical results - dispersive nature of the hydrogen–platinum interaction.

The experimental observation led to speculations concerning the influence of the hydrogen atom on the electronic structure of the Pt acceptor.

The latter has the $5d_z^2$ orbital pointing in the direction of the H atom, explaining the directionality of the interaction. Furthermore, it has been argued, as suggested in Ref. [38], that the Pt–H interaction is simply due to the space filling aspects of crystal packing [39]. Rizzato *et al.* have presented Hartree–Fock and MP$_2$ *ab initio* energies for the Pt atom interacting with a water molecule in both the H-ahead and the oxygen lone-pair ahead

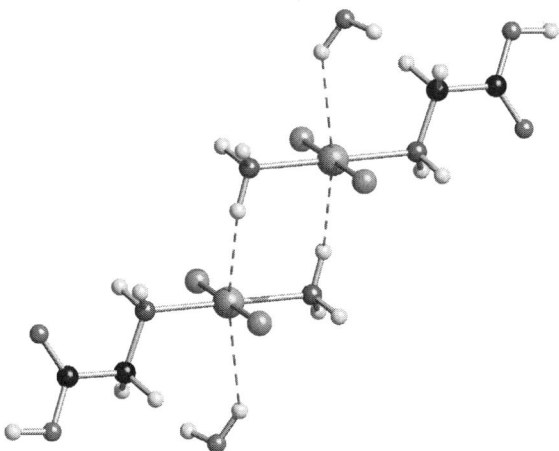

FIGURE 4.3 Coordination of the platinum atom in the [PtCl$_2$(NH$_3$)(N-glycine)]·H$_2$O complex, showing the interaction with a proton of a solvent water molecule. *Courtesy of Silvia Rizzato.* (For color version of this figure, the reader is referred to the online version of this chapter.)

configurations. A clear minimum in the MP$_2$ energy curve is found for the H-ahead configuration with an oxygen–platinum distance very close to the experimentally observed one (Figure 4.4a). The Pt-lone pair interaction is found to be substantially less stabilizing (Figure 4.4b). The curves obtained by HF calculations do not show any minima (Figure 4.4a and b). Given that HF calculations do not account for dispersion contributions whereas MP2 calculations do, it can be concluded that the stabilizing Pt...H...O interaction must be mostly dispersive in origin.

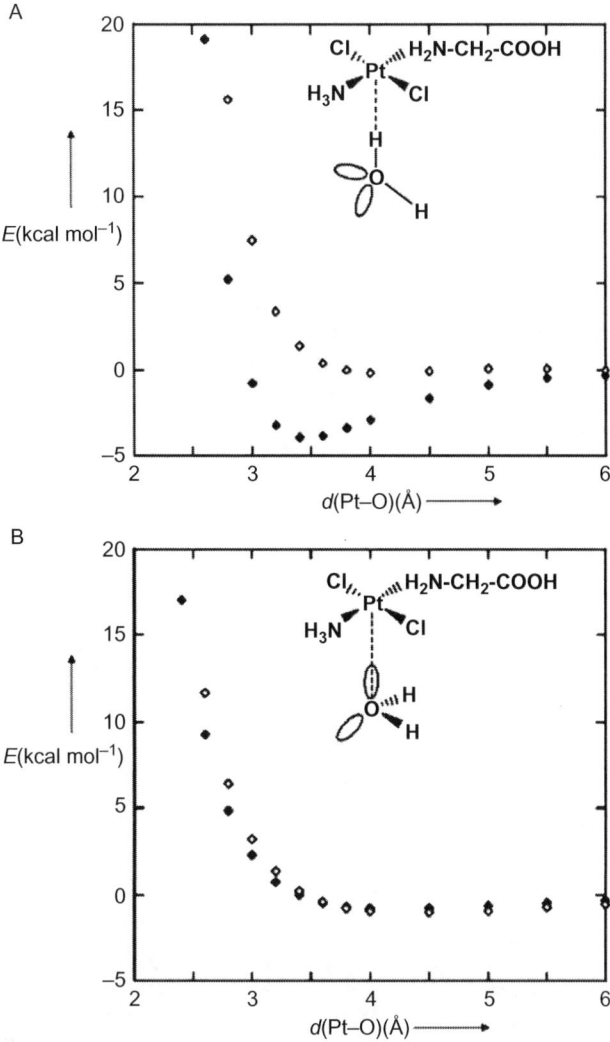

FIGURE 4.4 Calculated MP$_2$ (solid diamonds) and HF (open diamonds) interaction energies for the Pt complex with a water molecule in the "h-ahead" (a) or "lone-pair" ahead (b) configurations.

4.3.2 Proton Migration

Under the heading of *proton migration*, a long list of possible examples comes to mind. In biological systems, proton migrazion is often related to the mobility of water molecules through membranes in competition with ionic mobility [40]. In inorganic chemistry, the same term might be associated with proton conductivity in fuel cells and batteries [41]. A crystallographer might think of proton migration as a motion of protons between different minima of a potential energy surface. A typical example is the migration of a proton in a strong hydrogen bond, like the double proton exchange between the two carboxylate groups of the benzoic acid dimers [42]. It has been shown that large displacement of the hydrogen atoms are specific to system in which the strength of the H-bond is intermediate, while in systems involving strong and short interactions, the migration of the protons is of much smaller entities [43]. Many studies on proton migration in molecular crystals have been performed, in an effort of understanding which forces govern the reactivity of hydrogen bonds, their properties, and their influence on the processes they mediate, but also with the aim of designing molecular materials with tunable proton migration by effect of temperature, pressure, or chemical environment [44].

Although electron density maps obtained by accurate X-rays diffraction measurements have long been used in proton migration studies, together with a number of complementary techniques, the ability of neutron diffraction to give nuclear density is fundamental to evidence subtle proton migration effects involving atomic shifts of less than 0.1 Å. From a temperature-dependent study of proton migration in dimethylurea–oxalix acid complexes [45], the displacement of the proton within the hydrogen bond was found to be 0.031 Å in the 2:1 complex of N,N'-dimethylurea oxalic acid (Figure 4.5a) between 4 and 280 K, and 0.045 Å in the 2:1 complex of N,N-dimethylurea oxalic acid (Figure 4.5b) between 150 and 300 K.

Fourier difference maps obtained by X-ray diffraction data were inconclusive for determining the position of the proton involved in the hydrogen bond, due to the large smearing of the electron density in between the donor and acceptor oxygen atoms (Figure 4.6, left), while the negative neutron scattering length of the hydrogen atom, allowed to unambiguously determining the peak of the proton in the neutron Fourier difference maps (Figure 4.6, right).

The use of neutron diffraction for studying a shift in proton position of few hundreds of an angstrom might be considered inappropriate, but hydrogen bonds are very important structural features that play an important role on macroscopic physical properties of materials. The final aim of such studies is to identify the forces acting in proton migration, in order to design materials where hydrogen bonds can be used for "tuneable" transfer of charges.

Another form of proton migration is the diffusion of water into absorbing materials, either via channels and cavities or via some sort of pathway which involves a proton cascade. Already 200 years ago, Grotthuss, in trying to

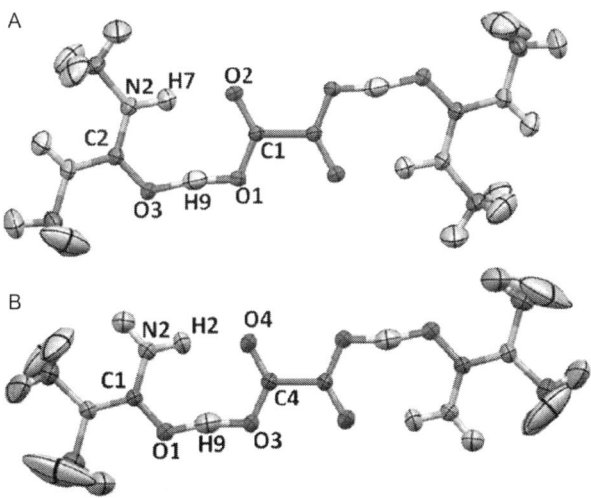

FIGURE 4.5 Crystal structure of the two 2:1 complexes of dimethylurea oxalic acid: the N,N'-dimethyl (a) and the N,N-dimethyl (b). (For color version of this figure, the reader is referred to the online version of this chapter.)

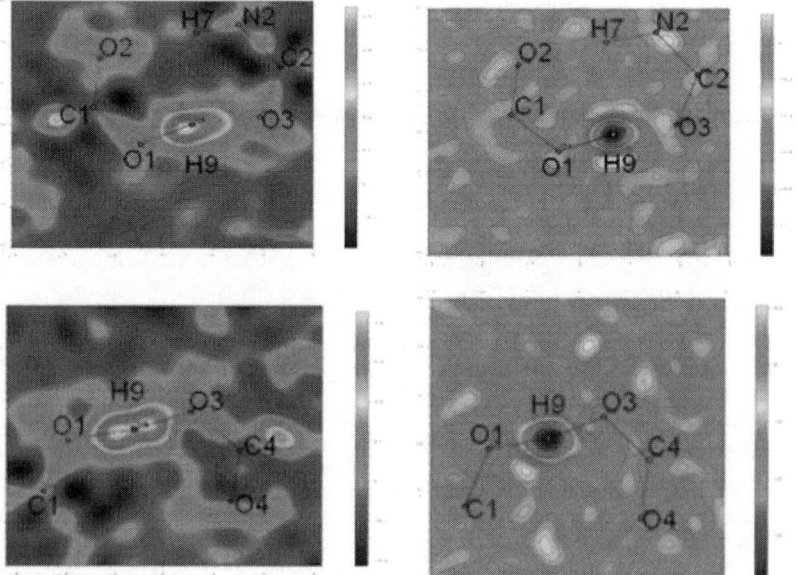

FIGURE 4.6 Fourier difference maps from X-rays (left) and neutron diffraction (right), centered on the hydrogen atom involved in the proton migration. Top : N,N'-dimethyl oxalic acid complex; bottom : N,N-dimethyl oxalic acid complex. (For color version of this figure, the reader is referred to the online version of this chapter.)

explain electrolysis in water solutions, proposed a mechanism in which the electric field was orienting the electrolytes into chains of alternating molecular polarity. When a charged atom at the end of a chain was reaching one electrode, reorientations along the chain would restore the original conformation and the process could be repeated [46].

Since then, such a hopping-reorientation mechanism has been invoked to explain proton mobility in a number of different processes, very often in solution, but has not always been supported by experimental evidences [47].

A nice experimental result that is in line with the Grotthuss handing over mechanism in the solid state, has been recently obtained with the essential contribution of neutron diffraction. A non-porous crystal of a manganese citrate cubane coordination polymer, $[Mn(H_2O)_6]_{2n}\{[Mn(H_2O)_5]Mn_4(citr)_4[\mu-Mn(H_2O)_4]\}_n \cdot 8nH_2O$, has been reported to undergo a complete exchange of protium for deuterium, when exposed to D_2O vapor at room temperature [48]. Metal-containing citrate cubanes have attracted attention because they manifest a broad range of chemical and physical reactivities: from simple physical solvent loss and reuptake via simple mechanisms, to solvent egress involving more complex chemical reactions. In addition, the Co-containing citrate cubane has been described as a single-molecule magnet [49, 50], opening up the study of analogous species.

The primary goal of studying the Mn-containing citrate cubane polymer was to understand the non-destructive, fully reversible single-crystal to single-crystal transformation that this non-porous system exhibit upon physical solvent loss and reuptake. The structure of the fully H_2O hydrated, dehydrated, and H_2O rehydrated compound, had been first measured by single-crystal X-rays diffraction in the lab. The determinations revealed the principal structural changes occurring at the solid-state transformations. However, because of the size of the structure and the small scattering power of hydrogen in X-ray diffraction, the hydrogen atoms of the water molecules could not be located. From neutron diffraction measurements at 20 K, the hydrogen atoms in the fully H_2O hydrated structure and in the D_2O rehydrated structure were located. The experiment showed that all the 30 water molecules in the structure, 21 of which bound to manganese centers in different arrangements of aquomanganese groups, undergo H/D substitution. All the H (or D) atomic positions were identified by neutron Fourier difference maps and fully refined with isotropic displacement parameters. Of the 30 water sites, 8 are found to have 3 hydrogen atoms at bonding distance from the oxygen atom, 2 of which half occupied. The fully occupied oxygen forms two chemically perfectly acceptable H_2O units with either of the half-occupied hydrogen. Two additional oxygen atoms are surrounded by four half-occupied hydrogen atoms. Again two water molecules with correct geometry can be formed by the oxygen atom and the two pairs of hydrogen atoms. The hydrogen atoms disorder implies a superposition of two orientations of water molecules frozen in place at the low temperature of the data collection. The disorder can therefore be

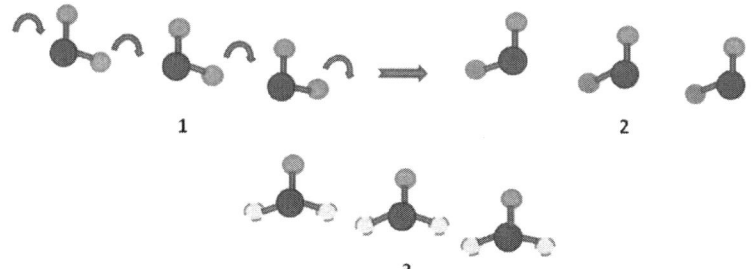

FIGURE 4.7 Schematic representation of a proton wire: a chain of double-minimum hydrogen positions which fit with the two states of a Grotthus mechanism. (For color version of this figure, the reader is referred to the online version of this chapter.)

interpreted in terms of two chains of double-minimum hydrogen potentials which fits with the two states of a Grotthus mechanism (Figure 4.7). Such a proton chain, or proton wire, allows the transport of hydrogen atoms inside the crystal in a direction parallel to the crystallographic a-axis.

4.3.3 Transition Metal Hydrides

For many years, the platinum group metals have played a significant role in the catalysis of hydrogenation processes [51]. In 1984, George Kubas demonstrated the existence of coordinated molecular hydrogen in some metallic complexes [52]. His discovery opened new frontiers both in the study of new chemical bonding to metals and in the understanding of the role of molecular hydrogen in coordination chemistry and catalysis. Since then, a number of studies on dihydrogen metal complexes have focused on various aspects of the chemistry of such systems (for a review see Ref. [53]). In the context of homogeneous hydrogenation, dihydrogen complexes of ruthenium have been shown to be more stable and easier to handle than those involving osmium or iridium [54], therefore, Ru complexes with substituted diphosphine ligands [55–57] that showed also bis-dihydrogen coordination [58, 59] were investigated.

Ru pincer complexes are another class of compounds that can coordinate hydrogen molecules [60], for which bis-coordination is relatively rare [61]. These complexes have been shown to be active as catalyst for C-H activation of arenes [62, 63], hydrogenation of nitriles [64] and as catalyst precursor for highly Z-selective hydroborations of acetylenes [65]. Among those systems, the complex $Ru(PCP)(H)(H_2)_2$ (PCP = 2,5-bis(di-*tert*-butyl-phosphanylmethyl)benzene), shows bis-dihydrogen coordination in the solid state [66]. Theoretical calculations predict two arrangements of minimum energy around the Ru atom: one with three and the other with five hydrogen atoms, that is, a mono- and a bis-dihydrogen-coordinated compound. Nuclear magnetic resonance measurements in solution are compatible with a nearly 1:1 mixture of the two complexes. From X-ray data the number of hydrogen atoms surrounding the

Chapter | 4 Structure of Complex Materials 333

FIGURE 4.8 Crystal structure of the complex the complex Ru(PCP)(H)(H$_2$)$_2$ (PCP = 2,5-bis(di-*tert*-butyl-phosphanylmethyl)benzene) as seen by neutron diffraction. The isotropic hydrogen atoms are those involved in the disorder (see text). (For color version of this figure, the reader is referred to the online version of this chapter.)

Ru atom could not be established unambiguously: a neutron diffraction structure of the complex determined at 18 K is shown in Figure 4.8. The ruthenium atom bears two di-hydrogen groups: one of them and the single hydrogen atom are disordered over two sites above and below the arene plane.

In addition to the structure at 18 K, Leitner and coworkers also measured the neutron structure of the ruthenium complex at 100 and 200 K and analysed the temperature dependence of the anisotropic displacement parameters of the hydrogen atoms not involved in the disorder. From such an analysis it is possible to determine low-frequency normal modes in the crystal [67]. A molecular Einstein model returns molecular translation, libration and deformation frequencies and their eigenvectors. The core of the Ru-complex has been described with the six rigid-body degrees of freedom (three librations and three translations), plus a libration around the molecular two fold axis affecting the dihydrogen group *trans* to the benzene ring. The frequency for the dihydrogen libration is 206(22) cm^{-1}. In the approximation of a dumbbell molecule in a double-minimum potential with one rotational degree of freedom, the barrier to rotation of this dihydrogen group may be estimated from this frequency to be around 1.8 kcal/mol, in agreement with calculated values and experimental results from inelastic neutron scattering (INS) for similar metal-dihydrogen complexes [68]. The barrier to rotation of the dihydrogen groups is a valuable piece of information on the metal-dihydrogen binding. The fact that such information can be attained from diffraction measurements on sample that are relatively small compared to those necessary for INS measurements and that do not require deuteration, represents an advantage in terms of economy of experiments. The method is also applicable to compounds bearing H and D atoms in different positions.

4.3.4 Porous Materials

In recent years, porous materials have attracted much attention for their physisorption and chemisorption properties. Metal-organic frameworks, covalent organic frameworks and hydrogen-bonded frameworks have been studied for applications as gas storage devices, for trapping small molecular solvents in their cavities, or as micro-reaction chambers of tunable-size, thereby responding to the need for solution of nowadays environmental issues. From a large choice of building blocks, tailored systems with pores or channel size adapted to specific applications can be synthesized [69]. One of the most widely envisaged applications for framework material is as gas storage systems, optimized for hydrogen storage and automotive applications. The requirements for such materials are very detailed and tend to be more and more internationally accepted (see, e.g., the criteria established for hydrogen storage systems by the U.S. Department of Energy, now commonly accepted worldwide [70]).

The potentially large scale applications of framework materials has intensified the search for purely organic porous materials either intrinsically or extrinsically porous ones.

Intrinsically porous materials are formed by building blocks containing a *pore*, that is, a cage or a bowl shape, that assumes a specific ordering during crystallization. In extrinsically porous materials, the pores are formed during the self-assembly process and the final characteristics of the porous structure are largely influenced by the crystallization solvent.

The most studied extrinsically porous material is probably the so-called Dianine's compound: 4-(*p*-hydroxy-phenyl)-2,2,4-trimethylchroman, which has been shown to accommodate a number of different solvents and gas molecules into its hexagonal, hydrogen-bonded channels [71]. This type of host-guest structure is a model system for studying the process of guest diffusion into the cavities of the host structure and consequently the nature of host-guest interactions [72].

A relatively new extrinsically porous organic material is 4-phenoxyphenol. In a combined X-ray and neutron study, the effect of the solvent on the final characteristics of the porous structure and of the interactions driving the self-assembly was investigated [73]. X-rays measurements of the crystal structures of 4-phenoxyphenol crystallized from methanol, acetone, diethyl ether, and acetonitrile, revealed hourglass-shaped channels running along the crystallographic *c*-axis. These channels form upon crystallization, and adapt to the steric demand of the solvent. In the limit of very bulky molecules no solvent is incorporated. The narrow part of the channel is defined by a hexagonal hydrogen-bonded ring about 5.4–5.6 Å wide. All the solvents used gave rise to the same overall structure with differences in the hydrogen-bonding distances and in the degree of disorder of the solvent in the channels: the smaller the solvent, the shorter the interaction distances and the higher the degree of disorder. The X-ray data did not allow to resolve the solvent disorder and gave only limited information on the intermolecular interactions involving H atoms. The neutron diffraction determination of the acetonitrile solvate, showed resolved positions of the disordered solvent molecules in the channels of 4-phenoxyphenol, including all the hydrogen atoms with their anisotropic

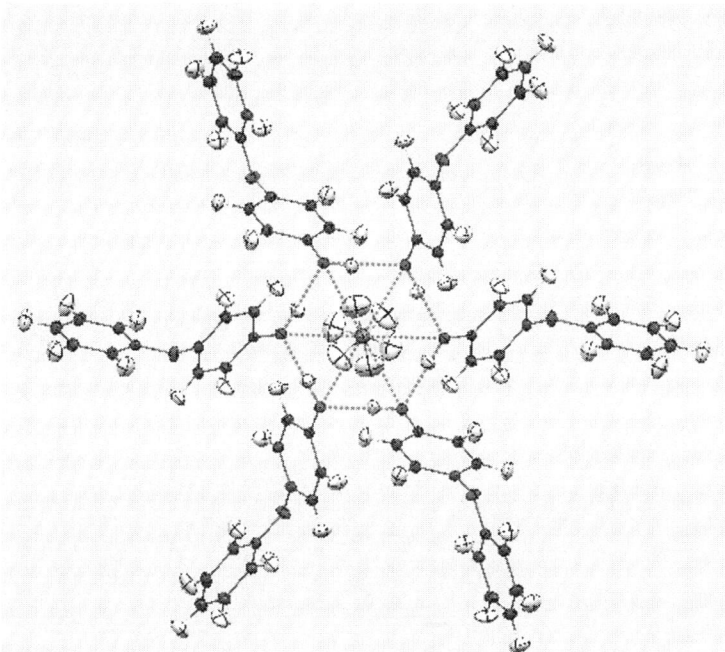

FIGURE 4.9 Crystal structure of the acetonitrile solvate of 4-pehnoxyphenol as determined by neutron diffraction, evidencing the hexagonal hydrogen-bonded channel and the twofold disorder of the acetonitrile molecule in the channel. (For color version of this figure, the reader is referred to the online version of this chapter.)

displacement parameters (Figure 4.9). The information obtained from neutrons was vital in order to fully characterize the interactions between solvent and channel-forming molecules.

4.3.5 Diffuse Scattering

Many crystallographers think that interpretation of diffuse scattering is presently the main unresolved problem in their field!

Diffuse scattering arises from any deviation from a perfectly ordered array of identical units. It can be static or dynamic in nature and can take many different forms: stacking faults of an otherwise ordered layer; disordered vacancies and ionic substitution; diffusion of atoms; disorder in molecular conformations or in crystalline domains; and phonon contributions (TDS). The analysis of diffuse scattering provides information about short range and local order in a crystalline material but needs accurate experimental data, preferably over a large range of reciprocal space. Laboratory X-rays have been the main probe for studying diffuse scattering because of the easy accessibility of the technique. It has been shown that such data can successfully model the short-range correlation in a structure but gives only an incomplete description of the thermal motion and has limited sensitivity to light atoms in the presence of heavier ones [74, 75]. High-energy synchrotron radiation can help to improve the situation, providing better data at higher resolution in reciprocal space. However, to recover the sensitivity for light atoms and in particular for hydrogen atoms, the best probe is still neutron scattering.

The diffuse scattering pattern for the $hh0$ plane of the cubic structure of a manganese Prussian blue analogue is given in Figure 4.10, for both neutron (left) and X-rays (right) [76].

The pictures nicely illustrate the complementarity between neutron and X-ray data: both patterns are calculated from the same structural model, and the differences between them are primarily due to the differences between neutron scattering lengths and X-rays atomic structure factors. The structure of $Mn_3(Mn(CN)_6)_2(H_2O)_6 \cdot xH_2O$ is full of water, but in the determination of the average X-ray structure, the hydrogen atom positions could not be determined and it was also difficult to locate the oxygen atoms of the x zeolitic water molecules [77]. Therefore, the model used to calculate the diffuse patterns contains only on the carbon, nitrogen, oxygen, and manganese atoms. As shown in Figure 4.2, the scattering length of Mn is negative and those of C, N, and O vary quite significantly, despite the small difference in atomic number. This means that in the neutron scattering pattern of this compound, each atom gives a significant contribution allowing to see the light atoms next to the heavy ones and to distinguish the light atoms among themselves, while the X-ray pattern is dominated by the scattering of the Mn atoms.

In order to take advantage of the information obtainable with the two scattering techniques, a simultaneous fitting of neutron and X-ray data represents the ideal approach for studying diffuse scattering. A nice example is given in

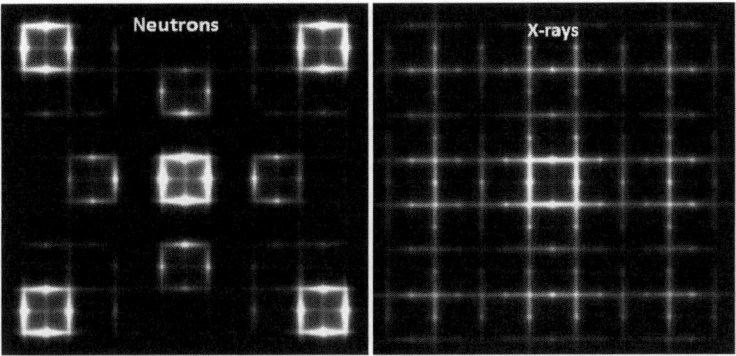

FIGURE 4.10 Comparison of calculated X-rays (right) and neutron (left) diffuse scattering patterns for $Mn_3(Mn(CN)_6)_2(H_2O)_6 \cdot xH_2O$. (For color version of this figure, the reader is referred to the online version of this chapter.)

Chapter | 4 Structure of Complex Materials

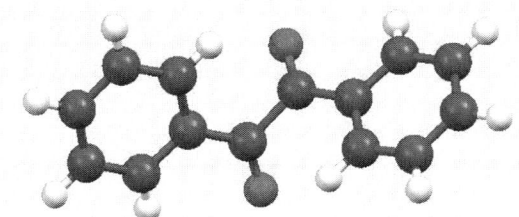

FIGURE 4.11 Drawing of the benzyl molecule. Conformation as observed in the crystal structure at 295 K [80]. (For color version of this figure, the reader is referred to the online version of this chapter.)

the work of Welberry on benzil [78]. A drawing of the molecular structure of benzil is given in Figure 4.11. The molecules have three internal degrees of freedom, namely, the torsions around the C-C single bond connecting the two phenyl rings, and form a two-dimensional network of hydrogen bonding. Benzil was initially selected for an X-ray diffuse scattering study aimed at understanding the influence of molecular flexibility on crystal packing and polymorphism [75]. In that study, a Monte Carlo model made of spring constants for describing effective intermolecular interactions, gave quantitative agreement with data obtained from four reciprocal lattice planes. This allowed to explain the diffuse lines observed in the diffraction pattern (Figure 4.12a) as due to strong longitudinal displacement correlations, affecting the oxygen and hydrogen atoms involved in the hydrogen-bonding contacts. A subsequent neutron study of the diffuse scattering of the same system [79] showed that the model derived from X-ray data was underestimating by a factor 2.25 the atomic displacements of each atom in the structural model.

Only the simultaneous refinement of laboratory X-ray diffuse scattering data on a crystal of size $0.5 \times 0.5 \times 2$ mm and neutron diffuse patterns collected on a crystal of size $5 \times 5 \times 10$ mm, have revealed the short-range correlation more accurately, and pointed out the dynamics of the molecules in the crystal [78].

In Figure 4.12, the observed and calculated patterns for both X-rays and neutrons for the $h00$ and $hk0$ reciprocal layer are shown. The comparison of the observed and calculated X-ray patterns (Figure 4.12a vs. c) shows that the large features around the Bragg peaks are well matched, while the comparison of the neutron observed and calculated patterns (Figure 4.12b vs. 12d) shows that the diffuse features are of the correct relative size, indicating that the atomic displacements are of the correct size. The simultaneous analysis concludes that the X-ray experiment provided the constraints with respect to the long-range correlations, while the neutron scattering one was giving the complementary information about atomic and molecular thermal motion. Furthermore, the model based on the X-ray data correctly determined the relative importance of the intermolecular interactions, but only the neutron data allowed the determination of the proper magnitude for these interactions.

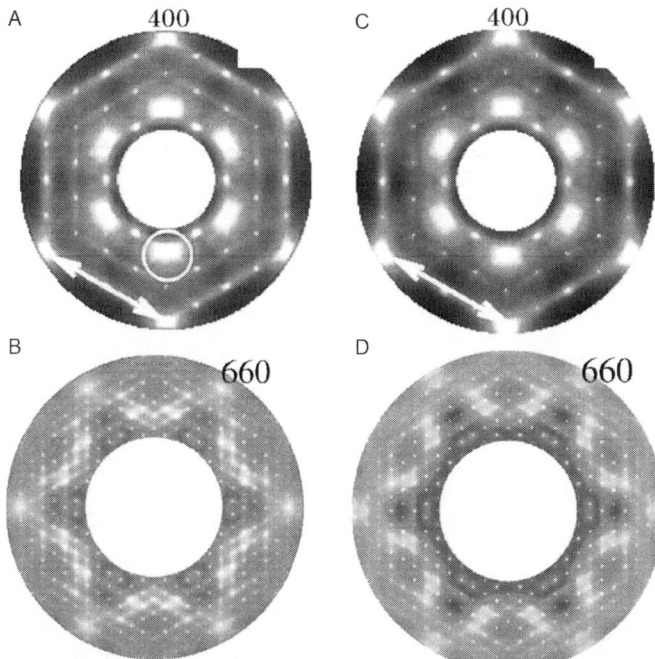

FIGURE 4.12 X-rays (a and c) and neutrons (b and d) observed (a and b) and calculated (c and d) diffuse scattering patterns for the $hk0$ reciprocal plane in the benzyl structure. (For color version of this figure, the reader is referred to the online version of this chapter.)

The benzil study shows that good refinement of a short-range correlation model requires high-quality low-background diffuse scattering data covering a large region in reciprocal space. Consequently, experiments involving diffuse scattering are really demanding especially when done with neutrons:

(a) the inherently low intensity of neutron sources has to be compensated with very large volumes of the sample (in the benzil case the volume of the sample used for the X-ray measurement was 0.5 mm^3, while that for the neutron one was 250 mm^3);

(b) the weak intensity of the diffuse scattering requires avoiding, or at least minimizing, any source of background, especially when using compounds containing hydrogens: the incoherent scattering of the H atoms can be of the same order of magnitude or even much larger than the diffuse features in the pattern; therefore, deuteration becomes imperative;

(c) neutron energies are comparable to phonon energies and effects due to inelastic scattering can appear in the diffraction pattern, giving rise to an artificial breaking of the diffraction symmetry (Figure 4.12b shows no data close to the origin of reciprocal space, due to the fact that data in that region have been rejected because of strong inelastic effects).

Chapter | 4 Structure of Complex Materials 339

4.4 OUTLOOK

In the previous section, some examples of using neutron diffraction techniques for getting more insights into the structure–property relationship have been presented. In particular, the example dedicated to diffuse scattering has shown both advantages and limitations in the use of neutrons.

In this section, a small overview of the different steps involved in a neutron diffraction experiment is given. The *fils rouge* guiding the discussion will be the relation that expresses the final intensity recorded after a neutron scattering process:

$$I = \varepsilon_{pr}\varepsilon_{sec}\varepsilon_{det}\sigma_s V_s I_0 \quad (4.5)$$

where ε_{pr} and ε_{sec} are the efficiencies of the primary and secondary spectrometers (respectively, monochromator and overall detection system including an analyzer module where applicable), ε_{det} is the detector efficiency, σ_s is the scattering power of the sample (depending on the scattering lengths of the elements in the unit cell), V_s is the sample volume, and I_0 is the intensity of the incident neutron beam [81]. In order to have a good quality experiment, it is necessary to optimize all the parameters involved in Equation (4.5).

Here we will go through a list of items that are related, some directly while others indirectly, to the parameters given in Equation (4.5). The list is not supposed to be exhaustive, nor the discussion will give the ultimate state of the art in the various fields. The aim of what follows is to look ahead at what could come in neutron crystallography and propose some possible developments.

4.4.1 Neutron Sources

As described in more details in Chapter 2, neutrons can be produced by radioactive-decay sources, by nuclear fission reactions or by collision of high-energy particles with heavy targets (spallation sources). In all cases, the emerging neutrons depart from the source in spherical geometry. For many years, because of the difficulties in focusing neutrons, the only possibility of increasing the intensity of a neutron beam - parameter I_0 in Equation (4.5) - was to increase the overall intensity, that is, the power of the source. High-flux reactors, like the ILL in Grenoble, can attain 10^{15} neutrons/cm^2 s, but to increase the production of neutrons even further the only possibility would be to build a more powerful reactor with increasing costs for fuel, safety of the installation, and radiation protection. Such high costs for building and running nuclear reactors, together with the long list of environmental issue connected to containment of fission reactions, and treatment of radioactive waste, have pushed research institutions in various countries in the direction of building spallation sources, like ISIS in the UK and more recently SNS in the USA and J-Park in Japan. At these facilities, the intensity of the outcoming neutron beam is of the order of 4×10^{12} neutrons/solid_angle · pulse for ISIS and

SNS, while J-Park has been constructed aiming at reaching 2×10^{13} neutrons/ solid_angle·pulse [82]. The idea of building a new generation spallation source like the European Spallation Source is in line with the need of increasing the beam intensity even further, with an expected neutron beam about 30 times brighter than the already running spallation sources [83]. This would allow the investigation of smaller and smaller samples in more complex environments.

Large scale facilities are governmental funded projects for which the accessibility of academic research groups is regulated by a system of peer-reviewed proposals. Industrial groups, on their side, have to pay for using these facilities and performing a neutron diffraction experiment could be very expensive. In order to offer an alternative to large scale facilities for industrial applications, some research has been conducted on portable neutron sources. At Berkley, for example, a high-intensity-pulsed neutron generator using the deuterium-deuterium fusion reaction has been developed [84]. Following this idea, a recent proposition for increasing the intensity of a neutron beam is a superconducting neutron source [85]. It is based on the collision of two deuteron beams travelling in the same direction, but staggered in time and energy, giving rise to a neutron beam that preserves much of the forward momentum of the colliding beams and a very narrow bandwidth. The idea is interesting, especially for the directionality of the resulting beam and for the use of a fusion reaction, that does not have the same drawbacks of a fission process. Such portable sources are not expected to attain the same brilliance of a high-flux reactor or of a last-generation spallation source, but they could help in increasing the accessibility of neutron techniques for fundamental research, leaving large scale facilities to deal with particularly difficult problems.

Further studies will tell about the technical feasibility and adaptability of portable neutron sources for research purposes, at the moment we can simply ask ourselves:

- Are stable, high-intensity, neutron laboratory sources coming soon?
- How neutron laboratory sources could change the impact of neutron crystallography in chemical, physical, and material science research?

4.4.2 Neutron Optics

In order to increase the performances of the already running neutron sources, one should consider which are the progresses in neutron optics that can improve the efficiency of the entire process of neutron transport and selection. The issue is related to the ε_{pr} parameter in Equation (4.5), if one considers the primary spectrometer as the ensemble of the neutron guide and the monochromator delivering the neutron beam to a specific instrument.

Neutrons are particles that cannot be stopped so easily, especially when they have a very high kinetic energy. When an instrument is built at some distance from the source, the need is to make as many neutrons as possible travel the expected distance. Neutron guides are especially conceived for such purpose, using the wave property of thermal and cold neutrons that undergo total external reflection as light waves do. Highly polished, very smooth metallic surfaces have been commonly used as mirrors for reflecting neutrons impinging at grazing angle. Supermirrors, that is, multilayers with adapted variation of thickness among the layers, had later substituted the initial monometallic surfaces, in order to increase the reflectivity with the aid of continuous Bragg reflections [86]. In this way, the acceptance angle of the impinging neutrons has increased and consequently the quantity of reflected particles. Nowadays, one can find ballistic supermirrors neutron guides at neutron research facilities [87], and some companies are producing super-polished glass surfaces for special neutron scattering application where smooth surfaces at atomic level are required [88].

Further down the guide tube, neutron energies can be selected with the aid of a monochromator. Probably, the hottest item in the field is diamond mosaic single crystals. After the first calculations [89], extensive work on the growth of single-crystal plates by chemical vapor deposition techniques has allowed to produce and test the first diamond single-crystal monochromator, working in transmission geometry using the (220) reflection [90].

By the time the incident beam reaches the sample, a large part of the neutrons have been lost on the way. Therefore, the best way not to lose any more neutrons is to try to focus the remaining ones. In many cases, a focussing monochromator can play the role, but in some others, and especially for small sample dimensions, focussing polycapillaries could represent a breakthrough [91, 92].

To the eye of a nonexpert, this quick snapshot of some of the progresses in neutron optics could be a real surprise. It indicates that the technology for processing materials for neutron applications has attained such a high level that a new generation of optical components for neutron scattering could come soon and disclose new possibilities. Maybe the time has come for neutron crystallography for a major technical upgrade?

4.4.3 Detectors

A neutron is a neutral particle. The most common way of detecting it, is though the production of a charged particle: for example, in ^{3}He wire detectors, a neutron is captured by a nucleus of ^{3}He, according to the reaction:

$$^3\text{He} + \text{n} \rightarrow {}^3\text{H} + \text{p} + 0.764\,\text{MeV} \tag{4.6}$$

The instable assembly decays producing a tritium atom and a proton, that ionize the stopping gas (usually CF_4) liberating the electrons that are then detected on the wires. This is, for example, the technology used for the detectors of the D9, D10, and D19 single-crystal diffractometers at the ILL [93].

Due to the fact that ^{3}He is slowing escaping the gravity of the earth and therefore the quantity of gas available for these types of detectors is finite, in recent years new ways of detecting neutrons have been developed, with particular emphasis on solid-state detection systems. Such detectors use the principle of a converter material that interacts with neutrons and produces charged particles detectable in a subsequent layer of the device. The most studied technology is probably the one based on boron ^{10}B (see, e.g., Ref. [94]), but other materials have been used, like the ^{6}LiF/ZnS:Ag scintillator screens installed on the SXD diffractometer at ISIS [95], and additional new technologies are under development. Very interesting is the use of single crystals as scintillators for neutrons, both from inorganic materials like lithium aluminates [96] and from completely organic ones like diphenylantracene [97].

It is impossible to anticipate here which type of new detectors will be installed in the future on neutron instruments, nor which will be the most efficient for diffraction work. In waiting for what will come, we can try to make the best use of what available. An example is the instrument CYCLOPS at the ILL [98]. Intensified cameras based on the well-known technology of cooled CCD cameras already developed for X-rays detectors have been combined with gadolinium scintillators for the detection of neutrons [99]. The result is a set of high-resolution, fast readout neutron CCD cameras that, combined in an octagonal arrangement with a white radiation incident beam, give a new type of neutron detector for quick reciprocal space investigation. Readout time is of the order of 300–500 ms; therefore, the exposure time becomes the limiting factor for time-resolved studies, but rapid temperature-dependent scans to allow the screening of phase transitions or the search of magnetic propagation vectors are ideal case studies to be performed with this instrument. In Figure 4.13, an example of a diffraction image taken with the new detector CYCLOPS is shown.

4.4.4 Samples and Sample Environment

4.4.4.1 Sample Dimensions

The "mith" in neutron crystallography has always been the size of the sample. Too many times, people renounce doing neutron diffraction

FIGURE 4.13 Diffraction image taken on CYCLOPS at the ILL of the spin crossover complex [Fe(L$_{222}$)(CN)$_2$]·H$_2$O [100] in the high-spin-7 configuration at RT. Note the eight parts of the image, each obtained with two CCD cameras. *Image courtesy of M.-H. Lemee-Cailleau and S. Lakhloufi.*

experiments because they need to grow "large crystals." It is true: a nonnegligible volume of sample is a physical requirement for a neutron experiments. Neutrons interact with nuclei, and in a solid, at atomic level, nuclei are far apart from each other; thus neutrons spend most of the time travelling through empty space; consequently, the only way to increase impact statistics is to increase the volume. It is clearly shown in Equation (4.5): the sample volume is a parameter directly proportional to the intensity of the scattered beam. In short: a large volume means good counting statistics in a reasonable time. But how large is large? And how long one needs to count for? Some people still think that for a neutron experiment, you need crystals of 2-3 cm^3, and to count for weeks before getting some results. This is no longer true. Few cubic millimeters of a good quality crystal with good diffracting power is usually enough, and exposure time can be of few seconds. According to the type of detector installed on the instrument used, the total time of an experiment can vary a lot. For example, using a large-area position-sensitive detector like the one installed on D19 at the ILL [101], for a medium-sized unit cell like that of β-alanine ($C_3H_7NO_2$, orthorhombic, Pbca, $a=6.030$ Å, $b=9.8982$ Å, $c=13.568$ Å, vol. $= 809.3$ Å^3) at a wavelength of 0.948 Å a full data collection at 15 K required 48 h, with 17,370 reflections measured, 2603 uniques, and average redundancy 5.7 ($R_{int}=4.2\%$, $R_\sigma=2.6\%$) [102]. In Figure 4.14, a screenshot of an image taken with the D19 area detector during the measurement of β-alanine at 15 K is shown.

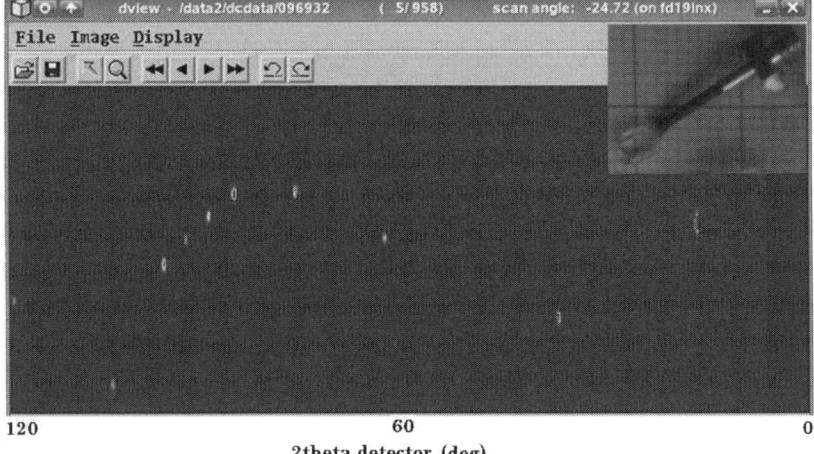

FIGURE 4.14 Single exposure of 3 s of a sample of β-alanine at 15 K on the PSD detector installed on D19 at the ILL. In the insert, the crystal used for the measurement (size ∼2.2 × 2.4 × 3.5 mm) mounted on a vanadium pin of 1 mm diameter. (For color version of this figure, the reader is referred to the online version of this chapter.)

4.4.4.2 Effects of Sample Environment

More and more often the investigation of materials has to be conducted in conditions as close as possible to those envisaged for possible applications. Furthermore, in studying a material at different temperatures, one can discover phase transitions or special conditions of meta-stability. There could be several reasons to study materials at their "extremes," and both home laboratories and large-scale facilities have developed tools and instrumentation in order to overtake new challenges.

Low/high temperature and high pressure are the most common conditions for crystallographic experiments. Low temperature is used in order to minimize effects of thermal motion and atomic vibrations, while high temperature or high pressure is used to study structural and/or magnetic phase transitions. Devices especially conceived to reach lower and lower temperatures have been developed for different types of diffraction experiments: for example, the liquid helium bath "Orange" cryostat developed at the ILL [103], working in a range of temperatures from 1.5 to 300 K, has become a standard for most neutron diffraction experiments at low-temperature both for powder and for single-crystal samples.

This is not the place where to give a list of devices for extreme-condition crystallography (for a comprehensive review, see Ref. [104]), the point to underline here is that experiments are becoming more and more complex, and in order to have good quality results, one has to take into account the additional effects on the diffraction pattern caused by the sample environment.

In the example of β-alanine given above, hydrogens represent 50% of the atoms in the structure; therefore, there is a huge inelastic scattering contribution to the background due to the hydrogens. As you can see in Figure 4.14, the diffraction pattern of β-alanine does not seem to be much affected, as there are well-defined reflections all over the surface of the detector, but special care had to be taken during the integration of the weak intensity reflections. As soon as you place your sample into a controlled environment such as a cryogenic device or a pressure cell, the walls of the device produce a scattering that will add up to the sample diffraction pattern. Heat screens made of aluminum produce powder diffraction lines on top of the diffraction pattern of the sample, and when the same screens are made of vanadium, the incoherent scattering produces a strong increase in the overall background. According to the type of detector used in the measurement, the effect on the final measurement can be different. For example, it is easier to avoid a powder diffraction line from an aluminum screen of a cryostat when measuring with a point detector rather than a 2D detector: in the first case, one can skip specific reflections affected by aluminum contamination with a selection based on the diffraction angle, while in the latter case, the powder lines will always be present on the detector images, with a consequent higher workload in order to remove specific regions of the detector in the integration process. In

Chapter | 4 Structure of Complex Materials   345

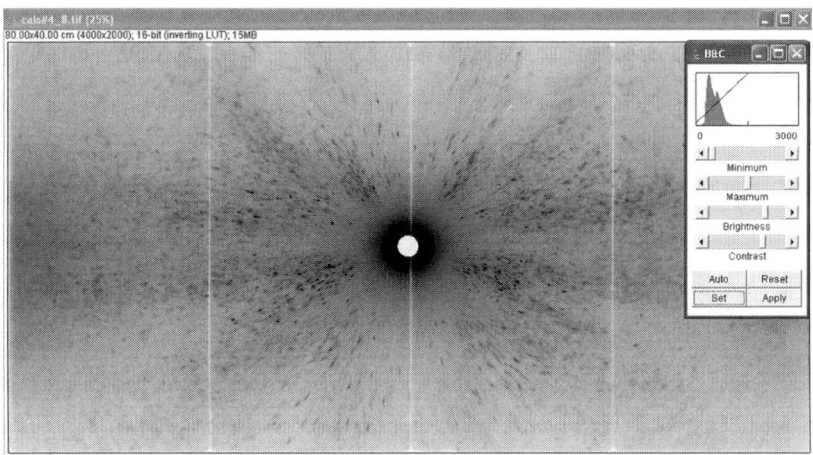

FIGURE 4.15 Diffraction image, taken on the instrument VIVALDI at the ILL, of an aluminum cylinder to be used as heat screen in a cryostat. It is evident that such a material is not well suited to be used in a sample environment for diffraction work as the diffraction spots coming from the aluminum will severely contaminate the diffraction pattern coming from the sample. *Courtesy of M.-H. Lemee-Cailleau.* (For color version of this figure, the reader is referred to the online version of this chapter.)

Figure 4.15, an example of the diffraction pattern from an empty calorimeter to be used in an orange cryostat, measured on the instrument CYCLOPS at the ILL. The radial lines in the picture are due to the mechanical treatment of the aluminum foil to shape it in a cylindrical form.

4.4.4.3 H/D Substitution

One more word has to be said about H/D substitution. In the study of the diffuse scattering in benzyl illustrated in Section 4.3.5, the inelastic scattering contribution due to hydrogen atoms had the same intensity of the diffuse features; therefore, for the neutron measurements, the fully detuterated sample had to be used in order to avoid disturbing effects in the final diffraction pattern. Deuteration is a common procedure for biological samples when to be studied with neutrons. The inelastic scattering contribution from hydrogen atoms in a large-size molecule like a protein or an enzyme that often crystallize with a huge amount of solvent water molecules can represent a strongly limiting factor in resolving the weaker intensity reflections.

On the other hand, when measuring a protein in a cryogenic device, the reduction in the background obtained by having exchanged for deuterium most of the hydrogen atoms present in the sample, is counterbalanced by the increase in background produced by the cryostat walls. It is true that a cryostat allowing to reach very low temperatures discloses new experimental opportunities for answering a broader range of question, especially in phase transitions and structural properties as a function of temperature, but often

FIGURE 4.16 The N_2-cryostream installed on the thermal neutron diffractometer D19 at the ILL, for data collection at 100 K. (For color version of this figure, the reader is referred to the online version of this chapter.)

in protein crystallography 100 K is the temperature at which most of the home-laboratory sources or the macromolecular synchrotron beamlines measure their samples. In Figure 4.16, a nitrogen cryostream installed on the thermal neutron diffractometer D19 at the ILL: it opens the possibility of performing neutron protein crystallography in sample conditions directly comparable to those common to X-ray work [105].

In conclusion, a little warning about H/D substitution : in both macromolecular and small-molecule crystallography deuteration is performed under the assumption that the use of the ^{2}H isotope does not change the final structure of the molecule under study. In Figure 4.17, one can see that this is not always true: the picture gives the reciprocal space reconstruction of the ($hk0$) layer of the structure of the manganese Prussian blue analogue $Mn_3(Mn(CN)_6)_2(H_2O)_6 \cdot xH_2O$ measured with X-ray diffraction. On the left side when

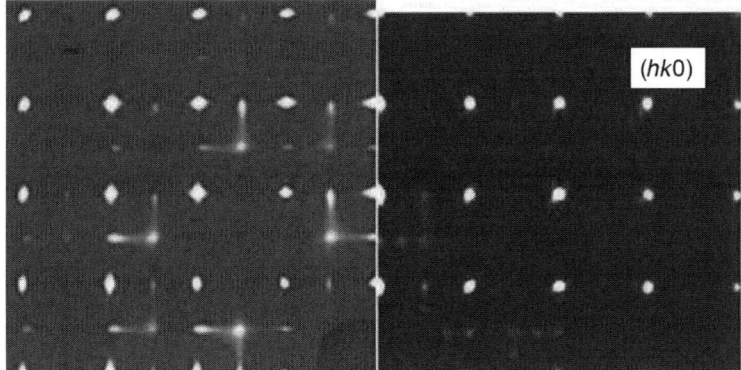

FIGURE 4.17 Reciprocal space reconstruction of the ($hk0$) layer of the structure of the manganese Prussian blue analogue $Mn_3(Mn(CN)_6)_2(H_2O)_6 \cdot xH_2O$ from X-ray diffraction measurements. On the left side diffraction obtained from a sample crystallized from hydrogenated water, on the right from deuterated water. Note the superstructure in the deuterated complex and the diffuse lines in the hydrogenated one. (For color version of this figure, the reader is referred to the online version of this chapter.)

the complex is crystallized from hydrogenated water, on the right when it is crystallized from deuterated water. In the first case, the structure shows diffuse features along the layers, while in the second one, it shows a superstructure that transforms the cubic unit cell into a tetragonal one, as the superstructure reflections appear only along the h and k reciprocal directions [106].

4.4.5 Software

As opposed to the X-ray diffraction scenario, in which single-crystal data collection protocols, instrumentations, sample environment, and setups are largely standardized and well known in a very large community of users, neutron single crystal is represented all over the world by a small set of instruments operated by specialists at large-scale facilities. Each of these instruments has a unique setup with dedicated hardware and software and, most of the time, an in-house made detector with in-house software for both data collection and data treatment. Often, single-crystal instruments at the same facility do not have the same software for data treatment and sometimes the data treatment software is composed of many different steps performed with different pieces of software. There are also cases in which the key piece of software for integrating reflection intensities as been written by a previous instrument responsible for a different detector and readapted to new needs, making almost impossible for anybody but the original programmer to follow the evolution of the code.

This situation is no longer sustainable. On the contrary, the risk is to keep away possible new neutron users, especially those who have experience of large X-rays facilities.

Some neutron single-crystal instruments, in particular, those with large coverage of reciprocal space, could perform very good complementary work to that done with X-rays, on the limit of becoming good competitors! Of course the large and easy accessibility of X-rays makes such an eventual competition with neutrons very unbalanced already at the starting line: as soon as commercial diffractometers have appeared, the X-rays community has started to share data, compare protocols and data analysis procedure, both informally and following directives of official projects [107]. Firmwares producing software have been stimulated for continuously improving data visualization and integration methods, pushing it to the limit that the complete automation of X-ray measurements became possible [108]. The importance of software in the development of X-rays work is attested by the CCP14 project [109]. It was the result of several software surveys between 1991 and 1993 that highlighted the need for a suite of programs to perform the steps necessary to the data analysis in X-ray diffraction in a more unified way. The project has been funded from 1995 to 2007, and the relative online repository, freely available to academics and students, still lists all the software codes deposited with the CCP14.

It is probably the time for the neutron diffraction community to engage in a collaboration effort aimed at developing standard protocols and procedures for single-crystal work, including data collection, data visualization, and analysis, in order to open the way to improved data comparisons and harmonization of the software in use.

4.5 CONCLUSIONS

The aim of this chapter was to illustrate the role of neutron diffraction in the study of complex materials. The structural studies given in Section 4.3 are typical examples belonging to the chemical crystallographic community, that is, studies aimed at understanding the properties of materials on the basis of the interrelations of the molecules and atoms composing it. The fact that in all materials proposed in the examples one of the "key" atoms is hydrogen shows, once more, the peculiarity of neutrons versus X-ray diffraction in detecting light atoms. The examples were chosen to show that crystallography in general, and neutron crystallography in particular, has still a role to play in understanding materials, not just in refining atomic positions but also in extracting physical properties from diffraction data. Such a "new role" for neutron crystallography can be pursued thanks to the technological advances in the field of neutron sources, optics, and detectors that put neutron experiments on a completely new stage as compared to those done until few years ago.

One should not forget that technological development has to advance in parallel with new approaches to neutron experiments: the use of neutron diffraction in combination or to complement X-ray work, as well as more standardize protocols, could open the way to new collaborations and new ways of approaching scientific questions.

REFERENCES

[1] Friedrich W, Knipping P, Laue M. Sitzungsber Math Phys Kl K Bayer Akad Wiss München; 1912. p. 303–22.
[2] Bragg WL. Proc Cambridge Philos Soc 1913;17:43–57.
[3] Chadwick J. Nature 1932;129:312.
[4] Von Halban H, Preiswerk P. Comptes Rendus 1936;203:73–5.
[5] Mitchell DP, Powers PN. Phys Rev 1936;50:486–7.
[6] Fermi E. Collected papers (Note e memorie). Vol. I: Italy, 1921–1938. Chicago, IL, USA: University of Chicago Press; 1962, *Vol. II: United States, 1939–1954*. University of Chicago Press, Chicago, IL, USA, 1965. Vol II edited by E. Amaldi, H. L. Anderson, E. Persico, F. Rasetti, C. S. Smith, A. Wattenberg, and E. Segrè..
[7] Mason TE, Gawne TJ, Nagler SE, Nestor MB, Carpenter JM. Acta Cryst 2012;A69:37–44.
[8] Hauptman HT, Karle J. Acta Cryst 1956;9:45–55.
[9] Cambridge Structure Database. Cambridge: CCDC; 2012.
[10] http:/www.bruker.com/products/x-ray-diffraction-and-elemental-analysis/single-crystal-x-ray-diffraction/smart-x2s/overview.html.
[11] Winter G, McAuley KE. Methods 2011;55:81–93.
[12] Arzt S, Beteva A, Cipriani F, Delageniere S, Felisaz F, Förstner G, et al. Prog Biophys Mol Biol 2005;89:124–52.
[13] http:/www.excillum.com/Technology/metal-jet-technology.html.
[14] https:/www.dectris.com/pilatus3_overview.html.
[15] Debye P. Ann Phys 1915;46:809–23.
[16] Koritsanszky TS, Coppens P. Chem Rev 2001;101:1583–628.
[17] Macchi P. Cryst Rev 2013;19:58–109.
[18] Ashcroft NW, Mermin ND. Solid state physics. Philadelphia: Saunders College; 1976.
[19] Dianoux A-J, Lander G. ILL Neutron Data Booklet. Philadelphia: OCP Science; 2003.
[20] Shull CG, Wollan EO. Science 1948;108:69–75.
[21] Pauling L. The nature of the chemical bond. Ithaca, N.Y: Cornell University Press; 1939.
[22] Moore TS, Winmill TF. J Chem Soc Trans 1912;101:1635–76.
[23] Latimer WM, Rodebush WH. J Am Chem Soc 1920;42:1419–33.
[24] Zachariasen WH. J Chem Phys 1933;1:634.
[25] Zachariasen WH. Z Kristallogr 1934;89:442.
[26] Robertson JM, Woodward I. J Chem Soc 1936;1817–24.
[27] Watson JD, Crick FHC. Nature. 171:737–8.
[28] Franklin R, Gosling RG. Nature 1953;171:740–1.
[29] Franklin R, Gosling RG. Nature 1953;172:156–7.
[30] Fonseca Guerra C, Bickelhaupt FM, Snijders JG, Baerends EJ. J Am Chem Soc 2000;122:4117–28.
[31] Desiraju GR. Acc Chem Res 1991;24:270–6.
[32] Thalladi VR, Weiss H-C, Bläser D, Boese R, Nangia A, Desiraju GR, et al. J Am Chem Soc 1998;120:8702–10.
[33] Brammer L. Dalton Trans. 3145–57.
[34] Arunan E, Desiraju GR, Klein RA, Sadlej J, Scheiner S, Alkorta I, et al. Defining the hydrogen bond: an account. IUPAC; 2011, Technical Report p. 1619–36..
[35] Rizzato S, Bergès J, Mason SA, Albinati A, Kozelka J. Angew Chem Int Ed 2010;49:7440–3.
[36] Baidina IA, Podberezskaya NV, Krylova LF, Borisov SV. J Struct Chem 1981;22:463–5.

[37] Kozelka J, Bergès J, Attias R, Fraitag J. Angew Chem Int Ed 2000;39:198–201.
[38] Falvello LR. Angew Chem Int Ed 2010;49:10045–7.
[39] Hamilton WC, Ibsen JA. Hydrogen bonding in solids. New York: W. A. Benjamin; 1968.
[40] Lehninger AL. Principles of biochemistry. USA: Worth Publishers; 1998.
[41] Fabbri E, Pergolesi D, Traversa E. Chem Soc Rev 2010;39:4355–69.
[42] Wilson CC, Shankland N, Florence AJ. Chem Phys Lett 1996;253:103–7.
[43] Wilson CC, Thomas LH. Compt Rendus Chem 2005;8:1434–43.
[44] Fontaine-Vive F, Johnson MR, Kearley GJ, Cowan JA, Howard JAK, Parker SF. J Chem Phys 2006;124:234503.
[45] Jones AOF, Lemée-Cailleau M-H, Martins DMS, McIntyre GJ, Oswald IDH, Pulham CR, et al. Phys Chem Chem Phys 2012;14:13273–83.
[46] Grotthuss CJT. Ann Chim 1806;58:54–74.
[47] Agmon N. Phys Lett 1955;244:456–62.
[48] Capelli SC, Falvello LR, Forcén-Vázquez E, McIntyre GJ, Palacio F, Sanz S, et al. Angew Chem Int Ed 2013; accepted for publication.
[49] Yang E-C, Hendrickson DN, Wernsdorfer W, Nakano M, Zakharov LN, Sommer RD. Appl Phys 2002;91:7382–4.
[50] Murrie M, Teat SJ, Stoeckli-Evans H, Güdel HU. Angew Chem Int Ed 2003;42:4653–6.
[51] Dedieu A. Transition metal hydrides. Germany: Wiley-VCH; 1992.
[52] Kubas GJ, Ryan RR, Swanson BI, Vergamini PJ, Wasserman HJ. J Am Chem Soc 1984;106:451–2.
[53] Kubas GJ. Metal dihydrogen and σ-bond complexes. New York: Kluwer Academic/Plenum Publishers; 2001.
[54] Sabo-Etienne S, Chaudret B. Coord Chem Rev 1998;178:381–407.
[55] Chaudret B, Poilblanc R. Organometallics 1985;4:1722.
[56] Borowski AF, Donnadieu B, Daran J-C, Sabo-Etienne S, Chaudret B. Chem Commun 2000;543–4.
[57] Abdur-Rashid K, Gusev DG, Lough AJ, Morris RH. Organometallics 2000;19:1652.
[58] Grellier M, Vendier L, Chaudret B, Albinati A, Rizzato S, Mason SA, et al. J Am Chem Soc 2005;127:17592–3.
[59] Grellier M, Mason SA, Albinati A, Capelli SC, Rizzato S, Bijani C, Coppel Y, Sabo-Etienne S, et al. Inorg Chem 2013;52:7329–37.
[60] Gruver BC, Adams JJ, Warner SJ, Arulsamy N, Roddick DM. Organometallics 2011;30:5133–40.
[61] van der Boom ME, Milstein D. Chem Rev 2003;103:1759–92.
[62] Prechtl MHG, Hölscher M, Ben-David Y, Theysen N, Loschen R, Milstein D, Leitner W. Angew Chem 2007;119:2319–22.
[63] Prechtl MHG, Hölscher M, Ben-David Y, Theysen N, Milstein D, Leitner W. Eur J Inorg Chem 2008;3493–500.
[64] Gunanathan C, Hölscher M, Leitner W. Eur J Inorg Chem 2011;3381–6.
[65] Gunanathan C, Pan F, Hölscher M, Leitner W. J Am Chem Soc 2012;14349–52.
[66] Gunanathan C, Prechtl MHG, Capelli SC, Englert U, Hölscher M, Leitner W. Eur J Inorg Chem 2013;5075–80.
[67] Bürgi HB, Capelli SC. Acta Cryst 2000;A56:403–12.
[68] Eckert J, Kubas GJ. J Phys Chem 1993;97:2378–84.
[69] Farrusseng D, editor. Metal-organic frameworks. West Sussex, UK: John Wiley & Sons; 2011.
[70] http:/www1.eere.energy.gov/hydrogenandfuelcells/hydrogen_publications.html.

[71] MacNicol DD, McKendrick JJ, Wilson DR. Chem Soc Rev 1978;7:65–87.
[72] Nemkevich A, Spackman MA, Corry B. Chem Eur J 2013;19:2676–84.
[73] Thomas LH, Cheung E, Jones AOF, Kallay AA, Lemee-Cailleau M-H, McIntyre GJ, et al. Cryst Growth Des 2012;12:1746–51.
[74] Welberry TR, Proffen T, Bown M. Acta Cryst 1998;A54:661–74.
[75] Welberry TR, Goossens DJ, Edwards AJ, David WIF. Acta Cryst 2001;A57:101–9.
[76] Chernishov D, Bosak A. Private communication; 2013.
[77] Franz P, Ambrus C, Hauser A, Chernyshov D, Hostettler M, Hauser J, et al. J Am Chem Soc 2004;126:16472–7.
[78] Goossens DJ, Welberry TR, Heerdegen AP, Gutmann MJ. Acta Cryst 2007;A63:30–5.
[79] Welberry TR, Goossens DJ, David WIF, Gutmann MJ, Bull MJ, Heerdegen AP. J Appl Cryst 2003;36:1440–7.
[80] More M, Odou G, Lefebvre J. Acta Cryst 1987;B43:398–405.
[81] Ioffe A. In: Brückel T, Richter D, Zorn R, Angst M, editors. Scattering methods for condensed matter research: towards novel applications at future sources, vol. 33. Schriften des Forschungszentrums Jülich/ Key Technologies. s.l.: Forschungszentrum Jülich Gmb; 2012.
[82] Maekawa F, Harada M, Oikawa K, Teshigawara M, Kai T, Meigo SI, et al. Nucl Instrum Methods Phys Res 2010;A620:159–65.
[83] http:/europeanspallationsource.se/unique-capabilities-ess.
[84] Williams DL, Vainionpaa JH, Jones G, Piestrup MA, Gary CK, Harris JL, et al. AIP conference proceedings 2009; vol. 1009: 936–9.
[85] Vaucher AR, McKinny KS. Superconducting neutron source; May 9, 2013, US 2013/0114773 A1 USA.
[86] Hayter JB, Mook HA. J Appl Cryst 1989;22:35–41.
[87] Abele H, Dubbers D, Häse H, Klein M, Knöpfler A, Kreuz M, et al. Nucl Instrum Methods Phys Res 2006;A562:407–17.
[88] http:/www.swissneutronics.ch/index.php?id=47.
[89] Freund AK. J Appl Cryst 2009;42:36–42.
[90] Courtois P, Freund AK, Fernandez-Diaz M-T, Nenert G. ICNS2013 - International Conference on Neutron Scattering. Edinburgh. IOP Institute of Physics; 2013.
[91] Gibson WM, Schultz AJ, Chen-Mayer HH, Mildner DRF, Gnäupel-Herold T, Miller ME, et al. J Appl Cryst 2002;35:677–83.
[92] http://www.xos.com/products/x-ray-optics-excitation-systems-x-beam/optics/neutron-optics/.
[93] http:/www.ill.eu.
[94] Robertson BW, Adenwalla A, Harken A, Welsch P, Brand JI, Dowben PA, et al. Appl Phys Lett 2002;80:3644–6.
[95] Rhodes NJ, Schooneveld EM, Eccleston RS. Nucl Instrum Methods 2004;A529:243.
[96] Fujimoto Y, Kamada K, Yanagida T, Kawaguchi N, Kurosawa S, Totsuka D, Fukuda K, et al. IEEE Trans Nucl Sci 2012;59:2252–5.
[97] van Loef EV, Glodo J, Shirwadkar U, Zaitseva N, Shah KS. Nucl Instrum Methods Phys Res 2011;A652:424–6.
[98] Ouladdiaf B, Archer J, Allibon JR, Decarpentrie P, Lemée-Cailleau M-H, Rodríguez-Carvajal J, et al. J Appl Cryst 2011;44:392–7.
[99] http://www.photonic-science.com/products/neutron-cameras.html#.
[100] Guionneau P, Le Gac F, Kaiba A, Sánchez Costa J, Chasseau D, Létard J-F. Chem Commun 2007;3723–5.
[101] Capelli SC, Mason SA, Forsyth VT, Allibon JR, Archer J. Acta Cryst A67: 2011;C35.

[102] Capelli SC, Moret M. ILL. Experimental report 5-12-289; 2013.
[103] Brochier D. Cryostat à température variable pour mesures neutroniques ou optiques; 1977. ILL. Technical report.
[104] Vincent L. Crystallography under extreme conditions: state of the art and perspectives, In: Benedict JB, editor. Recent advances in crystallography. Intech; 2012. 978-953-51-0754-5 http://dx.doi/10.5772/48608. Available from: http://www.intechopen.com/books/recent-advances-in-crystallography/crystallography-under-extreme-conditions-state-of-the-art-and-perspectives.
[105] Romoli F, Mossou E, Cuypers M, van der Linden P, Carpenthier P, Forsyth VT. submitted for publication to J Appl Cryst 2013.
[106] Capelli S.C., Larsern F.K., Buergi H.-B. 2013. [manuscript in preparation].
[107] Coppens P, Dam J, Harkema S, Feil D, Feld R, Lehmann MS, et al. Acta Cryst 1984;A40:184–95.
[108] Leslie AGW, Powell HR, Winter G, Svensson O, Spruce D, McSweeney S, et al. Acta Cryst 2002;D58:1924–8.
[109] http:/www.ccp14.ac.uk/.

Chapter 5

Large-Scale Structures

Jeffrey Penfold[*,†] and Ian M. Tucker[‡]
[*]ISIS Facility, Rutherford Appleton Laboratory, Chilton, Didcot, Oxon, United Kingdom
[†]Physical and Theoretical Chemistry Laboratory, Oxford University, Oxford, United Kingdom
[‡]Unilever Research and Development Laboratory, Port Sunlight, Bebington, Wirral, United Kingdom

Chapter Outline
5.1. Introduction 353
5.2. Experimental Details 356
   5.2.1. Fundamentals of Neutron Reflectivity 356
   5.2.2. Fundamentals of Small-Angle Neutron Scattering 358
   5.2.3. Experimental Details for Neutron Reflection 361
   5.2.4. Experimental Details for SANS 363
5.3. Thin Films, Interfaces, and Solutions 365
   5.3.1. Adsorption at the Air–Solution Interface 365
   5.3.2. Adsorption at the Liquid–Solid Interface 374
   5.3.3. Structure of Biological Membranes 378
   5.3.4. Micelles 382
   5.3.5. Lamellar Phases and Vesicles 385
   5.3.6. Colloidal Particles 392
   5.3.7. Polymers in Solution, Melt, and Thin Films 394
   5.3.8. Proteins and Biomacromolecules in Solution and at Interfaces 404
5.4. Summary and Future Prospects 409
References 409

5.1 INTRODUCTION

Large-scale structures encompass a broad area of science involving soft matter, biomaterials, and nanotechnology and are defined by relatively weak interactions and length scales that extend from nanometer to micrometer. In such systems, self-assembly and the development of hierarchical structures over a wide range of length scales in solution and in the solid phase are important characteristics. Thin films are also a prevalent feature of large-scale structures. In many of the systems, the properties of surfaces and interfaces dominate, and the structure of those interfaces and adsorption at those interfaces are key to understanding their diverse and important functions.

Characterizing large-scale structures is especially important, given the widespread industrial, technological, domestic, and medical applications, applications which impinge on every aspect of modern living. This covers applications such as home and personal care products, cosmetics, and many modern food formulations. In the context of advanced materials, it encompasses lubrication, adhesion, coatings, paints, sensors, photonics, and lightweight structural materials. In manufacturing and processing, this involves fuel oils, enhanced oil recovery, mineral flotation, and remediation. In biomedical applications, large-scale structures are implicated in the development of biosensors, encapsulates, drug delivery, the development of biocompatible and biofunctionalized surfaces, and synthetic biostructural materials. They are important to the fundamental understanding of some key biomedical issues, such as amyloid formation, interactions of drugs (antimicrobial and antibacterial) with cell membranes, protein structures (folding, unfolding), and bacterial structure and function.

Hence, it is not surprising that a wide range of direct and indirect experimental techniques have been applied to study the structural characteristics of such systems. These include light scattering; optical reflectivity and ellipsometry; optical imaging, such as microscopy and confocal microscopy; electron microscopy, including SEM, TEM, and cryo-TEM; surface probe techniques, such as AFM, STM, surface spectroscopic probes, diffraction, and X-ray; and neutron scattering techniques.

The focus of this chapter is how the neutron techniques of small-angle neutron scattering (SANS) and neutron reflectivity (NR) are used for the study of large-scale structures. The recent treatise on Elementary Scattering Theory by Sivia [1] provides an excellent introduction to the more mathematical aspects of SANS and NR. For reflectivity, the basic principles of X-ray and NR are covered in detail in some well-established review articles [2,3] and treated more fundamentally by Lekner [4]. Small-angle scattering is the more established technique, and the first detailed introduction to it was by Guinier and Fournet [5], and since then, for SANS, there have been many more tightly focused reviews and articles [6–10].

The X-ray and neutron scattering techniques of reflectivity and small-angle scattering have been extensively applied to many structural aspects of large-scale structures and provide fundamentally similar approaches. The wavelengths of X-rays and thermal/cold neutrons are on the order of nm and hence ideally matched to the scattering phenomena in large-scale structures. However, the nature of the interaction of X-rays and neutrons with matter is different and this has important implications. X-rays interact with electrons and so the scattering is dependent upon atomic number, and for many common elements, the adsorption coefficients are high and penetration is limited. This also means that, especially for modern synchrotron X-ray sources, radiation damage is an issue that requires special attention and penetration requires high-energy X-rays. For neutrons, the nuclear interaction dominates and the strength of the interaction is characterized by the neutron scattering length, b, which varies relatively randomly across the periodic table. Apart from known adsorbing materials (Cd, Gd, etc.), the adsorption cross sections are low and thermal neutrons are a penetrating probe that does little damage. Importantly, different isotopes have different scattering lengths, and in the context of this chapter, it is especially important for H and D, which have scattering lengths of different magnitudes and signs (see Table 5.1 for a list of the scattering lengths for some common elements).

This means that isotopic substitution (especially H/D substitution in organic materials) can be used to manipulate the neutron scattering length density and refractive index, which are defined as follows. The neutron refractive index, n, is related to the neutron scattering length density, such that (5.2),

$$n^2 = 1 - \frac{\lambda^2}{\pi}\rho \tag{5.1}$$

where λ is the neutron wavelength and ρ is the scattering length density (ρ is used elsewhere in neutron scattering as the particle-density operator) given by

$$\rho = \sum_i \bar{b}_i n_i \tag{5.2}$$

where $\bar{b}_i$ is the coherent scattering length and n_i is the number density of species i.

In this chapter, the fundamental principles of the two techniques, NR and SANS, are described in detail and the important experimental aspects

TABLE 5.1 Scattering Lengths for Some Common Elements (See Also Appendix)

Element	Scattering length ($\times 10^{-12}$ cm)
H	−0.374
D	0.6674
Si	0.4149
C	0.6648
O	0.5805
N	0.9300

summarized. Their application over a broad range of science will be presented, in a way that highlights their unique properties and their important contribution to the field.

5.2 EXPERIMENTAL DETAILS

5.2.1 Fundamentals of Neutron Reflectivity

The refractive index is not a convenient term to use in NR as it is ~ 1.0 (in contrast to light where the values for most materials are significantly different to 1.0), and the scattering length density is a more relevant term. As described earlier, for most materials, the coherent neutron scattering length, $\bar{b}$, is positive (hydrogen is one of the exceptions), and so n is generally <1.0. This highlights another major differences compared to light. Hence, in general, total external reflection occurs as most materials have a refractive index less than air. So, for example, at the air/D_2O or air/silicon interface, there will be total reflection from the interface below the critical glancing angle of incidence. The critical glancing angles are in general small. However, the information about the surface/interface structure occurs at grazing angles of incidence, θ, beyond the critical angle. The specular reflectivity can be conveniently described in the kinematic or Born approximation [2], as related to the square of the Fourier transform of the scattering length density profile, $\rho(z)$, in the direction, z, normal to the surface or interface,

$$R(Q) = \frac{16\pi^2}{Q^2} \left| \int \rho(z) e^{-iQz} dz \right|^2 \tag{5.3}$$

where Q is the wave-vector transfer perpendicular to the surface ($Q = 4\pi \sin\theta/\lambda$, where θ is the grazing angle of incidence). The reflectivity can be expressed in terms of the kinematic approximation, as above, or using the exact approach familiar in thin-film optics [11]. As such, the reflectivity for a thin film at an air–substrate interface (see Figure 5.1) can be written as

$$R = \frac{r_{01}^2 + r_{12}^2 + 2r_{01}r_{12}\cos(2\beta_1)}{1 + r_{01}^2 r_{12}^2 + 2r_{01}r_{12}\cos(2\beta_1)} \tag{5.4}$$

where the subscripts 0, 1, and 2 refer to the air phase, the thin film, and the substrate; $\beta = k_1 n_1 d_1$ is the optical path length, where n_1 is the refractive index of the thin film, d_1 is the film thickness, and k_0, k_1 are the neutron wave vectors in air and in the monolayer normal to the interface, $r_{ij} = (p_i - p_j)/(p_i + p_j)$, $p_i = n_i \sin\theta_i$.

Equation (5.4) can be simplified as

$$R(Q) = \frac{16\pi^2}{Q^4} \left[(\rho_1 - \rho_0)^2 + (\rho_2 - \rho_1)^2 + 2(\rho_1 - \rho_0)(\rho_2 - \rho_1)\cos Qd \right] \tag{5.5}$$

Chapter | 5 Large-Scale Structures

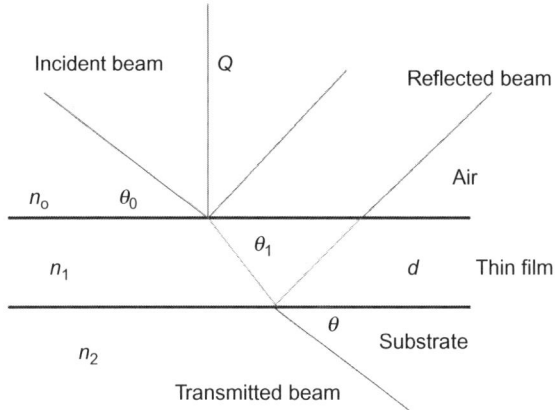

FIGURE 5.1 Neutron reflection geometry for reflection from a thin film with a refractive index n_1 on a substrate with a refractive index n_2. (For color version of this figure, the reader is referred to the online version of this chapter.)

At the interface between two bulk media (no thin film), Equation (5.4) reduces to Fresnel's law,

$$R(Q) = |r_{01}|^2 = \frac{16\pi^2}{Q^4}\Delta\rho^2 \tag{5.6}$$

where $\Delta\rho = (\rho_1 - \rho_0)$. This illustrates that for all interfaces, the reflectivity, in the absence of any other effects, decreases as $1/Q^4$. For a thin film at the interface where the air and substrate both have refractive indices $=1.0$ (a condition often used to study adsorption at the air-water interface as will be seen later), Equation (5.4) reduces to

$$R(Q) = \frac{16\pi^2}{Q^4}(2\rho)^2 \sin\left(\frac{Qd}{2}\right)^2 \tag{5.7}$$

For more complex surface structures than can be described as a single layer of uniform composition, there are two different and equivalent approaches, using the optical matrix method developed for thin-film optics [4,11] or the kinematic approximation [2,12]. The first of these relies on the surface structure being described as a series or stack of thin layers. This assumes that, in optical terms, an application of Maxwell's equations and the relationship between the electric vectors in successive layer lead to a characteristic matrix per layer, such that,

$$C_j = \begin{bmatrix} e^{i\beta_{j-1}} & re^{i\beta_{j-1}} \\ re^{-i\beta_{j-1}} & e^{-i\beta_{j-1}} \end{bmatrix} \tag{5.8}$$

For n layers, the matrices for each layer are multiplied together, $[C]=[C_1][C_2]\ldots[C_{n+1}]$. The resultant reflectivity is then given by the a and c elements of the final 2 × 2 matrix and their complex conjugates, such that, $R = cc^*/aa^*$. In practice, recurrence relationships between the Fresnel coefficients in the successive layers can be used to provide a more efficient calculation. Furthermore, a diffuse or rough profile characterized by a Gaussian roughness, σ, can be included at each interface in the stack, such that the reflectivity at the jth interface, r'_j, is given by

$$r'_j = r_j \exp(-0.5 k_j k_{j-1} \sigma^2) \quad (5.9)$$

In the kinematic approximation (see also Equation 5.3),

$$R(Q) = \frac{16\pi^2}{Q^2} |\hat{\rho}(Q)|^2 = \frac{16\pi^2}{Q^4} |\hat{\rho}'(Q)|^2 \quad (5.10)$$

where $\hat{\rho}(Q)$ and $\hat{\rho}'(Q)$ are the one-dimensional Fourier transforms of $\rho(z)$ and $\rho'(z)$. Dividing conceptually the interface region into the different components at the interface, the volume fraction distributions of those individual components can be obtained by

$$R(Q) = \frac{16\pi^2}{Q^2} \left[\sum_i b_i^2 h_{ii} + \sum_i \sum_{j<i} 2 b_i b_j h_{ij} \right] \quad (5.11)$$

The h_{ii} are the self-partial structure factors, $h_{ii} = |\hat{n}_{ii}(Q)|^2$, and the cross-partial structure factors, h_{ij}, are given by $h_{ij}(Q) = \text{Re}\{\hat{n}_i(Q)\hat{n}_j(Q)\}$, where $n_i(Q)$ is the one-dimensional Fourier transform of $n_i(z)$. The self-partial structure factors relate directly to the distributions of the individual components at the interface, whereas the cross-partial structure factors depend on the relative positions of pairs of components along the direction normal to the interface. This approach has been applied to determine the structure of adsorbed surfactant layers and membrane bilayers, where the different components are separated by making reflectivity measurements with the different components deuterium labeled (see later).

5.2.2 Fundamentals of Small-Angle Neutron Scattering

Small-angle scattering is scattering in the forward direction, and the pattern of the scattering at small scattering vectors, Q, provides information about structure over length scales from nanometer to fractions of a micrometer. The scattering vector, or wave-vector transfer, Q, is defined as $Q = 4\pi/\lambda \sin\theta$, where 2θ is the scattering angle, as shown in Figure 5.2.

The quantity measured in SANS is $d\sigma/d\Omega$, the differential scattering cross section, σ, per unit solid angle, Ω, as defined in Chapter 1. The scattering arises from the coherent superposition of the scattered waves and can be expressed as a summation of amplitude-weighted phase shifts or as an integral over the sample volume of the scattering length density distribution, $\rho(r)$, [13], such that

Chapter | 5 Large-Scale Structures

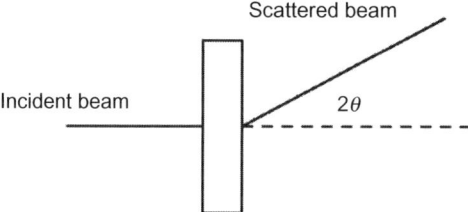

FIGURE 5.2 SANS scattering geometry.

$$\frac{d\sigma}{d\Omega} = \left\langle \left| \sum_j \overline{b}_j \exp(i\underline{Q}\cdot\underline{R}_j) \right|^2 \right\rangle = N \left\langle \left| \int_V \rho(r) \exp(i\underline{Q}\cdot\underline{r}) dr \right|^2 \right\rangle_Q \quad (5.12)$$

where $\overline{b}_j$ and R_j are the coherent neutron scattering length and position of the jth atom, the $\langle \rangle$ represents an average over all positions and orientations with respect to Q, $\rho(r)$ is the scattering length density, V is the sample volume, and N is the number of scattering objects.

For a dilute solution (which could be a liquid or a solid solution) of monodisperse scattering objects (e.g., micelles or other colloidal particles) then,

$$\frac{d\sigma}{d\Omega} = N\langle F^2(Q)\rangle_Q \quad (5.13)$$

and the form factor, which contains information about the particle shape and size, is defined as

$$F(Q) = \int_V [\rho(r) - \rho_s] \exp(iQ\cdot r) dr \quad (5.14)$$

ρ_s is the solvent (or matrix) scattering length density, the integral is over the particle volume, and $\langle \rangle_Q$ denotes an average of all orientations with respect to the Q vector.

For more concentrated solutions, where interparticle interactions exist, the situation is more complex,

$$\frac{d\sigma}{d\Omega} = \left\langle \sum_{n,m} \sum_{j(n)} \sum_{k(m)} b_j b_k \exp(i\underline{Q}\cdot\underline{r}_{jk}) \right\rangle \quad (5.15)$$

where $j=j(n)$ denotes the jth atom of the nth particle and r_{jk} is the relative separation of atoms, see Figure 5.3.

In the decoupling approximation [13], where it is assumed that there are no correlations between position and orientation, Equation (5.15) can be simplified as

$$\frac{d\sigma}{d\Omega} = N\left[S(Q)\left|\langle F(Q)\rangle_Q\right|^2 + \Delta(Q) \right] \quad (5.16)$$

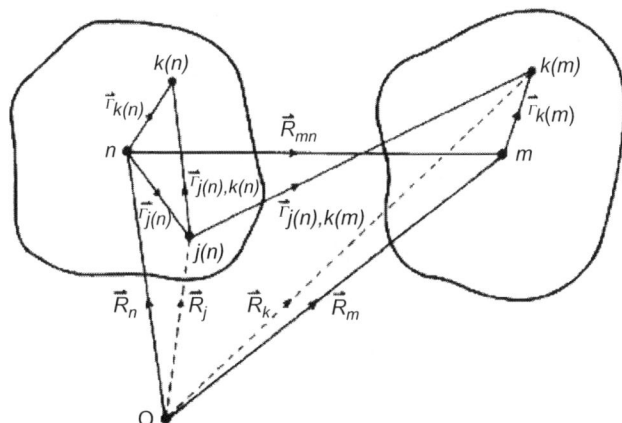

FIGURE 5.3 Geometrical notation of the intra- and interparticle relationships used to determine the "decoupling approximation." *Reproduced by permission of Ref. [13].*

where

$$\Delta(Q) = \left\langle |F(Q)|^2 \right\rangle_Q - \left| \langle F(Q) \rangle_Q \right|^2 \qquad (5.17)$$

and $S(Q)$ is the interparticle static structure factor, the Fourier transform of the real space interparticle pair distribution function $g(r)$. Analytic expressions exist for different model potentials, such as a hard-sphere interaction [14] and the soft repulsive Yukawa potential [15,16]. $\Delta(Q)$ can be considered as a coherent disorder term which comes into play when $S(Q) \neq 1.0$. For monodisperse spherosymmetrical particles, Equations (5.16) and (5.17) reduce to,

$$\frac{d\sigma}{d\Omega} \approx NS(Q)P(Q) \qquad (5.18)$$

where

$$\langle F^2(Q) \rangle_Q = \langle F(Q) \rangle_Q^2 = P(Q) = F_s^2(Q) \qquad (5.19)$$

A range of different form factors for different particle geometries, such as spheres, concentric shells, ellipses, rods, Gaussian coils, etc., exist and are frequently used [6–9,17]. For uniform spheres,

$$F_s(Q) = V_s(\rho - \rho_s)F_0(QR) \qquad (5.20)$$

$V_s = 4/3\pi R^3$, where R is the sphere radius, and $F_0(QR)$ the function,

$$F_0(QR) = 3[\sin(QR) - QR\cos(QR)]/QR^3 \qquad (5.21)$$

For a set of m concentric shells, Equation (5.21) can be extended to

$$F_s(Q) = V_1(\rho_1 - \rho_2)F_0(QR_1) + \cdots V_m(\rho_m - \rho_s)F_0(QR_m) \qquad (5.22)$$

Including polydispersity requires averaging over the form factor with an appropriate distribution function (a Schultz (log normal) [18] or Gaussian distribution is commonly used),

$$\langle F^2(Q)\rangle = \int f(r) F_s^2(Qr) dr \tag{5.23}$$

where $f(r)$ is the normalized probability distribution.

There are a number of approximate relationships which are frequently used to provide useful information when evaluating SANS data. The scattering cross section extrapolated to $Q=0$ and the Invariant provide an estimate of the volume and number density of the scattering objects, such that for a dilute solution,

$$\frac{d\sigma}{d\Omega}(Q=0) \approx NV^2 \Delta\rho^2 \tag{5.24}$$

where $\Delta\rho^2 = (\rho - \rho_s)^2$, and the Invariant is defined as

$$\int_0^\infty Q^2 \frac{d\sigma}{d\Omega} dQ = 2\pi^2 \phi(1-\phi)\Delta\rho^2 \tag{5.25}$$

where $\varphi = NV$ is the volume fraction.

Other expressions, such as Porod's law which determines surface area by extrapolation of the Q^{-4} dependence to large Q values, and the Guinier expansion which characterizes globular structures in terms of a radius of gyration, R_g, from the Q^{-2} region of the scattering are described in detail elsewhere [5–9].

5.2.3 Experimental Details for Neutron Reflection

A wide variety of neutron reflectometers now exist on reactor and pulsed neutron sources. The schematic in Figure 5.4 shows the layout of the time-of-flight (TOF) reflectometer SURF at the ISIS pulsed neutron source [19] and illustrates all the main features of a reflectometer.

The essence of a reflectivity measurement is to measure the specular reflectivity $R(Q)$ as a function of the wave-vector transfer, Q. This can be done with a horizontal or vertical sample geometry; D17 [20] at the Institut Laue-Langevin (ILL) has a vertical geometry, whereas CRISP, SURF, and INTER at ISIS [19,21,22] and FIGARO [23] at the ILL all have a horizontal geometry. The horizontal geometry is ideal for liquid surfaces, without precluding other surfaces and sample geometries. In most reflectometers, a narrow neutron beam, collimated by a pair of slits, is incident on the sample at grazing incidence and reflected into a single detector or a small area detector (see Figure 5.4). Measurements can be made by using a fixed wavelength and varying the angle of incidence, as on D17, or by using a fixed angle of incidence and TOF to sort a wide wavelength range, as on CRISP, SURF, INTER,

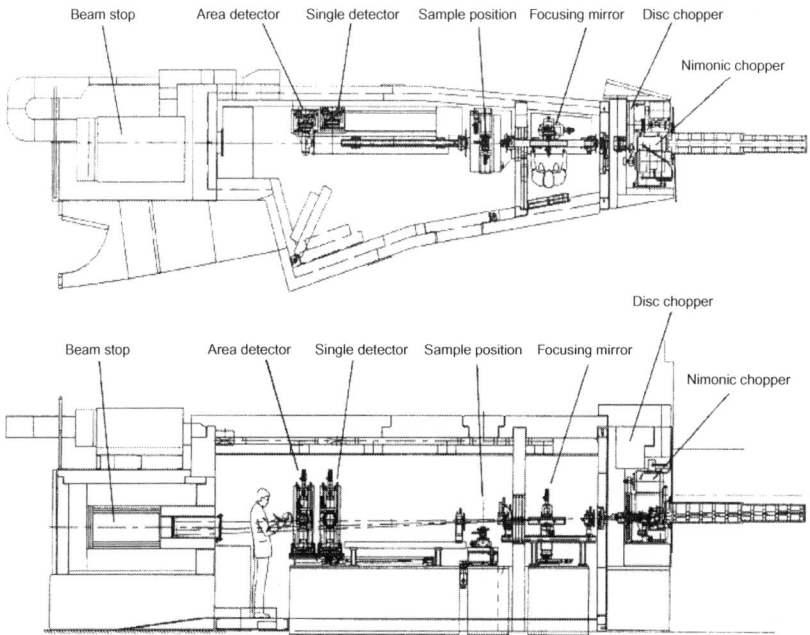

FIGURE 5.4 Schematic representation of the SURF reflectometer at ISIS. *Reproduced by permission of Ref. [19].*

and FIGARO. Typical angles of incidence range from a fraction of a degree to a few degrees and cover a Q range from ~ 0.002 to 0.5 Å^{-1}. Various different optical devices are incorporated into the instrument designs to ensure separation from the direct beam. As the grazing angles of incidence are small, the sample footprint is relatively large and is typically in the range of 10–1000 mm^2. The reflected beam is corrected for the incident spectral distribution, detector efficiencies, and solid angle by measuring the direct beam and calibrated to an absolute reflectivity (where total reflection corresponds to a reflectivity of unity) by reference to a standard surface (for liquid surfaces, this is usually D$_2$O). The minimum thickness, d, of adsorbed layer or thin film accessible is determined by the maximum Q value (where $d \sim 2\pi/Q$), and as $R(Q)$ decays as at least Q^{-4} (Fresnel's law), this is usually determined by the background signal which arises primarily from the incoherent scattering from the bulk of the sample, sample containers, and the instrument (typically $\sim 10^{-6}$ for most real samples). The maximum size accessible is determined by the instrumental resolution (δQ) and the sample quality and is typically ~ 3000 Å. The resolution can range from $\sim 1\%$ to 10% and has two major contributions, the wavelength resolution, $\delta \lambda$, and the collimation, $\delta \theta$, such that, $\delta Q^2 = (\delta \lambda / \lambda)^2 + (\delta \theta / \theta)^2$. On TOF instruments, $\delta \lambda$ is small and so the $\delta \theta$ term dominates. The ultimate signal to noise is determined by footprint size, resolution ($\delta \theta$), and sample quality.

In NR, data analysis is predominantly made using a model to compare with the data on a least-squares criterion, and a variety of approaches, as described earlier, exist, based on the exact optical thin-film approach [2–4,11] or the kinematic approximation [2,12]. An alternative approach is the determination of partial structure factors, as described above, and encapsulated in Equations (5.10) and (5.11). Various methods to directly invert the data by using reference layers have been developed [24] and are being increasingly used. These inversion techniques require high-quality data over as wide a Q range as possible; otherwise, it is easy to introduce artifacts into the interpretation.

5.2.4 Experimental Details for SANS

The design of most SANS diffractometers relies on pinhole collimation and a large-area detector. They are mostly located at the end of curved neutron guides to remove the sample and detector from the intense gamma-ray and fast neutron fluxes of the neutron sources. The designs are in essence all derived from the pioneering work of Schmatz *et al.* [25,26]. SANS has been most effectively exploited using cold neutrons, in the λ range from 4 to 20 Å. On reactor sources, a monochromatic beam, with a $\Delta\lambda/\lambda \sim 5$–20% defined by mechanical beam choppers, is used to provide a typical Q range of 0.002–0.5 Å^{-1} by means of an appropriate combination of neutron wavelengths and sample-to-detector distances. Lower values require longer sample-detector distances and correspondingly tighter collimation and hence reduce the incident flux. At high-Q sample or container, background scattering usually dominates and limits the accessible value of Q. Such instrumentation exists on most medium and high flux nuclear reactors and is epitomized by the D11 and D22 diffractometers at the ILL, France [27]. The advent of pulsed neutron sources has demonstrated the virtues of the white-beam-TOF method to provide simultaneously a wide Q range, an increasingly important feature for systems where there is a range of length scales and hierarchical structures present. The LOQ and SANS2D diffractometers at ISIS [28] and the D33 TOF diffractometer at the ILL [29] are prime examples, and the typical layout of a SANS diffractometer is shown in Figure 5.5.

Converting the scattered intensity into data that relate directly to the scattering cross section requires correction for instrumental factors and a normalization procedure. Taking such factors into account the scattered intensity is related to the scattering cross section of the sample by

$$I(Q) = C(\lambda)T(\lambda)t\sigma(\theta,\lambda) \quad (5.27)$$

where t is the sample thickness, $\sigma(\theta,\lambda)$ the scattering cross section of the sample, and $T(\lambda)$ the sample transmission (often measured separately). The wavelength-dependent instrumental factor is denoted by $C(\lambda) = \varphi(\lambda)A\Delta\Omega\varepsilon(\lambda)$, where A is the sample area, $\varphi(\lambda)$ the incident flux, $\Delta\Omega$ the solid angle, and

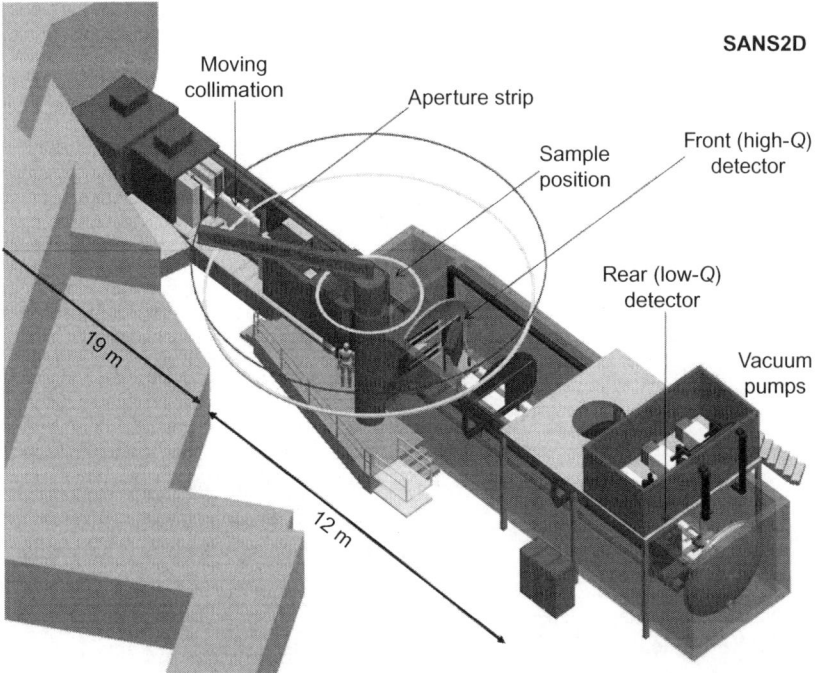

FIGURE 5.5 Schematic representation of a pulsed source SANS instrument. *Reproduced by permission of Ref. [28].* (For color version of this figure, the reader is referred to the online version of this chapter.)

$\varepsilon(\lambda)$ the detector efficiency. $C(\lambda)$ is usually evaluated and incorporated using a standard sample of known structure and cross section (such as H_2O, vanadium, or a D/H polymer blend) [30]. The absolute calibration of the scattering is an important powerful feature of SANS and is extensively exploited. Accurate transmission measurements and evaluation of sample and instrumental backgrounds are hence essential [31]. Although the corrections are made pixel by pixel on the detector, for isotropic scattering the corrected data are radially averaged to produce $d\sigma(Q)/d\Omega$ versus Q. For samples which have a preferred orientation, the scattering will be anisotropic and the corrected data are retained as a two dimensional array in Q, or cuts in specific directions are taken.

The detailed analysis of SANS data can take a number of forms, and many different packages exist [32]. In the simplest form, analysis of the data using a Guinier or Porod expansion provides an estimate of size and surface area. Extrapolation to $Q=0$ (Equation 5.27) is often used to get an estimate of the polymer molecular weight (MW) [33] and in combination with the Invariant (Equation 5.28) provides an estimate of the volume and number density of the scattering objects. For some systems, this can be sufficient information, especially when combined with the Q dependence of the scattering which can indicate shape: Q^{-1} for elongated and Q^{-2} for planar structures. However, the more common approach is to use modeling to compare with the data and provide an evaluation based on least-squares fitting procedures. For systems for which there is a suitable model for the form factor and structure factor, as described earlier for colloid particles and micellar structures and summarized by Equations (5.16)–(5.23), detailed modeling either using molecular constraints or in free form can be used [13,34]. For more complex form factors and structure factors where convenient analytic forms do not exist, numerical solutions or Monte-Carlo methods can be incorporated [35]. The alternative to modeling is the indirect Fourier transform (IFT) method pioneered by Glatter [36,37]. It provides a pair distribution function $P(r)$, which is a self-convolution of the excess scattering length density distribution multiplied by the distance squared. It is particularly effective at distinguishing between different shapes and in identifying key dimensions. It does, however, rely on good data over a wide Q range; otherwise, truncation effects can confuse interpretation. Although the effects of resolution smearing can be accounted for, it is difficult to incorporate polydispersity and interparticle interactions. However, in some cases, this approach provides a valuable alternative to modeling.

5.3 THIN FILMS, INTERFACES, AND SOLUTIONS

The theoretical background to the techniques of NR and SANS has been described and the key features of the associated instrumentation are presented. In the following sections, a broad range of scientific examples will be presented. For NR, they range from adsorption at the air–solution interface and the solid–solution interface to the study of biomembranes and thin solid films (polymeric). For SANS examples, these include the characterization of micellar phases, lamellar and vesicle phases, and the role of flow on self-assembly. It includes the study of colloidal particles and emulsions with an emphasis on characterization of the nature of adsorption onto the colloidal particles or droplets. Finally, the SANS examples will include the study of the structure of polymers in solution and in melts and the study of different biomacromolecules.

The examples are chosen in order to demonstrate the breadth of application and impact of the techniques, are predominantly taken from current research programs, and are relatively recently published. In addition, the case studies are chosen to demonstrate and highlight particular features and strengths of the techniques. Further experimental details are included where required in order to illustrate the flexibility in sample geometries and environments associated with the techniques.

5.3.1 Adsorption at the Air–Solution Interface

The air–solution interface is in many ways the ideal model hydrophobic interface to study the adsorption of surfactants, polymers, surfactant mixtures, and polymer–surfactant mixtures, and it is often where such studies start. Measurement of macroscopic properties, such as surface tension (ST), is still extensively used and provides invaluable information. However, the information about adsorbed amounts is obtained with the Gibbs equation [38], and the difficulties associated with its interpretation are well documented [12,39]. Furthermore, and most importantly, they provide no structural information and are difficult to apply to complex mixtures. NR, combined with D/H isotopic substitution, overcomes many of these shortcomings and has been discussed at length elsewhere [12]. In particular, adsorbed amounts can be obtained directly without any of the assumptions required for the interpretation of the Gibbs equation. It is straightforward to measure mixed adsorption and so determine the surface composition. Furthermore, it is possible to obtain detailed structural information, such as the thickness of the adsorbed layer, the relative positions of the different components at the interface, and the solvent distribution within the surface layer.

The facile determination of surfactant adsorption and mixed surfactant adsorption at the air–water interface using NR has provided a significant advance in our understanding of surfactant adsorption in recent years [12]. This is illustrated here with a recent study on the adsorption of biosurfactants and biosurfactant/conventional surfactant mixtures. There is increasing interest in the use of biosurfactants synthesized from a variety of different microorganisms, due to their lower toxicity, greater biodegradability, greater tolerance to pH, temperature and salinity, and the desire to reduce the dependence on petrochemicals. For measurements of a deuterium-labeled surfactant in null-reflecting water (nrw, 8 mol% D_2O/92 mol% H_2O, which has a refractive index of unity, the same as air), the reflectivity arises only from the adsorbed layer of deuterated surfactant and can be described for example by Equation (5.7). The adsorbed amount (Γ) is then given by

$$\frac{1}{\Gamma} = A = \frac{\sum b}{\rho d} \qquad (5.28)$$

where Σb is the scattering length of the adsorbed molecule, and ρ and d are the scattering length density and thickness of the adsorbed layer. It has been shown that treating the adsorbed layer of surfactant as a single layer of uniform composition is a good approximation [12]. Equation (5.28) is readily extended to n components such that,

$$\rho = \frac{\sum_n b_n}{A_n d} \qquad (5.29)$$

The glycolipids, such as the rhamnolipids and sophorolipids, are a promising biosurfactant in terms of applications for home and personal care formulations in that they have attractive surface properties and have relatively high yields. Smyth et al. [40] have adapted a strain of *Pseudomonas aeruginosa* in D_2O and glycerol-d_5 to produce rhamnolipids with >90% deuterium. Similarly, they have adapted *Candida bombicola* in isostearic acid-d_{35} to produce deuterium-labeled sophorolipids at a lower but acceptable level of deuteration. This is an important development as the availability of deuterium-labeled surfactants is a prerequisite for these NR measurements.

Chen *et al.* [41,42] have used NR and ST to investigate the adsorption of rhamnolipid and rhamnolipid/anionic surfactant mixtures at the air–water interface. The rhamnolipids consist mainly of one or two rhamnose molecules linked to one or two molecules of β-hydroxydecanoic acid. *P. aeruginosa* produces two predominant forms, R1 and R2, which have two C_{10} alkyl chains and a headgroup of one or two rhamnose groups, respectively. These have been extensively studied by ST where it is assumed that at low pH they are nonionic and at high pH they are anionic due to the protonation of the carboxyl groups. This has significant implications for the interpretation of their adsorption using the Gibbs equation:

$$\Gamma = \frac{1}{mRT}\frac{d\gamma}{dLnC} \qquad (5.30)$$

where C is the surfactant concentration and γ the interfacial tension, $m=1.0$ for nonionic surfactants and 2.0 for ionic surfactants. Chen *et al.* [41] measured the adsorption of R1 and R2 directly using NR and typical reflectivity data for R1 are shown in Figure 5.6.

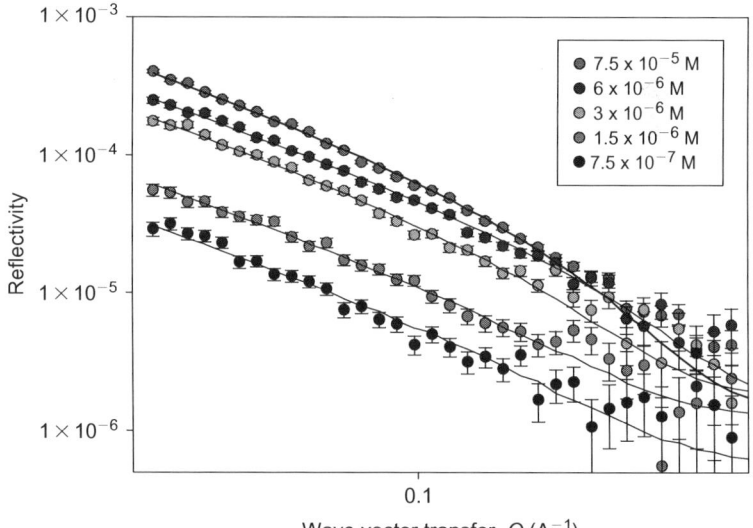

FIGURE 5.6 Specular reflectivity for R1 in nrw, at different concentrations as indicated in the figure legend. The solid lines are model calculations for a single layer of uniform composition. *Reproduced by permission of Ref. [41].* (For color version of this figure, the reader is referred to the online version of this chapter.)

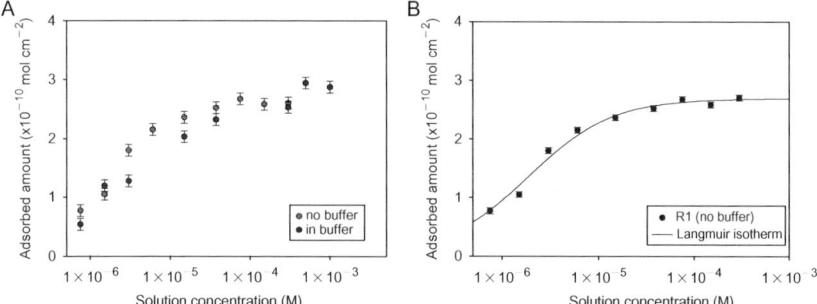

FIGURE 5.7 Adsorption isotherms: (a) R1 at pH 9 (in buffer) and in the absence of buffer, (b) adsorption isotherm for R1. The solid line has been calculated assuming a Langmuir isotherm, for $\Gamma_{max} = 2.7 \times 10^{-10}$ ml cm^{-2} and $k = 1.83 \times 10^{-6}$ mol l^{-1}. *Reproduced by permission of Ref. [41].* (For color version of this figure, the reader is referred to the online version of this chapter.)

The data are consistent with a thin monolayer of surfactant adsorbed at the interface, with a thickness $\sim 21 \pm 1$ Å. Measurements were made in buffer at pH 9 and 7; the resulting isotherms for R1 are shown in Figure 5.7 and are consistent with a Langmuir isotherm, $\Gamma = C/(C+k)$, where k is the adsorption coefficient.

The data for R1 and R2 show that R1 is more surface active than R2 (for R2 the saturation adsorption $\Gamma_{sat} \sim 2.7 \times 10^{-10}$ mol cm^{-2} compared to 2.1×10^{-10} mol cm^{-2} for R1). However, the overwhelming conclusion is that the adsorption is largely independent of pH, and in contrast to the ST data, R1 and R2 are only weakly anionic, even at high pH. Hence, previous interpretations based on the Gibbs equation have resulted in the misinterpretation of the adsorption properties. The difference between R1 and R2 can be explained by the greater steric (packing) constraints at the interface for the larger di-rhamnose headgroup of R2 and its greater hydrophilicity compared to the single rhamnose headgroup of R1. This has important consequences for the surface mixing properties of R1 and R2, as shown in Figure 5.8, where the adsorbed amounts and surface composition as a function of solution composition at a fixed surfactant concentration of 1 mM ($\gg$ critical micelle concentration (cmc)) are shown. The adsorbed amount of each component is estimated by making NR measurements with each component deuterated in turn and with both components deuterated, and using Equation (5.29) to evaluate the data. The cmc values for R1 and R2 are similar (0.36 and 0.2 mM, respectively), and from the cmc variation with solution composition, close to ideal mixing is expected [41]. However, the data in Figure 5.8 shows a pronounced departure from ideal mixing, which is also outside the expectations from the standard nonideal mixing approaches, such as regular solution theory (RST) [43,44]. Similar extreme departures from ideal mixing, in circumstance where ideal mixing might be expected, were also reported for the nonionic surfactant mixture of $C_{12}E_3/C_{12}E_8$ [45]. In this case, the surface composition

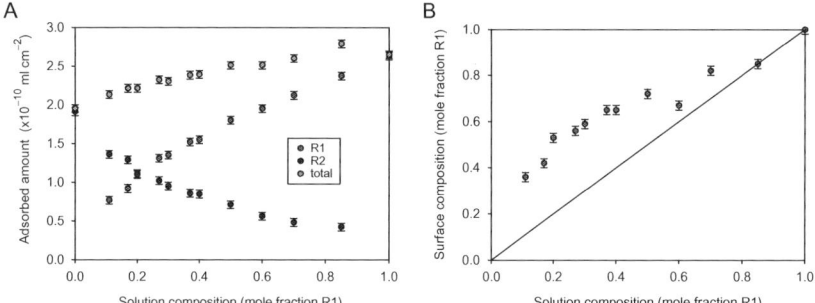

FIGURE 5.8 (a) Adsorption of R1, R2, and R1+R2 for 1 mM R1/R2 mixture at pH 9 (in buffer). (b) Surface composition (mole fraction R1) versus solution composition at a surfactant concentration of 1 mM, at pH 9 (in buffer) for the R1/R2 mixture. The solid line is a guide to the eye and represents the line of equal surface and solution composition. *Reproduce by permission of Ref. [41].* (For color version of this figure, the reader is referred to the online version of this chapter.)

is still dominated by the more surface active $C_{12}E_3$ at ~100 times the cmc. This extreme departure from ideality was attributed to the differences in surface activity and packing constraints, in particular, the steric hindrance of the larger EO_8 headgroup and the intrinsic difference in relative curvature of the two surfactants. Similar arguments apply here to the R1/R2 mixture, where the larger R2 headgroup means that R1 dominates the adsorption at concentrations well in excess of the cmc.

In terms of more immediate applications and incorporation into formulations, it is likely that the biosurfactants will replace some or part of the conventional components. To that end, Chen et al. [42] studied the surface adsorption of the ternary mixture of R1 and R2 and the anionic surfactant, sodium dodecyl benzene sulfonate, LAS. As in the R1/R2 mixture, ST measurements imply ideal mixing for the binary R1/LAS and R2/LAS mixtures. Although the NR measurements for the R1/LAS mixture show close to ideal mixing, the NR measurements for R2/LAS show that as in the R1/R2 mixture, the R2 component does not compete very well with the LAS for the surface. As shown in Figure 5.9a, four different NR measurements were made in nrw, for each component deuterated in turn and for all components deuterated.

Hence, the amount of each component at the interface and the total adsorption can be evaluated by solving by least squares the four simultaneous equations arising from Equation (5.29) for the four different labeling schemes. The amount of each component (in mole fraction) is shown in Figure 5.9b, and the adsorption is consistent with relative surface activities in the order LAS > R1 > R2. One of the key findings highlighted by Chen et al. [42] was that there is an enhancement in the total adsorption for the ternary mixture, which is not present in any of the associated binary mixtures. This enhancement was shown to be associated with a relief of the unfavorable

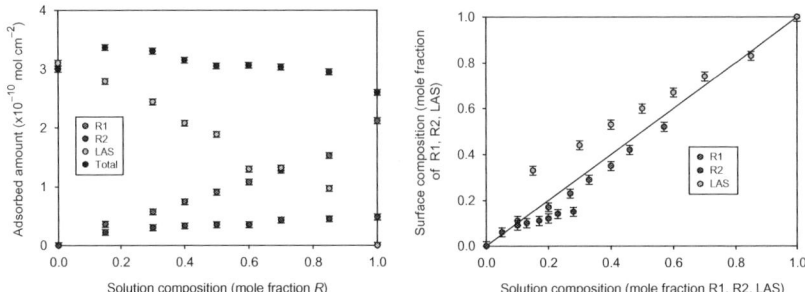

FIGURE 5.9 (a) Adsorption of R1, R2, LAS, and total adsorption versus solution composition (mole fraction of rhamnolipid) for R1/R2 2:1/LAS ternary mixture at 1 mM; (b) R1/R2/LAS surface composition versus solution composition for 2:1 R1/R2 composition. *Reproduced by permission of Ref. [42].* (For color version of this figure, the reader is referred to the online version of this chapter.)

packing constraints imposed by R2 in the ternary mixtures by the addition of the LAS.

A number of recent NR studies on surfactant adsorption in the presence of multivalent counterions [46,47] or polyelectrolytes [48–50] have revealed a surface structure more complex than a single monolayer, and which has not been revealed using other surface-sensitive techniques. Alargova et al. [46] showed how the strong complexation between the multivalent counterion Al^{3+} and the ethoxylated anionic surfactant, sodium dodecyl dioxyethylene sulfate, SLES, resulted in significant micellar growth at low surfactant and counterion concentrations at the point when the counterion charge is in excess. The addition of a nonionic cosurfactant with a bulky headgroup, such as $C_{12}E_{12}$, disrupted complexation and suppressed micellar growth. However, apart from a slight decrease in the ST, there was no indication in such systems that the form of the adsorption was other than a monolayer. Petkov et al. [47] used NR to investigate the surface adsorption of SLES and SLES/$C_{12}E_{12}$ mixtures in the presence of Al^{3+} ions, in the form of $AlCl_3$. Figure 5.10 shows the evolution in the NR patterns for a 1 mM 95/5 mol ratio SLES/$C_{12}E_{12}$ mixture with increasing amounts of Al^{3+}.

In the absence of Al^{3+}, the data are consistent with a simple monolayer adsorbed at the interface, as expected. At an $AlCl_3$ concentration of 0.24 mM, the NR data have a pronounced interference fringe at relatively high Q, and this is consistent with a three-layer structure (corresponding to an initial monolayer and a single bilayer beneath it) at the interface. At an $AlCl_3$ concentration of 0.36 mM, a double interference fringe appears, where the fringe at low Q is due to the total thickness of the adsorbed layer, and the fringe at higher Q is a broad Bragg peak arising from three bilayers at the interface. At the highest $AlCl_3$ concentration, 0.6 mM, a well-defined and relatively narrow Bragg peak, arising from >20 bilayers at the interface, has developed. The width of the Bragg peak is proportional to $1/N$, where N is

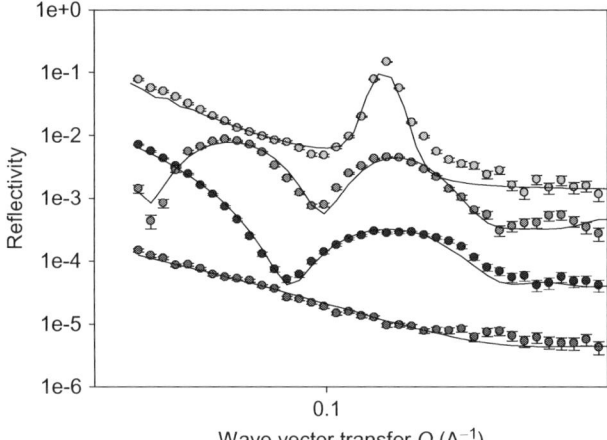

FIGURE 5.10 Reflectivity for 1 mM 95/5 mole ratio SLES/$C_{12}E_{12}$/$AlCl_3$/NaCl: red-filled circles, 0.0/3.0 mM $AlCl_3$/NaCl; dark-blue-filled circles, 0.24/1.8 mM; green-filled circles, 0.36/1.2 mM; and light-blue-filled circles, 0.6/0.0 mM. The different data are shifted vertically for clarity. *Reproduced by permission of Ref. [47].* (For interpretation of the references to color in this figure legend, the reader is referred to the online version of this chapter.)

the number of bilayers, convolved with the instrumental resolution. The data were analyzed using a multilayer model within the kinematic approximation such that [48],

$$R(Q) = \frac{16\pi^2}{Q^4} \left| \sum_0^{2N} (\rho_i - \rho_{i+1}) \exp(-iQd_i) \exp\left(\frac{-Q^2\sigma^2}{2}\right) \right|^2 \quad (5.31)$$

where ρ_i is the scattering length density of the ith layer counting from the subphase, $i=0$ represents the subphase, d_i is the distance between the ith and ith+1 layer, σ_i the roughness between ith and ith+1 layers, $\rho(2N+1)$ is the scattering length density of the air, and $2N$ the total number of layers. The bilayer thickness is ~44 Å when $N=3$, and ~48 Å when $N>20$. This approach has been shown in this study [47] and elsewhere [48] to provide a good description of these surface multilayer structures. From such data and corresponding evaluation, a surface phase diagram can be constructed, as shown in Figure 5.11a for SLES/$C_{12}E_{12}$/$AlCl_3$.

In particular, the impact of surfactant concentration and $AlCl_3$ concentration on the surface phase behavior is evident, and the role of the $C_{12}E_{12}$ in disrupting the surface multilayer formation is illustrated.

Similar observations were reported for LAS/$C_{12}E_8$ mixtures [48], but in that case, the divalent ion Ca^{2+} was sufficient to promote surface multilayer formation. In Figure 5.11b, the equivalent surface phase diagram obtained from NR data for LAS/$C_{12}E_8$/$CaCl_2$ is shown. The complexation with Ca^{2+} was not sufficient to promote multilayer formation for the single-chain SLES.

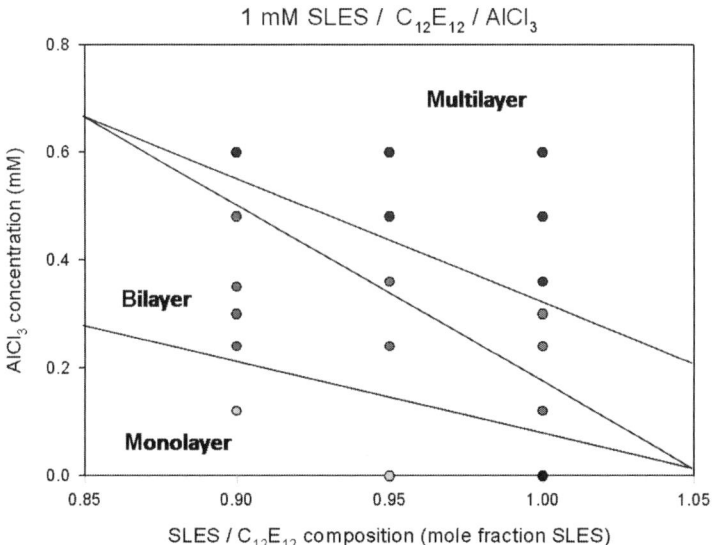

FIGURE 5.11 Surface phase diagram for 1 mM SLES/$C_{12}E_{12}$/$AlCl_3$/NaCl. *Reproduced by permission of Ref. [50].* (For color version of this figure, the reader is referred to the online version of this chapter.)

The comparison between the SLES and LAS anionic surfactants highlights the importance of the hydrophobic interaction between the alkyl chains in the multilayer formation. These surface structures are an increasingly important structural signature with potential properties for enhanced surface delivery of different components. The systems also exhibit extreme surface wetting properties and have implications in detergency and in bio- or soft-lubrication.

The other major area in which NR has transformed our knowledge and understanding is in the adsorption of polymer–surfactant mixtures at the air–water interface [49], in both weakly and strongly interacting polymer–surfactant mixtures. This is especially the case for strongly interacting polyelectrolyte–surfactant mixtures, where it is difficult to interpret the complex patterns of ST behavior in terms of adsorbed amounts and surface structure. Taylor *et al.* [49] summarized the variations in ST and adsorption behavior that arises from the competition between surface and solution complex formation. Without NR data to determine the amounts of surfactant, and in many cases polymer as well, adsorbed at the interface, and to determine the structure of the mixed surface layer, little progress had been previously made. This is especially epitomized in the recent studies on poly(ethyleneimine) (PEI)/sodium dodecyl sulfate (SDS) [50] and modified PEI/SDS [51] mixtures.

For a 25-kDa linear PEI, the ST behavior with SDS shows a strong pH dependence as the PEI goes from a highly charged polyelectrolyte at pH 3 to an essentially neutral polymer at pH 10 [50]. However, the NR data for PEI/d-SDS in nrw show that the adsorption is in the form of a monolayer over

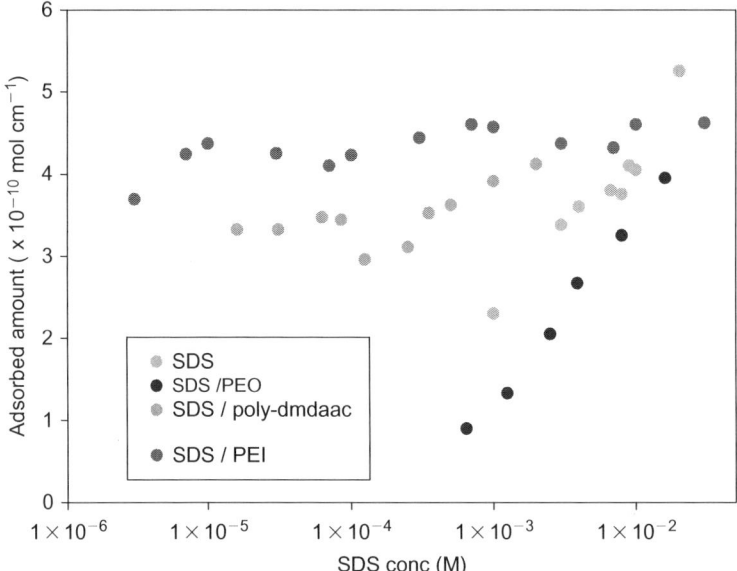

FIGURE 5.12 Comparison of the SDS adsorption for 20 ppm 25k MW linear PEI/SDS at pH 10 with SDS, SDS/polydmdaac, and SDS/PEO (see legend for details). *Reproduced by permission of Ref. [50].* (For color version of this figure, the reader is referred to the online version of this chapter.)

a wide SDS concentration range and, importantly, that the adsorption is independent of pH. Furthermore, even at an SDS concentration $\sim 10^4$ times below the cmc of SDS, the amount of SDS at the interface is similar to that at the cmc, as shown in Figure 5.12.

This remarkable adsorption pattern arises from the strong surface interaction and complexation between the PEI and SDS, which is just as strong at pH 10 as at pH 3. Similar patterns of adsorption exist for the branched form of PEI, except that at pH 7 and 10, the surface structure evolves from a monolayer to bilayer and multilayer structures, dependent upon the polymer MW, polymer and SDS concentrations, and pH. This is shown in Figure 5.13a for a 25-kDa branched PEI/SDS mixture at pH 10 and SDS concentrations from 0.3 to 5 mM.

This surface multilayer formation in polymer/surfactant mixtures is also shown in Figure 5.13b for an ethoxylated 2-kDa branched PEI, where a single ethylene oxide (EO) group has been attached to each nitrogen group of the PEI, PEI-EO$_1$ [51]. The data in Figure 5.13b are for 20 ppm PEI-EO$_1$/ 0.1 mM SDS at pH 10 for three different PEI-EO$_1$/SDS/solvent contrasts: d-SDS/PEI-hEO$_1$/nrw, h-SDS/PEI-dEO$_1$/nrw, and d-SDS/PEI-hEO$_1$/D$_2$O. This study shows the advantages of having data for different contrasts in refining the structural details of the surface layer, and such complementary

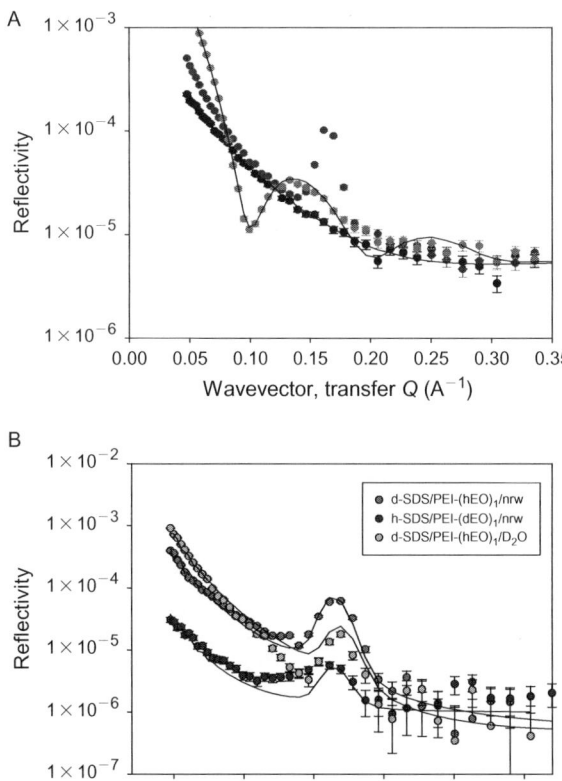

FIGURE 5.13 (a) NR data for 20 ppm branched 25k MW PEI/SDS at pH 10 and different SDS concentrations; (b) NR data for 20 ppm PEI-EO$_1$/0.1 mM SDS at pH 10 for different contrasts (see legend for details). *Reproduced by permission of Ref. [50].* (For color version of this figure, the reader is referred to the online version of this chapter.)

measurements are frequently employed. A structure in the form of 20 bilayers with a bilayer thickness of 36 Å and a scattering length density that reflects the deuterium-labeled components and hence different parts of the bilayer fits all the data simultaneously with the single model in which only the scattering length densities are changed.

5.3.2 Adsorption at the Liquid–Solid Interface

Adsorption at the liquid–solid interface is in many ways more important but more complex than the simpler air–liquid interface, but is also readily accessible to NR studies. Due to total scattering and the incoherent scattering cross sections, the transmission of thermal/cold neutrons through amorphous materials and liquids is relatively low. In contrast, provided that the combination of orientation and wavelength do not result in Bragg scattering, the transmission through crystalline materials such as silicon, quartz, and sapphire is much higher. For example, the transmission through a 100-mm path length of silicon single crystal is ~80%. Hence, making measurements by transmitting the neutron beam (at normal incidence) through a crystalline upper phase to grazing incidence at the solid–solution interface [52] (see Figure 5.14a) makes NR measurements at the liquid–solid interface tractable.

The scattering length densities of silicon, quartz, and sapphire (2.1, 4.2, and 5.4×10^{-6} Å^{-2}, respectively) are such that there is a good contrast with

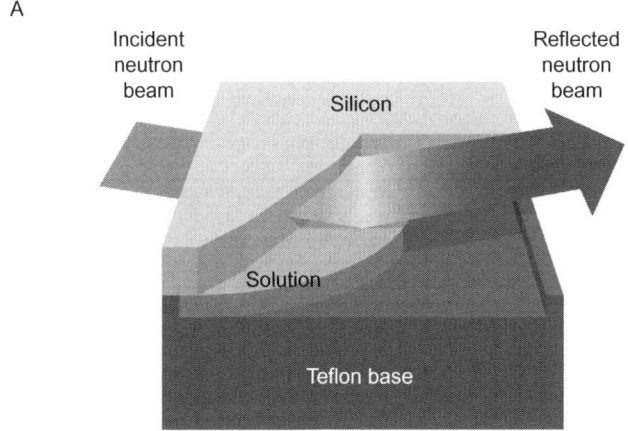

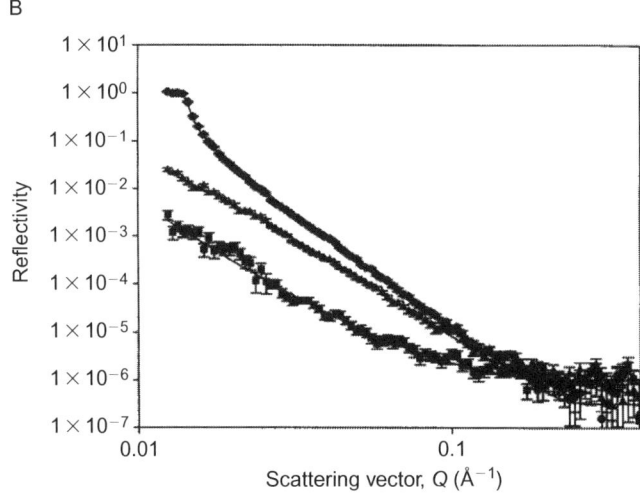

FIGURE 5.14 (a) Geometry for NR liquid–solid interface measurements; (b) NR data for (●) Si/D$_2$O, (□) Si/H$_2$O, and (■) Si/cmSi. *Reproduced by permission of Ref. [52].* (For color version of this figure, the reader is referred to the online version of this chapter.)

D_2O (6.4×10^{-6}) and H_2O (-0.56×10^{-6}), and H_2O/D_2O mixtures can be used to index match to the solid phase. In terms of studying surfactant adsorption at such interfaces, this means that both hydrogeneous and deuterated surfactant can be used. Optically flat surfaces are required and typical sample areas range from 20 to 100 cm^2. All such crystalline surfaces have a thin native oxide or gel layer at the surface and need to be characterized prior to adsorption studies. This is frequently done using NR measurements for different solvent contrasts, and some typical data are shown in Figure 5.14b for silicon/D_2O, silicon/H_2O and silicon/cmSi (cmSi is an H_2O/D_2O mixture index matched to silicon). The NR data in Figure 5.14b are consistent with a thin oxide layer $\sim$15 Å with a scattering length density $\sim 3.5 \times 10^{-6}$ Å^{-2}, which is close to that for SiO_2.

In recent years, a wide range of NR studies have been made on the adsorption of surfactants and mixed surfactants onto differently functionalized solid surfaces, some of which has been recently reviewed [53].

Recently, Penfold et al. [54–57] have used this approach to study surfactant adsorption [54], mixed surfactant adsorption [55], and polymer–surfactant adsorption [56,57] onto model hydrophobic and hydrophilic cellulose surfaces. The cellulose surfaces are prepared by Langmuir–Blodgett deposition of trimethylsilylcellulose, TMSC, onto a silanated silicon surface, to produce smooth layers $\sim$50–120 Å thick [54]. The characteristic structure of the cellulose films formed in this way is usually characterized by two layers which approximates to the density distribution within the film [54–57] and is obtained from NR measurements with different solvent contrasts. The films are initially hydrophobic, but are converted to hydrophilic surfaces by cleaving the surface methyl groups of the TMSC by exposure to HCl vapor. The NR data in Figure 5.15a show the evolution in the NR profiles for hexadecyltrimethyl ammonium bromide ($C_{16}TAB$) adsorption onto the hydrophilic cellulose surface with increasing $C_{16}TAB$ concentration [54].

The data are characterized by an interference fringe from the adsorbed bilayer of $C_{16}TAB$, with a thickness $\sim$30 Å, enhanced slightly by a weaker fringe, visible in the absence of surfactant, from the cellulose film, which in this case was $\sim$50 Å thick. The amount of surfactant at the interface can be estimated from

$$\Gamma = \frac{d(\rho_s - \rho_2)}{N_a V(\rho_s - \rho_a)} \quad (5.32)$$

where d is the adsorbed layer thickness; ρ_s, ρ_a, and ρ_2 are the solvent, adsorbate, and adsorbed layer scattering length densities; and V is the adsorbate volume. Figure 5.15b shows a comparison of the adsorption isotherms for $C_{16}TAB$ onto hydrophilic silica, and hydrophobic and hydrophilic cellulose. The adsorption of the $C_{16}TAB$ onto the anionic hydrophilic silica surface and onto the hydrophilic cellulose surface is similar, whereas the adsorption onto the hydrophobic cellulose surface is substantially smaller. This reflects

Chapter | 5 Large-Scale Structures 377

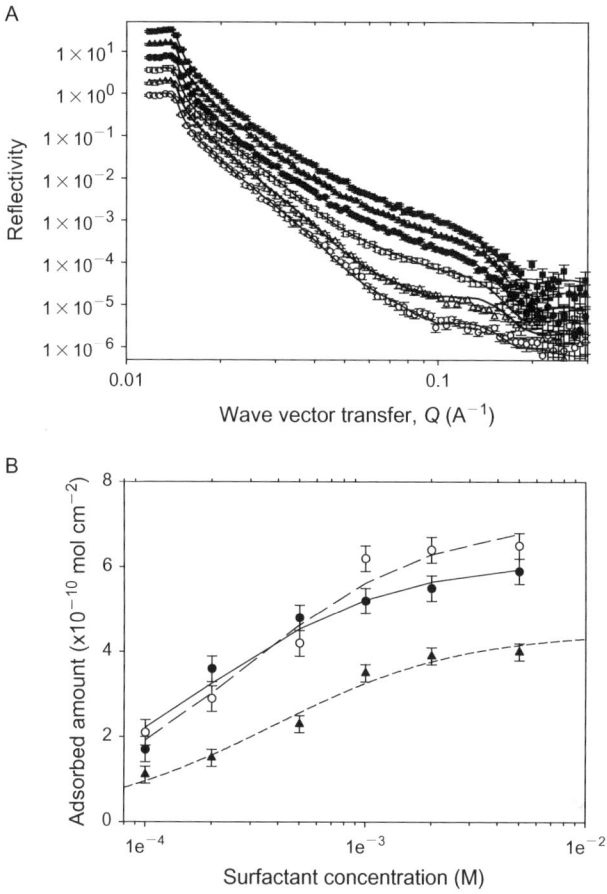

FIGURE 5.15 (a) Specular reflectivity versus wave-vector transfer, Q, for h-C_{16}TAB/D_2O at the hydrophilic cellulose surface, (○) 10^{-4} M h-C_{16}TAB, (Δ) 2×10^{-4}, (□) 5×10^{-4}, (●) 10^{-3}, (▲) 2×10^{-3}, and (■) 5×10^{-3}. The solid lines are model fits, and each profile is displaced for clarity. (b) Adsorbed amount of C_{16}TAB, $\Gamma(\times 10^{-10}$ mol cm$^{-2})$, as a function of surfactant concentration onto (●) hydrophilic silica, (○) hydrophilic cellulose, and (▲) hydrophobic cellulose. The lines are calculated curves, assuming a Langmuir isotherm. *Reproduced by permission of Ref. [54].*

the transition from bilayer adsorption on the hydrophilic surfaces to a monolayer on the hydrophobic surface. The detailed analysis shows the extent of solvent and surfactant penetration into the hydrophobic cellulose film, which does not exist for the hydrophilic cellulose film. Measurements for C_{16}TAB/hexaethylene monododecyl ether ($C_{12}E_6$) mixtures [55] show that the presence of the nonionic surfactant promotes a greater penetration. The interaction with the cellulose film is even more complex for polymer–surfactant mixtures, as was shown for SDS–poly(dimethyldiallyl ammonium chloride) (polydmdaac) mixtures at both hydrophilic and hydrophobic cellulose surfaces [56,57]. The NR measurements show that both polyelectrolyte (poly-dmdaac) and electrolyte promote swelling, penetration, and effects that are not reversible or only partially reversible. In such systems, the ability to change the solvent and adsorbate contrasts through deuterium labeling is essential, in order to unravel these complex patterns of adsorption behavior and to separate the contributions from the adsorbed layer and changes to the substrate.

5.3.3 Structure of Biological Membranes

Interest in biological membranes stems from a desire to understand the properties of cells, the structural and functional unit of most living organisms. Cells, in general, comprise a protoplasm of generic proteinaceous material plus electrolytes encased by a membrane which is only partially permeable and is itself composed a mixture of lipids and proteins. With relevance to drug delivery, and understanding complex transport in biological processes and membranes, the structure and composition of this cell wall membrane have attracted and continue to attract considerable interest. The membrane itself is complex, and our understanding is derived from a systematic process of simplification where the properties of individual membrane components were studied on their own and in mixtures, and there are many examples of model membrane studies [58–70]. A comprehensive review dedicated to NR was recently published by Wacklin [71].

Lipids are the scaffold of cell membranes and being amphiphiles are structurally similar to surfactants. Hollinshead *et al.* [59] have comprehensively studied the structural variations of a model lipid monolayer distearoylphosphatidylcholine (DSPC) under controlled surface pressure conditions using a Langmuir trough. Combining this approach with partial deuterium labeling of the headgroup and hydrocarbon chains on the model lipid membrane, the structure of the lipid monolayer at the air–water interface was obtained (Figure 5.16). Furthermore, it was possible to measure the changes in area per molecule, the hydration, and the tilt angle of the membrane lipid (Figure 5.17) as a function of surface pressure. Hollinshead also showed that above a certain pressure, the headgroup and alkyl chain regions begin to intermix.

In order to progress closer to real membranes and at the same time satisfy the strict optical requirements of NR, careful preparation of realistic model bilayer membranes is required. This task has developed into a sophisticated area of science itself, where membranes are readily cast from mixtures or formed against optically smooth surfaces which being crystalline are highly transparent to neutrons, and this together with selective deuteration enables studies of the arrangement of structural lipids, membrane proteins, and amphiphiles within the membranes. The "floating bilayer" method has been developed in different guises and continues to evolve, but at its simplest consists of a bilayer 2- to 3-nm thickness floating proximate to an identical bilayer

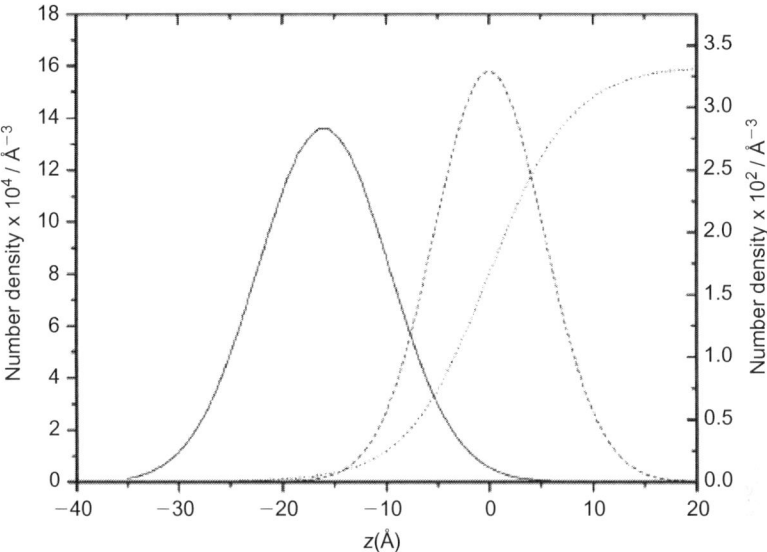

FIGURE 5.16 Number density profiles for DSPC monolayer at a surface pressure of 30 m Nm^{-1}. The profile for the alkyl chain of the lipid is the solid line, the solvent is the dotted line, and the headgroup is the dashed line. *Reproduced by permission of Ref. [59].*

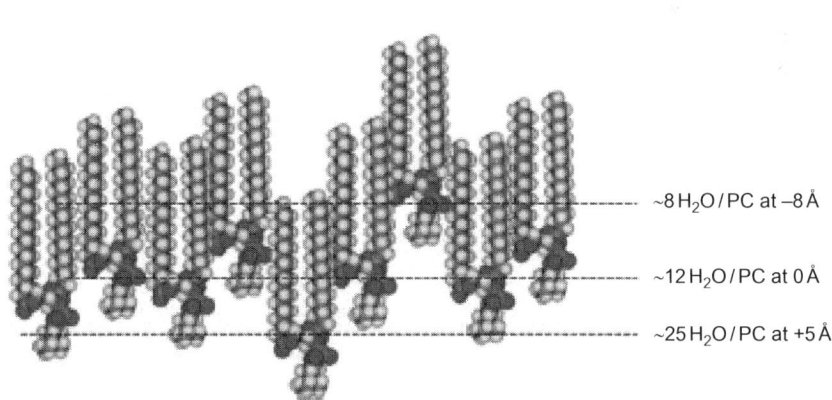

FIGURE 5.17 Schematic illustration of the structure of the DSPC monolayer formed at the air–water interface at a pressure of 30 m Nm^{-1}. Dashed lines denote distances measured along the interface normal, at $Z = -8$, 0, and $+5$ Å, measured with respect to the midpoint of the headgroup distribution. The number of water molecules per lipid is calculated from the ratio of the integrated number densities for the headgroups and solvent, with an integrated slab width of 10 Å. *Reproduced by permission of Ref. [59].* (For color version of this figure, the reader is referred to the online version of this chapter.)

directly attached to an optically flat surface which is either neutron or X-ray transparent [60–64]. Using this approach, the effects of lipid chain length, counterion charge, headgroup, and electric fields have been investigated. Fragneto [65] used NR to investigate the temperature dependence of the membrane behavior around the phase transition of double bilayers formed from saturated diacyl phosphocholines, varying the number of carbon atoms per chain from 16 to 20. Samples were prepared at room temperature (i.e., with the lipids in the gel phase), and measurements were performed at various temperatures so that the whole region of transition from gel to fluid phase was explored. In Figure 5.18, the results from NR measurements are reported, showing the existence of a very large swelling of the water layer between

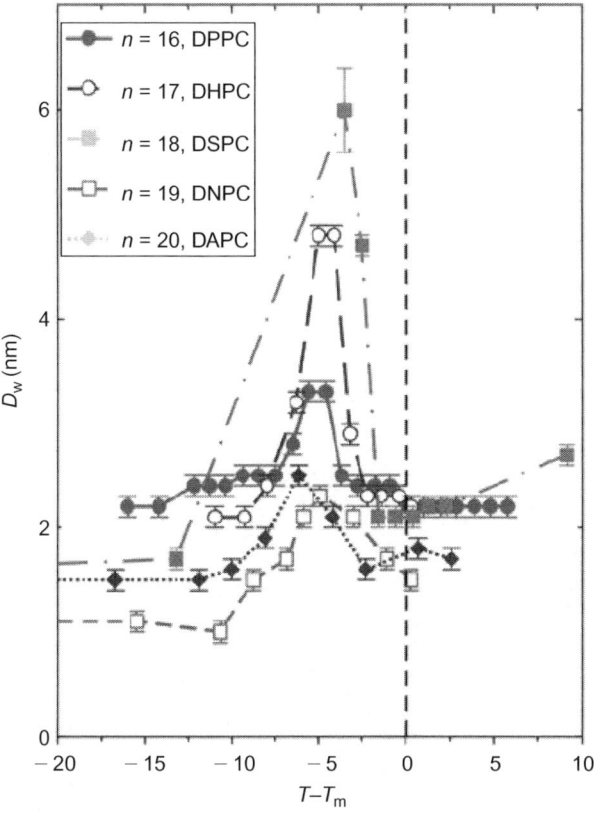

FIGURE 5.18 Distance, D_w, between the two bilayers derived from NR measurements of the hydrogenated phosphocholine lipids as a function of temperature, relative to the melting temperature of the lipid ($T-T_m$). Chain lengths varied from C16 (red filled circles), C17 (open blue circles), C18 (filled green square), C19 (open red square), and C10 (filled blue diamond). *Reproduced by permission of Ref. [65].* (For interpretation of the references to color in this figure legend, the reader is referred to the online version of this chapter.)

the two bilayers for lipids with 17 and 18 carbon atoms in the hydrophobic chain, in contrast to those with greater hydrocarbon content, suggesting that there is a change in membrane behavior between these structural homologues.

Petrache *et al.* [66] varied the headgroup charge on the lipid and using a combination of NR and scattering off the specular direction (off-specular measurements) compared 1,2-dipalmitoyl-*sn*-glycero-3-phosphoserine (DPPS) and dipalmitoylphosphatidylcholine (DPPC) bilayers. They showed that headgroup charge has an effect on bilayer structure, and light differences in bilayer thickness were observed. Off-specular scattering analysis suggests that membranes are strongly correlated over a large (optical) distance scale $\sim\mu$m, that over this distance a thin water layer ~ 1 nm is present between the two bilayers, [67], and that the headgroup charges are heavily screened due to counterion sharing between the bilayer membranes.

As discussed earlier, DPPC double bilayers exhibit large increases in the water layer separating the bilayers when heated to around the main lipid phase transition temperature, with corresponding increases in outer bilayer roughness. A more complex set of behaviors is displayed in mixed bilayers containing model lipid and cholesterol. Stidder *et al.* [68,69] observed two different sets of behaviors. Low levels (up to 2 mol%) of cholesterol have a softening effect on the membrane, vindicated by a reduced degree of swelling and increased span of temperatures over which the swelling occurred proximate to the gel–fluid transition temperature (Figure 5.19) and by a concomitant reduction in floating bilayer roughness. However, higher levels of added cholesterol stiffen the bilayer and ultimately (at ~ 10 mol%) lead to regions of phase separated

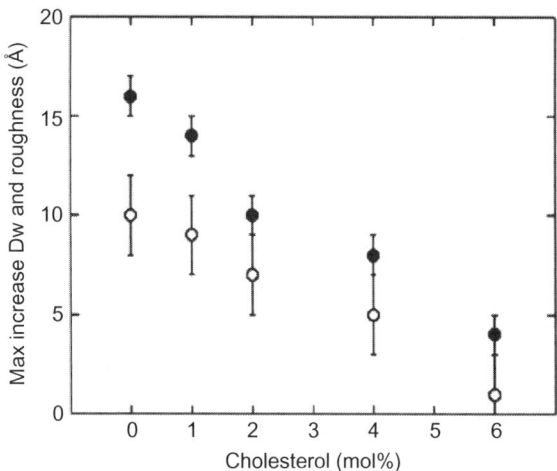

FIGURE 5.19 Variation in maximum increase in water layer thickness (D_w) between the bilayers (filled circles) and outer bilayer roughness (open circles) in the transition region as a function of cholesterol content for DPPC bilayers. *Reproduced by permission of Ref. [68].*

cholesterol included in a mixed matrix. The effect of cholesterol is therefore somewhat complex and modifies the interactions between the two bilayers.

Shen et al. [70] have observed similar behavior in DPPC bilayers adsorbed at the liquid–solid interface with the biosurfactant surfactin. An attractive interaction between the membrane protein and the surfactin is observed. Below the biosurfactant cmc, the surfactin adsorbs and becomes part of the bilayer structure, but at concentrations exceeding the surfactin cmc, DPPC is removed from the bilayers and complementary SANS measurements showed that it becomes included in a mixed aggregate.

5.3.4 Micelles

The Q range accessible in SANS is ideally matched to that required for the study of micellar structures. The detailed quantitative analysis pioneered by Hayter and Penfold [13], and encapsulated in Equations (5.16)–(5.23), gives access to the micelle structure, aggregation number (v), and surface charge (z). Using molecular constraints, a core–shell model can be used to describe the micellar structure in terms of an alkyl chain inner core radius of R1 and outer radius of R2, v, and z. The schematic diagram [72] in Figure 5.20 summarizes the main features of the model.

Some typical SANS data for 0.03 M $C_{16}TACl$ in D_2O are shown in Figure 5.21a [13].

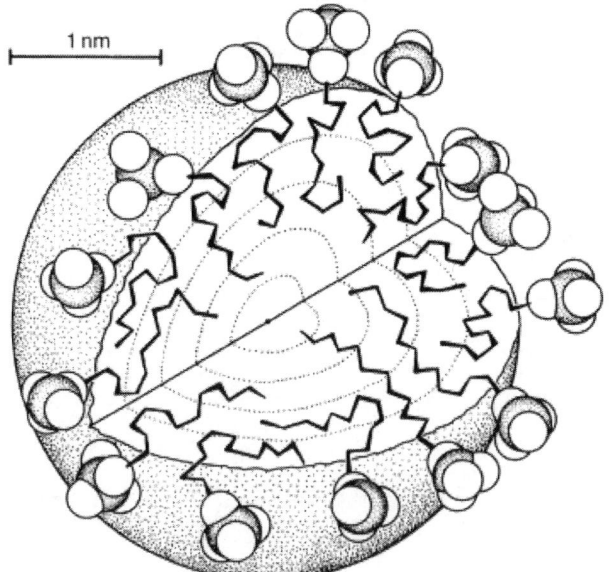

FIGURE 5.20 Schematic representation of the globular micellar model. *Reproduced by permission of Ref. [72].*

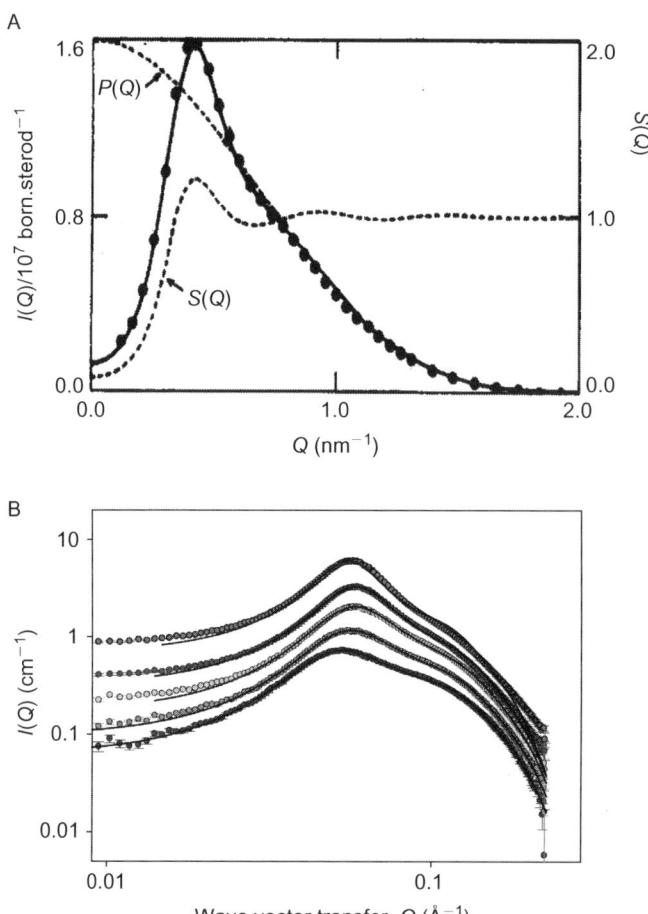

FIGURE 5.21 SANS data for (a) 0.03 M $C_{16}TACl/D_2O$. The solid line is a model fit, and the dashed lines are the contributions of $P(Q)$ and $S(Q)$ to the model (as indicated in the figure). (b) Scattered Intensity for LAS-6/D_2O for 35 mM (blue), 50 mM (pink), 75 mM (cyan), 100 mM (purple), and 150 mM (gray). The solid lines are calculated curves for interacting globular micelles, and the plots are shifted vertically for clarity. *Panel (a): Reproduced by permission of Ref. [13]. Panel (b): Reproduced by permission of Ref. [73].* (For interpretation of the references to color in this figure legend, the reader is referred to the online version of this chapter.)

The separate contribution from $S(Q)$ due to the intermicellar interactions and the form factor $P(Q)$ are shown in Figure 5.21. The data are modeled using the RMSA $S(Q)$ [15,16] and a globular core–shell form factor, with $v \sim 90$, R1 and R2 21 and 24 Å, respectively, and the degree of ionization, $\delta = z/v$ is ~ 0.3. This is typical of the analysis of SANS data for ionic surfactants at relatively low surfactant concentrations, where they are mostly globular, spherical, or elliptical. The degree of ionization, obtained from $S(Q)$, is consistent with the dressed-micelle model [74], which provides an analytic solution to the nonlinear Poisson–Boltzmann equation to describe the ionic distribution on the micellar surface. In Figure 5.21b, some recent SANS data from the micellization of the anionic surfactant LAS-6 in D_2O are shown [73]. Here, the evolution in the scattering profiles with increasing LAS concentration, from 35 to 150 mM, is shown. The data are consistent with relatively small elliptical core–shell micelles, in which the aggregation number increases from $\sim$30 to $\sim$130 and the ellipticity evolves from $\sim$2.0 to 4.0. At relatively low surfactant concentrations and for surfactants where the charge effects, such as for nonionic surfactants, are less dominant, the scattering is more dominated by the form factor. An example of this is illustrated in Figure 5.25 for the weakly anionic SLES at a surfactant concentration of 20 mM [47].

The data in Figure 5.22 show predominantly the evolution in the micelle form factor as $AlCl_3/NaCl$ is added. In the absence of $AlCl_3$, the micelles are relatively small globular micelles with an aggregation number $\sim$130, ellipticity $\sim$1.5, a minor axis radius $\sim$24 Å, and with a relatively low surface charge. With increasing $AlCl_3$ concentration, the scattering at low Q increases dramatically, consistent with micellar growth. For the highest $AlCl_3$ concentration shown in the figure, the micelle aggregation number is $\sim$1000 and the ellipticity is $\sim$10, consistent with significant elongation of the micellar structure.

Such SANS studies are relatively easily extended to mixed surfactant micelles, where the use of deuterium labeling enables both micelle structure and composition to be determined. This approach has been applied to a range of binary surfactant mixtures and provides the opportunity to directly compare

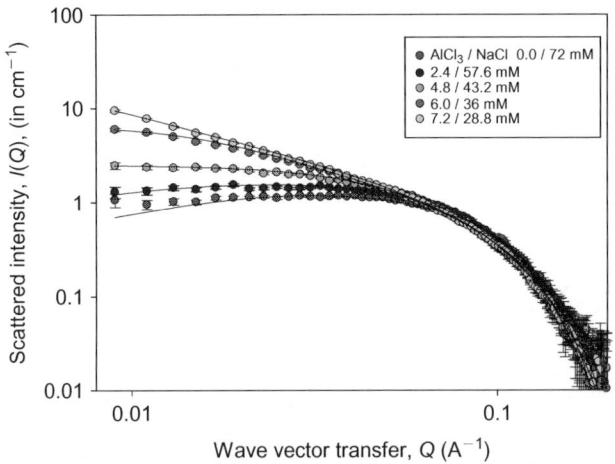

FIGURE 5.22 Scattered intensity, $I(Q)$, for 20 mM $SLES/AlCl_3/NaCl$, for different $AlCl_3/NaCl$ concentrations (see legend for details). $AlCl_3/NaCl$. The solid lines are model fits. *Reproduced by permission of Ref. [47].* (For color version of this figure, the reader is referred to the online version of this chapter.)

with the predictions of the thermodynamics of mixing, such as RST [76,77]. For a dilute solution of mixed surfactant micelles, in the limit of small Q where $S(Q) \sim P(Q) \sim 1.0$, the scattered intensity can be written as

$$I(0) \approx \sum_i N_i V_i^2 (\rho_{ip} - \rho_s)^2 \qquad (5.33)$$

where N is the micelle number density, V is the micelle volume, ρ_p and ρ_s are the micelle and solvent scattering length density, and Σ_i is a summation over the i components in the mixture. So for a binary mixture, Equation (5.33) reduces to two equations. If measurements are made for two different isotopic combinations, as shown in Figure 5.23a for h-SDS/h-$C_{12}E_6$/D_2O and d-SDS/h-$C_{12}E_6$/D_2O, then the ratio of the scattering is directly related to the micelle composition [76,77].

Figure 5.23b shows the evolution in micelle composition with solution concentration, for three different SDS/$C_{12}E_6$ solution compositions. The solid lines are the corresponding predictions from RST, which are in good agreement with the data but not for a single interaction parameter, β_m.

5.3.5 Lamellar Phases and Vesicles

The preferred structure and structural evolution in surfactant self-assembly can be broadly rationalized by the simple but elegant packing or curvature arguments of Israelachivili *et al.* [78], in which the aggregate packing parameter, pp, is defined as pp $= V/Al$, where V is the alkyl chain volume, A the area per molecule, and l the fully extended alkyl chain length. For pp $< 1/3$, the aggregates are small globular micelles; for $1/3 <$ pp $< 1/2$, they are more elongated structures from ellipses to rods; and for pp $> 1/2$, planar structures, lamellar or vesicular, exist. At low surfactant concentrations, most ionic and nonionic surfactants are globular micelles, as illustrated in the previous section. The addition of electrolyte reduces A, increases pp, and promotes micellar growth, as shown in the previous section for SLES/AlCl$_3$ in Figure 5.22. Increasing surfactant concentration also often promotes micellar growth, and this is illustrated for the dialkyl chain anionic surfactant LAS-6 in Figure 5.21b. At higher surfactant concentrations, LAS-6 transforms from a micellar to a lamellar structure.

In general, the dialkyl chain surfactants promote more planar structures, lamellar or vesicular, due to a larger pp associated with the large value of V, even at relatively low surfactant concentrations. The resultant curvature can be manipulated in mixed surfactant systems, where the different components have different preferred curvatures, and this was illustrated by Tucker *et al.* [79] as shown in Figure 5.24.

The SANS data in Figure 5.24a show the scattering from a 1.5 mM dihexadecyldimethyl ammonium bromide (DHDAB)/$C_{12}E_6$/D_2O mixture, from DHDAB rich to $C_{12}E_6$ rich. The data illustrate a transition from globular micelles ($C_{12}E_6$ rich) to bi-lamellar vesicles (DHDAB rich). The data

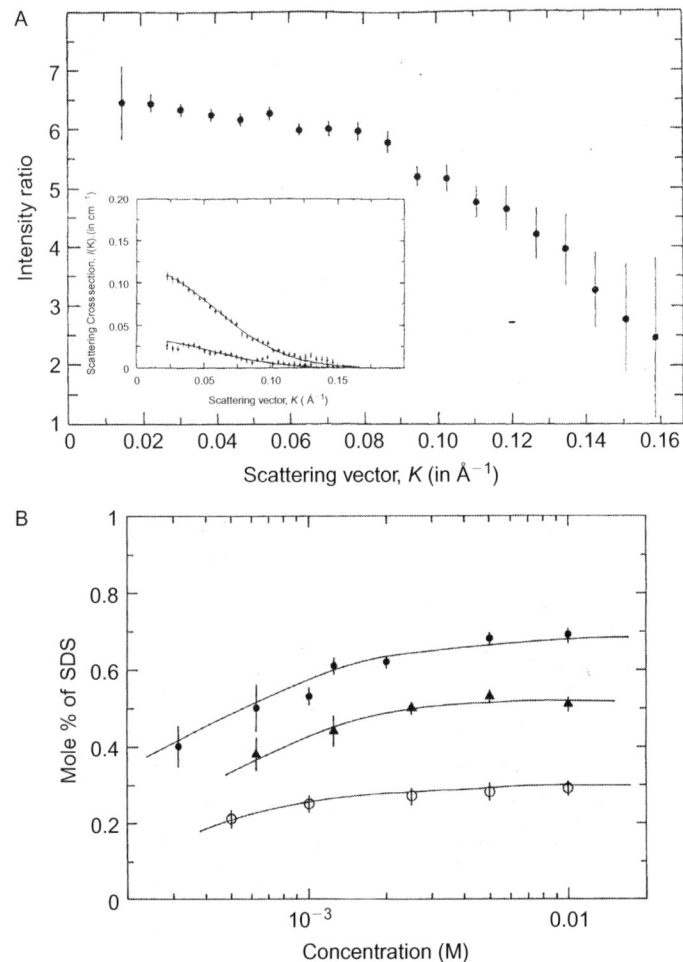

FIGURE 5.23 (a) Intensity ratio for h-SDS/h-$C_{12}E_6$/D_2O and d-SDS/h-$C_{12}E_6$/D_2O, the inset shows the individual scattering curves. (b) Variation in micelle composition for SDS/$C_{12}E_6$ for solution compositions of (●) 70/30, (△) 50/50, and (○) 30/70. The solid lines are for RST calculations with different interaction parameters. *Panel (a): Reproduced by permission of Ref. [76]. Panel (b): Reproduced by permission of Ref. [77].*

presented are multiplied by Q^2 to emphasize the transition from the Q^{-2} dependence of the scattering to the onset of the formation of the planar structures, and to more clearly identify the mixed phase region. From such measurements, the complex phase diagram was derived by Tucker *et al.* [79]. A quantitative analysis of the multi-lamellar and bi-lamellar vesicles as well as the micellar phase was presented.

The bi-lamellar and multi-lamellar vesicles and lamellar phase regions were analyzed using the approach developed by Nallet *et al.* [80], which takes

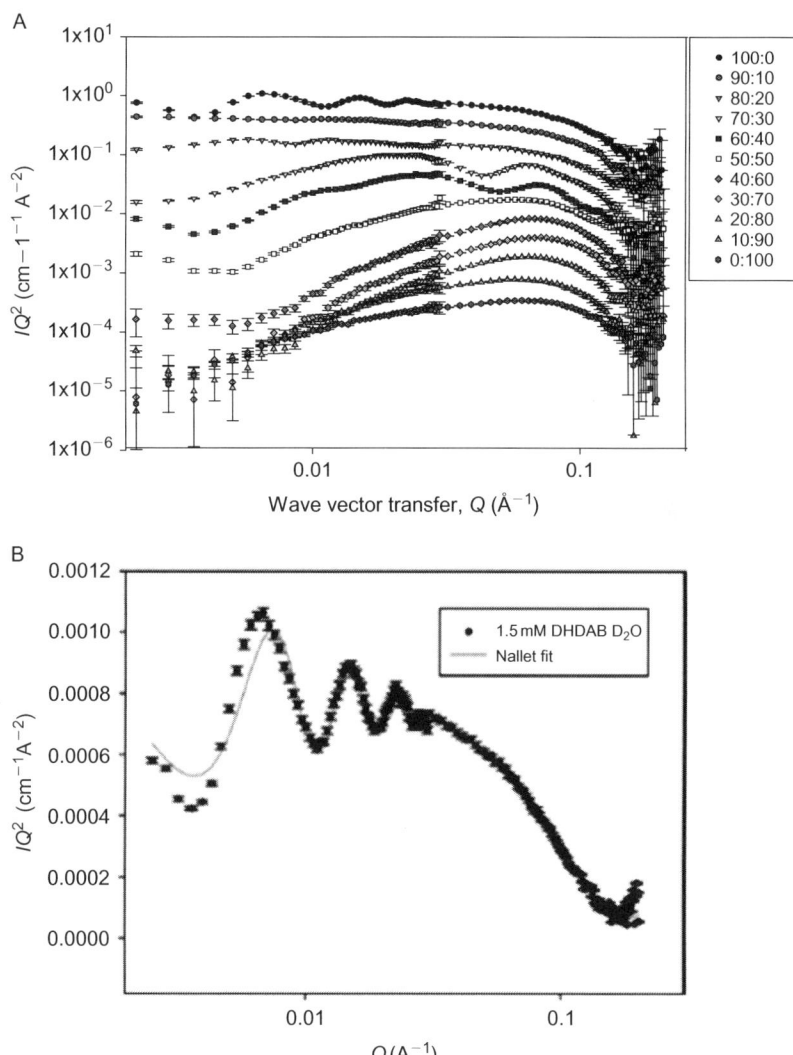

FIGURE 5.24 Variation in scattered intensity ($\times Q^2$) for (a) 1.5 mM DHDAB/$C_{12}E_6$/D_2O, see legend for details, (b) 1.5 mM DHDAB/D_2O. The solid line is a model fit to the Nallet model for the parameters described in the text. *Reproduced by permission of Ref. [79].* (For color version of this figure, the reader is referred to the online version of this chapter.)

into account the rigidity of the bilayer through the Caille parameter, η, which essentially determines the line shape of the scattering. It is defined as

$$\eta = \frac{Q_0^2 k_B T}{8\pi\sqrt{\kappa B}} \quad (5.34)$$

where B and κ are the bilayer compressibility and bending modulus, $Q_0 = 2\pi/d$, d is the bilayer spacing. The scattered intensity is then

$$I(Q) = 2\pi \frac{V}{d} \frac{1}{Q^2} P(Q) S(Q) \quad (5.35)$$

where $P(Q)$ is the lamellar form factor and $S(Q)$ is the structure factor which takes into account bilayer fluctuations and assumes powder averaging and a line shape width that takes into account the contribution from instrumental resolution. Figure 5.24b shows an application of the Nallet model (Equation 5.35) to a lamellar solution of 1.5 mM DHDAB/D$_2$O, which results in a bilayer spacing, d, ~850 Å, a bilayer thickness $\delta \sim 34$ Å, and $\eta \sim 0.07$. The low value of η implies that the membrane in the bi-lamellar vesicle is relatively rigid. A higher value of η would imply a more labile flexible structure, and this was observed by the addition of a nonionic cosurfactant.

Further manipulation of the preferred curvature and packing in vesicle structures by the addition of a third component, a short straight-chain alkanol, to the DHDAB/C$_{12}$E$_{12}$ mixture discussed above, resulted in the transformation from large bi-lamellar/multi-lamellar vesicles to small unilamellar vesicles, nanovesicles [75]. Similar observations were reported by Grillo et al. [81] in DDAB/C$_{12}$E$_4$ mixtures. The SANS data in Figure 5.25 show the scattering for 25 mM DHDAB/C$_{12}$E$_{12}$ for DHDAB/C$_{12}$E$_{12}$ mole ratios of 40/60, 55/45, and 70/30, with the addition of 50 mM octanol.

The data are consistent with small interacting unilamellar vesicles. In contrast to the scattering data for the larger bi-lamellar/multi-lamellar vesicles shown in Figure 5.24 for the DHDAB/nonionic surfactant mixtures where analysis required the more complex Nallet-type approach (Equations 5.34

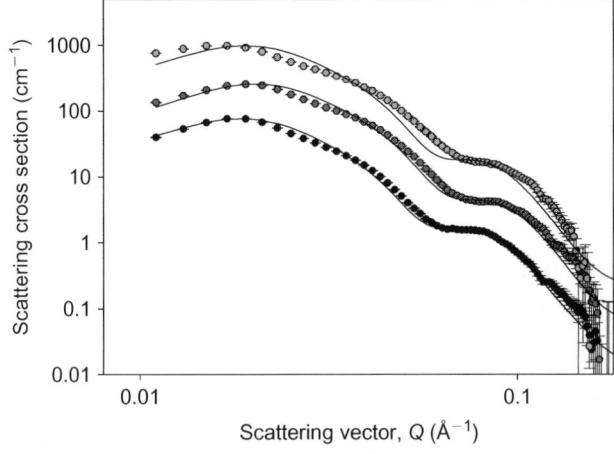

FIGURE 5.25 Scattering cross section, dσ/dΩ (in cm^{-1}) versus scattering vector, Q (in Å^{-1}) for 25 mM DHDAB/C$_{12}$E$_{12}$, +50 mM octanol (black-filled circles) 40/60 mole ratio DHDAB/C$_{12}$E$_{12}$, (red-filled circles) 55/45, and (green-filled circles) 70/30. The solid lines are model calculations. *Reproduced by permission of Ref. [75].* (For interpretation of the references to color in this figure legend, the reader is referred to the online version of this chapter.)

and 5.35), here the nanovesicles can be modeled by an adaptation of the interacting core–shell micelle model (Equations 5.16–5.23). In this case, the core has the same scattering length density as the surrounding solvent phase and the shell is a single bilayer of hydrogenous surfactant. The parameters from the core–shell model for the data in Figure 5.23 are an inner core radius R1 $\sim$ 24–30 Å, an outer radius $\sim$60 Å, a polydispersity, σ, $\sim$0.2, and a surface charge $\sim$15. The addition of the short-chain alkanol results in a manipulation of the preferred curvature, membrane flexibility, and packing anisotropy which result in a nonzero preferred curvature. The formation of such small unilamellar vesicles requires a relatively high bending energy, a finite preferred curvature, and a finite electrostatic contribution to the bending energy. The addition of octanol and decanol to the DHDAB/nonionic surfactant mixtures meets those requirements, but the addition of the longer dodecanol and hexadecanol does not.

There is much current interest in the effects of flow and different flow geometries on surfactant self-assembly, alignment of anisotropically shaped structures, such as rods and lamellar fragments, and shear-induced transformations [82]. The penetrating power of cold/thermal neutrons has enabled Poiseuille, Couette, and elongational flow geometries to be exploited for SANS studies [83,84]. In their pioneering study of the alignment of rod-like micelles in Couette flow, Hayter and Penfold [85] developed an analytical expression for the orientational distribution. As illustrated in Figure 5.26, measurements can be made with the neutron beam normal to the flow—vorticity plane or to the shear gradient—vorticity plane.

This was used by Penfold *et al.* [86] to show the transformation of an aligned lamellar phase of $C_{16}E_6$ from an orientation parallel to the flow—vorticity plane at low shear rates to the orthogonal orientation, parallel to the shear gradient—vorticity plane, at high shear, as shown in Figure 5.27a.

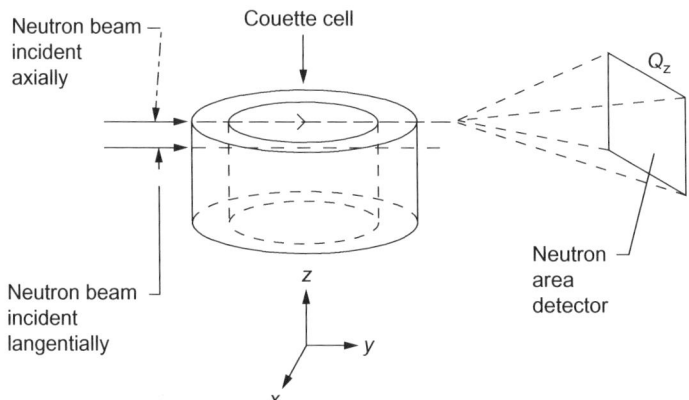

FIGURE 5.26 Couette flow cell scattering geometry.

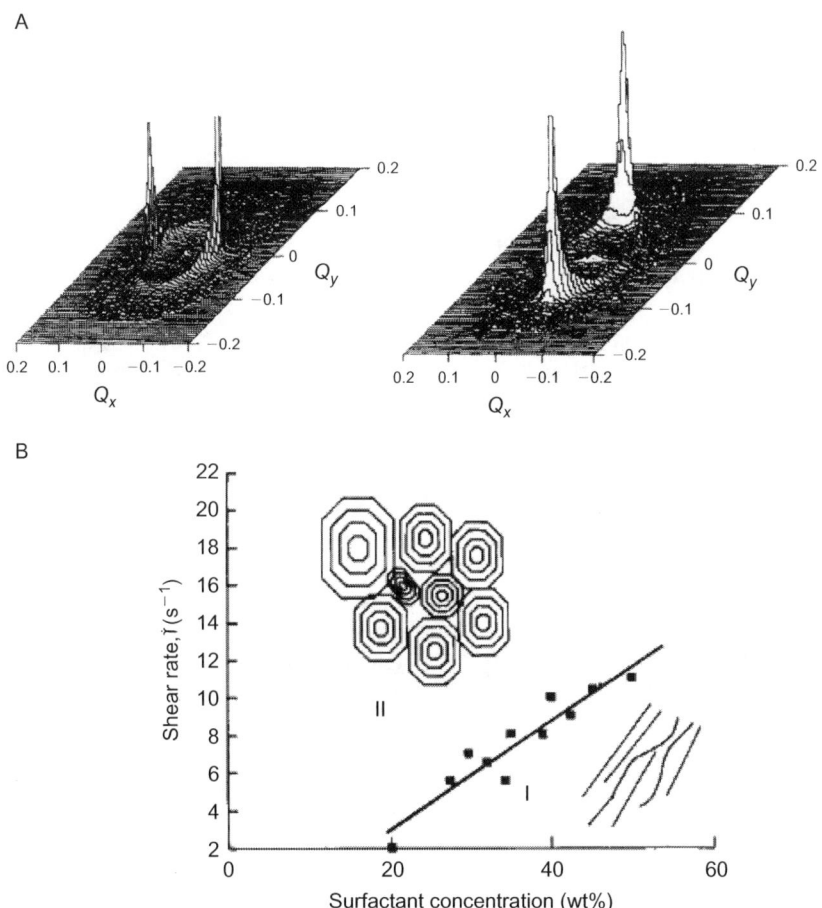

FIGURE 5.27 (a) SANS intensity maps for 50.6 wt% $C_{16}E_6$ in D_2O at $G=0$ and $5000\,s^{-1}$. (b) Shear diagram for DHDAC lamellar phase. *Panel (a): Reproduced by permission of Ref. [86]. Panel (b): Reproduced by permission of Ref. [87].*

Soubiran *et al.* [87] used the same scattering geometry and cell arrangement to investigate the shear-induced transformation from lamellar fragments to multi-lamellar vesicles with increasing shear for the dialkyl chain cationic surfactant (DHTAC) (see Figure 5.25b) and to map out the associated shear diagram.

More recently, Penfold *et al.* [84,88] have shown how spatially resolved SANS measurements (using a 1-mm diameter neutron beam to raster scan over an extended flow area) can be used to investigate the flow pattern and orientational distribution in lamellar phase solutions in a crossed flow elongational flow cell (see Figure 5.28a).

In Figure 5.28b, the SANS data, measured over the central region of the elongational flow cell for the lamellar fragments formed from the dialkyl

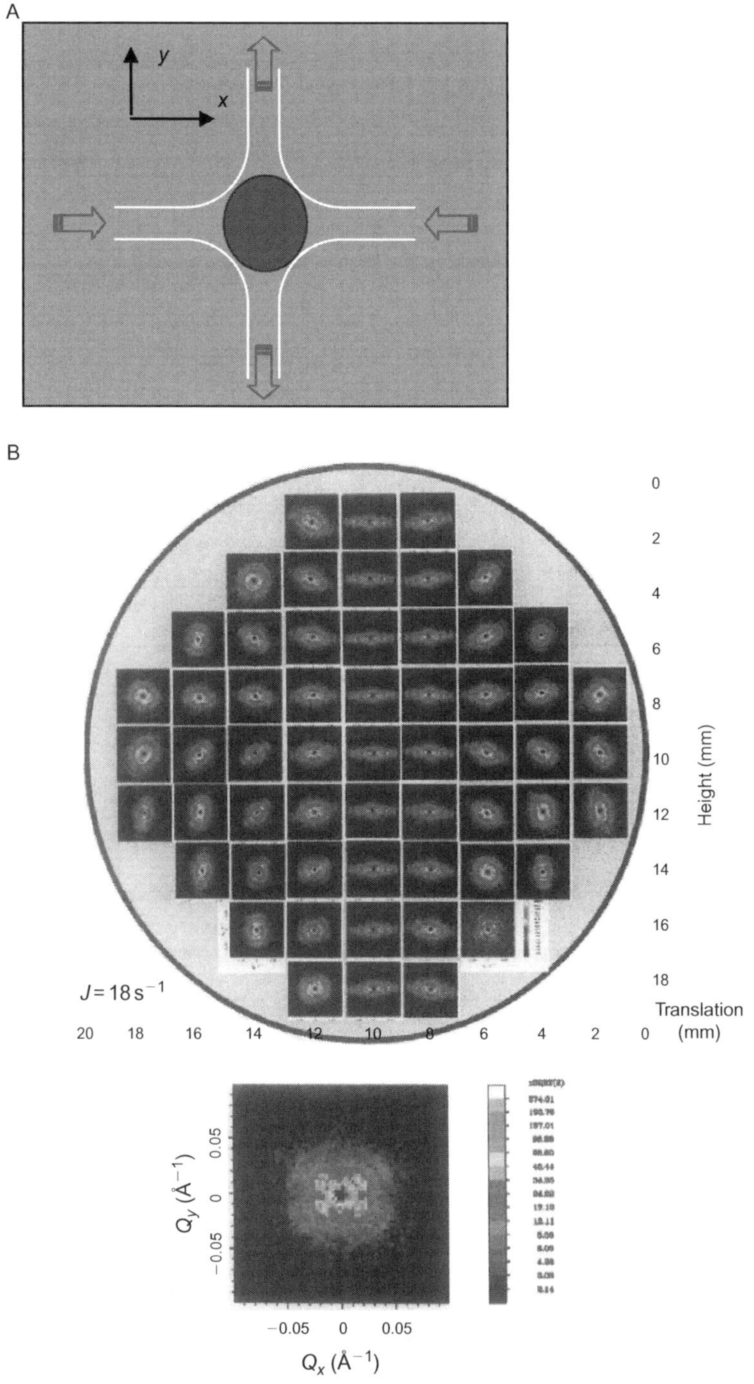

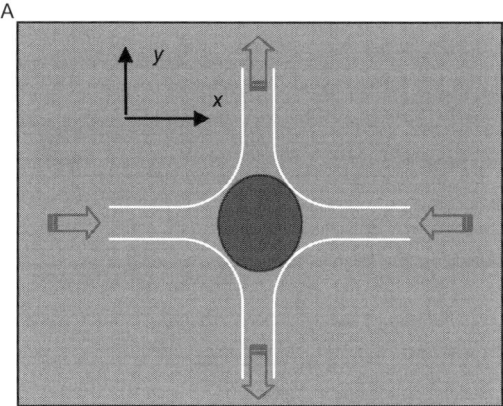

FIGURE 5.28 (a) Crossed slots elongational flow cell geometry. (b) Intensity contour maps (Q_y vs. Q_x) for 15 wt% DHTAC/0.5 wt% $C_{18}E_{10}$ in D_2O at a flow rate of 16 s^{-1}, measured at 2-mm intervals of the elongational flow cell aperture. *Panel (a): Reproduced by permission of Ref. [84]. Panel (b): Reproduced by permission of Ref. [88].* (For color version of this figure, the reader is referred to the online version of this chapter.)

chain cationic/nonionic surfactant mixture of DHTAC/$C_{18}E_{10}$, are shown. The data are consistent with highly ordered lamellar fragments aligned in the flow direction, and hence the variation in orientational ordering over that central region reflects the flow pattern. The more complex off-axis patterns (see inset in Figure 5.28b) reflect the competition between and demixing of the two principal flow directions within the cell.

These different examples of flow measurements described here represent some illustrative examples of what is feasible and a glimpse of a broad and active area of research [82].

5.3.6 Colloidal Particles

The adsorption of polymers and surfactants onto colloidal particles is a commonplace route to establishing and manipulating colloidal properties and especially stability [89]. SANS has played an important role in understanding the nature of adsorption onto colloidal particles and precedes and complements the earlier discussion on the use of NR to study the solid–liquid interface. Much of the early focus was on polymer adsorption, where the focus was on the adsorbed amount and the profile of the adsorbed polymer layer [90]. Subsequently, Cummins et al. [91,92] used SANS to explore systematically the adsorption of nonionic surfactants onto different silica sols, with typical diameters $\sim$80–200 Å, and to explore the effect of surfactant type, temperature, and pH. The measurements were made in a D_2O/H_2O mixture index matched to the silica, and hence, the scattering arises from the adsorbed layer of hydrogenous surfactant. Some typical data are shown in Figure 5.29a for $C_{12}E_6$ at 80% coverage adsorption onto a silica sol (diameter $\sim$150 Å, Dow Chemicals TM silica). The detailed modeling requires inclusion of the sol size and polydispersity, the adsorbed layer thickness and density, and the instrumental resolution. The adsorbed layer in that data could be well described by a layer of uniform density with a thickness $\sim$ 30 Å.

At much higher surfactant concentrations, the evidence of the contribution of "free" surfactant micelles, from the excess of surfactant in the system, is evident, as shown in Figure 5.29b.

More recent studies, which take advantage of the extended Q range and greater sensitivity to extend to lower scattering cross sections, and higher-quality sols, enabled Oberdisse et al. [93,94] to probe the structure of the adsorbed surfactant layer. As shown in Figure 5.30, they were able to demonstrate, in a region where there was no excess micelle scattering, that the scattering from the sol and its adsorbed layer, for the nonionic surfactant, $C_{12}E_6$, was consistent with a contribution from the lateral distribution, in the form of small surface micelles, of the surfactant adsorbed on the sol.

In Figure 5.30, the oscillations from the adsorbed layer are increasingly visible as the coverage increases from 25% to 100%. Furthermore, the broader

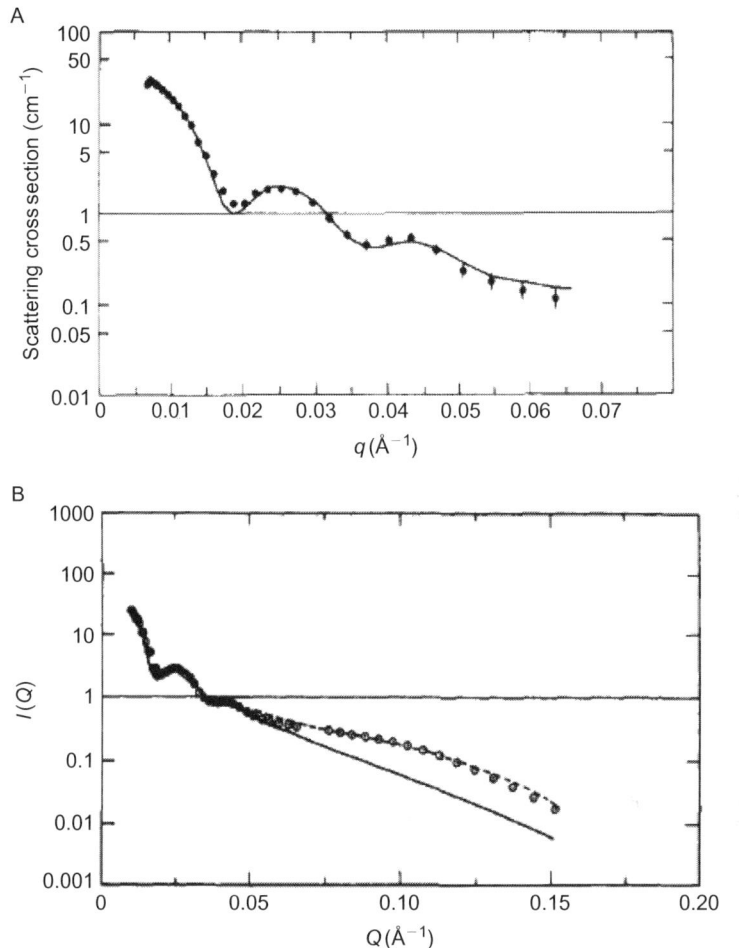

FIGURE 5.29 (a) Scattered intensity for 80% coverage of $C_{12}E_6$ on 4% TM silica sol, in H_2O/D_2O matched to silica. (b) Scattered intensity for 100% coverage $C_{12}E_6$ onto 3.2% TM sol in D_2O/H_2O index matched to silica, along with a model that includes adsorbed layer and free solution micelles. *Panel (a): Reproduced by permission of Ref. [91]. Panel(b): Reproduced by permission of Ref. [92].*

maximum in the scattering, mostly visible at higher Q values, due to the internal structure (surface micelles), increases as the coverage increases.

A similar approach has been applied to study surfactant adsorption at the oil–water interface by Staples *et al.* [95]. Stable oil-in-water emulsions, stabilized by a minimal amount of surfactant such as SDS, can be made using sonication, such that the emulsions are relatively monodisperse, $\sigma \sim 10\%$, sufficiently small, ~ 0.1 μm diameter, and at a few % concentration. When the oil, such as hexadecane, is index matched to D_2O, there is sufficient

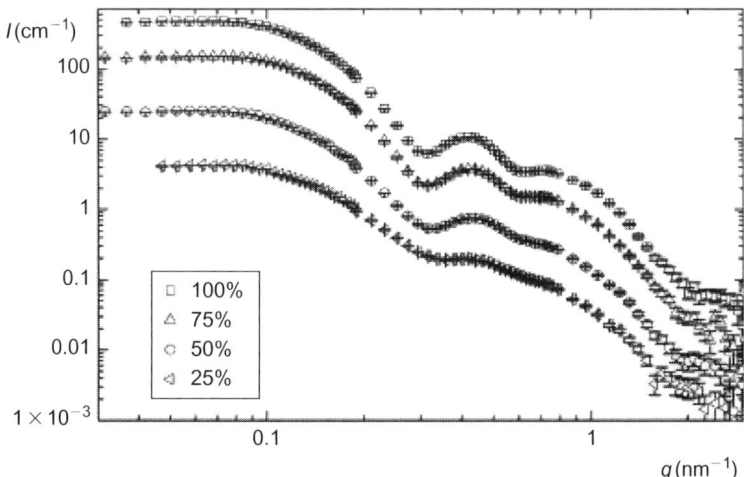

FIGURE 5.30 SANS profiles for 1.8 wt% silica in H_2O/D_2O contrast matched to silica for $C_{12}E_5$ adsorption at 25–100% coverage (see legend for details). *Reproduced by permission of Ref. [93].*

scattering from the adsorbed layer of hydrogenous surfactant that the layer thickness and adsorbed amount can be established using Equation (5.33). The emulsion droplets are sufficiently large that in terms of the adsorption properties, they can be considered as planar interfaces. Figure 5.31a shows the SANS data for h-SDS adsorbed at the oil–water interface of a 6.4 vol% hexadecane–water emulsion, d/h hexadecane index matched to D_2O, in the SDS concentration range of 1.25–20 mM.

The data are multiplied by Q^2 to emphasize the change from a thin absorbed layer, $\sim$12 Å, of surfactant for surfactant concentrations from 1.25 to 12 mM, to the adsorbed layer in equilibrium with free surfactant micelles at the higher surfactant concentrations. Figure 5.31b shows a typical adsorption isotherm derived from such measurements. Staples *et al.* [95] also showed how the emulsions once formed and stabilized with a minimal amount of surfactant can be used as a template for studying adsorption at that interface other than just the stabilizing species. They showed how mixed surfactant adsorption, in that case $SDS/C_{12}E_6$, can be studied by changing, index matching the different components to the solvent/oil phases. Subsequently, this approach has also been applied by Tucker *et al.* [96] to study the adsorption of polymer–surfactant mixtures.

5.3.7 Polymers in Solution, Melt, and Thin Films

One of the earlier applications of neutron scattering was the use of SANS to study polymer morphology in solution and in melts [9]. Since then, it has burgeoned into a much broader area. More recently, SANS studies have focused

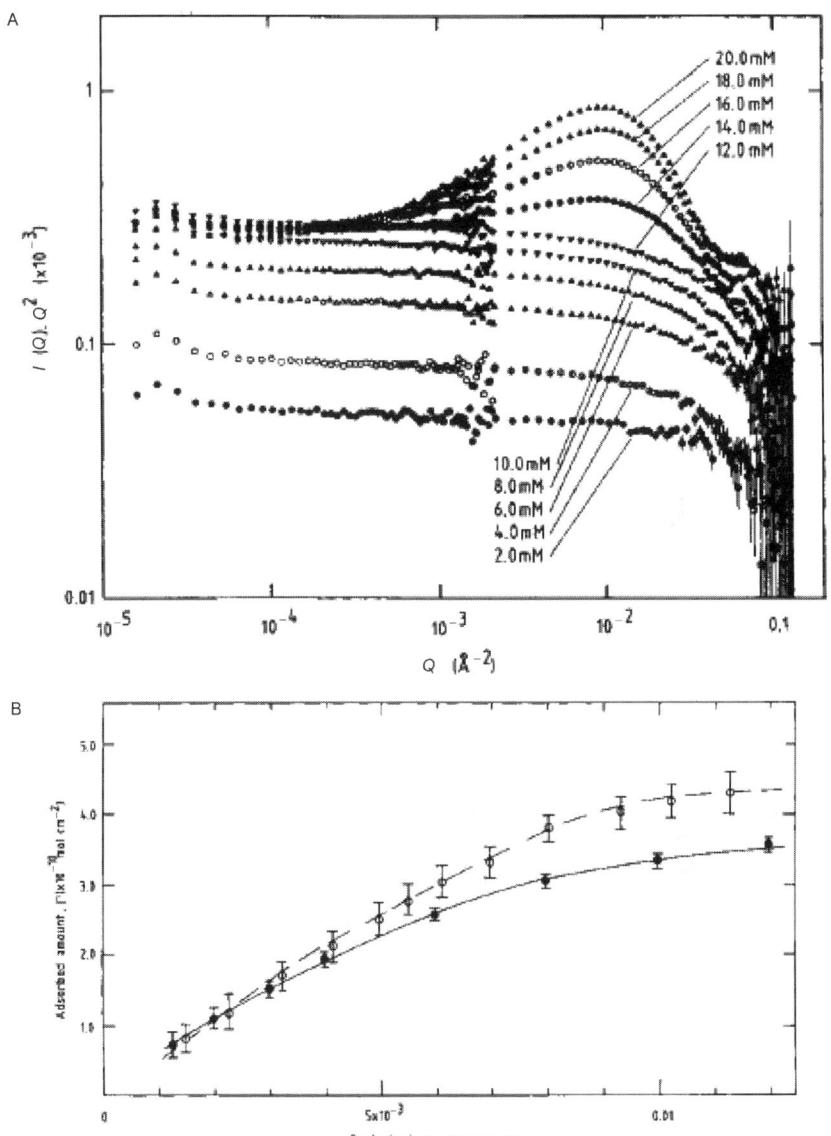

FIGURE 5.31 (a) Scattered intensity $\times Q^2$ for 6.4 vol% h/d hexadecane in D_2O emulsion with h-SDS in the concentration range of 1.25–20 M (see legend for details), (b) SDS adsorption isotherm at hexadecane–D_2O emulsion interface (●) SANS data, (○) surfactant electrode data. The lines are guides to the eye. *Reproduced by permission of Ref. [95].*

on issues such as the self-assembly properties of block copolymers. The application of NR to probe polymer thin films has emerged more recently [3]. An aspect of these current studies which embraces both techniques is the combination of the surface sensitivity with the bulk scattering to probe the role of surface and confinement on the polymer properties. A key feature of all these studies, and what made neutrons such an attractive probe in this area in the first place, is the ability to manipulate "contrast" using selective deuteration.

The Q range and resolution accessible by NR are ideal for the study of polymer thin films, which find application in, for example, photovoltaic devices and low-current microelectronics. In application, two or more polymeric films are used to form a multilayer coating as part of a device. Polymers however are inherently immiscible, and the degree of intermixing is commonly quantified using Flory–Huggins theory [97]. At the interface of two immiscible polymers of infinite MW, self-consistent mean-field theory gives the interfacial width as

$$W = 2a/\sqrt{6\chi} \tag{5.36}$$

where a is the segment length and χ is the mixing parameter [98]. The degree of mixing can be calculated from the measurements of W which can be obtained directly using NR. In the NR studies, enhanced sensitivity or selectivity is usually obtained by deuterium labeling or partially labeling one or part of the polymers. Samples are usually prepared by spin casting from dilute evaporative solution or casting from the melt. In the latter instance, polymer films have been studied where the film is cast against an optically smooth (usually silicon) substrate. Typical data for melt cast polymers are shown in Figure 5.32. For instance, Bucknall et al. [99] have quantified the degree of mixing and demixing in blends of hydrogen and deuterium-labeled polystyrene (h-PS–d-PS$_1$), contrasting the reflectivity from the original spin cast film with the reflectivity from the annealed melt cast film. By repeatedly scanning the reflectivity variation with time, and using the unique ability to change the reflectometer geometry, observations of the reduction of interference fringe amplitude enabled the quantification of the diffusion mixing processes as a variation in the interfacial width with time (Figure 5.33). This relates directly to the transport properties in the melted thin film.

The interfacial width varies with time in the melt state for this polystyrene blend, and the data presented in Figure 5.33 show that the characteristic timescales are directly correlated to diffusion processes but only when the diffusive timescales exceed the reptation rates of the individual polymer species present in the films [99]. A similar approach was used by Kanaya et al. [100] to study the effect of annealing on the thickness of deuterated polystyrene thin films. Their studies revealed a complex set of behaviors depending on the annealing history of the sample. In contrast, Mayes et al. [101] studied the structure of annealed films of polystyrene-b-methylmethacrylate diblock

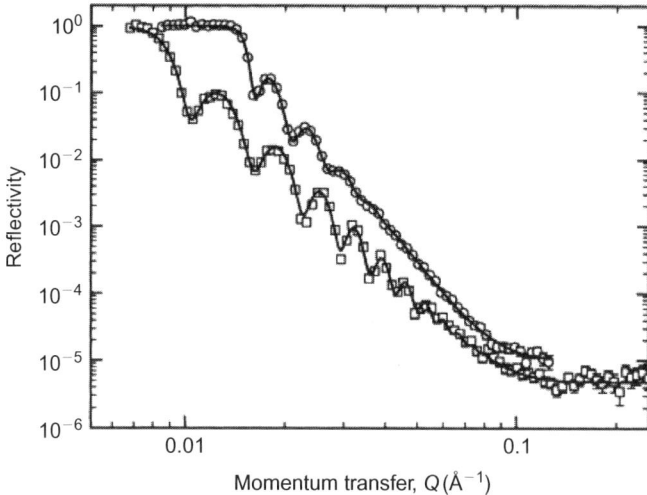

FIGURE 5.32 The variation in reflectivity versus momentum transfer for d-iPP-PE (squares) and d-aPP-PE (circles) measured *in situ* at 185 °C. The solid lines are model fits to the experimental data. *Reproduced by permission of Ref. [99].*

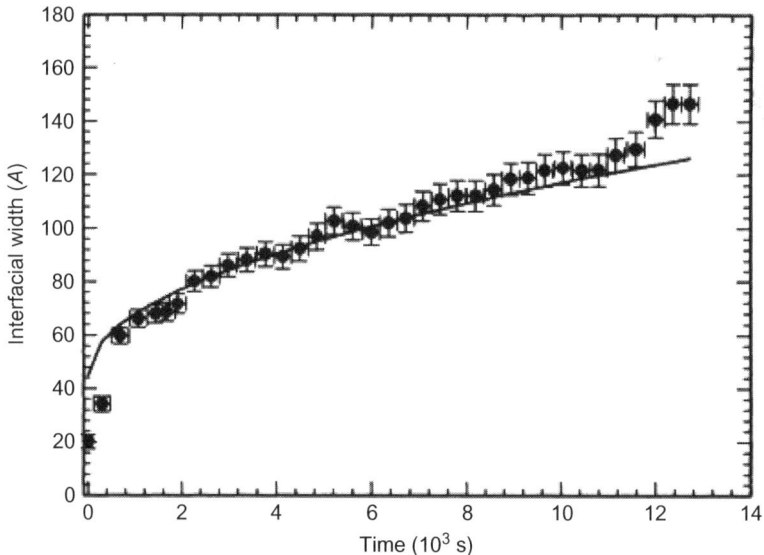

FIGURE 5.33 Variation in interfacial width with time for h-PS–d-PS$_1$ mixtures. The solid line is a calculation based on a $t^{0.5}$ dependence. *Reproduced by permission of Ref. [99].*

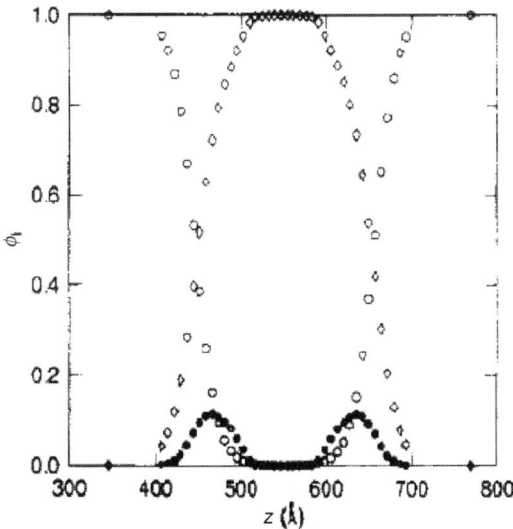

FIGURE 5.34 Volume fraction profiles of PS (open circles) d-PS (filled circles) and PMMA (open diamonds). *Reproduced by permission of Ref. [101].*

copolymers. These polymers have very different interactions with the substrate and can in one sense be regarded as large-scale amphiphiles. Annealing studies combined with selective deuteration led to the determination of the spatial variation in volume fraction for both the polystyrene and PMMA in these mixtures (Figure 5.34) and showed that the mixtures organized themselves into alternating lamellar microdomains of either polystyrene or polymethylmethacrylate.

Studies of polymers in their melt state are often useful in revealing the preferred structure of macromolecules. As NR provides information about the 1D spatial variation in scattering length density orthogonal to an interface, it is common practice to combine the information from other depth profiling techniques with NR. Clarke et al. [102] have used NR and He^3 nuclear reaction analysis (NRA) to measure adsorbed amounts and concentration profiles of carboxy-terminated polystyrene chains in the melt adsorbed at the native oxide surface of a silicon polymer interface. By following the changes in NRA as a function of annealing (Figure 5.35a) and correlating with the NR studies (Figure 5.35b), they showed that the carboxylate termination predominately locates at the oxide surface and by careful modeling using self-consistent field theory (SCFT) determined a sticking energy of $\sim$8.5 kT per molecule.

In similar manner, Nicolai et al. [103] have used NR to follow the change in conformation of physically end-adsorbed polystyrene chains in polystyrene melt matrices and compared their results to predictions of numerical SCFT. They varied the amount and MW of the adulterant polymer, deuterated

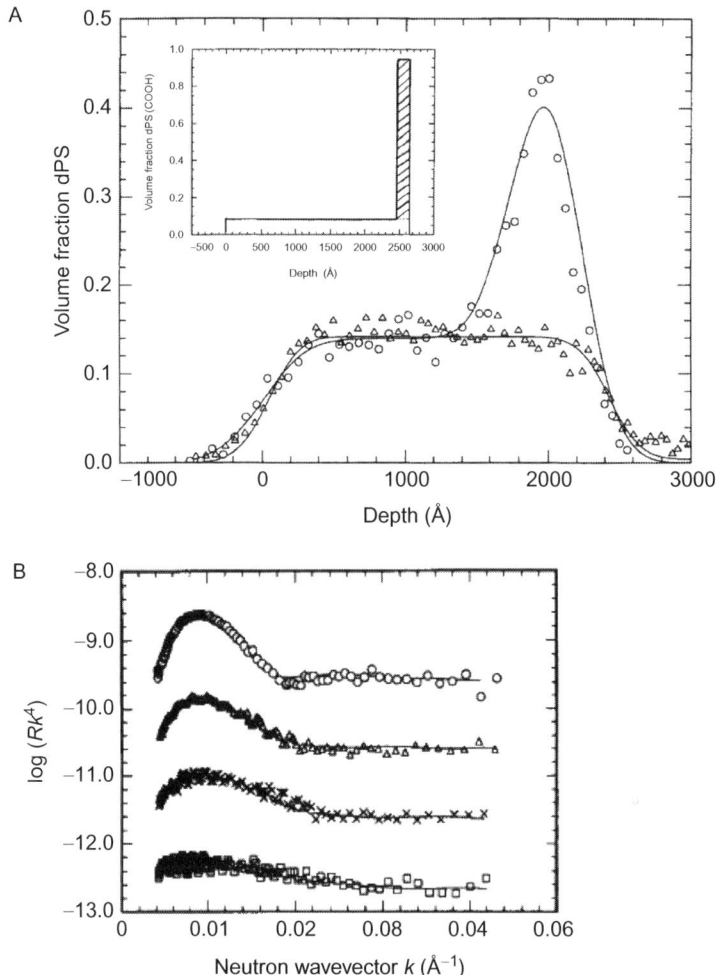

FIGURE 5.35 (a) NRA profiles obtained from a sample of 15% dPS(COOH) before annealing (triangles) and postannealing at 184 °C for 24 h (circles). The solid lines are fits to the data. The inset shows the model used in the fitting. The surface excess, z^*, is represented by the shaded area. (b) Neutron reflectivity data for samples annealed for 1 h (squares), 4 h (crosses), 24 h (triangles), and 7 days (circles). Data vertically shifted for clarity. *Reproduced by permission of Ref. [102].*

polystyrene containing the carboxy termination. Figure 5.36 shows the variation in interfacial excess (z^*) of carboxyl-terminated chains (normalized by the unperturbed brush chain radius of gyration (R_g)) as a function of bulk volume fraction. The solid and dotted lines are the predictions of SCFT for sticking energy of 6.3 and 5.7 kT per molecule, respectively. The experimentally derived interfacial excess data are consistent with their theoretical predictions

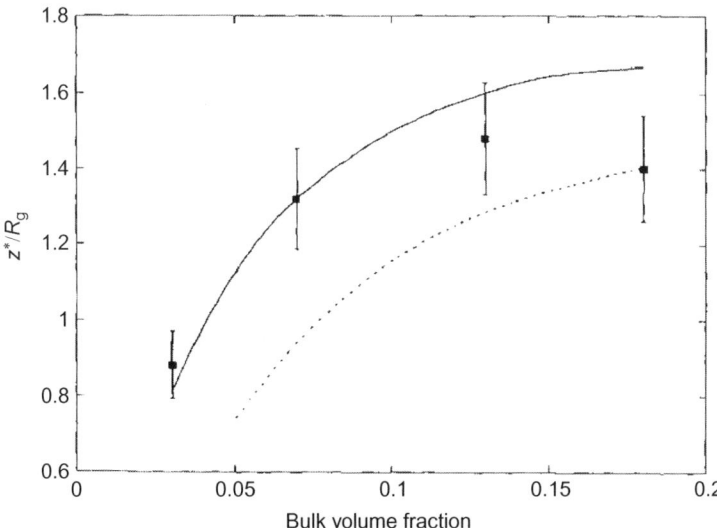

FIGURE 5.36 The variation in interfacial excess (z^*) of carboxyl-terminated chains (normalized by the unperturbed Brush chain R_g) as a function of bulk volume fraction. The solid and dotted lines are the predictions of their SCF theory for sticking energies of 6.3 and 5.7 kT per molecule, respectively. *Reproduced by permission of Ref. [103].*

with a sticking energy of ~6 kT per molecule. They concluded that the end-adsorbed chains formed weakly stretched brushes and the widths of the interface between free and brush chains are broadly in agreement with, but slightly less than, the widths predicted by theory.

Recently, there has been considerable interest in the structure and composition of mixed polymer films particularly for photovoltaic applications [104,105]. Lu et al. [105] described the morphological characterization of a crystalline bulk heterojunction structure in a blend. Blends of a low-bandgap silole-containing conjugated polymer, poly[(4,4-bis(2-ethylhexyl)dithieno[3,2-*b*;2,3-*d*]silole)-2,6-diyl-alt-(4,7-bis(2-thienyl)-2,1,3-benzothiadiazole)–5,5-diyl] (PSBTBT) with [6,6]phenyl-C61-butyric acid methyl ester (PCBM), were investigated under different processing conditions. The surface morphologies and vertical segregation of the as-spun, preannealed, and postannealed films were studied by scanning force microscopy, contact angle measurements, X-ray photoelectron spectroscopy, near-edge X-ray absorption fine structure spectroscopy, dynamic secondary ion mass spectrometry, and NR. The NR results (Figure 5.37) showed that PSBTBT was enriched at the cathode interface in the as-spun films and thermal annealing increased the segregation of PSBTBT to the free surface, while thermal annealing after deposition of the cathode increased the PCBM concentration at the cathode interface. Grazing-incidence X-ray diffraction and SANS showed that the crystallization of PSBTBT and segregation of PCBM occurred during

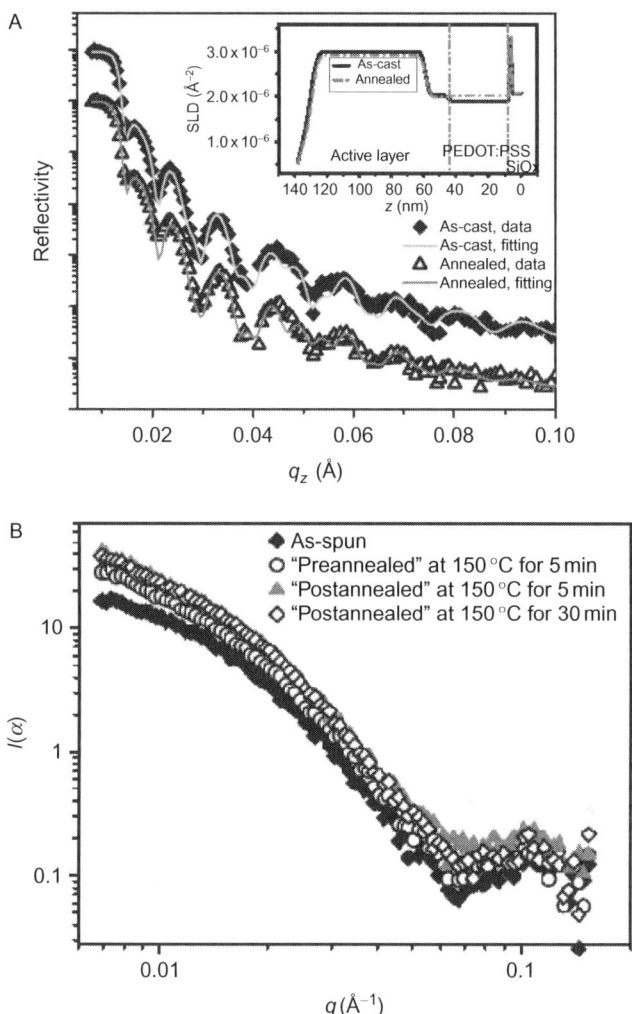

FIGURE 5.37 (a) Comparison of pre- and postannealed NR profiles and (b) corresponding SANS data. *Reproduced by permission of Ref. [105].*

spin coating, and thermal annealing increased the ordering of PSBTBT and enhanced the segregation of the PCBM, forming domains ~10 nm, and were subsequently correlated with an improvement in photovoltaic performance.

Polyelectrolytes are an important class of molecules and can be used as models for interactions with biopolymers and biomembranes, an area recently reviewed in detail by Kilbey and Anker [106]. NR provides direct information regarding the nuclear scattering length density variations normal to a surface, which can be directly related to the average segment density variation of a polymer film or grafted layer attached to that same interface. The studies have been essentially on two generic families of polyelectrolytes. The first referred to as "strong" polyelectrolytes which are sufficiently structured and charged so that their structure is essentially invariant with solution conditions and "weak" polyelectrolyte where the polymer conformation and hence adsorbed layer structure can be manipulated via changes in solution pH and electrolyte. Mir *et al.* [107] studied the structure of strong polyacids forming polystyrene sulfonate *in situ* on *a priori* grafted polystyrene layers. Their results were complicated by incomplete conversion to PSS and a resultant random block structure persisted throughout the network. Effort in this area has focused on improved methods of forming uniform charged layers usually based on grafting a preformed charged polymer to a surface [108]. Parnell *et al.* [109] have found a unique way to probe the changes in free and tethered segment density fluctuations in their studies of polymethylacrylic acid stretching by combining NR studies with high-resolution force spectroscopy.

The studies of polymers in solution and melts using SANS have developed substantially from the early pioneering studies [9] and now embrace more complex polymer morphologies and resulting structures, the impact of flow and the relationship between structure and bulk properties such as the rheology.

A relatively recent example of the complexity in structure and components is the study of the interaction of polyethylene glycol star polymers with the anionic surfactant SDS by Wesley *et al.* [110]. Using SANS and contrast variation, they were able to demonstrate how the binding of SDS micelles to the star polymer has a profound effect on the structure of the polymer, as shown in Figure 5.38.

It results in the star adopting and increasingly elongated conformation and is likely to have a significant impact upon the rheological properties of the solution. Habas *et al.* [111] probed the link between structure and rheological properties more specifically in their study of the nanostructure of PEO–PPO triblock copolymers. They demonstrated a richness in microstructure, from unimers to micelles, to body-centered cubic (bcc) and lamellar structures, dependent upon polymer concentration and composition. Furthermore, they showed how these structures evolved with flow. The study by Bent *et al.* [112] showed how SANS can be used to map polymer flow through a restriction. They used a fine neutron beam to scan the variation in the polymer scattering through the restriction. From the analysis of the anisotropy in the scattering, the SANS data provided a direct indication of the evolution in the polymer conformation with flow and spatial position.

The optical flow birefringence, shown in Figure 5.39, correlates with the neutron data and the calculated flow profile, to provide a direct link between the flow properties and the molecular properties of the melt.

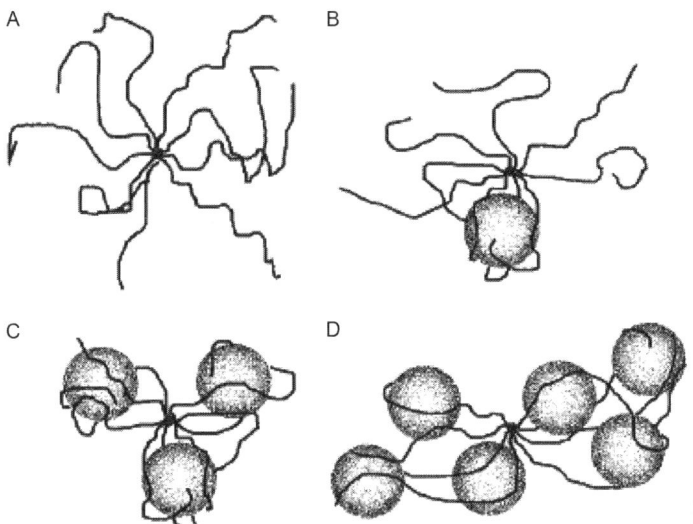

FIGURE 5.38 Schematic representation of the star polymer as a function of SDS concentration: (a) no surfactant, (b) above T_1, (c) at the cmc, and (d) at saturation. *Reproduced by permission of Ref. [110].*

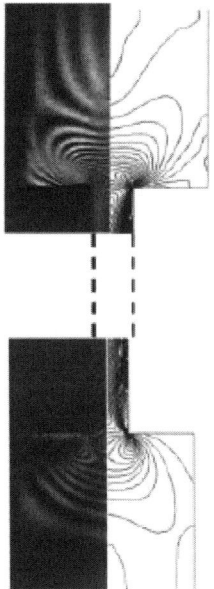

FIGURE 5.39 Experimental and predicted contours of principle stress difference at a flow rate of 4.7 mm s^{-1}. *Reproduced by permission of Ref. [112].*

5.3.8 Proteins and Biomacromolecules in Solution and at Interfaces

In recent years, an increasingly wide range of biomacromolecules have been studied in solution and at interfaces. These studies provide important information that relate directly to biological function, disease formation and prevention, and in more technological areas such as processed foods, and biofouling.

Drawing analogy with the previous section, proteins can be regarded as weak polyelectrolytes, but are also generally surface active. Lu [113] has recently reviewed the area of protein adsorption at interfaces using NR. More broadly than proteins, many biomacromolecules also have amphiphilic properties and are currently receiving much attention. It is here that the power of NR can be used to determine important physical properties of such large macromolecules without the complexity often associated with conventional techniques such as the value of the Gibbs prefactor in ST measurements.

A good example of the information that can be obtained is in the work of Su et al. [114,115], who studied the adsorption of lysozyme at the hydrophilic and hydrophobic silica–water interfaces. Su [114,115] showed that the adsorption of lysozyme was entirely reproducible but unidirectional, in that once adsorbed it did not desorb. At low concentrations, a 3-nm-thick monolayer was obtained, but at higher concentrations, the adsorbed layer structure was consistent with a bilayer of lysozyme with a thickness ~6 nm. Adsorption was unaffected by low levels of added electrolyte but became less at electrolyte concentrations in excess of 50 mM NaCl. pH had an effect which was completely reversible, but where lowering the pH to 4 from 7 resulted in a reduced adsorption of the protein due to enhanced electrostatic repulsion between adjacent lysozyme molecules.

The naturally occurring amphiphilic protein hydrophobin HFBII is another good example and has recently been shown to adsorb strongly at the air–water interface to form a close-packed monolayer approximately 30 Å thick with a mean limiting area per molecule of 400 Å^2 [116]. The adsorption is approximately constant at concentrations in excess of 0.01 g dm^{-3} and is invariant with solution pH. Zhang et al. [116] produced an adsorption isotherm (see Figure 5.40) and also observed that at low concentrations, a period of several hours was needed for the surface to come to equilibrium. It was interesting to note that the hydrophobin used was in the natural state without deuterium labeling as the molecule has an MW of approximately 7000 Da and the scattering length density is dominated in this instance by the carbon and oxygen content.

Zhang et al. [116] also reported the competitive adsorption of HFBII with the cationic, anionic, and nonionic surfactants: $C_{16}TAB$, SDS, and $C_{12}E_6$. In competition with the conventional surfactants $C_{16}TAB$, SDS, and $C_{12}E_6$, the HFBII adsorption totally dominates the surface for surfactant concentrations less than the critical micelle concentration, cmc. Above the cmc of the

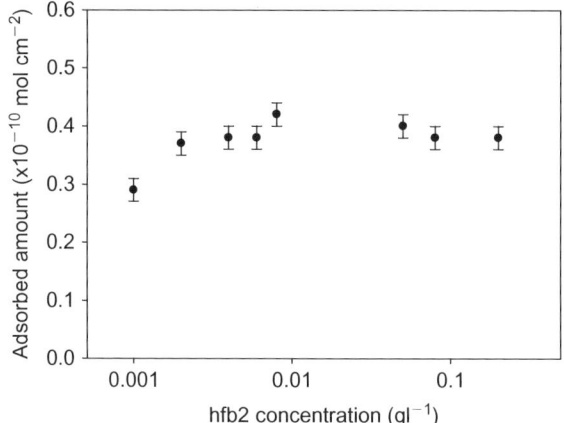

FIGURE 5.40 Adsorption isotherm for hydrophobin at the air–water interface. *Reproduced by permission of Ref. [116].*

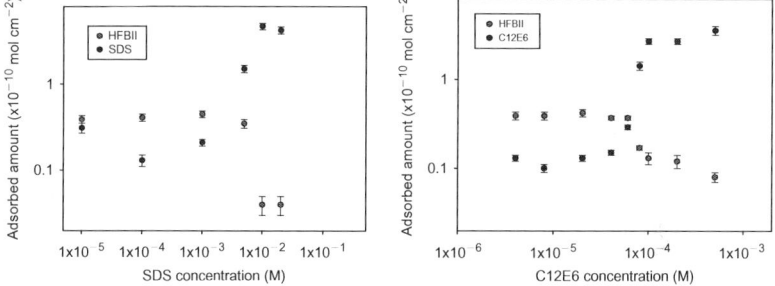

FIGURE 5.41 Variation in surface composition with surfactant concentration for (a) hydrophobin/SDS mixtures and (b) hydrophobin/$C_{12}E_6$ mixtures. *Reproduced by permission of Ref. [116].* (For color version of this figure, the reader is referred to the online version of this chapter.)

conventional surfactants, the hydrophobin is displaced by the surfactant ($C_{16}TAB$, SDS, or $C_{12}E_6$). For $C_{12}E_6$, this displacement is only partial and some hydrophobin remains at the surface for concentrations greater than the $C_{12}E_6$ cmc, as shown in Figure 5.41.

There was some variation in the pattern of adsorption at low pH for hydrophobin/SDS and hydrophobin/$C_{12}E_6$. At concentrations just below the surfactant cmc, there is now mixed hydrophobin/surfactant adsorption for both the SDS and $C_{12}E_6$. For the hydrophobin/SDS mixture, the structure of the adsorbed layer is more complex in the region immediately below the SDS cmc, resulting from the hydrophobin/SDS complex formation at the interface. Complementary studies using SANS confirmed the formation of mixed

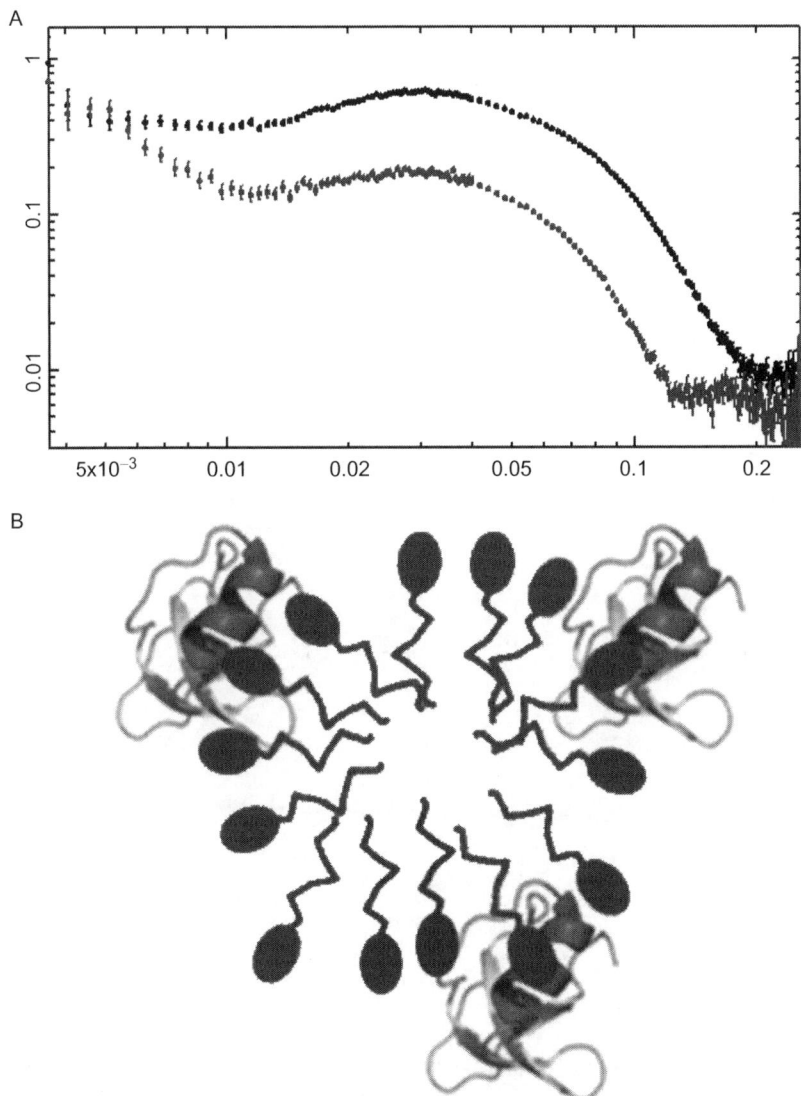

FIGURE 5.42 Scattering cross section versus wave-vector transfer for 5 mg/ml hydrophobin/ 5 mM h-C_{16}TAB (black) and hydrophobin/d-C_{16}TAB (blue). *Reproduced by permission of Ref. [117].* (For interpretation of the references to color in this figure legend, the reader is referred to the online version of this chapter.)

surfactant aggregates at concentrations in excess of the surfactant cmc, and the use of selective deuteration and a relatively simple core+shell model enabled the determination of the solution structure of the mixed aggregates [117]. Figure 5.42a shows the SANS data for hh and dh C_{16}TAB/hydrophobin mixtures at a concentration of 5 mg/ml hydrophobin and 5 mM surfactant, and in Figure 5.42, the solution aggregate structure derived is presented.

Many biomacromolecules derived from human sources are also frequently studied. The scattering studies frequently provide the crucial link between structure and function. Perkins [118] has studied the solution structure of the human blood clotting enzyme, factor VIIa (FVIIa). FVIIa is a soluble four-domain plasma serine protease coagulation factor that forms a tight complex with the two extracellular domains of the transmembrane protein tissue factor in the initiating step of blood coagulation. X-ray and neutron scattering data in solution for free FVIIa and the complex between FVIIa and soluble tissue factor (sTF) had been obtained for comparison with crystal structures of the FVIIa–sTF complex and of free factor IXa (FIXa). The solution structure of free FVIIa as derived from scattering data was shown to be consistent with the extended domain arrangement of FVIIa seen in the crystal structure of its complex with sTF, but not consistent with the bent, less extended domain conformation seen in the FIXa crystal structure. Analytical ultracentrifugation data and the modeling of these data were consistent with the scattering results. It is concluded that in solution, FVIIa has an extended or elongated domain structure, which allows rapid interaction with sTF over a large surface area to form a high-affinity complex.

Recently, Perkins *et al.* [119,120] have also determined the self-assembled microstructure of the mucosal protein Immunoglobulin A (IgA) [119,120]. Secretory IgA (SIgA) is the principal antibody for immune defense. This is formed from a dimer of IgA in complex with the secretory component (SC). Its heavy glycosylation precludes a detailed crystallographic study.

By a process of conversion of the molecular structure to Debye spheres on a preset lattice [121], the IFT approach becomes more computationally viable, and in Figure 5.43, a comparison is made between the different protein scattering data and the constrained fits for four antibody-related structures for serum human IgA1 (Figure 5.43a), recombinant human IgA2m (Figure 5.43b), human myeloma IgD (Figure 5.43c), and recombinant human SC (Figure 5.43d). Using a similar approach, a constrained modeling of neutron and X-ray scattering data, Perkins located the SC relative to dimeric IgA within SIgA. First, the solution structures for the individual monomers and dimers of IgA as well as for the free SC were determined, followed by mixtures. Figure 5.44 shows that the IgA dimer adopts an extended solution structure of two monomers connected end to end, and the arrow indicates that the SC is most probably located along the outermost convex edge of the central region in the IgA dimmer. This shows that it is able to bind to FcαRI receptors and offers protection against bacterial proteolytic action, and the formation and transportation of SIgA into mucosa.

The constrained and combined modeling of SANS and SAXS data, constrained by fragment crystallographic data and data from ultracentrifugation, and the use of the IFT method provide a sophisticated approach to unraveling

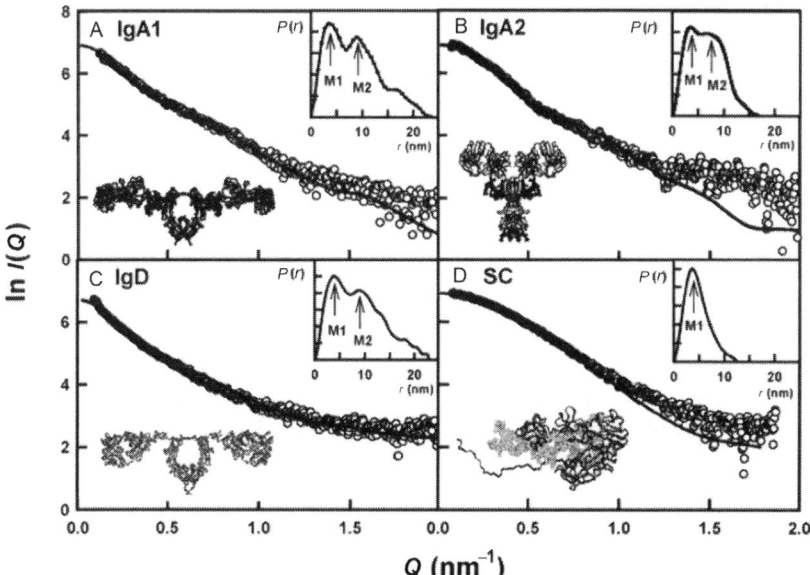

FIGURE 5.43 Constrained scattering curve $I(Q)$ fits for four antibody-related structures. Each model fit is shown as a black line and compared with the experimentally observed X-ray curve (circles). A ribbon view of the model used for the fit is shown (lower left) and the inset shows the modeled $P(r)$ curve with one or two maxima M1 and M2. (a) Serum human IgA1; (b) recombinant human IgA2m(1); (c) human myeloma IgD; and (d) recombinant human SC. *Reproduced by permission of Ref. [119].*

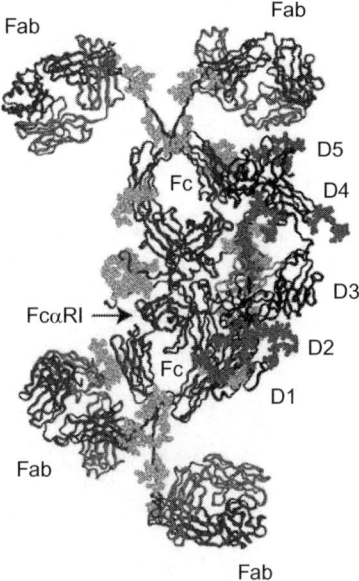

FIGURE 5.44 Model of the dimer structure of IgA combined with the secretory component protein. The arrow shows the location of the latter component in the structure. *Reproduced by permission of Ref. [121].*

the structure of these complex macromolecules. It was pioneered for such systems, but is not yet widely used in the simpler amphiphilic self-assembly systems discussed earlier where simpler constrained modeling is often adequate.

Understanding the structure–function relationship in this instance will aid in understanding and developing effective medicines and drugs that can preserve, protect, and repair human body mucus membranes.

5.4 SUMMARY AND FUTURE PROSPECTS

The basic principles of NR and SANS, in the context of their application to large-scale structures, have been described. With a wide range of current or relatively recent examples in soft matter, biomaterials, and nanomaterials, the importance and key contributions of the techniques in this field are demonstrated. In particular the ability to manipulate contrast remains a powerful advantage over other surface and solution characterisation methods.

The current and emerging generation of neutron sources and neutron instrumentation provide exciting opportunities to extend the studies in a number of key areas. These are notably in the study of more complex multi-component systems, and in some cases there is a reduced reliance on deuteration to enhance the signal. There is now also the real opportunity to probe kinetic processes, both at interfaces and in solution, on relevant timescales. The increased fluxes available open up the accessibility to a wider range of interfaces, different solid–liquid interfaces, and the liquid–liquid interface on a routine basis. The developments in instrumentation are being accompanied by developments in data analysis and modeling methods, and these are key aspects for future exploitation.

In this chapter, we have sought to illustrate the power of the techniques by recourse to a limited selection of scientific examples. However, the application of NR and SANS in the field of large-scale structures is much broader and is a growing and vibrant area of research with a rich future. This is especially true as it encompasses the key areas of nanotechnology, advanced materials, health care, and the environment.

REFERENCES

[1] Sivia DS. Elementary scattering theory for X-ray and neutron users. Oxford: Oxford University Press; 2011.
[2] Penfold J, Thomas RK. J Phys Condens Matter 1990;2:1368–412.
[3] Russell TP. Mater Sci Rep 1990;5:171–271.
[4] Lekner J. Theory of reflection. Dordrecht: Martinus Nijhoff; 1987.
[5] Guinier A, Fournet G. Small angle scattering of X-rays. New York: Wiley; 1955.
[6] Kostorz G, editor. Neutron scattering treatise on Materials Science and Technology, vol. 15. New York: Academic Press; 1982.
[7] Lindner P, Zemb Th, editors. Neutron, X-ray and light scattering. Amsterdam: North Holland; 1996.

[8] Penfold J. In: Hubbard AT, editor. Neutron scattering studies of micellar structure. Encyclopedia of surface and colloid science, vol. 4. New York: Marcel Dekker; 2002.
[9] Richards RW. J Macromol Sci Chem 1989;A26:757–815.
[10] Perkins SJ. X-ray and neutron scattering. In: Neuberger A, van Deenen LLM, editors. Modern physical methods in biochemistry. Part B. Amsterdam: Elsevier; 1988 [chapter 6].
[11] Heavens OS. Optical properties of thin films. London: Dover; 1955.
[12] Lu JR, Thomas RK, Penfold J. Adv Coll Int Sci 2000;84:143–304.
[13] Hayter JB, Penfold J. Coll Polym Sci 1983;261:1022–30.
[14] Ashcroft NW, Lekner J. Phys Rev 1966;145:83–90.
[15] Hayter JB, Penfold J. Mol Phys 1981;42:109–18.
[16] Hansen JP, Hayter JB. Mol Phys 1982;46:651–6.
[17] Hayter JB. Determination of the structure and dynamics of micellar solutions by SANS. In: Proceedings of Enrico Fermi school of physics, Varenna, Italy, North Holland, Amsterdam; 1983.
[18] Schultz GV. Z Phys Chem 1935;43:25–9.
[19] Penfold J, Richardson RM, Zarbakhsh A, Webster JRP, Bucknall D, Rennie AR, et al. J Chem Soc Faraday Trans 1997;93:3899–917.
[20] Cubitt R, Fragneto G. Appl Phys 2002;A74(Suppl. 1):5329–31.
[21] Penfold J, Ward RC, Williams WG. J Phys E 1987;20:1411–8.
[22] INTER reflectometer at the ISIS facility, http://www.isis.stfc.ac.uk/instruments/SURF/ [last accessed March 2103].
[23] FIGARO reflectometer at the ILL, http://www.ill.eu/instruments-support/instruments-groups/instruments/figaro/ [last accessed March 3013].
[24] Berk NF, Majkrzak CF. Langmuir 2009;25:4132–44.
[25] Schmatz W, Schelten J. J Appl Cryst 1971;4:410–2.
[26] Schmatz W, Springer T, Schelten J, Ibel K. J Appl Cryst 1974;7:96–116.
[27] Lieutenant K, Lindner P, Gahler R. J Appl Cryst 2007;40:1056–63.
[28] Heenan RK, King SM, Turner D, Treadgold JR. Proc ICANS-XVII 2006;1:780–8.
[29] http://www.ill.eu/instruments-support/instruments-groups/instruments/d33/, [last accessed March 2013].
[30] Cotton JP. In: Lindner P, Zemb T, editors. Neutrons, X-rays and light scattering. Amsterdam: North Holland; 1991.
[31] Horbay F, Hecht AM, Mallam S, Geissler E, Rennie AR. Macromolecules 1991;24:2896–902.
[32] Wignall GD, Bates FG. J Appl Cryst 1987;20:28–40.
[33] Higgins JS, Benoit H. Polymers and neutron scattering. Oxford: Oxford University Press; 1994.
[34] Jerke G, Pedersen JS, Egelhaaf SU, Schurtenberger P. Langmuir 1999;14:6013–24.
[35] Pedersen JS. Curr Opin Coll Int Sci 1999;4:190–6.
[36] Glatter O. J Appl Cryst 1977;10:415–21.
[37] Glatter O. J Appl Cryst 1980;13:577–84.
[38] Chattaraj DK, Birdi KS. Adsorption and the Gibbs surface excess. New York: Plenum Publishing; 1984.
[39] Xu H, Li P, Ma K, Thomas RK. Penfold J. Langmuir: Lu JR; 2013. http://dx.doi.org/10.1021/la401835d.
[40] Smyth TJP, Perfume A, Marchant R, Chen ML, Thomas RK, Penfold J, et al. Appl Microbiol Biotechnol 2010;87:1347–54.
[41] Chen ML, Penfold J, Thomas RK, Smyth TJP, Perfume A, Marchant R, et al. Langmuir 2010;26:18281–92.

[42] Chen ML, Penfold J, Thomas RK, Smyth TJP, Perfume A, Marchant R, et al. Langmuir 2010;26:17958–68.
[43] Holland PM. Coll Surf 1986;19:171–83.
[44] Penfold J, Thomas RK. Annu Rep Prog Chem C 2010;106:14–35.
[45] Penfold J, Staples E, Thompson L, Tucker I. Coll Surf A 1995;102:127–32.
[46] Alargova RG, Petkov JT, Petsev DN. J Coll Int Sci 2003;261:1–11.
[47] Petkov JT, Tucker IM, Penfold J, Thomas RK, Petsev DN, Dong CC, et al. Langmuir 2010;26:16699–709.
[48] Penfold J, Thomas RK, Dong CC, Tucker I, Metcalfe K, Golding S. Langmuir 2007;23:10140–9.
[49] Taylor DJF, Thomas RK, Penfold J. Adv Coll Int Sci 2007;132:69–110.
[50] Penfold J, Tucker I, Thomas RK, Zhang J. Langmuir 2005;21:10061–73.
[51] Zhang XL, Taylor DJF, Thomas RK, Penfold J. Langmuir 2011;27:2601–12.
[52] Lee EM, Thomas RK, Cummins PG, Staples EJ, Penfold J. Chem Phys Lett 1989;162:196–202.
[53] Penfold J, Thomas RK. Probing the liquid-solid interface by neutron reflectivity. In: Imae I, editor. Advanced chemistry of monolayers and interfaces. Amsterdam: Elsevier; 2007 [chapter 1].
[54] Penfold J, Tucker I, Petkov J, Thomas RK. Langmuir 2007;23:8357–64.
[55] Tucker I, Petkov J, Penfold J, Thomas RK. Langmuir 2010;26:8036–48.
[56] Tucker I, Petkov J, Penfold J, Thomas RK. Langmuir 2012;28:10773–80.
[57] Tucker I, Petkov J, Penfold J, Thomas RK. Langmuir 2012;28:10223–9.
[58] Mortitsen O. Life as a matter of fat. Berlin: Springer; 2005.
[59] Hollinshead CM, Harvey RD, Barlow DJ, Webster JRP, Hughes AV, Weston A, et al. Langmuir 2009;25:4070–7.
[60] Charitat T, Bellet-Amalric E, Fragneto G, Graner F. Eur Phys J B 1999;8:583–93.
[61] Hughes AV, Goldar A, Gerstenberg MC, Roser SJ, Bradshaw J. Phys Chem Chem Phys 2002;4:2371–8.
[62] Chen J, Kohler T, Gutberlet T, Mohwald H, Krastev R. Soft Matter 2009;5:228–33.
[63] McGillivray DJ, Valincius G, Heinrich F, Robertson JWF, Vanderah DJ, Febo-Ayala W, et al. Biophys J 2009;96:1547–53.
[64] Fragneto G, Charitat T, Daillant J. Eur Biophys J 2012;41:863–74.
[65] Fragneto G, Chariatat T, Bellet-Amalric E, Cubitt R, Graner F. Langmuir 2003;19:7695–702.
[66] Petrache HI, Tristam-Nagle S, Gawrisch K, Harries D, Parsegian VA, Nagle JF. Biophys J 2004;86:1574–86.
[67] Malaquin L, Charitat T, Daillant J. Eur Phys J E 2010;31:285–301.
[68] Stidder B, Fragneto G, Roser SJ. Langmuir 2005;21:9187–93.
[69] Stidder B, Fragneto G, Cubitt R, Hughes AV, Roser SJ. Langmuir 2005;21:8703–10.
[70] Shen H-H, Thomas RK, Penfold J, Fragneto G. Langmuir 2010;26:7334–42.
[71] Wacklin HP. Curr Opin Coll Int Sci 2010;15:445–54.
[72] Gruen DWR. Prog Coll Polym Sci 1985;70:6–16.
[73] Bradbury R, Penfold J, Thomas RK, Tucker IM, Petkov JT, Jones C, Grillo I. Langmuir 2013;29:3234–45.
[74] Hayter JB. Langmuir 1992;8:2873–6.
[75] Tucker I, Penfold J, Thomas RK, Bradbuty R, Grillo I. Langmuir 2009;25:4934–44.
[76] Penfold J, Staples E, Thompson L, Tucker I, Hines J, Thomas RK, et al. Langmuir 1995;11:2496–502.

[77] Penfold J, Staples E, Thompson L, Tucker I, Hines J, Thomas RK, et al. J Phys Chem B 1999;103:5204–14.
[78] Israelachivili J, Mitchel DJ, Ninham BW. J Chem Soc Faraday Trans 2 1976;72:1525–68.
[79] Tucker I, Penfold J, Thomas RK, Grillo I, Barker JG, Milner DFR. Langmuir 2008;24:7674–87.
[80] Nallet F, Laversanne R, Roux D. J Phys II 1993;3:487–503.
[81] Grillo I, Cousin F, Penfold J, Tucker I. Langmuir 2009;25:3932–43.
[82] Butler P. Curr Opin Coll Int Sci 1999;4:214–21.
[83] Cummins PG, Staples E, Millen B, Penfold J. J Meas Sci Technol 1990;1:179–83.
[84] Penfold J, Staples E, Tucker I, Carroll P, Clayton J, Cowan JS, et al. J Phys Chem B 2006;110:1073–82.
[85] Hayter JB, Penfold J. J Phys Chem 1984;88:4589–93.
[86] Penfold J, Staples E, Tucker I, Tiddy GT. J Appl Cryst 1997;30:744–52.
[87] Soubiran L, Staples E, Tucker I, Penfold J, Creeth A. Langmuir 2001;17:7988–94.
[88] Penfold J, Tucker I. J Phys Chem B 2007;111:9496–503.
[89] Ottewill RH, Rochester CH, Smith AL, editors. Adsorption from solution. London: Plenum Press; 1987.
[90] Cosgrove T, Crowley TL, Ryan K, Webster JRP. Coll Surf 1990;51:255–69.
[91] Cummins PG, Staples E, Penfold J. J Phys Chem 1990;94:3740–5.
[92] Cummins PG, Staples E, Penfold J. J Phys Chem 1991;95:5902–5.
[93] Despert G, Oberdisse J. Langmuir 2003;19:7604.
[94] Lugo D, Oberdisse J, Karg M, Schweins R, Findenegg GH. Soft Matter 2009;5:2928–36.
[95] Staples E, Penfold J, Tucker I. J Phys Chem B 2000;104:606–14.
[96] Penfold J, Tucker I, Petkov J, Jones C, Thomas RK, Rogers S, et al. Langmuir 2013;28:15474–981.
[97] Flory PJ. J Chem Phys 1941;9:660–1.
[98] Helfand E, Sapse AM. J Chem Phys 1975;62:1327–32.
[99] Bucknall D, Butler SA, Higgins JS. J Phys Chem Sol 1999;60:1273–7.
[100] Kanaya T, Miyazaki T, Watanabe H, Nishida K, Yamano H, Tasaki S, et al. Polymer 2003;44:3769–73.
[101] Mayes AM, Russell TP, Deline VR. Macromolecules 1994;27:7447–53.
[102] Clarke CJ, Jones RAL, Edwards JL, Clough AS, Penfold J. Polymer 1994;35:4065–72.
[103] Nicolai T, Clarke CJ, Jones RAL, Penfold J. Coll Surf A 1994;86:155–63.
[104] Staniec PA, Parnell AJ, Dunbar ADF, Yi H, Pearson AJ, Wang T, et al. Adv Energy Mater 2011;1:499–504.
[105] Lu H, Akgun B, Russell TP. Adv Energy Mater 2011;1:870–8.
[106] Kilbey II SM, Anker JF. Curr Opin Coll Int Sci 2012;17:83–9.
[107] Mir Y, Auroy P, Auvray L. Phys Rev Lett 1995;75:2863–6.
[108] Muller F, Romet-Lemonne G, Delsanti M, Mays JW, Daillant J, Geunoun P. J Phys Condens Matter 2005;17:S3355–S3361.
[109] Parnell AJ, Martin SJ, Dang CC, Geoghegan M, Jones RAL, Crook CJ. Polymer 2009;50:1005–14.
[110] Wesley RD, Cosgrove T, Thompson L. Langmuir 1999;15:8376–82.
[111] Habas JP, Pavie E, Perreur C, Lapp A, Peyrelasse J. Phys Rev E 2004;90:061802.
[112] Bent J, Hutchings LR, Richards RW, Gough T, Spares R, Coates PD, et al. Science 2003;301:1691–5.
[113] Lu JR, Zhao X, Yassen M. Curr Opin Coll Int Sci 2009;12:9–16.
[114] Su TJ, Lu JR, Thomas RK, Cui ZF, Penfold J. J Coll Int Sci 1998;203:419–29.

[115] Lu JR, Su TJ, Thirtle PN, Thomas RK, Rennie AR, Cubitt R. J Coll Int Sci 1998;206:212–23.
[116] Zhang XL, Penfold J, Thomas RK, Tucker IM, Petkov JT, Bent J, et al. Langmuir 2011;27:11316–23.
[117] Zhang XL, Penfold J, Thomas RK, Tucker IM, Petkov JT, Bent J, et al. Langmuir 2011;27:10514–22.
[118] Perkins SJ, Nealis SA, Sim RB. Biochemistry 1990;2:1167–75.
[119] Perkins SJ. The Biochemist 2009;31:32–5.
[120] Perkins SJ, Okemefuna AI, Nan R, Li K, Bonner KA. J R Soc Interf 2009;6:S679–S696.
[121] Perkins SJ, Bonner A. Biochem Soc Trans 2008;36:37–42.

Chapter 6

Dynamics of Atoms and Molecules

Mark R. Johnson* and Gordon J. Kearley[†]
*Institute Laue Langevin, Grenoble, France
[†]The Bragg Institute, Australian Nuclear Science and Technology Institute, Kirrawee DC, New South Wales, Australia

Chapter Outline
6.1. Introduction 416
6.2. Brief Review of Theoretical Concepts 418
6.3. Modeling 419
   6.3.1. Mapping Potential Energy Surfaces 419
   6.3.2. Molecular Dynamics Simulation 420
   6.3.3. Empirical and *Ab Initio* Energy Calculation 420
6.4. Instrumentation 422
   6.4.1. Three-Axis Spectrometers 422
   6.4.2. Time of Flight 422
   6.4.3. Neutron Compton Scattering Spectrometers 423
   6.4.4. Molecular Spectrometers 424
   6.4.5. Backscattering Spectrometers 425
   6.4.6. Neutron Spin-echo Instruments 426
   6.4.7. The Measured Neutron-Scattering Signal 427
6.5. Oscillatory Motion, Incoherent Scattering 427
   6.5.1. Molecular Vibrations of Benzene 429
   6.5.2. Hydrogen-Bonded Systems 430
   6.5.3. Complex Hydrides 431
   6.5.4. Polymers 433
6.6. Oscillatory Motion, Coherent Scattering 434
   6.6.1. Classic Phonons and Soft Modes in $SrTiO_3$ 435
   6.6.2. Negative Thermal Expansion 435
   6.6.3. Nanostructured Materials 438
   6.6.4. Oxygen-Ion Conductors—Brownmillerites 439
   6.6.5. Thermoelectrics—Skutterudites 441

6.6.6. Pnictides	442	6.8.3. Coherent QENS, Rotation	456
6.6.7. Strontium Gallium Oxides	443	6.8.4. Dynamical Transitions from Elastic Scans	457
6.6.8. Deoxyribonucleic acid	445		
6.7. Tunneling	**446**	6.8.5. Diffusion of Coherent Scatterer CO_2	459
6.7.1. Rotational Tunneling	446		
6.7.2. Translational Tunneling	453	6.8.6. Water and Complex Diffusion	460
6.8. Stochastic Relaxation/ Dynamics	**453**	6.8.7. Ionic Liquids	463
6.8.1. Complex Diffusion	455	**6.9. Conclusion and Perspectives**	**464**
6.8.2. Ligand Water Rotation	456	**References**	**466**

6.1 INTRODUCTION

Understanding of the link between structure, dynamics, and function is of central importance in the development of both smart and sustainable materials that are becoming commonplace in neutron-scattering studies and in our everyday lives. While neutron scattering plays an obvious role in structure determination, interactions at the atomic and molecular scale are often the key to understanding not only the materials' mechanical properties but also their electrical and magnetic properties, and these are best probed via the dynamics at an atomistic level. Although there are many techniques capable of measuring these dynamics, the complementary techniques of inelastic (INS) and quasielastic neutron scattering (QENS) can give key insight. Many of the generic strengths of neutron scattering are similar across the entire range of neutron techniques, but of particular importance to this chapter are sensitivity to H-atom motion, the simultaneous spatial and temporal information that is available for coherent (collective) and incoherent (local) scattering processes, and finally, the straightforward connection to atomistic simulations.

This is different from other spectroscopies that have very limited potential for measuring geometrical, phase, and spatial aspects of the dynamics. We first consider the timescale of the dynamics that are accessible to the neutron-scattering technique, and for clarity, we distinguish between kinetics and dynamics as "kinetics" for the macroscopic scale of, for example, a phase transition or chemical reaction, while we use "dynamics" for motion at the atomic or molecular scale. In this chapter, we are concerned entirely with dynamics, although this may of course be used to follow a kinetic process on the macroscopic scale. Faster dynamics corresponds to larger energy transfer to/from the neutron, which is in

Chapter | 6 Dynamics of Atoms and Molecules   417

general easier to measure up to a few tens of femtoseconds, while the current limit for slow dynamics of a few hundred nanoseconds is achieved by neutron spin-echo (NSE). Compton scattering (or deep inelastic scattering) achieves even higher energy transfers, and we will only discuss this briefly here.

The overwhelming majority of dynamical studies made with neutron scattering are in the condensed phases or in species close to a surface, which means that interactions will be occurring over a variety of length scales. Neutrons are sensitive to both spatial and temporal (equivalently energy) aspects of dynamics and we need to recognize the regions of the space–time surface that are accessible to neutron spectrometers. In practice, small energy changes originate from weak interactions that tend to be long range, and it follows that low energy transfer and low momentum transfer often go together so that in Figure 6.1, we concentrate on the scientific areas that can be studied in the accessible time/energy regions.

In complex systems that characterize much of current neutron scattering, there are different dynamical processes occurring on different time and length scales, for example, those discussed in Section 6.8.6, and it is essential that we are able to attribute the measured signal to the correct dynamical process. Here, the interplay between the energy (inverse temporal) and momentum transfer (inverse spatial) ranges, which is almost unique to neutron scattering, provides vital information for deciphering the underlying dynamics, particularly with the help of molecular modeling methods.

The examples chosen to illustrate this chapter generally come from, or are related to, the field of smart or functional materials, as this is a major growth

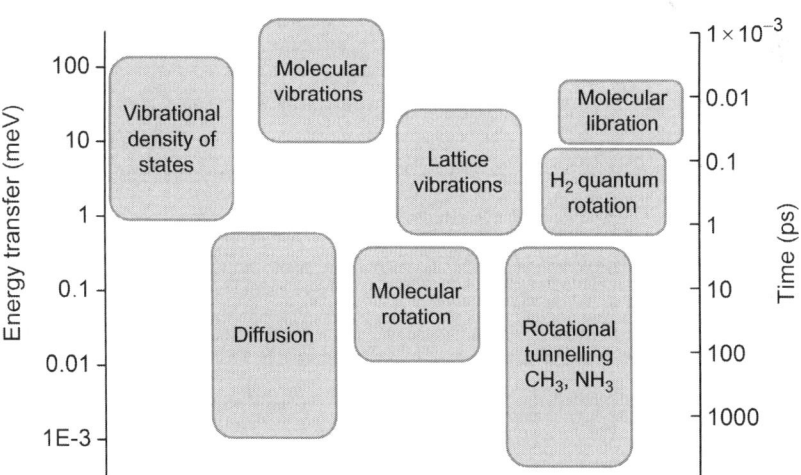

FIGURE 6.1 Types of dynamics that can be measured over different time and energy ranges with instrument types. As a general rule, times shorter than ~50 ps are accessed using time-of-flight methods, and longer times by backscattering and eventually spin-echo. (For color version of this figure, the reader is referred to the online version of this chapter.)

area not only for neutron scattering but also for X-ray synchrotron applications [1]. A few examples have been selected for completeness or academic interest, but these serve to illustrate a concept, like the idea of single-molecule calculations as the starting point to study solid-state effects in the case of benzene (Section 6.5.1). Together, the examples also demonstrate what can be learnt from samples and systems in different forms: crystals, fibers, powders, liquids, surfaces, molecules, polymers (including proteins), inorganics, etc. These examples originate mainly from the authors work or that of their colleagues, which we hope will provide a broad overview. The full scientific context of these examples can be gleaned from the references at the end of the chapter. The keen reader is also encouraged to read reference books [2,3] and reviews on inelastic and quasielastic scattering to study the dynamics of atoms and molecules to gain a fuller perspective on this aspect of neutron scattering.

6.2 BRIEF REVIEW OF THEORETICAL CONCEPTS

The foundations of neutron scattering theory is given in the first chapter, so we will concentrate here only on those aspects that are required to understand what can be extracted from an INS or QENS experiment. We will do this by developing the function that describes how neutrons "follow" the way in which atoms explore accessible space within the sample at a given temperature. It is particularly relevant that atomistic modeling provides exactly the same function furnishing a convenient link between modeling, theory, and experiment.

Conceptually, INS or QENS measures the decay (or variation) of correlations of atomic position with time, which is the van Hove correlation function [4], $G(\mathbf{r},t)$. We will develop this from the basic case through to its formal generalization. In favorable cases, such as diffusion of water, all H-atoms are equivalent and their incoherent scattering dominates the measured signal so that we need only consider $G(\mathbf{r},t)$ for a single atom. We would expect $G(\mathbf{r},t)$ to be at least approximately an exponential decay in time (t), with a rate constant that increases with r, although, as we will show in the examples, current studies are increasingly concerned with deviations from exponential decay. The vast majority of current experiments involving water investigate $G(\mathbf{r},t)$ for its dynamics in (or on) a variety of environments (on proteins, in pores, etc.). The first extension to the basic case is to accommodate several inequivalent incoherent scatters such as the H-atoms in hydrated proteins that we will cover in Section 6.8.6.3. The signals from the different H-atoms in water, on the protein back-bone, and terminal CH_3 groups give rise to the sum of their $G(\mathbf{r},t)$ weighted appropriately for the composition. We will deal with the problem of separating these components from the experimental signal later in this chapter, but clearly, these components of $G(\mathbf{r},t)$ are scaled by their scattering cross sections, so selective deuteration (or more generally isotopic substitution) can be very powerful in separating the partial contributions.

For incoherent scatterers, $G(\mathbf{r},t)$ is the evolution of atomic position with respect to the atoms' own original position, but where there is interference of the waves scattered coherently by different atoms $G(\mathbf{r},t)$ includes the cross-terms between these atoms scaled by their coherent scattering cross sections. The total $G(\mathbf{r},t)$ including all of these extensions, but without the scattering cross sections, then reads

$$G(\mathbf{r},t) = \frac{1}{N}\left\langle \sum_{j=1}^{N}\sum_{k=1}^{N} \delta\big(r + r_k(0) - r_j(t)\big) \right\rangle \qquad (6.1)$$

The quantities actually measured by the neutron scattering instruments are either the space or space and time, Fourier transforms of $G(\mathbf{r},t)$, and we will cover this formalism in the following sections.

6.3 MODELING

Atomistic modeling has become an integral part not only of the understanding of inelastic and quasielastic neutron experiments but also in the planning and the choice of instrument. An overview of the general modeling methods that are available can be found in Ref. [5] and specific examples of their use in Ref. [6]. The obvious connection between modeling and the experiment is the evolution of relative atomic position with time, embodied in $G(\mathbf{r},t)$, which in turn depends on how kinetic and potential energy interchange at the atomistic level. The potential energy of a given atomic configuration is the fundamental grain of modeling methods, and we will describe the methods for its determination later.

6.3.1 Mapping Potential Energy Surfaces

The "traditional" data-analysis approach is to use a model function that characterizes the dynamics and is parameterized according to a theoretical or intuitive prediction, with the best values of the parameters being obtained by fitting the model function to the experimental data. Much of the arbitrariness of this approach can be removed by using calculations to map the potential energy of the system via a set of conformations or structures that characterize the supposed dynamics. This potential energy expression can then be used in a Hamiltonian or dynamical matrix that characterizes the geometry and kinetic energy to obtain the calculated dynamics of the system from which the neutron spectrum can be calculated. A common example of this is in the calculation of phonons or molecular vibrations where the force constants that are adjusted (or assumed) in a traditional analysis are replaced by values obtained from successive atomic displacements and energy calculations. The vibrational dynamics are then determined by the normal lattice-dynamics calculations as illustrated by the examples of benzene and $SrTiO_3$ in Sections 6.5.1

and 6.6.1, respectively. Because a relatively small number of energy calculations are required, this mapping or matrix method has the advantage that the calculations can be made at very high precision.

6.3.2 Molecular Dynamics Simulation

A second method that makes even fewer assumptions relies on putting the requisite quantity of kinetic energy into the model and, by allowing the system to evolve under Newtonian dynamics, it follows the way in which the atoms explore their environment with time—$G(\mathbf{r},t)$. This method is the well-known molecular dynamics (MD) simulation, and in many cases, it gives the most complete information since we can easily store atomic positions and energies for as much of the temporal evolution of the system that we require. An example of the difference between assumed harmonic potentials and MD with no assumptions is given for hydride dynamics in Section 6.5.3. At each time step, the trajectories of the atoms are calculated from the forces acting upon them and the system will obviously diverge if the time between steps (recalculating new forces and trajectories) is too long. In practice, this step is ~ 1 fs which clearly limits the total time and number of degrees of freedom that can be used. A further limitation that is of importance for light atoms in low-energy processes is the neglect of the quantum aspects of the nuclear dynamics. In the section on tunneling, we show that a good approximation to the potential energy surface can be obtained and then treated quantum mechanically, but it is becoming increasingly attractive to use path integral MD to take account of quantum effects.

6.3.3 Empirical and *Ab Initio* Energy Calculation

The keystone of both mapping and MD methods is a calculation of the potential energy of a given atomic arrangement, and there are two main approaches for this calculation, as illustrated in Figure 6.2. Classically, the interactions between atoms are characterized by two-body terms: bond, van der Waals and Coulomb, with three- and four-body terms restricted to angular and dihedral coordinates. Expressions for these are parameterized and summed to obtain the total energy of a given atomic (or molecular) arrangement, with the parameters being tabulated for various atoms, fragments, and species. This method is known as empirical or force field and has the advantage of being rapid, allowing large systems to be studied over periods of many nanoseconds. The main limitations are that electron density and wavefunctions are absent so that the choice of parameters becomes crucial and it is difficult to handle distributed properties such as charge and polarizability and the limits of two-body interactions.

The second, more rigorous, approach is based on calculating the wavefunctions (collectively known as *ab initio*) that minimize the electronic energy

Chapter | 6 Dynamics of Atoms and Molecules

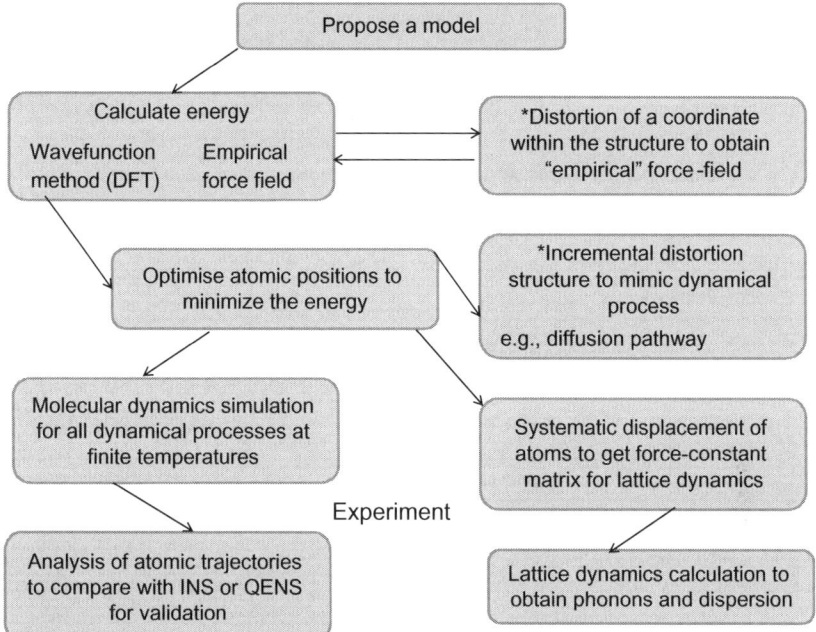

FIGURE 6.2 Schematic illustration of the common simulation strategies used to understand experimental neutron scattering data. The route on the left is basically for molecular dynamics, while the right is for the dynamical matrix approach. (For color version of this figure, the reader is referred to the online version of this chapter.)

of a given distribution of atoms, the most popular at present being density functional theory, DFT, a good introduction being available in Ref. [7]. These methods overcome most of the difficulties of parameterization (although there are choices that can seem arbitrary) and naturally deal with changes of interaction type such as bond breaking/making. They provide high-quality results and would normally be the method of choice wherever possible, but unfortunately, they require far more computer resources than empirical methods.

The choice of method is frequently a balance between the size and time limits of wavefunction methods and the approximations of a parameterized force field. These force fields are generated on the basis of types of atoms and bonds and will normally be optimized for a particular class of compound. Because the transferability of force-field parameters to a particular system may be subject to question, paradoxically, these parameters are often determined for specific systems by wavefunction methods, particularly where it is impractical to get these from experiment.

All combinations of matrix/MD methods with empirical/DFT methods are used in the neutron-scattering community depending on the nature of the system and the experiment to be analyzed, as illustrated in Figure 6.2. The most common strategy is to first ensure that the model can successfully reproduce any structural data by energy minimization, followed by the calculation of the dynamics. The dynamic results are then transformed into the neutron-scattering signal, including resolution functions, etc., to produce a calculated signal which can be compared with that measured. Good agreement is taken as validation of the model which is then used for a more complete analysis of the system's dynamics and other related properties such as electron–phonon coupling. In some cases (see the example in Section 6.8.7), the final step is to use the simulation results to develop an analytical model that is more amenable to theoretical use, closing the loop between experiment, simulation, and theory.

6.4 INSTRUMENTATION

More complete information on instrumentation is given in Chapter 3, and here we will only summarize particular aspects that are of relevance to this chapter. Although the quantity being measured in INS and QENS is well defined, the analysis of the experimental data is often quite challenging. Atomistic modeling can be of considerable help in attributing the observed peaks and profiles to the correct dynamical processes, but this is of little help if the experimental data themselves are flawed. Provided that backgrounds, calibrations, sample-shape effects, and multiple scattering are treated properly, neutron-scattering techniques can be used to obtain unique information about subtle phenomena and complex samples. A variety of instrumentations are available, and in this section, we give an overview of these characteristics and uses.

6.4.1 Three-Axis Spectrometers

Three-axis spectrometers (TASs) were the first neutron scattering spectrometers, being invented at Chalk River by Bertram Brockhouse who was Nobel Laureate in 1994 for this work. The basic technique allows, within kinematic limitations, complete, point-wise access to the reciprocal and energy space of a crystal. They are very flexible in use, but in practice, other types of instruments (below) are now better optimized, in the same energy-transfer range, for molecular spectroscopy and density-of-state measurements. The three axes (monochromator, sample, and analyser) allow the incident energy, scattering angle, and final energy to be chosen, while the orientation of the sample determines the accessible part of reciprocal space. This makes them ideal for lattice dynamics (see examples in Section 6.6.1), and although this was their stronghold for several decades, these instruments are now largely used for measurements of excitations in magnetic and strongly correlated electron systems that are described in Chapter 5.

6.4.2 Time of Flight

Neutron pulses can be produced naturally by spallation sources, or by a rotating device (usually a chopper) at a reactor source, providing a convenient means by which it is possible to analyze neutrons scattered by the sample over a wide angular range simultaneously. A good overview of general time of flight (TOF) is available in Ref. [8]. Initially, this was the method of choice for incoherent scattering, but the method has now evolved, particularly as a result of spallation neutron sources, to exploit advantages for single-crystal and coherent-scattering experiments.

Two basic methods are used. In the first, direct TOF, a monochromatic neutron pulse, is scattered by the sample and the energy of the scattered neutron is determined from its flight time from the sample to the detectors. Alternatively, a polychromatic neutron pulse is generated at several meters before the sample, where it is scattered to an array of constant-energy analysers before arriving at the detectors. In this case, the TOF provides the incident energy, while the analyser determines a single scattered energy, this technique being known as "inverted TOF" for obvious reasons. In direct and inverted instruments, the TOF spectra collected at different scattering angles can be mapped onto the momentum- and energy-transfer (Q,E) plane using slightly different algorithms.

The energy transfer E is the difference between the incident energy and the scattered neutron energy, and because the detectors cover a large angular range, the momentum transfer can also be measured quite accurately if required. This coverage of $\mathbf{Q},E$ space is required for a number of experiment types, but for some vibrational density-of-states (VDOS) experiments, the $\mathbf{Q}$-dependence is of secondary concern, the main advantage being the high resolution of these instruments at low energy transfer. In these cases, spectra from detectors at different scattering angles can generally be summed together to give higher count rates because the integral over $\mathbf{Q}$ does not cause any significant loss of information. In practice, the more straightforward generalized density of states (GDOS) is often used because it ignores the different scattering cross sections of the nuclei involved and the details of the inherently Q-dependent coherent scattering intensity by using the incoherent approximation [9]. Molecular vibrations, particularly those involving hydrogen and strongly bonded light atoms, tend to arise at energy transfers above ~ 30 meV and better optimized techniques described below are most efficient to study these. In contrast, where the system is less molecular in nature the vibrational energies will be lower, and in the absence of strong scattering due to H-atom displacement, TOF instruments described here are more appropriate.

The relative energy resolution generally improves with increasing neutron energy loss, while it degrades with increasing neutron energy gain. For this reason, neutron energy loss is the preferred choice, but sometimes, the desired energy transfer is close to, or exceeds, the incident neutron energy and neutron energy gain is used within the limits of the sample temperature.

6.4.3 Neutron Compton Scattering Spectrometers

This technique was originally intended for measuring the condensate fraction in superfluid ^{4}He [10], but has subsequently been used successfully for a wide range of studies (see Ref. [11] for review). This type of spectrometer is exemplified by VESUVIO at ISIS (UK), the current status, and instrumental details of which are given in Ref. [12]. This instrument has been used in the measurement of proton momentum distributions in materials such as water in different environments and hydrogen-bonded systems. The principle of operation is similar to inverted TOF described above, but because the neutron energies are so high (>1 eV), the final energy is determined by a sequence of gold-foil transmissions with a combination of neutron and γ detection in the back and forward scattering directions, respectively. At these high neutron energies, the interaction between the neutron and the struck atom is assumed to satisfy the impulse approximation so that the single-particle properties of the atom are probed. For an atom of mass M, this leads to a scattered neutron peak maximum at

$$E = \frac{\hbar^2 Q^2}{2M} \tag{6.2}$$

This peak is then broadened by the Doppler effect to a width that is proportional to the kinetic energy of the atom and with a shape that depends on the distribution of atomic momenta. In principle, this can be used to obtain the Born–Oppenheimer potential of the atom, which is particularly interesting in H-bonded systems [13].

6.4.4 Molecular Spectrometers

We use the term "molecular spectroscopy" in instrumentation to denote spectrometers that were conceived mainly for the study of molecular vibrations and with the defining characteristics of comparatively high energies, high flux, high energy resolution, and relaxed **Q**-resolution. Because molecular vibrations are normally considered to be undispersed (not always true), these instruments can trade a very coarse Q-resolution, and a fixed trajectory through $\mathbf{Q},\omega$ space (Figure 6.3), in exchange for higher count rates. These conditions are met in two different ways that allow us to introduce some of the elements of TAS (Section 6.4.1) and inverted TOF (Section 6.4.2). While a full TAS machine controls both the incoming and analyzed wavevectors, for molecular spectroscopy, we need only analyze the scattered energy, which enables us to integrate over the angular range of the scattered neutrons. At reactor sources, this is achieved by the "beryllium filter technique" that dates back to 1960 [14], and although the instruments have developed enormously, the basic technique is unchanged. The three-axis analyser system is replaced with a beryllium/graphite fixed-energy filter that covers a large angular range so that only neutrons with a specific energy will pass to the detectors. With the sample between the monochromator and the fixed-energy, its energy spectrum is obtained by recording the number of neutrons detected after the filter as a function of the scanned incident neutron energy from the monochromator. Examples of this

Chapter 6 Dynamics of Atoms and Molecules

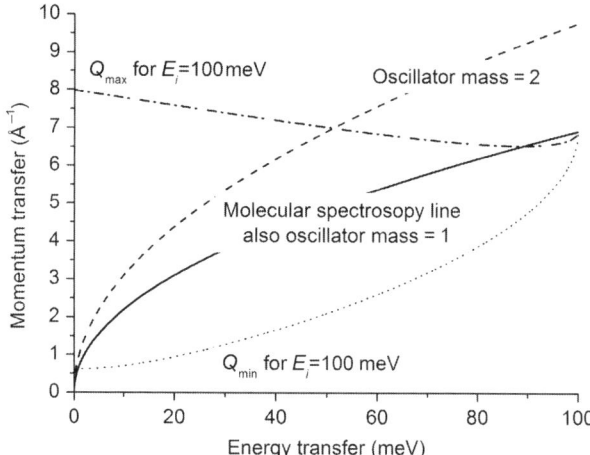

FIGURE 6.3 The solid line shows the cut that molecular spectrometers take through $\mathbf{Q},E$, this being dictated by the constant low final energy. This line coincides exactly with the maximum intensity for vibrational peaks from oscillator mass of 1.0, while the dashed line shows the corresponding maximum for mass of 2.0. The dotted and dot-dashed lines show the bounds of $\mathbf{Q},E$ for neutron energy loss on a TOF spectrometer with $E_i = 100$ meV.

type of instrument are IN1 (now Lagrange) at ILL (France), FANS at NIST (USA), and TAIPAN (with filter option) at ANSTO (Australia).

The inverted TOF approach for molecular spectroscopy arranges the analyser banks so that neutrons undergoing the same energy transfer, irrespective of the initial and final energies (within limits), arrive at the detector at the same time. This time focusing greatly increases the count rate without altering the energy resolution, but, in common with the filter instrument above, has little control over the momentum transfer. The principle examples of this type of instrument are TOSCA at ISIS (UK), FDS at LANSCE (USA), and VISION that is currently under commissioning at SNS.

Samples are frequently powdered and in a simple slab geometry that is cooled to about 20 K to remove the thermal motion that would otherwise blur the signal. Data treatment is minimal and usually involves normalization by the incident flux, background subtraction, and conversion/correction for the energy scale. The majority of molecular spectra are now analyzed using DFT methods as described in Section 6.3 and examples in Sections 6.5 and 6.6.

6.4.5 Backscattering Spectrometers

Backscattering instruments were first conceived in the late 1960s [15] and have been extensively used to measure rotational tunneling of molecular species and slow diffusive processes (examples in Sections 6.7.1 and 6.8, respectively). These are inverted TOF machines which scan the spectrum as a varying incoming energy is analyzed at fixed energy after scattering at the sample. These are a subset of the inverted TOF instruments described in Section 6.4.2. Their defining characteristic is the use of monochromator and/or analyser crystals in backscattering to obtain the best resolution from the crystals, without going to very long wavelengths that would otherwise limit the **Q**-range. The incident neutron energy is modulated either by the velocity distribution in a pulse (as described in Section 6.4.2) or by imposing a periodic variation in the neutron wavelength, usually by means of a Doppler drive. In the latter, the crystal monochromator (usually identical to the analysers) is driven back and forth along the direction of the incident beam causing a shift in the neutron energy. The time/velocity of the monochromator can then be correlated with the arrival time of scattered neutrons at the detector to determine the incident energy as described for inverted TOF instruments above. Some of these instruments have rather limited energy transfer ranges so that in practice they are frequently used in conjunction with experiments on TOF spectrometers to cover an adequate energy-transfer range.

Those backscattering instruments that modulate the incoming energy in a controllable way can also be used to make purely elastic scans by simply removing the incident-beam modulation so that only elastically scattered neutrons (within the instrumental resolution) will be detected. This provides a very high count rate, and the initial idea was to monitor this rate as a function of sample temperature to determine the temperature range in which the dynamics could be measured on the instrument. An inflection in the temperature dependence of these elastic measurements signals the onset of dynamics within the range of the instrument and full quasielastic measurements can then be made around this temperature. This method has since been developed to make use of the **Q**-dependence of the purely elastic scattering to determine an effective force constant that is the resultant of all motions contributing to the Debye–Waller factor. Changes in the effective force constant with temperature signal dynamical transitions, an example for a biological system being given in Section 6.8.4. Many backscattering instruments can be used in this way, but IN13 at ILL (France) is on a thermal neutron beam and has the advantage of reaching high **Q**.

6.4.6 Neutron Spin-echo Instruments

In a magnetic field, B, the neutron spin precesses at a frequency $\omega = \gamma_n B$, where γ_n is the gyromagnetic ratio, and because ω is quite high even in modest fields, the time the neutron spends in the field and therefore its velocity can be determined very precisely. This endows all NSE instruments with very good resolution, but does require polarized beams and magnetically "quiet" environments. Traditionally, this type of instrument used the spin-echo principle [16] to compare the velocity of incident and scattered beams, but an alternative use of the precession to determine trajectories has led to other applications such as powder diffraction, reflection, and small-angle scattering. Neutron-scattering data are usually considered in terms of the scattering law, because most instruments work with defined values $S(\mathbf{Q},\omega)$, but for QENS/INS, the measured quantity in NSE is the average polarization of the scattered beam which is proportional to the Fourier transform of the energy spectrum. NSE instruments are now in routine operation at most major facilities, providing the largest length scales and the longest timescales, typically for the study of slow dynamics in polymers, biological systems, confined media, and complex fluids. A review of the method and its use can be found in Ref. [17].

6.4.7 The Measured Neutron-Scattering Signal

If we are only interested in differences, for example, diffusion as a function of temperature or composition, it may be possible to extract the required information directly from the measured signal. To understand the underlying dynamics however, it is only in the simplest case that $G(\mathbf{r},t)$ for a particular atom (or species) can be obtained directly from the data and in systems of any complexity modeling has become an essential element of the interpretation. The model can be phenomenological (e.g., atoms jumping between different sites that may be described by combinations of exponential time functions) or molecular and lattice dynamics simulations—all provide an expected $G(\mathbf{r},t)$ that can be tested and/or refined against the measured signal.

However, we do not usually measure $G(\mathbf{r},t)$, but its double Fourier transform. In the best case, this is only the space transform that is usually expressed as the intermediate scattering function $I(\mathbf{Q},t)$ which happens naturally with NSE methods (where it is commonly known as $F(\mathbf{Q},t)$), but otherwise, we are always confronted with a frequency spectrum in reciprocal space $S(\mathbf{Q},\omega)$. In most of this chapter, we refer to frequency, ω, and use $S(Q,\omega)$, which is equivalent to $S(\mathbf{Q},E)$, where E is the neutron energy transfer:

$$S(Q,\omega) = \frac{1}{2\pi\hbar} \int\int G(\mathbf{r},t) e^{i(Q \cdot r - \omega t)} \, d\mathbf{r} dt \qquad (6.3)$$

While back transformation of this to the time domain involves truncation errors, models developed in the time domain can be evaluated over large ranges enabling forward transformation for direct comparison with the measured $S(\mathbf{Q},\omega)$. The exponential-type decay of positions in the time domain Fourier transforms to Lorentzian-type broadening of the elastic peak in the energy domain which gives rise to the term quasielastic scattering. In practice, the term "quasielastic" is currently used for any change in the spectral profile around the foot of the elastic peak, regardless of its origin.

6.5 OSCILLATORY MOTION, INCOHERENT SCATTERING

For oscillatory motions, $G(\mathbf{r},t)$ is of course an oscillating function whose Fourier transform contains peaks at the vibrational frequencies with intensities determined by the amplitudes. Within the harmonic approximation, and for small displacements, the incoherent scattering from a molecule (or unit cell) dictates that each normal-mode frequency with its amplitude makes a contribution to the measured spectrum $S(\mathbf{Q},\omega)$:

$$S(\mathbf{Q},\omega) = \sum_{i=1}^{n} \frac{h|\mathbf{Q}\cdot e_i^j|^2}{2m_i\omega_j} \exp(-2W_i)\delta(h\omega - h\omega_j) \quad (6.4)$$

where n is the number of atoms in the molecule (or unit cell), j is the normal mode, e_i^j is the displacement along the direction of $\mathbf{Q}$ for atom i in mode j, m is the atomic mass, and ω_j is the frequency of the normal mode j. The summation effectively includes $3n$ normal modes. The Debye–Waller factor, exp$(-2W_i)$, deserves some comment. The spectral intensity for a particular frequency of an atom along a given vector is reduced by the atomic Debye–Waller factor, which is given by the projection all modes of the atom along that vector. Consequently, a low-frequency large amplitude displacement will effectively "steal" the intensity from any higher-frequency mode that has a significant fraction of an atomic displacement along the same direction. This militates against analysis of spectral fragments, but ensures that a complete set of vibrations should provide consistency with all measured frequencies and intensities in a single calculation.

The intensity of a normal-mode peak will increase with $\mathbf{Q}$ according to the first term in Eq. (6.4) which is proportional to $\mathbf{Q}^2 U^2$, where the amplitude U is related to the oscillator mass and frequency. However, the Debye–Waller factor causes the intensity to decrease exponentially with $\mathbf{Q}$ so that the maximum intensity for a given frequency arises at a Q-value that is related to the oscillator mass, as shown in Figure 6.3, although the oscillator mass is rarely extracted by this means. VDOS are usually obtained by summing energy spectra within selected $\mathbf{Q}$-ranges that will sample different parts of this Q-dependent intensity function for each peak. Clearly, the $\mathbf{Q}$-range should contain as much of the peak as possible, and it is convenient that optimizing a spectrometer for VDOS measurements using a low fixed final energy imposes a Q,ω trajectory that is ideal for a mass ~1 oscillator (i.e., hydrogen, see Figure 6.3). The expression above is for a set of fundamental vibrations, but INS measures the whole phonon expansion, which can be calculated fairly easily within the harmonic approximation. While this "extra data" has the advantage of further constraining any model, the experimental spectrum becomes increasingly crowded at high energy and momentum transfer and often becomes completely intractable above ~200 meV. Ethylene has low molecular mass and intermolecular interaction and thus provides an extreme example where peaks from many high-order phonons merge to an almost featureless spectrum [18].

In disordered systems, there is often an anomalous peak in the VDOS that is typically attributed to an excess density of vibrational states and referred to as the boson peak [19]. It originates from quasilocalized dynamics that are caused by static or dynamic distributions that are characteristic of amorphous materials and, interestingly, also seen in proteins. While the boson peak does vary with temperature and hydration of the protein, it is not affected by the dynamical transition.

6.5.1 Molecular Vibrations of Benzene

Benzene is a good example of molecular spectroscopy because it has been studied at several stages of the technique's development [20,21] and can be used to show both molecular and crystal contributions to the spectrum. Figure 6.4 shows the experimental spectrum (lower curve) taken at 20 K [20] using TOSCA at ISIS (UK). Below about 30 meV we see the lattice modes and then a good separation to the molecular modes starting at about 45 meV. The top spectrum in Figure 6.4 is that calculated using DFT for a gas-phase benzene molecule, in which there are no lattice modes, but it can be seen that the overall agreement between the observed and calculated spectra is quite good.

The calculation provides the atomic displacements for each of the modes so that when there is good agreement between calculation and observation, it is straightforward to assign the observed peaks to specific normal modes of vibration of the molecule. There are, however, some discrepancies and although these can be due to either local crystal effects or dispersion, only calculations including these can provide an unambiguous answer. The second spectrum from the top is from a DFT calculation using the full periodic

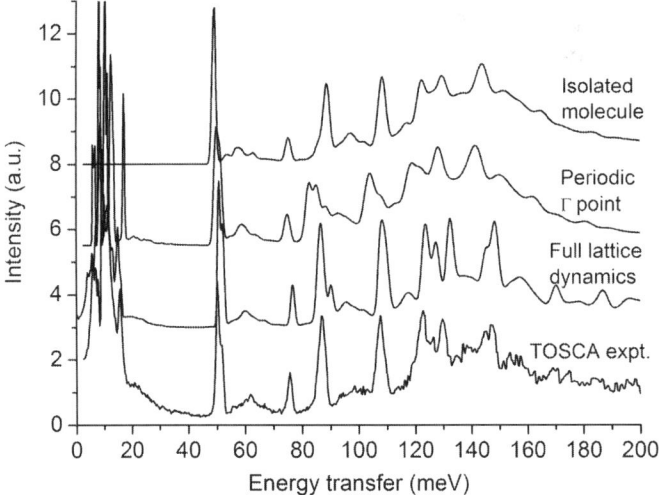

FIGURE 6.4 Observed INS spectrum of solid benzene (bottom) compared with spectra calculated at different levels of approximation: isolated, or gas-phase molecule (top), at the long-wavelength limit (as for IR and Raman) (second), calculation taking full account of dispersion (third).

structure, and it can be seen that the agreement with experiment is certainly no better, possibly worse. The best result is the second spectrum, directly above the experimental profile, and it is evident that the agreement, including the lattice modes, is very good because the calculation included the full lattice dynamics [22] with complete dispersion, but it was also the most expensive in computational effort. If the interest is only in molecular modes, the first calculation would be adequate for the assignment of the peaks, but if the interest is in intermolecular interactions, the full calculation would be required to understand the more subtle changes in the spectrum.

6.5.2 Hydrogen-Bonded Systems

The benzene example above illustrates that the INS spectrum of a condensed-matter molecular system is close to that of an isolated molecule in the gas phase so that a single calculation can normally be used to obtain a reliable assignment of the vibrational modes. Information about weak intermolecular interactions can be obtained as explained above, but the interplay between inter- and intramolecular interactions is more striking when the former are stronger—this is the case of intermolecular hydrogen bonds. There have been many studies of hydrogen-bonded systems [23], involving both inter- and intramolecular hydrogen bonds, because they are ubiquitous in molecular systems in a very wide range of scientific fields and they are challenging for theoretical modeling using quantum-chemical methods. Intermolecular hydrogen bonds are particularly challenging from a modeling point of view since they can form extended solids and are of interest as examples of the simplest chemical interaction—proton transfer.

In this context, benzoic acid has been studied by INS [24], also using TOSCA. In the solid state, the larger carboxylic acids tend to form dimers, leading to the possibility of studying quantum and classical aspects of proton transfer (see translational diffusion Section 6.8). The purpose of this investigation was to identify vibrational modes that couple to the proton transfer coordinate and enhance or hinder that process. Spectra were measured on the four isotopologues (Molecule_H:Hbond_H, Molecule_H:H bond_D, Molecule_D:Hbond_D, Molecule_D:Hbond_H), allowing molecule and hydrogen bond modes to be identified, highlighted, or hidden following the isotopic mass change and scattering power of H/D, which is a particular advantage of INS. Three spectra are shown in Figure 6.5 and are compared with solid state, DFT-based, phonon calculations. Hydrogen bond modes are clearly evident, and the agreement between experiment and calculation is generally good, even for small deuteration-related spectral shifts at low frequency. This analysis clearly identified the lower frequency, intradimer, intermolecular modes that modulate proton transfer.

Short, strong hydrogen bonds (SSHBs) increase the ratio of inter- to intramolecular interactions further still and are of scientific interest since the

Chapter | 6 Dynamics of Atoms and Molecules   431

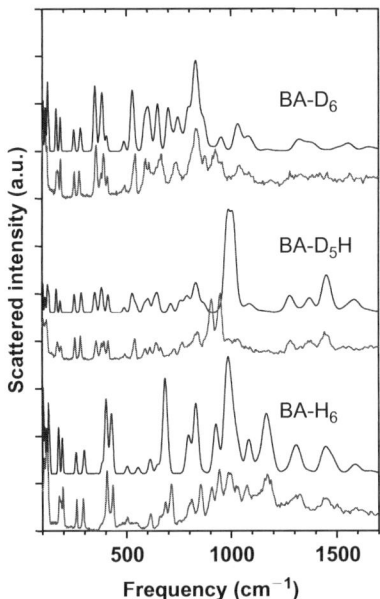

FIGURE 6.5 INS spectra of benzoic acid for three isotopologues: fully protonated (bottom), fully deuterated (top), and hydrogen-bond protonated (middle). The lower spectrum in each pair was measured on the TFXA (now TOSCA) spectrometer at ISIS, UK. The upper spectrum is the result of a DFT-based normal-mode calculation. All three calculated spectra are derived from a single Hessian matrix of force constants. *Reprinted with permission from Ref. [24].* (For color version of this figure, the reader is referred to the online version of this chapter.)

proton transfer reaction in a longer hydrogen bond can become an instability between molecular species, showing temperature-dependent, proton migration from donor to acceptor atoms. In this way, SSHBs are thought to mediate important biomolecular processes. Two systems, identified by variable temperature diffraction, have been studied by INS: urea phosphoric acid (UPA) [25] and 3,5-pyridine dicarboxylic acid (PDCA) [26]. As for benzoic acid, the purpose of combined INS and computational investigations of the vibrational modes is to identify those which modulate the hydrogen-bond geometry in such a way as to drive the proton as a function of temperature from donor to acceptor atom. Figure 6.6 shows the spectra and the partial density of states of UPA for the SSHB since this cannot be so easily highlighted experimentally by deuteration due to the multitude of labile protons in hydrogen bonds. The vibrational investigation of PDCA has shown that vibrational entropy is a driving force in proton migration.

6.5.3 Complex Hydrides

We are often interested in the vibrational dynamics of a part of the molecule, and INS is particularly useful in this respect when signals from other parts of the molecule can be suppressed by deuteration. A recent example is a di-metal

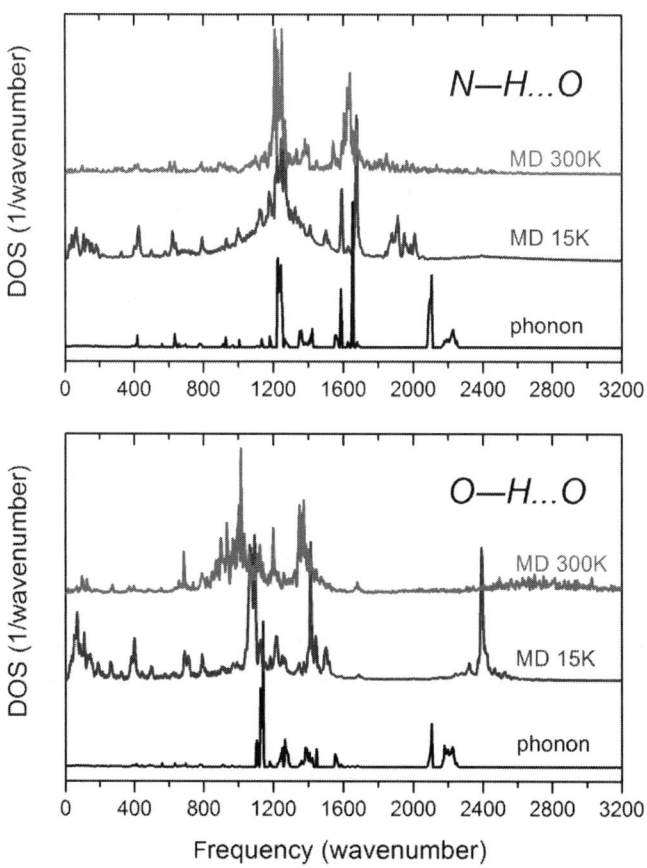

FIGURE 6.6 Normal-mode calculations and AIMD simulations for UPA reproduce measured spectra as shown for benzoic acid above. From these calculations, the partial density of states for the proton in the short N—H...O bond (upper panel) and the normal O—H...O bond (lower panel) are shown in this figure. The N—H stretch mode at ~2000 cm^{-1} softens with increasing temperature which is not the case for the O—H mode in the O—H...O bond. *Reprinted with permission from Ref. [25].* (For color version of this figure, the reader is referred to the online version of this chapter.)

mono-hydride where the interest in the hydride environment is a model for catalytic systems [27]. As is often the case, the molecule is only stable with bulky organic ligands, but by deuterating these (as far as possible), the signal from the hydride can be studied. This example is particularly interesting due to the spectacular disagreement between the observed INS spectrum and that calculated using DFT with the matrix method (see Section 6.3.1) to obtain the normal modes as shown in Figure 6.7.

The lowest spectra are those measured for the selectively deuterated molecule with the hydride ligand (black) and the deuteride (red). Perhaps surprisingly, there are only slight differences between the two. The top spectra were

Chapter | 6 Dynamics of Atoms and Molecules   433

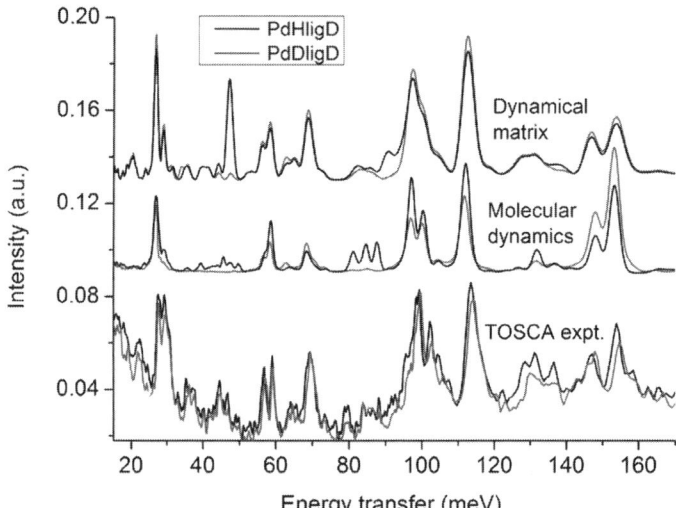

FIGURE 6.7 Spectra of Pd_2HLigD (LigD = deuterated mesityl-imidazo) where the red spectra are where the hydride H has also been deuterated (Pd_2DLigD). The bottom spectra are experimental and the top spectra are calculated using the harmonic approximation. The middle spectra are calculated from an MD simulation. (For interpretation of the references to color in this figure legend, the reader is referred to the online version of this chapter.)

calculated using DFT and the matrix method and clearly show a strong peak for the hydride (black) at 46 meV that is absent in the deuteride (red). It might appear that the experimental deuteration procedure failed, but when the calculation is made using DFT and MD, it indeed shows that there are only slight differences between the hydride and deuteride spectra. This effect arises because the hydride is in a highly anharmonic environment where its motions are strongly determined by those of the surrounding organic ligands, which effectively hold the hydride in place. The experimental spectra now provide a validation of the MD procedure that can be used to examine the processes that are of catalytic interest.

6.5.4 Polymers

It has often been suggested that vibrations can be used to obtain structural information in a general way, although this clearly only makes sense when diffraction methods cannot be used. This is particularly common with polymeric materials in which it is the local structure rather than the long-range structure that is of interest.

Structure-based analysis of INS spectra can be used to determine local structure on the nanometer length scale when diffraction fails. Kevlar provides such an example [28]. It exists in partially crystalline, fiber form which limits the accuracy of diffraction measurements but allows oriented samples to be measured on the spectrometer. Spectra were measured as a function of fiber orientation on IN1 at ILL, France, and were then compared with computed spectra based on structures in which the relative alignment of polymer chains, which interact via hydrogen bonds or simply Coulomb and van der Waals' interactions, was varied. Remarkably, the INS analysis correctly identified the intersheet alignment, in agreement with the most recent diffraction data based on partially deuterated samples, through an accurate modeling of the hydrogen bonds and the weaker intermolecular interactions.

Unfortunately, INS spectra of adequate quality are notoriously difficult to obtain for polymers, but this difficulty can sometimes be circumvented by the use of oligomers, which provide much better spectra. Note that because these materials are soft, they can also be partially aligned by extrusion, which allows the polarization of the different spectral peaks to be determined [29].

An example of using oligomers is provided by the popular model system for polymer electrolytes, polyethylene oxide, which becomes conducting when doped with salts. This raises questions of how the local structure adapts to the presence of ions and, in turn, affects their mobility. For the INS study, this polymer has been simplified by using diglyme, $(CH_3OCH_2CH_2)_2O$. It turns out the INS is quite sensitive to the *cis/trans* orientations along the chain and what happens when salt is added [30]. Addition of nanocrystalline materials to these polymer electrolytes significantly increases their conductivity, and in a further INS study, it was shown that this addition actually has little effect on the backbone conformations [31].

6.6 OSCILLATORY MOTION, COHERENT SCATTERING

Solid-state physicists quickly realized that neutron scattering provides a unique way to determine the forces holding a crystal together, which are entirely responsible for its bulk physical properties and are related to many of its electronic and magnetic properties too. Fundamental vibrations propagate in different ways along different directions in the crystal lattice and within certain approximations provide the matrix of force constants between the constituent atoms. This has provided the basis for understanding microscopic properties, phase transitions, and thermodynamics in crystalline materials.

Compared to the examples given above for incoherent scattering, the phase-related dynamics of coherently scattering atoms gives rise to richly structured dispersion and intensity variations in $\mathbf{Q},\omega$ space. While this spectroscopic information could traditionally be analyzed for data from single crystals of simple systems on TASs, more complex systems and even powder data measured over a wide Q-range on TOF spectrometers can be understood with DFT-based phonon calculations. For thermodynamic quantities, measurements performed over a wide Q-range are important to determine a reliable GDOS. The combination of *ab initio* determination of force constants with lattice dynamics calculations first showed its potential by giving excellent agreement with experimental dispersion curves for GaAs as early as 1990 [32].

6.6.1 Classic Phonons and Soft Modes in $SrTiO_3$

Although strontium titanate is not multiferroic, it is a model system among the perovskite systems, some of which are of current interest for smart devices due to their actual or potential multiferroic properties. $SrTiO_3$ remains cubic at room temperature, but at $\sim$105 K, it undergoes a phase transition to a tetragonal phase that has become the classic example of a soft-mode-driven transition [33]. Soft modes are evident experimentally as their energies tend to zero as the phase transition is approached. In the case of the cubic-to-tetragonal transition in $SrTiO_3$, this is a zone-boundary optical mode that is readily followed using a triple-axis spectrometer. These modes are also evident by lattice-dynamics calculations which show the mode at an imaginary frequency due to the lattice instability.

There are many interesting aspects to the lattice dynamics of perovskites, and for $SrTiO_3$, there is a further low-temperature phase transition from a paraelectric to ferroelectric phase that can be induced by stress or an electric field. However, the atomic displacements in the transition are comparable with their zero-point motion so that under normal conditions the ferroelectric phase is only seen when the heavier ^{18}O isotope is used. The transition is readily seen by the dielectric properties, and lattice dynamics calculations clearly show a soft mode (Figure 6.8), but the changes in dispersion curves or the VDOS calculated from these are very challenging to measure [34].

6.6.2 Negative Thermal Expansion

Negative thermal expansion (NTE), which is also known as "contraction upon heating," is of technological interest to develop materials with specific expansion properties that can compensate the more normal positive thermal expansion, which causes obvious problems in many applications. The main aim of neutron studies is to understand the detailed mechanisms, and because this generally requires geometrical and temporal information, the **Q**-dependence of the spectra is analyzed. While positive thermal expansion is due to anharmonicity, the most common mechanism of NTE is via low-energy transverse vibrations of atomic or molecular bridges that cause a decrease in the average distances as the temperature is increased. These low-energy dynamics are commonly modeled by using rigid unit modes, where polyhedra are considered to be completely rigid, with the only flexibility being via linkages. In simple terms, peaks in the VDOS that are associated with NTE vibrational modes would be expected to shift to higher energy (harden) on heating, and although this general trend has sometimes been observed, in general, the

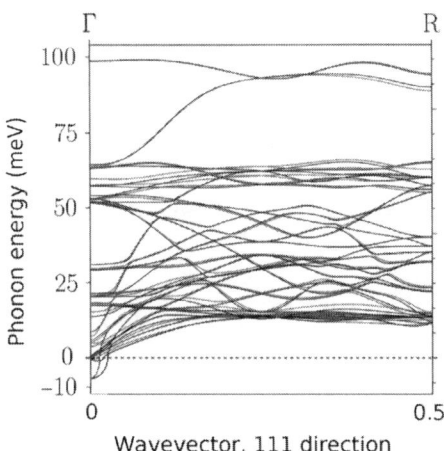

FIGURE 6.8 Phonon dispersion curves, calculated for the tetragonal phase of SrTiO$_3$ (black) showing the negative (imaginary) energies close to the zone center. The red curves are for the orthorhombic structure that was found by following the atomic displacements of the eigenvector of the soft mode. *Reprinted from Bartowiak [35].* (For interpretation of the references to color in this figure legend, the reader is referred to the online version of this chapter.)

NTE mechanism is spread over several normal modes. There have been a number of studies using INS, which combined with modeling have revealed a variety of mechanisms.

Metal organic frameworks (MOFs) have been the subject of considerable activity over the past decade due to their large pore volumes and flexible structures [36]. They consist of metal ions (usually first-row transition elements) linked by modestly large organic molecules, such as carboxylic acids, to form a three-dimensional nanoporous structure. An example of the classic MOF5 (Zn$_4$O(1,4-benzenedicarboxylate)$_3$) is shown in Figure 6.9 in which the metal atoms are zinc, and the organic spacers are benzenedicarboxylates. We will use these as example systems in three sections of this chapter because their large pore volume, which is evident in Figure 6.9, allows the study of diffusive motions of H$_2$, CO$_2$, and other gases, while the presence of both a metal site (Zn in MOF5) and an aromatic site in the framework provides interactions that are strong enough to localize adsorbates at low temperatures so that we can study local motion. Further, the cage itself is very flexible, showing interesting dynamics such that many MOFs show NTE over some temperature range.

We will consider the example of CuBTC (Cu$_3$(1,3,5-benzenetricarboxylate)$_2$), the basic structural unit being shown in Figure 6.10. In the three-dimensional structure, all of the carboxylate oxygen atoms (red) are linked to copper in the same way as at the center of Figure 6.10. VDOS spectra were measured at several temperatures over a large energy range and were in good agreement with those predicted by MD [37]. However, analysis of the MD

Chapter | 6 Dynamics of Atoms and Molecules   437

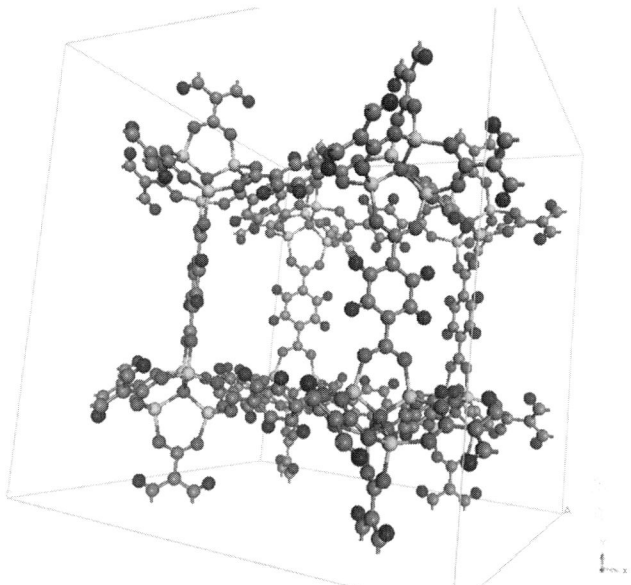

FIGURE 6.9 Crystal structure of MOF5 showing open and flexible cage structure. Atom colors are Zn (green), O (red), C (gray), and H (blue). (For interpretation of the references to color in this figure legend, the reader is referred to the online version of this chapter.)

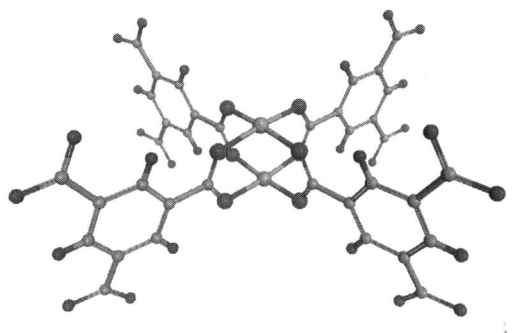

FIGURE 6.10 Fragment of CuBTC showing the paddle-wheel unit that is found at each corner of the cage linked by the acetyl oxygens. Atom colors are Cu (pink), O (red), C (gray), and H (blue). (For interpretation of the references to color in this figure legend, the reader is referred to the online version of this chapter.)

trajectories revealed that the NTE was associated with a bimodal distribution of twist angles between the upper and lower CuO_4 units that are highlighted in Figure 6.10, and that the frequency of this motion was at $\sim$5 meV. To reach this energy, the spectrum of CuBTC was measured at 20 and 100 K using a direct-TOF spectrometer on a cold neutron beam (NEAT at HZB, Germany).

The measured spectra are compared with those predicted by MD and clearly show the main NTE mode at ~5 meV associated with the "paddle-wheel" motion. Interestingly, there is a downward energy shift of this mode in the experimental spectrum of ~1 meV on cooling from 100 to 20 K that seems also to be present in the MD-derived spectra. This softening on cooling is often seen in modes that play a role in NTE.

Because there are fairly large organic spacers between the paddle-wheel units in the crystal structure, the MD simulation found very little correlation between the paddle-wheel motions at different sites. Although the TOF data were difficult to obtain, they are of adequate quality to look for signs of dispersion in the **Q**-dependence of the peak at ~5 meV. None was found, which supports the MD results, showing that this is a material in which the NTE modes are local, rather than collective.

NTE also occurs in some liquids. Water, just above its melting point, is possibly the best known example, but there are also some liquid alloys composed of atoms in the vicinity of Ge, Sb, Te, etc. (GST) in the periodic table. These elemental mixtures (GST) have been widely studied in the context of phase-change materials which depend on the phase (amorphous–solid) variation of optical properties for recording data. GST liquids do not show NTE, but some Ge–Te mixtures do as does pure Te. The most pronounced NTE occurs for $GeTe_6$. GeTe in the solid state (α-phase) exhibits a Peierls distortion which results in alternating short- and long-Ge–Te bonds, minimizing enthalpy. At higher temperature, there is a transition to the β-phase in which all bonds have the same average length. Neutron diffraction and DFT-based MD have demonstrated that the same mechanism occurs in the liquid phase for $GeTe_6$ and underpins the NTE [38], the thermally averaged bonds at higher temperature being shorter than the average of the short and long bonds at lower temperature. But what drives this evolution toward shorter bonds, that is, NTE? The DFT MD simulations revealed a softening of the VDOS over the temperature range of the NTE. TOF measurements revealed a corresponding but much more pronounced softening in the VDOS (see Figure 6.11). These data were used to calculate thermodynamic quantities from which it was concluded that vibrational entropy is the dominant contribution to NTE, that is, 70% of the total entropy change estimated from specific heat data [39].

6.6.3 Nanostructured Materials

Even crystalline nanomaterials can be difficult to study using diffraction because of defects and small particle sizes. Determination of crystal structure by INS is too difficult, but it is possible to distinguish between possible structures and to provide phase information.

An interesting example is MgH_2 which is of interest as a hydrogen-storage material. It has high capacity, but is limited by the kinetics, and to some degree the thermodynamics, of the reaction between Mg and H_2. It was found that the rate at which H_2 reacts with Mg, and can subsequently be removed to return to the metal, increases dramatically when the metal is nanostructured

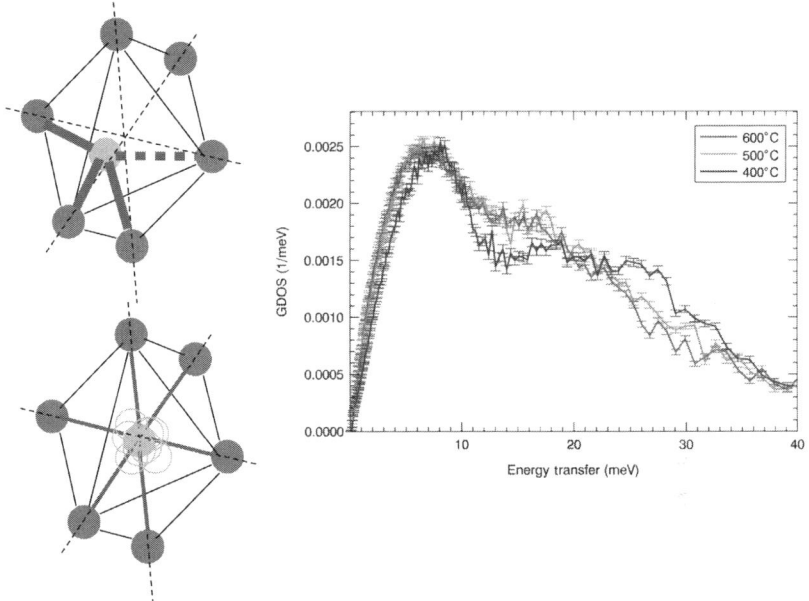

FIGURE 6.11 Octahedral environment of tellurium atoms (orange) around germanium atoms (purple) at low temperature (top left) and high temperature (bottom left) illustrating mechanism of NTE. GDOS spectra (right) show pronounced softening between 400 and 500 °C from which the vibrational entropy contribution to NTE is determined. (For interpretation of the references to color in this figure legend, the reader is referred to the online version of this chapter.)

by ball milling [40]. Diffraction methods could only provide poor information, but because the length scale probed by phonons is much shorter, the VDOS neutron spectra are more informative. These were calculated for the normal β-phase and the high-pressure γ-phase of MgH_2 shown in Figure 6.12. The measured INS spectrum [41] of the bulk crystalline material is shown at the top of Figure 6.12 and has many non-Gaussian features that are characteristic of dispersive modes, but on ball milling, the spectrum marked "bm" was obtained. The loss of crystallinity is clear from the reduction in spectral quality, but the spectrum is also different, and it turns out to be fitted quite well by a mixture of 75% β-MgH_2 and 25% γ-MgH_2 plus a simple background function.

From a technological point of view, the interest is that the spectrum returns to pure β-MgH_2 on subsequent unload/load of H_2, showing that any strain has been relaxed. It follows that the improved kinetics on nanostructuring is due not to strain but probably to increased surface area.

6.6.4 Oxygen-Ion Conductors—Brownmillerites

Materials that conduct light ions like H^+, Li^+, and O^{2-} are of considerable and growing importance in the context of new solutions for energy production. Oxygen-ion conductors have potential applications in membranes for

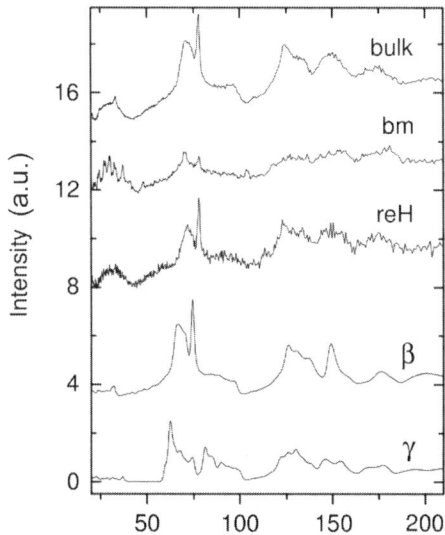

FIGURE 6.12 Spectra of MgH_2 in different phases. Working from the top, the first 3 are experimental spectra of the bulk MgH_2, the freshly ball-milled material, and the ball-milled material after desorption and the readsorption of hydrogen, respectively. Note that the rehydriding reestablishes the main features of the bulk spectrum. The lower spectra are calculated for the normal and high-pressure phases of MgH_2, respectively. *Reprinted with permission from Ref. [41].*

solid oxide fuel cells. While the diffusion (see below) of O^{2-} through a lattice is the fundamental process of interest, understanding this process requires knowledge of structural fluctuations that may trigger or facilitate this, so do vibrational modes play a role in diffusion? In this context, the Brownmillerite family of materials has been studied [42], with the $Sr_2Fe_2O_5$ compound being a much better conductor than the $Ca_2Fe_2O_5$ compound. These materials present alternating layers of FeO_6 octahedra and FeO_4 tetrahedra, the tetrahedral layers being disordered in the Sr compound and ordered in the Ca compound. On heating, the Ca lattice develops disorder in the tetrahedral layers before oxygen loss changes the stoichiometry of the compound. A key structural parameter that highlights the difference between the Sr and Ca compounds is the spacing between the tetrahedral and octahedral layers, it being larger for Sr than for Ca and it expands normally with increasing temperature. A TOF instrument (IN6 at ILL, France) was therefore used to measure the inelastic spectrum of both compounds over a wide Q-range to determine the GDOS as a function of temperature. DFT-based phonon calculations could be performed on the ordered Ca structure and showed good agreement with the low-temperature spectra, enabling the vibrational modes to be reliably identified (see Figure 6.13). Shell model calculations using General Utility Lattice Program [43] were then used to investigate the dependence of the vibrational modes with interlayer spacing. The vibrational mode measured at 10 meV, which was

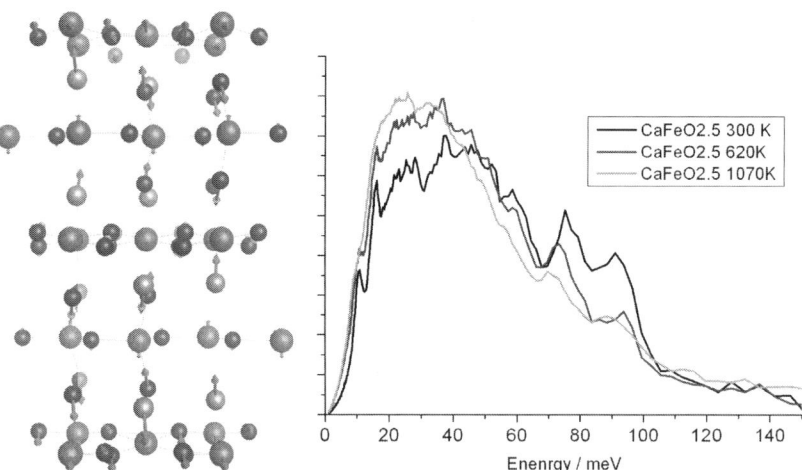

FIGURE 6.13 GDOS of $Ca_2Fe_2O_5$ as a function of temperature (right). The first prominent mode in the spectrum at 10 meV is shown on the left. The octahedral layers are at 0, $b/2$, b, and the tetrahedral layers are at $b/4$ and $3b/4$, where b is the vertical cell axis. (For color version of this figure, the reader is referred to the online version of this chapter.)

observed to soften at high temperature, was found to soften significantly with the interlayer separation. This mode is polarized in the direction perpendicular to the layers and triggers a structural instability of the apical oxygen atoms of the polyhedra that are common to the octahedral and tetrahedral layers. DFT-based MD simulations show clearly this instability and enhanced oxygen-ion mobility in the Sr compound. More detailed studies of this mode in the Ca compound have subsequently been performed on crystals using TASs.

6.6.5 Thermoelectrics—Skutterudites

Also on the energy theme, thermoelectric materials, which transform thermal energy into electrical energy, are studied by neutron spectroscopy. The Seebeck coefficient characterizes the efficiency of these materials as the ratio of electrical potential to temperature gradient. While the electrical conductivity of a material cannot be significantly modified, thermal conductivity and therefore the temperature gradient across a material can be modified more easily, for example, by including heavy atoms in cage-like structures to scatter the heat-carrying phonons. Skutterudites present this type of structure and have been studied by INS to investigate to what extent the heavy guest atoms rattle independently of their cages [44]. In this case, the Q-range of a TOF spectrometer is required to investigate the dispersion of the vibrational modes, the rattling heavy-atom mode should show no dispersion, and spectra were measured on IN6 at ILL. The GDOS for the La and Ce guests in the

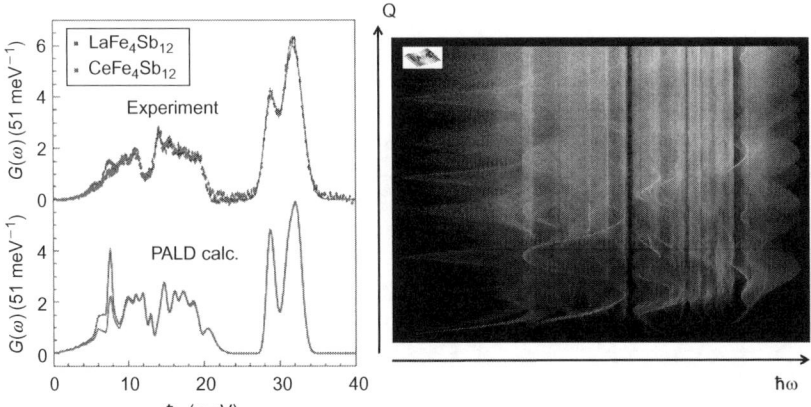

FIGURE 6.14 Left: Measured and calculated GDOS for LaFe$_4$Sb$_{12}$ and CeFe$_4$Sb$_{12}$ showing the excess GDOS at 6–8 meV for the more strongly scattering La. Right: Calculated, high-resolution $S(Q,\omega)$ map for LaFe$_4$Sb$_{12}$ and showing rich structure due to the coherent scattering. *Reprinted with permission from Ref. [44].* (For color version of this figure, the reader is referred to the online version of this chapter.)

iron-antimony cages clearly identify the low-frequency modes of the heavy atoms. The La compound shows clear spectral intensity at 5–7 meV which is absent in the Ce sample, Ce having a similar mass to La but a very low scattering cross section. Despite the iron-antimony cages being isotropic, the spectral intensity in the La-GDOS has a doublet structure indicative of dispersion of the heavy atom mode (see Figure 6.14). DFT-based phonon calculations are in good agreement with the experimental data, both for $S(\mathbf{Q},\omega)$ and for GDOS, in particular, the richly-structured $\mathbf{Q}$-dependent scattering function being well reproduced by the harmonic model. The data and simulations therefore show that the heavy atoms are coherently coupled to the cages, giving rise to dispersive modes which do not therefore scatter heat-carrying phonons. These materials have low thermal conductivity because the phonons have a small group velocity in a large part of phase space so that phonon transport is hindered by Umklapp scattering. This type of scattering occurs when two phonons create a new phonon whose wavevector is outside the first Brillouin zone, effectively limiting thermal conductivity in crystalline materials.

6.6.6 Pnictides

The field of superconductivity (SC) has been dominated in recent years by the discovery of iron-based superconductors, which is iron-arsenide pnictides. There are several families of these materials, the most widely studied being the 122-AFe$_2$As$_2$ and 1111-AFeAsF groups, where A = Ca, Sr, Ba. Doped variants of these two groups show SC up to 50–60 K. The parent compounds show structural and magnetic phase transitions on cooling. Electron or hole doping suppresses the long-range magnetic order and induces SC, suggesting an interplay between spin degrees of freedom and SC. The GDOS were measured on a wide range of powder samples of the parent and doped compounds [45] but no strong evidence was found for electron–phonon coupling as being the dominant mechanism for Cooper-pair formation. However, comparison with DFT-based phonon calculations revealed a systematic discrepancy between measured and calculated spectra that was surprising given the structural simplicity of the materials: Fe–As layers separated by alkali metals and fluorine (see Figure 6.15). Initial calculations were performed in the usual way without considering spin polarization. Subsequent calculations included spin–spin contributions in the interatomic interactions and provided much better agreement with experimental spectra both in the magnetic and in the nonmagnetic phases, demonstrating the importance of spin degrees of freedom both over long range at low temperature and locally at higher temperatures. Now that this field is maturing and better samples become available, the initial work on powders is being complemented by detailed studies on crystals.

6.6.7 Strontium Gallium Oxides

Magnetic systems, some of which have potential applications in microelectronics, are widely studied by neutron scattering. Of particular interest are frustrated magnetic systems which are typically based on triangular lattices and for which the ground state is not well defined. There may be a set of

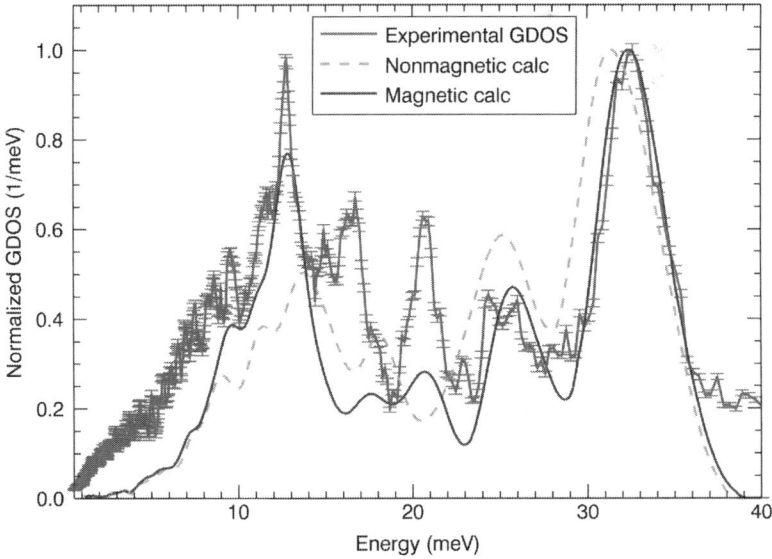

FIGURE 6.15 Measured and calculated GDOS for $BaFe_2As_2$ showing the importance of magnetic interactions in correctly reproducing the phonon mode observed at 20 meV. (For color version of this figure, the reader is referred to the online version of this chapter.)

states closely spaced in energy that can be modulated by low-energy structural excitations. This situation was realized in the Kagome bilayer, pyrochlore slab antiferromagnet $SrCr_{9x}Ga_{(12-9x)}O_{19}$ (SCGO) in which spin-echo measurements revealed magnetic relaxation with an activation energy of 7 meV, which prompted an investigation of the low-frequency phonon spectrum [46]. The isostructural, nonmagnetic sample $SrGa_{12}O_{19}$ was therefore measured on TOF spectrometers since the low Q–E spectrum of SCGO is dominated by magnetic excitations. In addition, DFT-based phonon calculations are problematic on the SCGO structure due to the disorder on the Ga site, whereas they could be performed on strontium gallium oxides (SGO), despite the size of the system, 256 atoms in the periodic model. The experimental and calculated data reveal a phonon mode close to 7 meV which the calculations showed to be dominated by the Ga sites, in particular, those sites that host Cr atoms. These data were also consistent with Raman measurements. However, inspection of the gamma-point modes at 7 meV could not explain the magnetic relaxation in terms of vibrationally modulated spin–spin interactions. Exploring further reciprocal space and the dispersion of the lattice modes revealed approximately flat phonon branches close to the M point (1/2, 1/2, 0) that modulate the magnetic sublattice and could drive magnetic relaxation (see Figure 6.16).

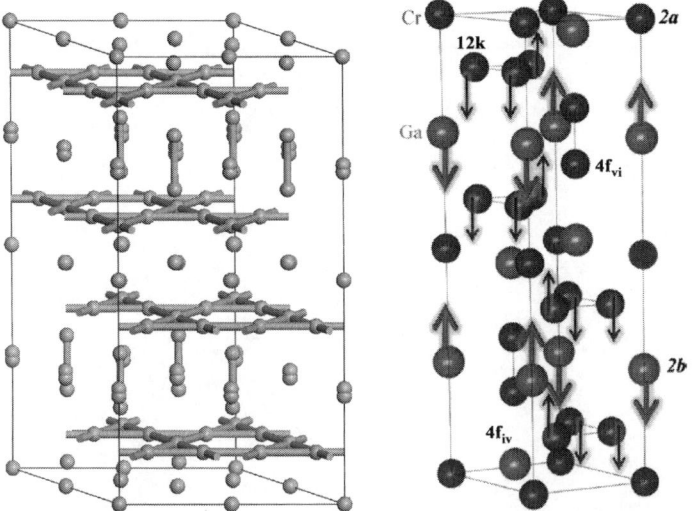

FIGURE 6.16 Left: Ga/Cr sites in the SG(C)O system showing the Kagome layers which pair-up to form pyrochlore slabs at 0, $c/2$, and c, where c is the vertical cell axis. Right: One of the M-point modes which modulates the Cr sublattice and can therefore drive magnetic relaxation. *Reprinted with permission from Ref. [46].* (For color version of this figure, the reader is referred to the online version of this chapter.)

6.6.8 Deoxyribonucleic acid

Deoxyribonucleic acid (DNA), the molecule of life, is one of the systems, like water, that will be studied over and over again as science and instrumentation evolve. DNA has been investigated with neutrons in the more distant past, but interest was renewed due to access to large, aligned, fiber samples and new TOF spectrometers with unprecedented Q–E resolution. On complex molecular systems, like DNA and proteins, there is an issue on the experimental side of what data can be measured and will be able to advance our understanding, this depending on how the data can be analyzed. In this context, new TOF experiments were performed on aligned fibers to study the acoustic and low-frequency optic phonons [47]. Inelastic X-ray scattering has also been used to study phonons along the DNA helix, reporting acoustic phonon dispersion in the Brillouin zone defined by the base-pair separation of $\sim$3.3 Å. The low Q dispersion gives an estimate of the sound velocity along the DNA helix ($\sim$3 km/s) and the stiffness of the molecule, while at higher Q the dispersion seems to be liquid-like. INS measurements could not be performed at low Q and were restricted to the vicinity of the base-pair Bragg peak at 1.9 Å^{-1}. The high Q–E resolution clearly shows acoustic dispersion, but now in the Brillouin zone defined by the helix, that is, the zone boundaries are observed at 1.8 and 2.0 Å^{-1} (see Figure 6.17); these are the only spectroscopic data to date that unequivocally displays the helical character of DNA. On this longer

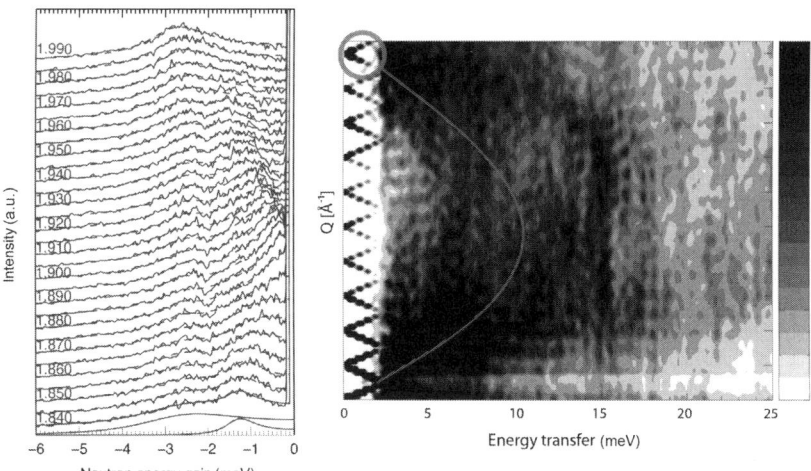

FIGURE 6.17 Left: Spectra of DNA measured on IN5 at ILL, showing the acoustic mode going to zero frequency at $\mathbf{Q} = 1.9 \pm 0.1$ Å^{-1} and the optic mode at $\sim$2.5 meV. Right: $S(Q,w)$ map from the atomistic phonon calculation showing the acoustic phonons in the helix Brillouin zone at $w \sim 0$ and the acoustic branch in the base-pair Brillouin zone which has a maximum of $w \sim 10$ meV at $\mathbf{Q} \sim 1$ Å^{-1}. *Reprinted with permission from Ref. [47].* (For color version of this figure, the reader is referred to the online version of this chapter.)

length scale, the sound velocity (4.2 km/s) and stiffness are higher. In addition, the data reveal the lowest-lying optic mode which shows weak dispersion. Large-scale, force-field phonon calculations reproduce these experimental data and show that the optic mode is an intramolecular mode, rather than an intermolecular mode as proposed from early optical spectroscopy of DNA. New measurements could therefore be reconciled with older spectroscopic measurements from a range of techniques: INS, IXS, and IR/Raman. However, one significant discrepancy remained: structure-based measurements used to determine molecular stiffness indirectly tend to report values 10–100 times weaker than those obtained from spectroscopy. MD simulations using the phonon models were used to explore the effective force constant as the helix is stretched and compressed. The value obtained by this method, which allows the water around DNA to relax on the longer, "structural" timescale, is much lower than the "short-time" value, without water relaxation, obtained from spectroscopy, and is consistent with the structure-based determinations.

6.7 TUNNELING

Neutron scattering has been particularly successful in the study of tunneling of atoms and molecules, this being almost entirely restricted to quantum motion of H and D. The utility of the method is limited by two principal factors. First, as the temperature increases, the coherence between the tunneling sites is usually lost rapidly and the motion becomes more classical-like and diffusive. Second, the exponential dependence of the tunneling frequency to the barrier height means that many examples may only be studied at high resolution on backscattering instruments, if at all. When tunneling can be observed however, it gives a very precise measure of the potential-energy surface that controls the quantum motion as illustrated in the review article [48].

6.7.1 Rotational Tunneling

The monograph by Press [49] gives a good introduction to the relation between classical and quantum single-particle rotations. Low-temperature quantum rotation of molecular entities such as CH_4, CH_3, NH_3, H_2, attracted considerable attention over a period of several years, during which a surprisingly large number of examples were studied [50]. The central questions of temperature dependence with the eventual transition to classical dynamics, rotor/rotor coupling, and coupling of rotational and translational degrees of freedom have now been resolved, and the phenomenon is regarded as well characterized. Nevertheless, calculation of the quantum motion of atoms from first principles is far from straightforward because both mapping the potential correctly and constructing a full Hamiltonian are challenging tasks.

6.7.1.1 One-Dimensional Threefold Rotors

The most common example is the one-dimensional rotor of a CH_3 group (the tunneling particle) bound to a molecule at 0 K—the rotor will have a minimum of three equivalent orientations. If the barrier to rotation is low enough the zero-point motion of the CH_3 group will allow overlap between the wavefunctions of the three orientations of the particle and hence tunneling through the barrier. The Hamiltonian for this motion is as follows:

$$H = \frac{\hbar^2}{2I}\frac{\partial^2}{\partial \emptyset^2} + \sum_{n \geq 1} \frac{V_{3n}}{2}(1 + \cos(3n(\emptyset + \alpha_{3n}))) \quad (6.5)$$

where I is the moment of inertia and the potential energy is expressed as a Fourier series where in practice n is limited to <3. The evolution of the energy-level diagram is illustrated in Figure 6.18, and the crucial point is that the tunneling (rather than quantum free rotation) is almost exponentially dependent on the barrier and hence a sensitive probe of interatomic interactions. In practice, it is often too sensitive and even modest changes to the system reduce the tunnel splitting to energies <1 μeV which is too small to measure. Because these transitions require a nuclear spin flip, they are not directly accessible to other spectroscopies, making neutron scattering the best technique to study this phenomenon.

The technique has evolved to a spectroscopy that has had its greatest recent impact in measuring the interaction between hydrogen molecules and

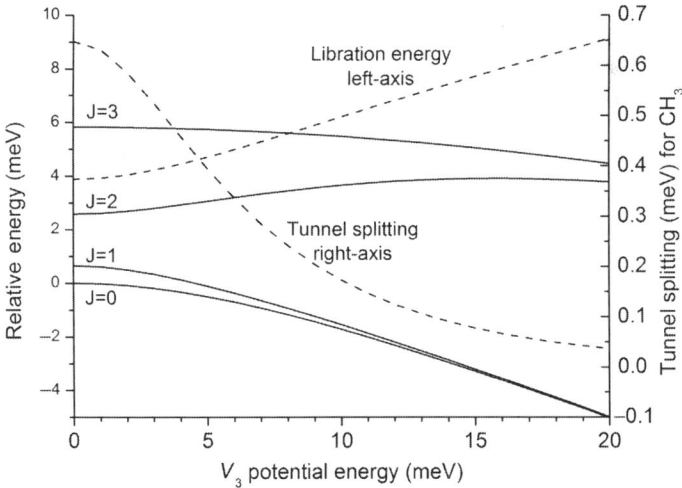

FIGURE 6.18 Energy levels and transitions as a function of the amplitude of a $\cos(3\phi)$ potential. Solid lines are the energy levels and dashed lines are the transitions: rotational tunneling and libration as indicated. The libration is taken as the separation between the average of the $J=0$ and the $J=1$ to the average of the $J=2$ and the $J=3$ because in practice the separate components are rarely resolved experimentally.

potential hydrogen-storage materials. INS peaks from quantum rotation and tunneling are quite intense and because the H_2 molecule is short and of low mass the rotational constant is high $\sim$6.35 meV, thus falling in a spectral range that is easily accessible to many spectrometers. Similarly, methane clathrates have also been studied as samples that occur naturally on the sea floor and may liberate catastrophic quantities of methane, a powerful greenhouse gas, if global warming results in a sufficient rise in sea temperature.

6.7.1.2 The Two-Dimensional Rotor H_2

One of the most straightforward experiments on a TOF instruments is to measure the quantum rotation of hydrogen molecules on a substrate that scatters only weakly. Although for *para*-H_2, scattering below the rotational bandhead is purely coherent (and therefore weak); incoherent QENS experiments are still possible with *ortho*-H_2 (if it can be kept as such at low temperatures below ca. 150 K) [51]. The sensitivity of neutron scattering to the quantum motion is so high that we can present a study made on a very simple, but interesting, rotating crystal spectrometer at the 1 MW reactor in Delft. This TOF instrument combined the neutron-pulse generation with monochromation into a single element—a rapidly rotating crystal. The neutron pulse arises when the Bragg condition is fulfilled at the crystal for the angle between the incident white beam and the monochromatic beam to the sample. Despite its simplicity, this instrument was used very successfully to follow-up claims made in the late 1990s that single-walled carbon nanotubes (SWNT) had exceptionally high hydrogen-storage capacities.

The crucial point is the interaction of the H_2 molecule with the nanotube wall, which must be significantly higher than in other forms of carbon for this claim to be true. The rotational spectrum of hydrogen on four different forms of carbon was measured [52]: two activated carbons, fish-bone carbon nanofibers, and SWNT. The results are shown in Figure 6.19.

A striking feature is the lack of a significant elastic peak which reflects the insignificant scattering from the carbons and the quantum dynamics of the hydrogen nuclei. The important point, however, is that the inelastic peak from the $J=0$ to $J=1$ transition of the H_2 rotor is just below 15 meV for all four carbons, showing clearly that the same weak interaction exists between the H_2 molecule and the carbon surface in all cases. The difference in intensity relates only to the effective surface area, not the strength of interaction. A number of other works demonstrated that claims of high H_2 capacity for SWNT were due to adsorbates other than hydrogen being present in the original gravimetric studies.

MOFs have a large internal surface of around 4500 m^2 that provides sites for H_2 with the adsorption interaction strength for MOF5 being $\sim$70 meV per H_2 molecule. This interaction is too weak for practical H-storage applications, but the organic or metal sites in MOFs are especially interesting because they

Chapter | 6 Dynamics of Atoms and Molecules   449

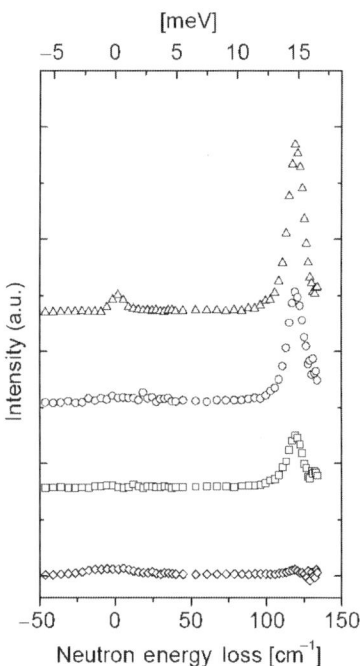

FIGURE 6.19 TOF spectra of free rotation of H_2 on different forms of carbon. The top spectrum is NoritAC990293, single-walled nanotubes, then Norit GSX, and the bottom spectrum is graphitic nanofibers. *Reprinted with permission from Ref. [52].*

can be modified via changes either to the metal or to the linker to enhance the adsorption properties. Because of the need to understand the weaker interactions, and eventually modify them, it is the lower-energy region of the molecular spectra (<100 meV) that is of most interest.

The strategy here is to understand the interaction of molecular hydrogen with different sites within the pores so that it can be increased for H-storage purposes, but the spectral features corresponding to the $J=0$ to $J=1$ rotational transition of H_2 have to be identified and correctly attributed to different absorption sites [53]. These are complex materials so that a large number of vibrations occur as shown in Figure 6.20a, but in general, mixing of vibrations between the guest and host is slight so that the spectral features arising from the H_2 rotations are easily recognized.

The INS spectra of MOF5 shown in Figure 6.20b illustrate how the rotational part of the spectrum varies as a function of H_2 loading, providing evidence of several different adsorption sites. In this work, structural data on H_2 sites, INS, DFT calculations, and adsorption isotherms were brought together to positively assign INS subspectra to specific crystallographic sites and to elucidate how H_2 molecules on different sites influence each other as the loading is varied. Binding energies for each site were also assigned and varied

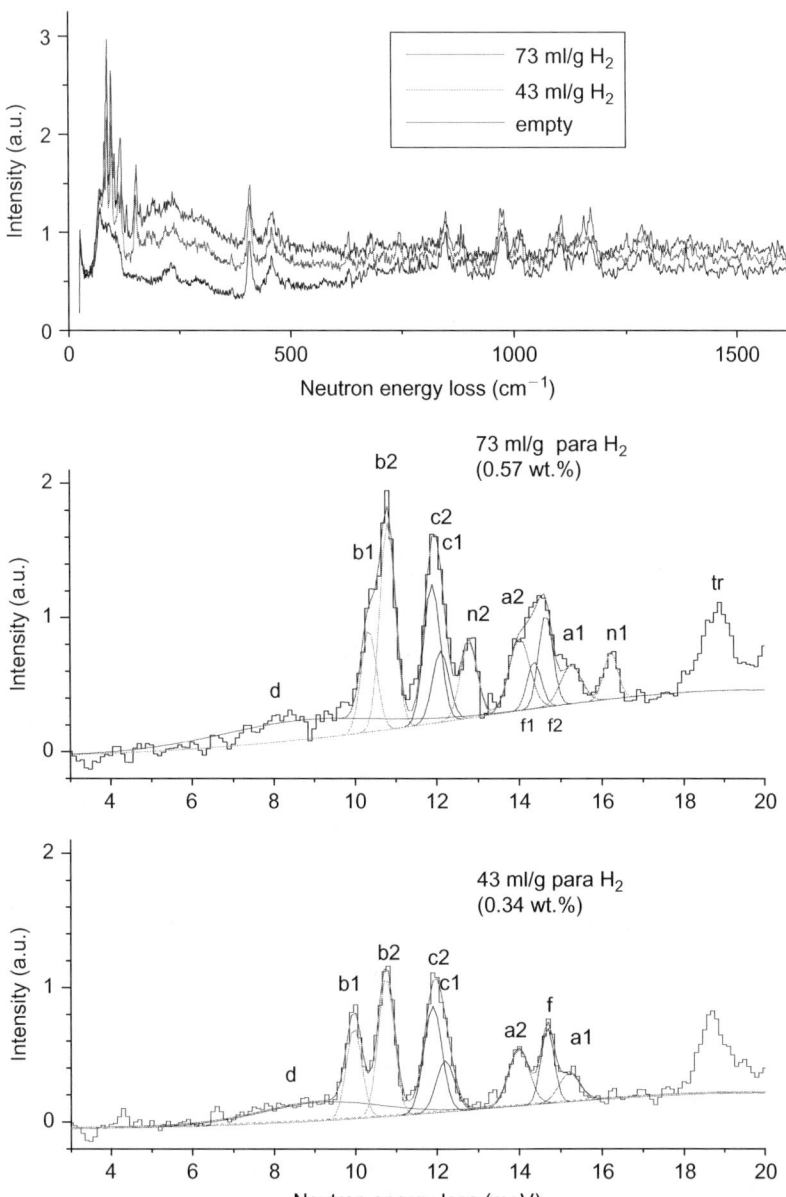

FIGURE 6.20 (a) Spectra of pure MOF5 and with two different loading of H. Most of the spectrum above 20 meV arises from the organic linker molecule. (b) The low energy region where the H_2 rotations arise showing how the peak intensities vary with loading. *Reprinted with permission from Ref. [53].* (For color version of this figure, the reader is referred to the online version of this chapter.)

between 100 and 800 K for different sites. Very recently, in an iron MOF, there has been renewed interest in INS for understanding interactions between neighboring adsorption sites [54]. Some MOFs, such as CuBTC, provide exposed metal sites for H_2 binding, and by combining structural data and INS spectra as a function of loading, with *ab initio* calculations, a very detailed understanding of the binding can be obtained [55], showing surprisingly that despite the strength of the binding to the exposed metal site, there is no elongation of the H–H distance. Although there is a general correlation between binding energy and the energy of the rotational transitions, the details are more complex, including possible interactions between neighboring H_2 rotors, which has recently been excluded for some MOFs by using gas adsorption measurements, neutron diffraction, and INS [55].

The H_2-ice clathrate is not only a fascinating example of a potential hydrogen-storage material but also from a more theoretical standpoint raises questions of both rotation/translation coupling and guest/host coupling which are recurring themes throughout MD and are often resolved by INS or QENS. The well-known ice clathrates that form hydrogen-bonded cages around other molecules to form solids at temperatures and pressures where these would normally exist as free gas provide an interesting environment for the H_2 quantum rotor and potentially a hydrogen-storage medium. This possibility arose because it has been found that the pressure at which the clathrate forms can be drastically reduced by incorporating tetrahydrofuran [56], although the potential maximum capacity is then reduced. In this system, there is a large volume for the quantum delocalization of the H_2 nuclei among the two rotational and three translational degrees of freedom, in contrast to the MOFs above in which the translational and rotational degrees of freedom can be considered separately. The more isotropic environment of the ice-clathrate cage means that the rotational and translational degrees of freedom become intricately coupled and the energies and eigenfunctions for transitions between the various rotational/translational levels must be calculated from the full five-dimensional Hamiltonian to obtain a comparison with the measured INS spectrum (Figure 6.21).

Simulation of host/guest systems for any dynamics always raises the question of coupling between host and guest motions. In the case of the H_2-ice clathrate, it would be a formidable task to construct the Hamiltonian including motion of the ice cages, and indeed, the agreement between the observed and calculated spectra would suggest that any coupling is negligible [57]. This is surprisingly given that the ice-lattice modes arise in the same general spectral region as the H_2 quantum transitions, which are deliberately suppressed in the INS spectrum by use of D_2O. Although DFT MD simulations treat the nuclei classically, at low temperature, they still provide a reasonable approximation of the free-energy surface in which the quantum dynamics of the H_2 molecule occurs, knowing that the simulated H_2 dynamics will be classical. These simulations show significantly different classical rotational/translational dynamics

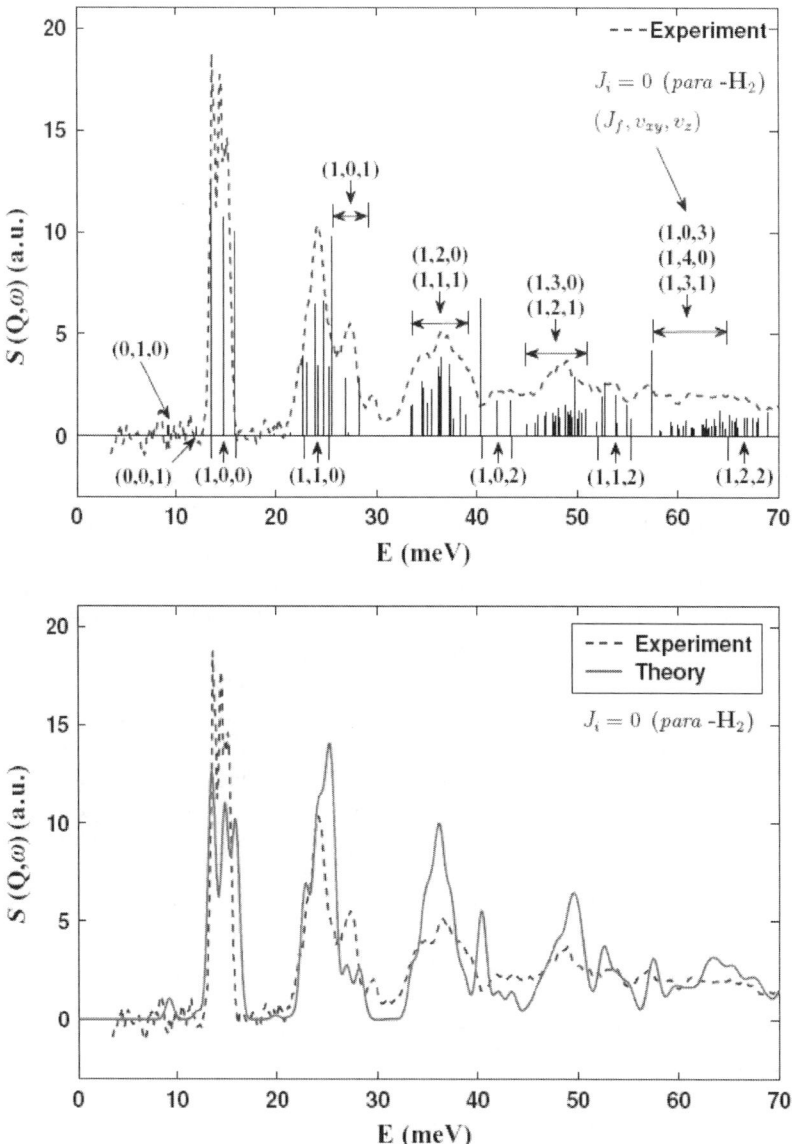

FIGURE 6.21 Observed and calculated spectrum of H_2 in ice clathrate. The top spectrum shows how the envelope of intensity is derived from the stick spectrum, while the lower spectrum shows the additional instrumental effects such as resolution. *Reprinted with permission from Ref. [57].* (For color version of this figure, the reader is referred to the online version of this chapter.)

of the H_2 molecule when the ice cage is held fixed, or left flexible, which is clearly in conflict with the good agreement shown in Figure 6.21. With this coupling, the H_2 quantum dynamics will be sensitive to deuteration of the ice cage that would lead to a fortuitous cancellation of errors [58]. This hypothesis can be resolved by an accurate study of the INS spectrum of samples with partially deuterated ice.

6.7.2 Translational Tunneling

Polarons are common throughout dynamics and arise from the tendency of the system to become "stuck" by relaxation around a local configuration. Tunneling is extremely sensitive to changes in local potential so that it is unlikely to be observed unless the motion is between indistinguishable conformations. This equivalence is difficult to achieve with translation, so tunneling between states is rare for translational motion [59,60]. One system that stands out for translational tunneling is the bridging hydrogen bond arrangement in which a proton is formally between a hydrogen-bond donor and acceptor. As the distance between the donor and acceptor decreases, the distinction between their roles is progressively lost and the proton transfers rapidly between them. Ultimately, the bimodal free-energy function for the proton within the bridge merges into a broad flat-bottomed profile.

Benzoic acid dimers were studied [61] with a view to optimizing the symmetry of the two-well potential governing proton transfer since the asymmetry only arises from the weak, interdimer interactions. A QENS study on aligned crystals of partially deuterated benzoic acid (Molecule_D: Hbond_H) minimized the elastic scattering intensity and allowed the temperature dependence of the QENS signal to be measured accurately. Data on the proton dynamics were consistent with NMR data and enabled the low-temperature activation energy to be determined. This activation energy was found to be consistent with vibrational modes studied by INS that enhances proton transfer by quantum tunneling. Similar experiments have been attempted on SSHBs, but the short proton-jump distance moves the maximum amplitude in the QENS signal out to prohibitively large values of Q.

6.8 STOCHASTIC RELAXATION/DYNAMICS

Local motion differs from diffusion because there is not only a quasielastic component but also an elastic peak arising from the finite probability of the particle returning to its original position, which is expressed by

$$S(Q,\omega) = A_0(Q)\delta(\omega) + S_{qe}(Q,\omega) \tag{6.6}$$

The term $A_0(Q)$ describes the Q-dependence of this probability and is referred to as the elastic incoherent structure factor (EISF), which is widely used to characterize the geometry of the underlying motion, such as for a threefold jump rotation of a molecular group:

$$S(Q,\omega) = A_0(Q)\delta(\omega) + (1 - A_0(Q))\frac{1}{\pi}\left(\frac{\tau}{1+\omega^2\tau^2}\right) \quad (6.7)$$

with

$$A_0 = \frac{1}{3}\left(1 + 2j_0\left(Qr\sqrt{3}\right)\right) \quad (6.8)$$

where r is the radius of rotation, τ is the time between jumps, and j_0 is a Bessel function of the first kind. A number of dynamical models have been developed over the years [3,62], and in many cases, one or more of these can be used for an analytical approach to experimental data, some standard EISFs being illustrated in Figure 6.22.

The early period of quasielastic scattering was the heyday for molecular rotation, and although current interest in this type of motion is usually in combination with other types of motion, there are still pedagogic examples where rotation alone is important. Changes in rotational dynamics often characterizes phase transformations, the classic example being the ammonium halides, and these were indeed successfully studied by QENS many years ago [63]. Diffusion is characterized by a continuous decay of position with time (albeit that the volume explored may be finite), and in the energy

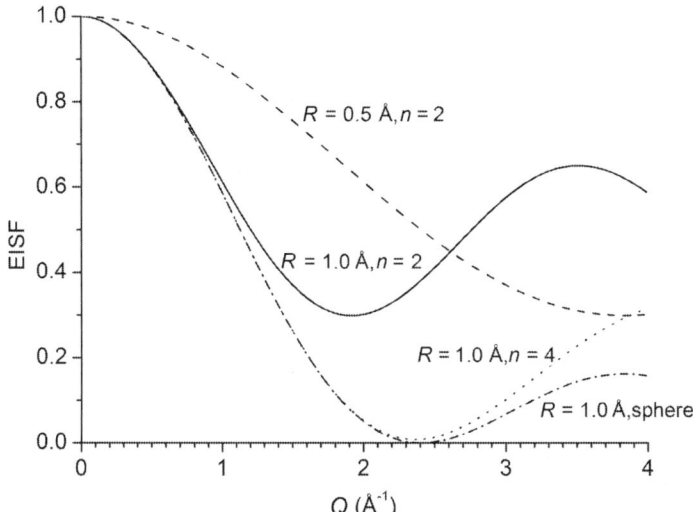

FIGURE 6.22 EISFs for motions of various geometries. The dashed line is for interchange of two atoms separated by 1.0 Å (radius 0.5 Å), showing that the minimum is at rather high **Q**. The solid line is for the same process, but with a separation of 2.0 Å were it can be seen that the minimum is at the upper limit of the **Q**-range of most instruments. The other curves show how the EISF at the minimum depends on the geometry of the motion.

domain, this gives rise to a quasielastic broadening of the elastic peak. We emphasize that what is measured is self-diffusion over a distance that is given by the **Q**-range on the measurement, and this is not always compatible with bulk (or driven) diffusion measured by other techniques. For all quasielastic scattering (not only diffusion), it is important to know the approximate timescale of the motion so that the appropriate instrument (or range) can be selected. Often this timescale can be inferred from other techniques, or modeling, but unless the dynamics are already well characterized, some adjustment of the sample temperature is often required to bring the motion within the timescale accessed by the instrument. This is particularly true for backscattering machines that often have a rather limited energy range and has led to the elastic-scan technique for finding accessible dynamics using backscattering machines (see Section 6.4.5).

6.8.1 Complex Diffusion

Measurement of diffusive processes by QENS has evolved considerably in recent years with far more complex systems and diffusion mechanisms becoming increasingly common. This has placed more emphasis on larger dynamic ranges, better data-correction, and models based on atomistic MD simulations, and the reader is referred to an excellent review by Jobic and Theodorou [64], covering the many aspects of diffusion in zeolites for more examples on this topic. Simple Brownian motion in which space is explored isotropically and without driving force leads to the well-known Lorentzian quasielastic broadening of the elastic peak with a full-width-at-half-maximum (FWHM) that increases as $2D\mathbf{Q}^2$, where D is the diffusion constant. At low **Q**, this equation often works reasonably well, but as we will discuss later, deviations from simple continuous diffusion are very common, and the $2D\mathbf{Q}^2$ rule is probably most useful in constraining the initial data-analysis procedure. A number of models (Chudley–Elliot [65], Singwi–Sjolander [66], Hall–Ross [67]) have been developed to take account of the microscopic mechanisms that cause deviations from Fickian behavior. They are mostly based on the idea of diffusion by successive jumps and lead to functions for the FWHM that flattens and decreases at higher Q, this being related to the treatment of the jump trajectory and the residence time between jumps.

Deviation from simple diffusion also affects the shape of the quasielastic broadening because the decay of position may no longer be purely exponential and appear as subdiffusion. Trial functions for decay in the time domain may have no analytical form in the frequency domain and become intractable if decay functions are multiplied because these will be convoluted in the frequency domain of the measurement. These difficulties can be minimized by analyzing quasielastic spectra over a range of momentum transfers with a single function, rather than the more traditional "one spectrum at a time" approach. A flexible function that is commonly used for nonexponential decays is the phenomenological stretched exponential, or Kohlrausch–Williams–Watts (KWW), function in which the exponent is raised to a power β:

$$I(Q,t) = \exp\left[-\left(\frac{t}{\tau}\right)^{\beta}\right] \tag{6.9}$$

Numerous diffusion processes and distributions of diffusion lead to decay that is well approximated by the KWW function, and it is only by obtaining data over a very wide range of timescales and/or carefully analyzing the **Q**-dependence of the parameters that possible underlying mechanisms can be distinguished.

6.8.2 Ligand Water Rotation

A more recent example that shows how this can be extended to more complex systems is the evolution of rotational dynamics through the phase transitions of transition-metal hexaquo ions [68] using IN5 at ILL (France). In these salts, a divalent first-row transition metal ion is octahedrally coordinated by six water molecules. Four phases are found below about 360 K, and these have been characterized by studying the EISF, activation energies and correlation times. In the lowest temperature phase IV (below $\sim$200 K), no rotational dynamics is observed, although librational motion would of course persist. In phase III, the hydrogen-bonding interactions are still quite strong, but there is an onset of rotational jumps of the water ligands, with the EISF being consistent with 180° jumps. There is an abrupt decrease in the activation energy and correlation time of the water rotational jumps (still 180°) at the transition to phase II at $\sim$280 K, but perhaps surprisingly, only slight changes at the transition to phase I at $\sim$350 K. The EISF for rotational diffusion is very characteristic, but the observed EISF was clearly still that of 180° jumps, the slight change from phase II most probably arising from the effect of dynamical changes of the molecular counterions (ClO_4^-). The EISFs at different temperatures are shown in Figure 6.23.

6.8.3 Coherent QENS, Rotation

The overwhelming majority of QENS studies are concerned with incoherent scattering, almost always hydrogen atoms. Coherent quasielastic scattering arises from correlated rotational (or diffusive) motion and a convenient example, which also leads us to diffusion in the next section, is the two-stage melting of cesium/lead alloys [69]. In the solid, this alloy is formulated as Cs_4Pb_4, reflecting the ordered structure of the tetrahedral ions and shows no broadening of the elastic line (Figure 6.24a) measured on IN6 at ILL, France. The experimental coherent QENS of the phase after "first melting" was successfully modeled as a plastic crystal in which there is rotational freedom of the metal tetrahedron, but with translational order, which accounted not only for the

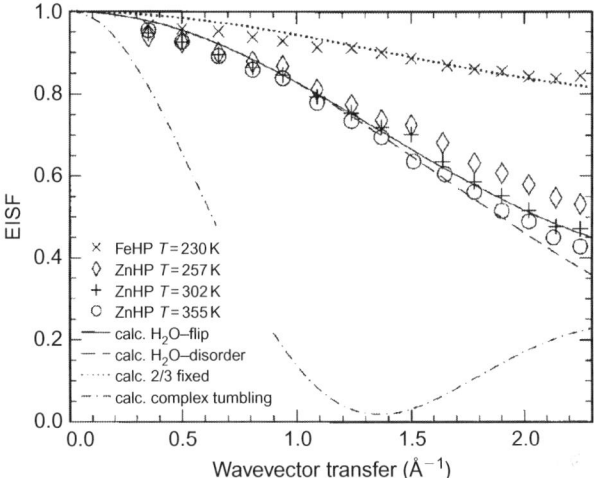

FIGURE 6.23 Symbols are observed EISFs for hydrogen dynamics in [Fe(H$_2$O)$_6$])ClO$_4$)$_2$ at different temperatures and the lines are the EISFs calculated for the different types of motion in the different phases at the corresponding temperatures. The experimental EISF always decreases with increasing temperature due to the overall increase in amplitude of motion. *Reprinted with permission from Ref. [68].*

QENS signal (Figure 6.24b) but also for the "not quite liquid" static structure factor. Above the final melting temperature, a liquid-like QENS signal was found (Figure 6.24a), with no trace of strictly elastic scattering, as illustrated in Figure 6.24c.

6.8.4 Dynamical Transitions from Elastic Scans

There has been considerable interest over the past decade or so in the microscopic nature of dynamical transitions in proteins by using elastic neutron scattering. The basic idea is to measure the Q-dependence of the elastic scattering (within the energy resolution of the backscattering instrument) of the sample to obtain the overall mean-square amplitude of the average motion. Within the harmonic approximation, the amplitude will increase linearly with temperature with a slope that is inversely proportional to the force constant, so that any changes in slope on heating indicate the onset of a new dynamical process. The classic experiment on purple membrane [70] showed a force constant of 2 N/m below 220 K and a sharp decrease to 0.3 N/m above this temperature, reflecting the onset of a new rather-free motion. Selective deuteration of the extracellular part of the bacteriorhodopsin reduced the contribution from this part of the molecule from the scattering, and no transition was then seen, confirming that the transition was characterized by a sudden onset of motion of this entity.

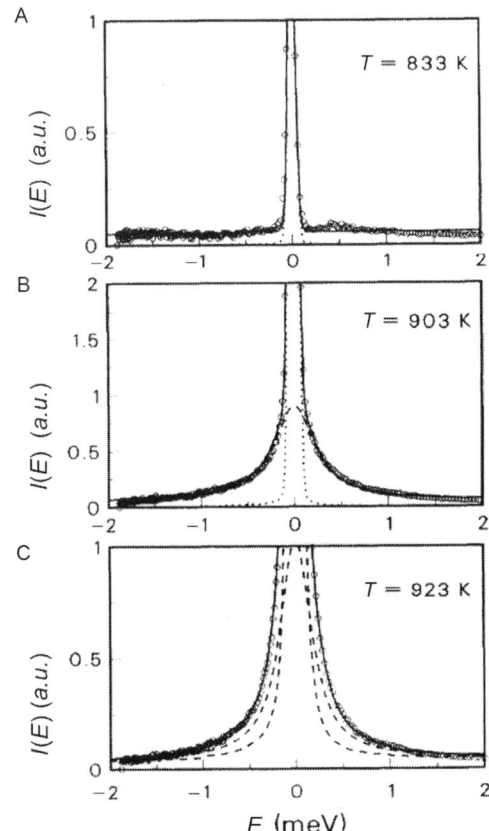

FIGURE 6.24 Examples of QENS spectra of CsPb showing how purely elastic scattering (dotted line) in the solid gives way partly to a quasielastic signal in the intermediate phase and then completely quasielastic, as is normal for a liquid phase. *Reprinted with permission from Ref. [69].*

This has become a very useful neutron-scattering tool for biologists with one instrument (IN13 at ILL, France) now being almost entirely dedicated to this type of experiment. Usually, it is difficult (if not impossible) to deduce any atomic-level details of the underlying dynamical process without also collecting and analyzing the quasielastic signal, unless some fairly drastic assumptions are made. Indeed, it is clear that some caution is required to avoid confusing the effect of the finite instrumental energy resolution with a fictive dynamical transition [71]. Recently, however, it has been shown that a combination of selective deuteration, varying hydration levels, and temperatures can successfully be used to address the question: "How do local thermal motions relate to specific protein activity?" [72] In this work, deuterium labeling of the core of bacteriorhodopsin (shown in Figure 6.25) showed that its dynamics were also subject to the effects of hydration.

6.8.5 Diffusion of Coherent Scatterer CO_2

Functional materials such as MOFs seem to be becoming ubiquitous in potential applications that exploit their porous and flexible nature, as in the potential storage of the green-house gas CO_2, which is a serious, economic, global issue. For QENS, CO_2 is very different from the more usual hydrogen-containing adsorbates being a purely coherent scatterer. Nevertheless, successful experiments are now being made, leading to a good understanding of CO_2 diffusion mechanisms [73]. A recent example demonstrating the complexity that can now be understood examines the transport diffusivity of CO_2 in an MOF with two polymorphs (large pore and narrow pore) whose relative abundance depends on the CO_2 loading. CO_2 is a coherent scatterer so that the dynamic structure factor is scaled by the static structure factor, all of which has to be rolled into the modeling. The authors are able to show that the

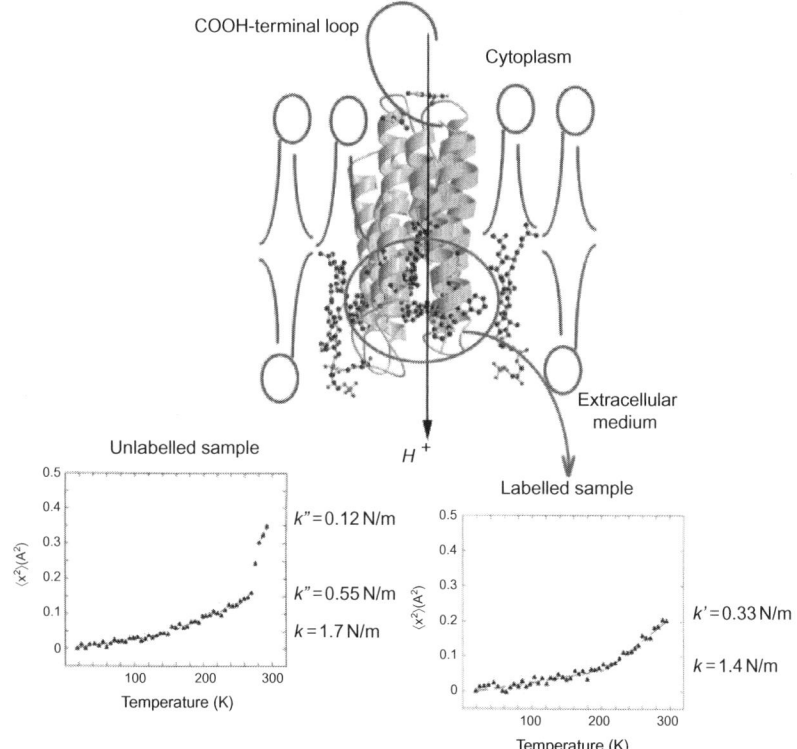

FIGURE 6.25 Illustration of how the breakaway of amplitude at a particular temperature can be assigned to the correct fragment of the protein by selective deuteration. The picture shows bacteriorhodopsin with the deuterated sites in red. *Reprinted with permission from Ref. [72].* (For interpretation of the references to color in this figure legend, the reader is referred to the online version of this chapter.)

diffusion is always one-dimensional and is very rapid in the large pore and slower and single-file in the narrow pores.

6.8.6 Water and Complex Diffusion

Analysis of QENS spectra frequently consists of disentangling the signals from different motions and/or different species within the sample. The common example of this difficulty is water, which, because of its importance, has been studied in a wide variety of environments, and because of its multiple interactions (H-bonding, polarity, phyla and phobia), the resulting dynamics are rarely simple. The ubiquitous occurrence of both rotation and translation can simply be additive, from, for example, fixed and mobile water in cement, or convoluted when it is the same species involved in a combined motion. This is difficult to analyze because the convolution makes the fitting procedure less sensitive to the parameters of the function and it is tempting to approximate the convolution by a sum. While this is a reasonable approximation when the two dynamical processes occur on very different timescales, it will give a seductively good but insensitive fit if the timescales are similar.

We will examine examples in which water interacts with a complex environment so that the diffusion process develops with time and can be punctuated by bound or caged periods between the longer-range excursions. The analysis of this complex dynamics usually requires interplay between the experimental data, MD simulation, and the development of a consistent analytical model.

The potential ambiguity in interpretation is highlighted in a recent QENS study of supercooled water by Qvist *et al.* [74], this system being of fundamental interest due to the strong slowing down of diffusion in a three-dimensional hydrogen-bonded network on cooling. Previous interpretation involved a jump process that was inferred from flattening of the Q-dependence of the quasielastic width at high Q, this being a common signature of finite jumps. A second component had been attributed to continuous rotational diffusion, allowing the QENS spectra to be described with a small number of parameters. Unfortunately, the jump process of this "standard model" had never been observed in MD simulations, and the rotational timescales were in conflict with nuclear-spin relaxation rates. Qvist *et al.* used a careful QENS experiment and then a substantial MD simulation to deduce the origin of the two-component signal. The simulations revealed two types of structural dynamics: local fluctuations within a group of molecules (termed a basin) and slower dynamics between the basins. These dynamics account quantitatively for the measured QENS and its temperature dependence and resolve the difficulties that persisted in the "standard model."

6.8.6.1 Water in Nafion®

Fuel cells using H_2 and air to produce electricity are important to replace fossil-fuel energy applications, particularly in transport. The Nafion®

proton-exchange membrane used in current low-temperature fuel cells has poor performance and short lifetime as well as high fabrication cost. Despite considerable work, there is still a need for a detailed understanding of the water dynamics and the proton conductivity. Nafion® has a hydrophobic polytetrafluoroethylene backbone terminated by a hydrophilic sulfonic group and decorated by fluorether side chains (see Figure 6.26), but despite considerable effort, the structural details of the polymer organization are still not known in detail. The various studies of water in Nafion® by QENS illustrate how ever more detailed dynamical information of water within the ill-defined porous system can be obtained.

The first study [75] in the early 1980s was analyzed with a model for diffusion inside a sphere and then much later with the Hall–Ross jump-diffusion model [66] with a distribution of jump lengths. These provided the basic information for the time and length scales of the dynamics, but only little information about the local microscopic mechanism within the pores and between the pores. By combining QENS data over a greater range of timescales with SANS for samples at different hydrations, a consistent model was derived that involves two populations of diffusing protons [76]. The slow population is attributed to protons of the hydronium ion moving over a length scale of 2–4 Å, while the fast population of water moves over the same scale but with an additional long-range diffusion through poorly defined walls to neighboring domains.

It is a general observation that dynamical properties in a confined space are different from those of the bulk, and this is clearly seen in the behavior

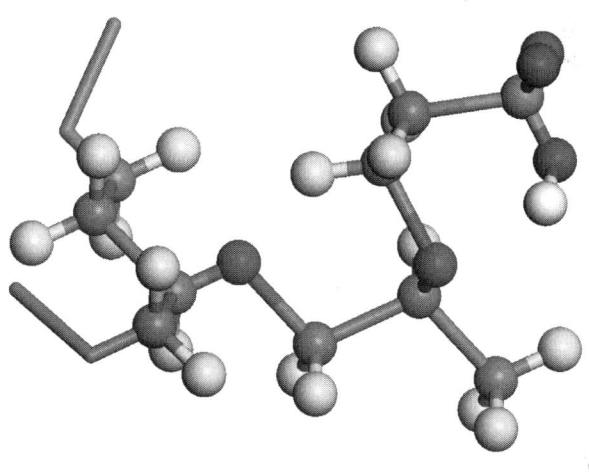

FIGURE 6.26 Schematic illustration of a fragment of the Nafion® polymer showing types of branching, linkages and terminal groups. Carbon is gray, oxygen red, hydrogen white, fluorine blue, and sulfur yellow. (For interpretation of the references to color in this figure legend, the reader is referred to the online version of this chapter.)

of water in Nafion®, but even the form of the confining structure in Nafion® is not known in any detail. In order to disentangle the effects of the confining media, one approach is to make a simple structural analogue of Nafion® that self-assembles into a well-defined mesoporous structure with water: perfluorooctanesulfonic acid (PFOS) [77]. This molecule, illustrated in Figure 6.27, resembles an oligomer of Nafion, comprising a hydrophobic backbone with seven CF_2 units terminated on one end by a hydrophilic sulfonic group and on the other by CF_3.

Depending on the concentration in water, this surfactant forms a mesoporous hexagonal phase (60 wt% PFOS) or a lamellar phase (90 wt% PFOS). The QENS experiment confirmed the basic findings of water in confinement in Nafion®, showing similar confinement sizes, residence times, and diffusion coefficients for comparable levels of hydration. The real advantage was that the well-defined confinements of PFOS allowed an analysis of how the geometry of confinement affects the diffusive process.

6.8.6.2 Water in Concrete and Silica Materials

QENS is being used to study increasingly complex systems, and few are more complex than the hydration of cement. Concrete and cement are hugely important in modern society, and until recently, very little was known about the setting process, its acceleration, and retardation, at the atomistic level. Neutron-scattering techniques—diffraction, inelastic, and QENS—have played a crucial role in unraveling the complex sequences involved in setting processes (see review by Peterson [78]), and in these systems, there are several types of water with characteristic dynamics that evolve as the cement sets. The identity of the free and bound fractions of water can be determined by including, for example, calorimetric measurements, which have been used to understand the acceleration of setting caused by $CaCl_2$ and the retardation caused by sugars [79]. The water shows characteristics of confinement, as in

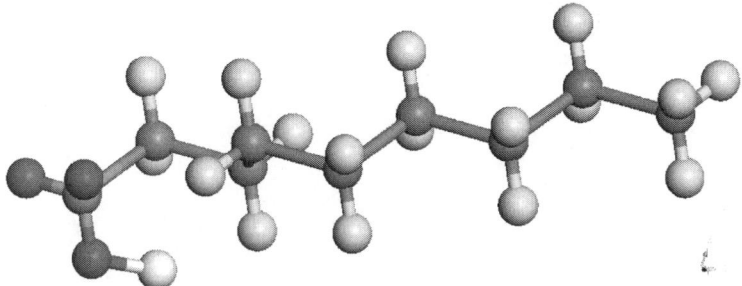

FIGURE 6.27 Illustration of the PFOS molecule showing it as a simplified model for Nafion® in Figure 6.26. Carbon is gray, oxygen red, hydrogen white, fluorine blue, and sulfur yellow. (For interpretation of the references to color in this figure legend, the reader is referred to the online version of this chapter.)

mesoporous silica materials, and the QENS is conveniently understood using the relaxing cage model (RCM) [80] in which the water is considered to be bound and librating at short times, but the position then decays via rotation and translation at longer times [81]. The RCM is an important concept and is written as

$$I_T(Q,t) = I_t^l(Q,t)\exp\left[-\left(\frac{t}{\tau_T}\right)^b\right] \quad (6.10)$$

where I^l is the local motion and the stretched exponential decay is for the translation, which signals a complex diffusion that is fundamentally different from free water.

6.8.6.3 Water in Biological Materials

Water accounts for almost 3/4 of the mass of cells and plays a host of different roles in the workings of biology, but detailed analysis of the dynamics at an atomistic level is in the early stages of development. The main difficulty has been complexity and the simplifying approach of elastic scans (see Section 6.4.5) has laid the groundwork for quasielastic scattering, many excellent examples being given in the book "Neutron Scattering in Biology" [82]. Here, we will outline the challenges for protein dynamics in which water plays a central role.

We return to the question of the dynamical transition in proteins that we introduced in Section 6.8.4. The interest arises because the changes in the QENS, or elastic signal and the onset of biological activity often (not always) occur in similar temperature ranges, but are absent in dehydrated proteins. For this reason, the transition has been loosely attributed to dynamics of the hydration shell, but there is no consistent picture from using elastic scattering studies alone. By combining this with VDOS, QENS, and NSE measurements, however, it is possible to separate the internal motions and global diffusion in a consistent way [83]. Other efforts to disentangle the rather homogeneous contributions to the QENS signal use variable hydration level, deuteration, temperature, and aligned membranes [84].

6.8.7 Ionic Liquids

Room-temperature ionic liquids (RTILs) are of interest as clean solvents plus an increasing range of other applications, but despite the obvious connection for these materials between QENS and understanding their macroscopic properties, there have been only a few studies. They are of interest here as an example of how to understand the combined internal and external motions of the molecules. RTILs have at least one large, molecular, polarizable ion with a more modest-sized charge-balancing counterion. The example that we will consider here is 1-ethyl-3-methyl-imidazolinium bromide (emimBr)

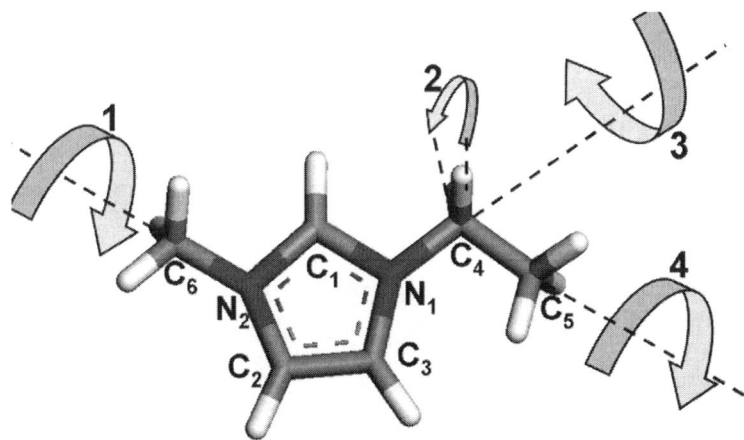

FIGURE 6.28 Different reorientational motions possible for the molecular entities in emim$^+$. *Reprinted with permission from Ref. [85].* (For color version of this figure, the reader is referred to the online version of this chapter.)

[85], the imidazolium-based cation being illustrated in Figure 6.28. In cases like this, it is common to try separating any rotational and translational components in the quasielastic spectrum by fitting with Lorentzian functions, as was done in this work. It is also then common to find that either Lorentzians provide a poor description, usually due to sub-diffusion, or the EISF of the rotational component cannot be interpreted unambiguously. The way forward is to perform an MD simulation to find which motions are possible on the timescale covered by the QENS experiment.

The difficulty with emim$^+$ was the rotational component, and the MD shows clearly that reorientation of the methyl and methylene groups is too fast to be discernible and that whole molecule reorientation is too slow. Ethyl group reorientation, however (see Figure 6.28), was plausible, and subsequent fitting of the appropriate model function was successful as shown in Figure 6.29.

It is important to understand that several approximate analytical models will give reasonable fits to data from complex systems in which several motions on different timescales may be present, and that care should be taken to ensure the correct choice. In general, a good MD simulation will reproduce the observed QENS well and even when errors in empirical force fields require temperature scaling to match the experimental data, it should be possible to identify the motions occurring on the approximate timescale of the measurement.

6.9 CONCLUSION AND PERSPECTIVES

The examples shown in this chapter illustrate that considerable insight can be gained from studying the dynamics of atoms and molecules using INS and/or

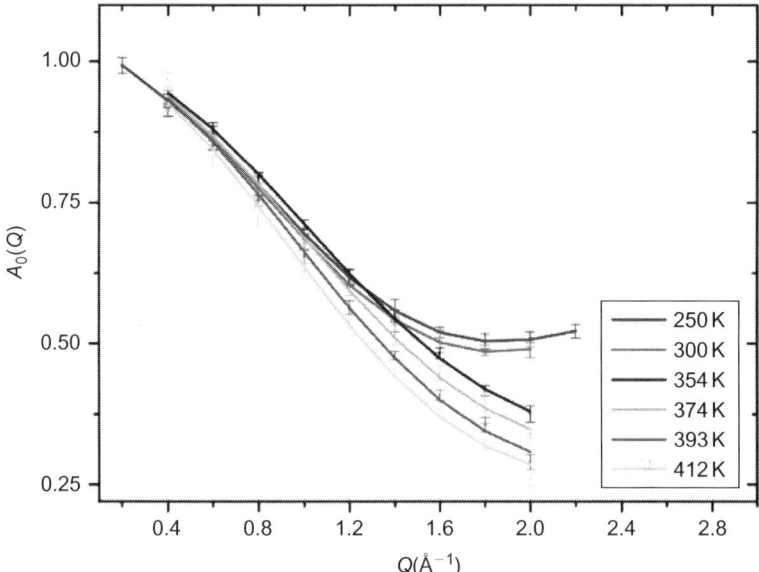

FIGURE 6.29 Observed and calculated EISFs for emim$^+$ as a function of temperature. *Reprinted with permission from Ref. [85].* (For color version of this figure, the reader is referred to the online version of this chapter.)

QENS. In virtually all cases cited here, atomistic simulations, based on empirical force-fields or *ab initio* methods, have been essential in fully exploiting the experimental data. Most experimental studies start with structure determination, and it is not surprising that this is also the starting point for most atomistic modeling that is increasingly used to support and understand the experimental results. Structure is a time-averaged property of a molecular-dynamics trajectory, while dynamics are the time-dependent structural fluctuations, and this extension with additional insight that it brings is often crucial to the property/function relation of the material. We have seen in this chapter how early work on highly idealized systems paved the way to the study of more complex materials.

The study of dynamics at central neutron facilities currently constitutes about 25% of the experiments (excluding magnetism). Structural studies are frequently the starting point of an investigation, and they are facilitated by long tradition of software support. Although the study of dynamics has developed along a different path, it too now relies on good software, not only to analyze experimental data but also to produce numerical models that can be compared with the data. There is a realization at central facilities that analysis software is a key component in the production of scientific results and, at the same time, the providers of simulation software know that their investment in producing high-performance codes can be rewarded through ease-of-use facilitating large user communities—the major DFT codes now have user bases of the order of 1000 research groups.

Future interest in studying dynamics will continue to be stimulated as new materials of both academic and applied interest are devised or discovered, which will provide an incentive for new types of instrumentation that will also open new doors. For example, new or novel materials are normally produced in small quantities and in powdered form. At present, in the most favorable case for a hydrogenated sample, ~ 10 mg of sample may be sufficient to obtain some experimental information, and this points to a major problem for the future. When sample quantities become this small, it is difficult to spread the sample out to use the entire neutron beam, which typically has an area of a few cm^2. Current instrumentation developments are already addressing this difficulty by focusing the neutron beam to reduce the required sample area to less than a cm^2. Smaller samples also tend to facilitate novel sample environments, one example being container-less levitation devices that allow very high temperatures (>2000 K) to be reached [86].

We have shown that INS and QENS have a deceptively strong link to applied materials, which is achieved partly through modeling, but also by better instrumentation allowing a shift from model systems more towards real systems. Cement, energy-storage, thermoelectric, and NTE materials are examples of this trend, and this has encouraged neutron scattering centers to participate in whole projects of this type, in addition to the more normal "punctual" experiments.

REFERENCES

[1] Dosch H, Van de Voorde M. Gennesys white paper. A new European partnership between nanomaterials science & nanotechnology and synchrotron radiation and neutron facilities. Stuttgart: Max-Planck-Institut für Metallforschung; 2009. ISBN 978-3-00-027338-4.
[2] Mitchell PCH, Parker SF, Ramirez-Cuesta AJ, Tomkinson J. Vibrational spectroscopy with neutrons. Singapore: World Scientific; 2005. ISBN 981-256-013-0.
[3] Bée M. Quasielastic neutron scattering: principles and applications in solid state chemistry, biology and materials science. Bristol: Adam Hilger; 1988.
[4] van Hove L. Phys Rev 1954;95(1):249.
[5] Leach AR. Molecular modelling principles and applications. Harlow, UK: Longman; 2001.
[6] Gonzalez M, Johnson M, Kearley GJ, Mounir T, editors. Special issue. Chemical physics "Combining Simulations and Neutron Scattering Experiments: from Models to Systems of Biological and Technological Interest, vol. 317; 2005. p. 111–318, Issues 2–3.
[7] Koch W, Holthausen MC. A chemists' guide to density functional theory. Weinheim, Germany: Wiley-VCH; 2000.
[8] Newport RJ, Rainford BD, Cywinski R. Neutron scattering at a pulsed source. Bristol: Adam Hilger; 1988. ISBN 0-85274-366-1.
[9] Squires GL. Introduction to the theory of thermal neutron scattering. Cambridge: Cambridge University Press; 1978.
[10] Sosnick TR, Snow WM, Silver RN, Sokol PE. Phys Rev B 1991;43:216.
[11] Watson GL. J Phys Condens Matter 1996;8:5955–75.
[12] Mayers J, Reiter G. Meas Sci Technol 2012;23:04902–20.

[13] Homouz D, Reiter G, Eckert J, Mayers J, Blinc R. Phys Rev Lett 2007;98:115502.
[14] Brockhouse BN. Bull Am Phys Soc 1960;5:373.
[15] Alefeld B, Birr M, Heidemann A. Naturwissenschaften 1969;56:410.
[16] Mezei F. Z Phys 1972;255:146.
[17] Mezei Ferenc, Pappas Catherine, Gutberlet Thomas, editors. Neutron spin echo spectroscopy: basics, trends and applications. Berlin: Springer-Verlag; 2003.
[18] Jobic H. Chem Phys Lett 1984;106:321–4.
[19] Zorn R. Physics 2011;4:44–8.
[20] Kearley GJ, Johnson MR, Tomkinson J. J Chem Phys 2006;124:044514.
[21] Fitch AN, Jobic H, Renouprez A. J Chem Soc Chem Commun 1985;5:284–6.
[22] Parlinski K. Software PHONON, Cracow; 2010.
[23] Kohen A, Limbach H-H. Isotope effects in chemistry and biology. Boca Raton, FL: Taylor and Francis; 2006. ISBN 0-8247-2449-6.
[24] Plazanet M, Fukushima N, Johnson MR, Horsewill AJ, Trommsdorff H-P. J Chem Phys 2001;115:3241.
[25] Fontaine-Vive F, Johnson MR, Kearley GJ, Howard JAK, Parker SF. J Am Chem Soc 2006;128:2963.
[26] Fontaine-Vive F, Johnson MR, Kearley GJ, Cowan JA, Howard JAK, Parker SF. J Chem Phys 2006;124:234503.
[27] Boyd PDW, Edwards AJ, Gardiner MG, Hol CC, Lemée-Cailleau M-H, McGuinness DS, et al. Angew Chem Int Ed Engl 2010;49(36):6315–8.
[28] Plazanet M, Fontaine-Vive F, Gardner KH, Forsyth VT, Ivanov A, Ramirez-Cuesta AJ, et al. J Am Chem Soc 2005;127:6672.
[29] Jobic H. J Chem Phys 1982;76:2693–6.
[30] Kearley GJ, Johansson P, Delaplane GG, Lindgren J. Solid State Ion 2002;147:237.
[31] van Eijck L, Best AS, Kearley GJ. Macromolecules 2004;37:9591–5.
[32] Strauch D, Dorner B. J Phys Condens Matter 1990;2:1457.
[33] Stirling WG. J Phys C Solid State Phys 1972;5:2711.
[34] Bartkowiak M, Kearley GJ, Yethiraj M, Mulder AM. Phys Rev B 2011;83(6):064102.
[35] Bartowiak M. PhD Thesis, University of New South Wales, Sydney, Australia; 2012.
[36] Kepert CJ. Chem Commun 2006;695.
[37] Peterson VK, Kearley GJ, Wu Y, Ramirez-Cuesta AJ, Kemner E, Kepert CJ. Angew Chem Int Ed Engl 2010;49(3):585–8.
[38] Bichara C, Johnson MR, Raty J-Y. Phys Rev Lett 2005;95:267801.
[39] Otjacques C, Raty J-Y, Coulet M-V, Johnson MR, Schober H, Bichara C, et al. Phys Rev Lett 2009;103:245901.
[40] Huot J, Liang G, Boily S, Van Neste A, Schulz R. J Alloys Compd 1999;495:293–5.
[41] Schimmel HG, Johnson MR, Kearley GJ, Ramirez-Cuesta AJ, Huot J, Mulder FM. J Alloys Compd 2005;393:1.
[42] Paulus W, Schober H, Eibl S, Johnson MR, Berthier T, Hernandez O, et al. J Am Chem Soc 2008;130:16080.
[43] Gale JD, Rohl AL. Mol. Simul. 2003;29:291–341.
[44] Koza MM, Johnson MR, Viennois R, Mutka H, Girard L, Ravot D. Nat Mater 2008;7:805.
[45] Zbiri M, Mittal R, Rols S, Su Y, Xiao Y, Schober H, et al. J Phys Condens Matter 2010;22:315701.
[46] Zbiri M, Mutka H, Johnson MR, Schober H, Payen C. Phys Rev B 2010;81:104414.
[47] van Eijck L, Merzel F, Rols S, Ollivier J, Forsyth VT, Johnson MR. Phys Rev Lett 2011;107:88102.
[48] Johnson MR, Kearley GJ. Annu Rev Chem 2000;51:297–321.

[49] Press W. Springer tracts in modern physics, Berlin: Springer-Verlag; 1981. ISBN 0-387-10897-1.
[50] Prager M, Heidemann A. Chem Rev 1997;97(8):2933–66.
[51] Fernandez-Alonso F, Cabrillo C, Fernández-Perea R, Bermejo FJ, Gonźalez MA, Mondelli C, et al. Phys Rev B 2012;86:144524.
[52] Schimmel HG, Kearley GJ, Nijkamp MG, Visser CT, de Jong KP, Mulder FM. Chem Eur J 2003;9:4764.
[53] Mulder FM, Dingemans TJ, Schimmel HG, Ramirez-Cuesta AJ, Kearley GJ. Chem Phys 2008;351:72–6.
[54] Queen WL, Bloch ED, Brown CM, Hudson MR, Mason JA, Murray LJ, et al. Dalton Trans 2012;41:4180–7.
[55] Brown CM, Liu Y, Yildirim T, Peterson VK, Kepert CJ. Nanotechnology 2009;20:204025.
[56] Florusse LJ, Peters CJ, Schoonman J, Hester KC, Koh CA, Dec SF, et al. Science 2004;306:469–71.
[57] Xu M, Ulivi L, Celli M, Colognesi D, Bačić Z. Phys Rev B 2011;83:241403.
[58] Peterson VK, Shoko E, Kearley GJ. Faraday Discuss 2011;151:37–46.
[59] Rush JJ, Udovic TJ, Berk NF, Richter D, Magerl A. Europhys Lett 1999;48(2):187–93.
[60] Antonov VE, Dorner B, Fedotov VK, Grosse G, Ivanov AS, Kolesnikov AI, et al. J Alloys Compd 2002;330–332:462–6.
[61] Neumann M, Brougham DF, McGloin CJ, Johnson MR, Horsewill AJ, Trommsdorff H-P. J Chem Phys 1998;109:7300.
[62] Hempelmann R. Lovesey SW, Mitchell EWJ, editors. Quasielstic neutron scattering and solid state diffusion, Oxford series on neutron scattering in condensed matter, vol. 13. Oxford: Clarendon Press; 2000.
[63] Livingston RC, Rowe JM, Rush JJ. J Chem Phys 1974;60:4541.
[64] Jobic H, Theodorou DN. Microporous Mesoporous Mater 2007;102:21059.
[65] Chudley CT, Elliot RJ. Proc R Soc Lond 1961;77:353.
[66] Singwi KS, Sjolander A. Phys Rev 1960;119:863.
[67] Hall PL, Ross DK. Mol Phys 1981;42:673.
[68] Noldeke C, Asmussen B, Press W, Buttner H, Kearley GJ, Lechner RE, et al. J Chem Phys 2000;113:3219.
[69] Price DL, Sabougni ML, Reijers R, Kearley G, White R. Phys Rev Lett 1991;66:1894–7.
[70] Zaccai G. Science 2000;288:1604–7.
[71] Magazù S, Migliardo F, Benedetto A. J Phys Chem B 2011;115:7736–43.
[72] Wood K, Lehnert U, Kessler B, Zaccai G, Oesterhelt D. Biophys J 2008;95:194–202.
[73] Salles F, Bourrelly S, Jobic H, Devic T, Guillerm V, Llewellyn P, et al. ACS Nano 2010;4:143–52.
[74] Qvist J, Schober H, Halle B. J Chem Phys 2011;134:144508–27.
[75] Volino F, Pineri M, Dianoux AJ, De Geyer AJ. J Polym Sci Polym Phys Ed 1982;20:481.
[76] Perrin J-C, Lyonnard S, Volino F. J Phys Chem C 2007;111:3393–404.
[77] Lyonnard S, Berrod Q, Brüning B-A, Gebel G, Guillermo A, Ftouni H, et al. Eur Phys J ST 2010;189:205–16.
[78] Peterson VK. Studying the hydration of cement systems in real-time using quasielastic and inelastic neutron scattering. In: Studying kinetics with neutrons Springer series in solid-state sciences, vol. 161; 2009. p. 19–75.
[79] Peterson VK, Garci Juenger MC. Chem Mater 2006;18:5798–804.
[80] Chen S-H, Liao C, Sciortino F, Gallo P, Tartaglia P. Phys Rev E 2001;59:020201.
[81] Faraone A, Liu L, Copley JRD, Chen S-H. J Chem Phys 2003;119:3963–71.

[82] Fitter J, Katsaras J, Gutberlet T. Neutron scattering in biology. Berlin, Heidelberg: Springer; 2006.
[83] Wood K, Caronnac C, Fouqueta P, Haussler W, Natali Francesca, Ollivier J, et al. Chem Phys 2008;345:305–14.
[84] Fitter J, Lechner RE, Büldt G, Dencher NA. Proc Natl Acad Sci USA 1996;93:7600–5.
[85] Aoun B, González MA, Ollivier J, Russina M, Izaola Zunbeltz, Price DL, et al. J Phys Chem Lett 2010;1(17):2503–7.
[86] Hennet L, Pozdnyakova I, Bytchkov A, Cristiglio V, Palleau P, Fischer HE, et al. Rev Sci Inst 2006;77:53903.

Appendix

Neutron Scattering Lengths and Cross Sections

Javier Dawidowski, José Rolando Granada, Javier Roberto Santisteban, Florencia Cantargi and Luis Alberto Rodríguez Palomino
Comisión Nacional de Energía Atómica, Consejo Nacional de investigaciones Científicas y Técnicas, Centro Atómico Bariloche and Instituto Balseiro, Bariloche, Río Negro, Argentina

Chapter Outline

A.1. Introduction	471	A.3.6. Neutron Gravity Refractometer 490
A.2. Theoretical Background	472	A.3.7. Neutron Interferometry 491
A.2.1. Scattering Length	472	A.3.8. Small-Angle Scattering 492
A.2.2. Spin-Dependent Scattering Lengths	476	A.3.9. Total Reflection 493
A.2.3. Neutron–Atom Interactions	477	A.3.10. Pseudomagnetic Method 493
A.3. Methods of Measurement of Scattering Lengths	482	A.3.11. High-Energy Experiments 494
A.3.1. Transmission	482	A.4. Tables of Neutron Scattering Lengths and Cross Sections 495
A.3.2. Bragg Diffraction	484	
A.3.3. Dynamical Diffraction	486	
A.3.4. Prism Refraction	488	
A.3.5. Christiansen Filter	489	References 527

A.1 INTRODUCTION

Low-energy neutron scattering techniques are widely employed nowadays as a tool to investigate the structure and dynamics of matter, as well as the fundamental interactions between neutron and matter.

Practically in all the scattering experiments with thermal neutrons, the main interactions involved are the nuclear force and those arising from the interaction of the neutron spin with unpaired electron density in the material under investigation, when it exists [1]. While the latter can be calculated from the electronic structure of the atoms, the former involves the strong forces that depend on the detailed structure of the nuclei. Therefore, the basic quantity that describes the slow neutron–nucleus interaction (scattering length) must be measured and tabulated for the different nuclear species. In 1947, Fermi and Marshall [2] conducted the first comprehensive set of measurements of scattering lengths, whose validity extends to a large extent to the present day. However, since then, numerous techniques have been developed to increase the accuracy and precision of these measurements.

Nuclear absorption processes (comprising radiative-capture, neutron–proton, neutron–alpha, and neutron–fission processes) are also involved in the neutron–nucleus interaction. In the formal theory of neutron scattering, the scattering length is a complex number, where the imaginary part is related to absorption. As in the previous case, there is no precise theory that describes this phenomenon, so the values of absorption cross sections must also be measured and tabulated.

For those reasons, scattering length tables are a necessary reference source for those who use low-energy-neutron experimental techniques. In the past, several tables of scattering lengths have been published, those of Sears [3] and Koester *et al.* [4] being the two most recent and outstanding works.

The aim of this work is to present the most updated values of the scattering lengths. We first present a summary of the theory of scattering lengths, based on the fundamental work of Sears [5], who accurately stated the terminology and basic definitions, helping to correct ambiguities. We also include a paragraph on the neutron–electron interaction, whose order of magnitude is larger than presently available experimental uncertainties, so it may require consideration in many experiments, in order not to confuse the neutron–nucleus and neutron–atom scattering lengths.

The third section is devoted to describe the main experimental techniques used in the determination of scattering lengths. To conclude, we present a comprehensive table of scattering data, along with a brief explanation of its use, and associated bibliography.

A.2 THEORETICAL BACKGROUND

A.2.1 Scattering Length

In this section, we introduce the neutron–nucleus interaction governed by the strong nuclear forces. In this context, we develop the theoretical background of neutron scattering lengths, valid in the framework of slow neutron scattering, in which the spatial range of neutron–nucleus interactions is negligibly small relative to thermal neutron wavelengths. We leave for Section A.2.3 a more complete picture of the various neutron–atom interactions.

We first pose the scattering problem of neutrons of energy E (wave vector of modulus k), propagating along the z-axis, interacting with a nucleus of infinite mass. As a consequence, the scattering processes we will describe are elastic and we can write $|\mathbf{k}_i|=|\mathbf{k}_f|$, and as a result, we will obtain an expression for the bound scattering length.

The wave function that formally solves the scattering problem is

$$\psi(\mathbf{r}) = \exp(ikz) + \frac{e^{ikr}}{r} f(\mathbf{k}), \qquad (A.1)$$

where first term is the incident wave and the second term is the scattered wave whose propagation vector is $\mathbf{k} = k(\mathbf{r}/r)$. The *scattering amplitude* $f(\mathbf{k})$ is directly related to the interaction potential U through the expression [6]

$$f(\mathbf{k}) = -\frac{m_n}{2\pi\hbar^2} \int e^{-i\mathbf{k}\cdot\mathbf{r}'} U(r') \psi(\mathbf{r}') d\mathbf{r}'. \qquad (A.2)$$

For thermal neutrons, the range of nuclear forces is negligible compared with the neutron wavelength, so in Equation (A.2) $\mathbf{k}\cdot\mathbf{r}' \ll 1$ (S-wave scattering approximation), and the scattering amplitude admits an expansion in powers of k:

$$f(k) = -b + ikb^2 + O(k^2), \qquad (A.3)$$

where b is called the *scattering length* and is normally determined by experiments. For thermal neutrons $k \approx 2\,\text{Å}^{-1}$, and since typical values of b are about 5 fm, then $|kb| \approx 10^{-4}$ and normally a good approximation is to set

$$f(k) = -b. \qquad (A.4)$$

We will assume that the atom is in its ground (nuclear and electronic) state and remains unexcited after the collision. Under those conditions, the scattering length $f(k)$ can be calculated within the Born approximation, bearing in mind the short range of the nuclear forces and the weakness of the long-range electromagnetic forces:

$$f(\mathbf{k}) = \frac{m_n}{2\pi\hbar^2} \int U(r) e^{-i\mathbf{Q}\cdot\mathbf{r}} d\mathbf{r}, \qquad (A.5)$$

where $\mathbf{Q} = \mathbf{k}_i - \mathbf{k}_f$ is the difference between the incident and the scattered wave vectors.

The scattering length is in general a complex number [5]

$$b = b' - ib'', \qquad (A.6)$$

and its imaginary part (related to neutron absorption) is usually of the same order of magnitude as the second term in Equation (A.3). In the cases where Equation (A.4) is valid, and the scattered wave is described in the first Born approximation, the Fermi pseudopotential is adequate to describe $U(r)$

$$U(r) = \frac{2\pi\hbar^2 b}{m_n} \delta(r). \qquad (A.7)$$

The Fermi pseudopotential is constructed as the potential with scattering length b for which the first Born approximation is exact. It is widely employed in thermal neutron scattering, but in cases where the imaginary part of $f(\mathbf{k})$ is important, it is no longer valid. In most practical cases, the scattering amplitude is uniquely determined by the scattering length, although it is not always described by the Fermi pseudopotential [5].

The scattering amplitude of an atom is directly related to the differential cross section for scattering into the solid angle $d\Omega$ about the direction $\mathbf{\Omega} = \mathbf{k}/k$

$$\frac{d\sigma}{d\Omega} = |f(\mathbf{k})|^2, \qquad (A.8)$$

equivalent to

$$\frac{d\sigma}{d\Omega} = |b|^2 \left(1 - 2kb''\right) \qquad (A.9)$$

neglecting terms of order higher than k. For S-waves, this cross section is independent of the angle.

Important properties are derived from $f(\mathbf{k})$ regarding the total cross section

$$\sigma_t = \sigma_s + \sigma_a, \qquad (A.10)$$

$$\sigma_t = \frac{4\pi}{k} \mathrm{Im}[f(\mathbf{k})]_{\theta=0} \qquad (A.11)$$

equivalent to

$$\sigma_t = \frac{4\pi}{k} b'' + 4\pi \left(b'^2 - b''^2\right). \qquad (A.12)$$

- Integrating Equation (A.9) in all directions, we obtain the total scattering cross section

$$\sigma_s = 4\pi |b|^2 \left(1 - 2kb''\right). \qquad (A.13)$$

- Subtracting Equation (A.13) from Equation (A.12), we obtain an expression for the absorption cross section

$$\sigma_a = \frac{4\pi}{k} b'' |1 - 2kb''|. \qquad (A.14)$$

The term kb'' is of the order of 10^{-4} and is usually neglected, so the following expression is normally used for the scattering total cross section

$$\sigma_s = 4\pi |b|^2 \quad (A.15)$$

and for the absorption cross section

$$\sigma_a = \frac{4\pi}{k} b''. \quad (A.16)$$

Equation (A.16) is known as the "$1/v$" law for the absorption cross section, because of its inverse proportionality with neutron velocity.

The above expressions involve the bound cross sections. For free nuclei, the scattering problem must be posed in the center-of-mass frame. Thus, the potential of Equation (A.7) must include the free-atom scattering length a, and the reduced neutron mass $\mu = \frac{m_n M}{m_n + M}$, where M is the mass of the scattering atom. The problem is formally the same, and the following expression holds between the free and bound atom scattering length [5]

$$a = \left(\frac{A}{A+1}\right) b \quad (A.17)$$

(both for its real and imaginary parts). Thus, Equation (A.15), valid for a bound atom, is related to the free-atom total cross section through

$$\sigma_{s,\text{free}} = \left(\frac{A}{A+1}\right)^2 \sigma_s, \quad (A.18)$$

whereas the free-atom absorption cross section is

$$\sigma_{a,\text{free}} = \frac{4\pi}{k_i} a'', \quad (A.19)$$

where k_i is the incident neutron wave vector in the center-of-mass system. Since $k/k_i = (A+1)/A$ then,

$$\sigma_{a,\text{free}} = \sigma_a. \quad (A.20)$$

In the discussion developed so far, we implicitly assumed that there are no nuclear resonances in the thermal region. In cases where (n,γ) resonances (due to internal nuclear particle rearrangements) influence the thermal zone, the scattering length must be rewritten by adding the Breit–Wigner terms [7]

$$b = R + \frac{\Gamma_{n,r}/2k}{E - E_r + i\Gamma_r/2} = R + b_r(E) \quad (A.21)$$

where R is the scattering length due to potential scattering (i.e., the direct scattering term due to the presence of an interaction potential) and $b_r(E)$ is the resonance scattering length. Here, E_r is the energy of the rth resonance, $\Gamma_r = \Gamma_{n,r} + \Gamma_{\gamma,r}$ is its total width, and E is the incident neutron energy in the center-of-mass system. Examples of isotopes with thermal neutron resonances are shown in Table A.4.

A.2.2 Spin-Dependent Scattering Lengths

So far we have not discussed the dependence of the scattering length with the neutron spin. The nucleus of spin $\mathbf{I}$ and the neutron spin $\frac{1}{2}\boldsymbol{\sigma}$ couple to the total spin

$$\mathbf{J} = \mathbf{I} + \frac{1}{2}\boldsymbol{\sigma}. \quad (A.22)$$

The scattering length takes the values b^+ and b^- depending on the coupling. This is conveniently described through the operator $\hat{b}$ that has the eigenvalues $b^{\pm}$ for the two possible states of the total spin $I \pm \frac{1}{2}$ [8]

$$\hat{b} = \frac{(I+1)b^+ + Ib^-}{2I+1} + \frac{b^+ - b^-}{2I+1}\boldsymbol{\sigma}\cdot\mathbf{I}. \quad (A.23)$$

The *coherent scattering length* is defined as the mean $\bar{b}$ value

$$\bar{b} = \frac{(I+1)b^+ + Ib^-}{2I+1}, \quad (A.24)$$

whereas the *incoherent scattering length* is defined as its variance

$$b_i = \frac{\sqrt{I(I+1)}}{2I+1}(b^+ - b^-). \quad (A.25)$$

With these definitions, Equation (A.23) can also be written as

$$\hat{b} = \bar{b} + \frac{1}{\sqrt{I(I+1)}}b_i\boldsymbol{\sigma}\cdot\mathbf{I}. \quad (A.26)$$

Conversely,

$$b^+ = \bar{b} + \sqrt{\frac{I}{(I+1)}}b_i \quad (A.27)$$

and

$$b^- = \bar{b} - \sqrt{\frac{(I+1)}{I}}b_i. \quad (A.28)$$

The scattering and absorption cross sections for bound nuclei are

$$\sigma_s = 4\pi\left\langle |\hat{b}|^2 \right\rangle \quad (A.29)$$

and

$$\sigma_a = \frac{4\pi}{k}\langle \hat{b}''\rangle, \quad (A.30)$$

where the brackets mean the average over neutron and nuclear spins. For unpolarized nuclei and neutrons, $\langle \hat{b}\rangle = \bar{b}$ and the bound scattering cross

Appendix | Neutron Scattering Lengths and Cross Sections

section can be expressed as the sum of the bound coherent and the bound incoherent cross section

$$\sigma_s = \sigma_c + \sigma_i, \qquad (A.31)$$

where

$$\sigma_c = 4\pi |\bar{b}|^2 \qquad (A.32)$$

and

$$\sigma_i = 4\pi |b_i|^2. \qquad (A.33)$$

The absorption cross section can be written as

$$\sigma_a = \frac{4\pi}{k} \bar{b}''. \qquad (A.34)$$

The expressions showed so far depend on the nuclide considered. When we consider an isotopic distribution of nuclides, the average expressions (A.29) and (A.30) must include both the isotope and spin distributions. If C_I is the fraction of isotopes of type I so $\sum C_I = 1$ then

$$\sigma_s = \sum_I C_I \sigma_{sI}, \quad \sigma_a = \sum_I C_I \sigma_{aI}, \quad \bar{b} = \sum_I C_I \bar{b}_I. \qquad (A.35)$$

Now the incoherent cross section can be written as

$$\sigma_i = \sigma_i(\text{spin}) + \sigma_i(\text{isotope}), \qquad (A.36)$$

where

$$\sigma_i(\text{spin}) = 4\pi \sum_I C_I |b_{iI}|^2 \qquad (A.37)$$

and

$$\sigma_i(\text{isotope}) = 4\pi \sum_{I<I'} C_I C_{I'} |\bar{b}_I - \bar{b}_{I'}|^2. \qquad (A.38)$$

A.2.3 Neutron–Atom Interactions

In the preceding sections, we discussed only the neutron–nucleus interactions dominated by strong nuclear forces. To complete this description, it is necessary to discuss additional neutron–atom interactions.

The scattering length is dominated by the strong nuclear force and the contribution due to the magnetic–dipole coupling with the magnetic electrons when it exists. In addition, other electromagnetic interactions of the neutron with the nucleus and the atomic electrons may contribute to the cross section of the scattering process, including the spin–orbit and Foldy interactions, as well as those related to the electric polarizability and the finite intrinsic charge radius of the neutron [1].

Those electromagnetic interactions are typically a few orders of magnitude smaller than the nuclear interaction, yet depending on the neutron energy, some of them may need to be accounted for while performing high-precision data analysis or calculations. We briefly review here the interaction potentials for those electromagnetic contributions, following the formulation given by Sears [1].

Let us consider the scattering of a neutron by a single, isolated atom placed at the coordinate's origin. We will assume that the atom is in its nuclear ground state and remains unexcited after the collision, which is then described in terms of an effective interaction potential $U(r)$. Under those conditions, the scattering length $b(\mathbf{Q})$ can be calculated within the Born approximation through Equation (A.5).

The total potential can be written in the form

$$U = U_N + U_M + U_E \qquad (A.39)$$

where U_N is the potential that represents the strong nuclear force, U_M is the coupling of the neutron dipole moment with the electromagnetic field of the atom, and U_E is the electrostatic energy that arises from the charge structure of the neutron.

In Equation (A.21), we presented the expression for the scattering length, where thermal resonances are present, and in Equation (A.26), we presented the expression for the scattering length of a nucleus without considering the contribution due to resonances $b_r(E)$. Thus, the nuclear interaction between the neutron and the atomic nuclei, within the Fermi pseudopotential approximation and for S-wave neutrons, leads to

$$b_N(E) = \overline{b} + b_i + b_r(E), \qquad (A.40)$$

where $b_N(E)$ represents the nuclear scattering length. This expression includes all the effects due to nuclear forces in the interaction of unpolarized neutrons.

As already mentioned, the neutron is a spin-1/2 particle which, during a collision with an atom, is moving inside an electromagnetic field ($\mathbf{E}(\mathbf{r})$ and $\mathbf{B}(\mathbf{r})$) that leads to Zeeman and spin–orbit interactions:

$$U_{M1}(\mathbf{r}) = -\boldsymbol{\mu} \cdot \mathbf{B}(\mathbf{r}), \qquad (A.41)$$

$$U_{M2}(\mathbf{r}) = -\frac{1}{m_n c} \boldsymbol{\mu} \cdot (\mathbf{E}(\mathbf{r}) \times \mathbf{p}), \qquad (A.42)$$

where $\boldsymbol{\mu}$ and $\mathbf{p}$ are the magnetic moment and momentum of the neutron, respectively. In addition to the above two terms, it is necessary to include residual contributions of relativistic quantum origin, as the Foldy term:

$$U_{M3}(\mathbf{r}) = -\frac{2\pi\hbar}{m_n c} \mu \rho_e(\mathbf{r}), \qquad (A.43)$$

where $\rho_e(\mathbf{r})$ is the electronic charge density.

Appendix | Neutron Scattering Lengths and Cross Sections

The internal structure of the neutron can be described in terms of quark models; however in our present analysis, it is simply represented by an effective charge density $\rho_n(\mathbf{r})$ [1]. Then, the scattering length corresponding to the electrostatic energy can be derived from

$$U_{E1}(r) = \int \varphi(\mathbf{r}+\mathbf{r}')\rho_n(\mathbf{r}')d r', \quad (A.44)$$

where $\varphi(\mathbf{r})$ is the atomic electrostatic potential. In U_{E1}, we have treated the neutron as having a rigid charge distribution $\rho_n(\mathbf{r})$, whereas it can be expected to have an electric polarizability α which gives rise to an interaction energy in the presence of the atomic electric field $E(\mathbf{r})$:

$$U_{E2}(\mathbf{r}) = \frac{1}{2}\alpha E(\mathbf{r})^2. \quad (A.45)$$

According to Equations (A.4) and (A.5) and the potentials associated to the different terms of the interactions (Equations A.41–A.45), for unpolarized neutrons, we can write in general:

$$b(E) = b_N(E) + b_M + b_E + b_P(E), \quad (A.46)$$

where $b_N(E)$ is the nuclear scattering length given by Equation (A.40) and $b_P(E)$ is the contribution from the neutron electrostatic polarizability. The other terms, corresponding to total magnetic and electric interactions, include **Q**-dependent form factors that reflect the atomic electromagnetic field and nuclear charge structure and can be rearranged in order to show their explicit dependence with spin. However, for the interaction of unpolarized neutrons with diamagnetic atoms, we essentially end up with contributions to coherent, incoherent, and spin–orbit or Schwinger cross sections (the latter very small for thermal neutrons) [1], as well as the absorption cross section.

Leaving aside the incoherent resonance contributions and rearranging terms in Equation (A.46), we can define the *atomic coherent* scattering length as

$$b_c(\mathbf{Q}) = b_c(0) - b_e Z[1 - F(\mathbf{Q})], \quad (A.47)$$

where $b_c(0)$, *the zero-energy nuclear coherent scattering length related to the purely nuclear interaction*, is given by

$$b_c(0) = \overline{b} + b_P(0) \quad (A.48)$$

and the constant b_e in Equation (A.47) is the neutron–electron scattering length, given by

$$b_e = b_F + b_I. \quad (A.49)$$

In principle, b_P depends on the neutron energy through a nuclear form factor, but for thermal neutrons, it is a constant related to the neutron electrical polarizability α [9–11]. In Equation (A.49), b_F is the Foldy scattering length [12] and b_I is the *intrinsic* neutron–electron scattering length associated to the (neutron charge separation) electrostatic interaction. The value of b_e can be experimentally determined, the presently accepted value being [11,13–15]:

$$b_e = -(1.33 \pm 0.03) \times 10^{-3} \, \text{fm}. \tag{A.50}$$

We must, however, emphasize that precise determinations of that quantity may involve complex corrections [16].

The atomic form factor $F(\mathbf{Q})$ in Equation (A.47) is measured by X-ray scattering for each element of atomic number Z or can be calculated using a relativistic Hartree–Fock formalism. In any case, it will be adequate for our purposes to employ an approximation due to Sears [1]:

$$F(Q) = \left[1 + 3\left(\frac{Q}{Q_0}\right)^2\right]^{-1/2}, \tag{A.51}$$

where $Q = 2k \sin(\varphi/2)$ and $Q_0 = 1.9 Z^{1/3} \, \text{Å}^{-1}$. The angular average of this expression is as follows:

$$\overline{F} = \frac{2}{x}\left(\sqrt{1+x} - 1\right) \tag{A.52}$$

with $x = 12(k/Q_0)^2$.

In Figure A.1, we display the angular averaged atomic form factor of Pb, as obtained from a full calculation and that resulting from the use of the approximation (A.52), over the slow-neutron energy range.

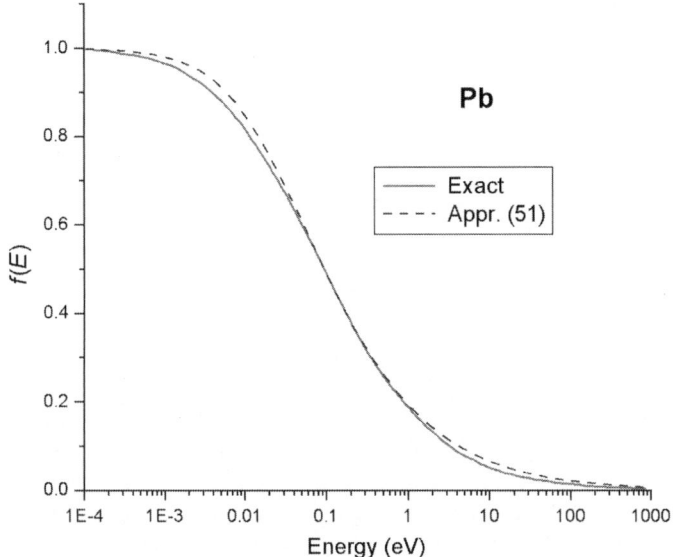

FIGURE A.1 The atomic form factor for Pb. See text for details. (For color version of this figure, the reader is referred to the online version of this chapter.)

Correspondingly, the scattering length which determines the total coherent cross section

$$\sigma_c(E) = 4\pi |b_c(E)^2| \qquad (A.53)$$

is obtained from Equation (A.47):

$$b_c(E) = b_c(0) - b_e Z\left[1 - \overline{F}(E)\right]. \qquad (A.54)$$

The latter expression is sometimes written in the form

$$b_c(E) = b_v + b_e Z \overline{F}(E), \qquad (A.55)$$

where

$$b_v = b_c(0) - b_e Z = \overline{b} + b_P - b_e Z \qquad (A.56)$$

is the *nuclear* coherent scattering length, containing contributions due to purely nuclear forces, the electrical polarizability of the nuclear charge, and the neutron interaction with the electric field of the Z protons in the nucleus.

The conclusion is that in scattering processes involving thermal neutrons, the coherent scattering length varies according to Equation (A.47), while in transmission experiments (where total cross section as a function of energy is measured), it does according to Equation (A.54) or Equation (A.55). If the neutron energy tends to zero, we will measure $b_c(0) = b_c$, the *bound atom coherent scattering length*, whereas in the free-atom energy region and beyond the coherent cross section is determined by b_v. In Figure A.2, we

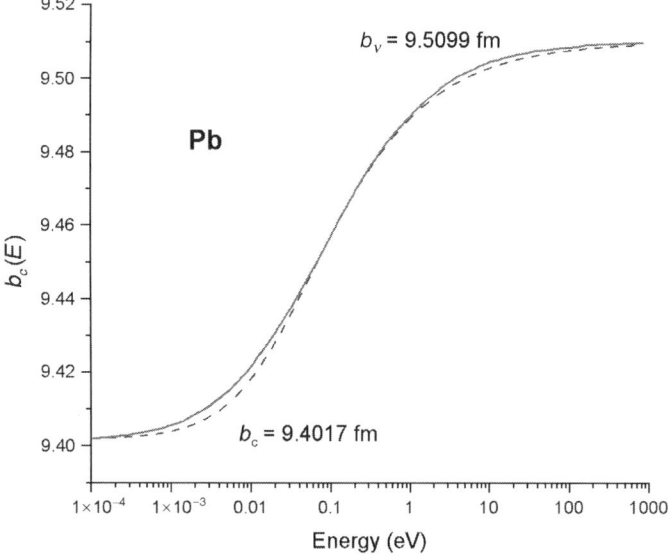

FIGURE A.2 Variation of the coherent scattering length of Pb over the thermal energy region. The exact calculation is shown in full line, while the dotted line shows the approximate formalism. (For color version of this figure, the reader is referred to the online version of this chapter.)

illustrate the variation of the atomic coherent scattering length, Equation (A.55), for the case of elemental Pb, evaluated according the exact and approximate (Equation A.52) formalisms.

A.3 METHODS OF MEASUREMENT OF SCATTERING LENGTHS

In this section, we show the main methods for measuring scattering lengths. The different techniques that we will refer to are in general complementary and are designed to determine the coherent and incoherent scattering length, as well as its imaginary part, related to the absorption cross section.

In Table A.1, we show a summary of the main features of the experimental techniques that will be treated in this section. Although most of the techniques are based on neutron beams incident on the samples to be studied, the high-energy technique employs the neutrons produced in nuclear reactions, to derive its fundamental interactions with neutrons and protons.

Most of the coherent scattering lengths were firstly determined by Bragg and mirror reflections. However, they are seldom used today due to their relatively low accuracy. Among the methods we will mention, the most versatile are neutron interferometry, transmission, and Christiansen-filter technique. It is worth noting that all the scattering length data shown in Table A.2 of this work were measured by one (or several) of the techniques mentioned in this section.

A.3.1 Transmission

Transmission experiments are designed for the measurement of total cross sections. In the case of unpolarized neutrons and atoms, Equation (A.31) shows that the scattering total cross section is simply the sum of the coherent and the incoherent components. Traditionally, this technique has been used complemented with some of the others described in the paragraphs below to determine both components.

A transmission experiment is perhaps the most basic of all neutron techniques [17]. It consists on a collimated neutron beam that hits perpendicularly on a flat sample that can be removed from the beam. The "sample-in, sample-out" method allows to measure the ratio of transmitted neutrons. In pulsed neutron sources, a white neutron beam is used that allows to discriminate the neutron energy through the time-of-flight (TOF) method. The transmission factor is directly related with the total cross section through

$$T(E) = e^{-Nt\sigma_t(E)}, \qquad (A.57)$$

where N is the number of atoms per unit volume in the sample and t its thickness. It can be seen from Equation (A.10) that the total cross section is the sum of the scattering and the absorption components. For epithermal neutrons in most nuclei, the absorption component is negligible. In the absence of

TABLE A.1 Summary of the Main Features of the Experimental Methods Employed in the Measurement of Scattering Lengths

Method	Typical accuracy (%)	Neutron beam	Special requirements/ suitable for
Transmission	0.1	Polychromatic, thermal and epithermal	All kind of samples/ suitable for gases
Bragg diffraction	1	Monochromatic, thermal	Polycrystalline samples
Dynamical diffraction	0.03	Monochromatic, thermal	Large perfect crystal samples
Prism refraction	0.03	Monochromatic, cold	Very small deflection angles measured
Christiansen filter	0.1	Monochromatic, cold	Allows small amounts of sample
Neutron gravity refractometer	0.02	Monochromatic, cold	Large liquid samples
Neutron interferometry	0.1	Monochromatic, cold	Very accurate knowledge of incident wavelength
Small-angle scattering	1	Monochromatic, cold	Allows small amounts of sample
Total reflection	1	Monochromatic, cold	Plane surface samples. Liquids
Pseudomagnetic method	3	Monochromatic, cold, polarized	Polarized nuclei in samples
High-energy experiments	3	Neutrons generated as product of nuclear reaction	Large accelerator facilities

low-energy nuclear resonances, the total cross section has the asymptotic expansion [6]

$$\sigma_t(E) = \sigma_{s,\text{free}} \left[1 + \frac{k_B \overline{T}}{AE} + \cdots \right], \quad (A.58)$$

where $\overline{T}$ is the effective temperature of the system [18] from which $\sigma_{s,\text{free}}$ can be determined. Figure A.3 shows the total cross section of iron [19] along with a theoretical calculation. Its main features are governed by the Bragg edges, and at epithermal energies, it is asymptotically described by Equation (A.58). Accuracies of 0.1% can be attained in the determination of

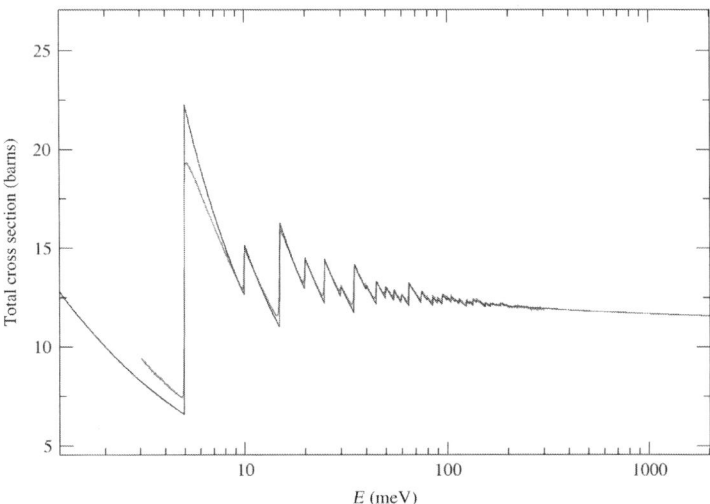

FIGURE A.3 Experimental total cross section of iron powder (after Santisteban et al. [19], red symbols). The theoretical calculation using the code CRIPO [20] (blue curve) tends asymptotically to Equation (A.58) at epithermal energies. The difference in intensity in the first Bragg edge is due to extinction effects. (For interpretation of the references to color in this figure legend, the reader is referred to the online version of this chapter.)

$\sigma_{s,\text{free}}$ by employing this technique, which is compatible with the accuracies of the techniques to determine the coherent scattering length.

Although examples are rare in the literature, a direct measurement of the incoherent scattering length can be attained by polarized neutron transmission in a sample with polarized atoms [21]. Also the absorption cross section has been preferably explored with thermal neutron transmission. From the measured total cross sections, the imaginary component of the scattering length can be determined by fitting the "$1/v$" law in Equation (A.16). With respect to the spin dependence of σ_a, in most of the nuclides, absorption is significant either in the $I+1/2$ or in the $I-1/2$ state of spin. In such cases, the determination of b'' from unpolarized neutron transmission is straightforward. In ^{6}Li where the absorption is large both for $I\pm 1/2$, polarized neutron transmission must be employed.

A.3.2 Bragg Diffraction

The majority of coherent scattering lengths have been determined from the Bragg intensities of monochromatic neutrons scattered on polycrystalline samples using the kinematical theory of diffraction to evaluate the diffracted intensities [2,22]. A typical arrangement of a modern instrument is shown in Figure A.4. For a cylindrical sample of volume V completely immersed in a

Appendix | Neutron Scattering Lengths and Cross Sections

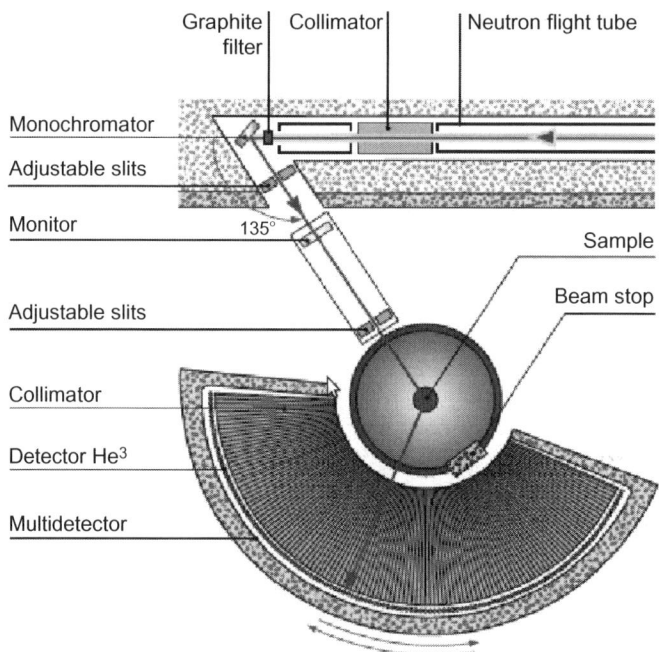

FIGURE A.4 Experimental arrangement used to measure Bragg diffraction from polycrystals. The setup corresponds to instrument D2B at Institut Laue Langevin (Grenoble, France). Modern instruments use a position-sensitive detector array. The original experiments by Shull and Wollan used a single detector rotated around the sample. *Reprinted with permission of ILL, Grenoble, France.* (For color version of this figure, the reader is referred to the online version of this chapter.)

neutron beam of wavelength λ and intensity I_0, the area of the diffraction peak appearing at an angle $2\theta_{hkl}$ from the incident neutron beam is

$$I_{hkl} = I_0 \frac{\lambda^3 V N_c^2 m_{hkl} F_{hkl}^2}{\sin\theta_{hkl} \sin 2\theta_{hkl}} A_{hkl}, \tag{A.59}$$

where N_c is the number of unit cells per unit volume, m_{hkl} the multiplicity of the hkl lattice plane, A_{hkl} an absorption factor, and F_{hkl} the scattering amplitude per unit cell at the temperature T at which the measurement is made, given by

$$F_{hkl} = \sum_{j \in \text{atoms in unit cell}} \overline{b}_j \exp\left[2\pi i\left(hx_j + ky_j + lz_j\right)\right] \exp\left[-W_j(T)\sin^2\theta_{hkl}/\lambda^2\right]$$

$$\tag{A.60}$$

with $W_j(T)$ the mean square displacement of the atom located at the position (x_j, y_j, z_j) of the unit cell. For monatomic lattices, the geometrical structure factor can be separated out and the intensity becomes proportional to $\overline{b}^2$.

For crystalline compounds, the coherent scattering amplitude for the individual elements involved can be determined by measuring several hkl peaks, for which the contributions to F_{hkl} by the elements in question are different. For ferromagnetic and antiferromagnetic materials, there could be a sizeable contribution to the coherent scattering from the magnetic interaction of the neutron, and the observed data must be resolved into its nuclear and magnetic components.

A direct measurement of the absolute intensity of the monochromatic incident beam offers considerable technical difficulties, so it has been found more convenient and accurate to compare the diffracted intensities of the unknown sample with those from a substance of known coherent scattering cross section. So, I_0 is evaluated from the diffracted intensities of the calibrating substance, and this value is used in subsequent studies. Ideally, the above calibration should be performed with an element having zero nuclear spin (ThO_2, C, Ni^{58}), so that the coherent scattering cross section is identical to the total bound scattering cross section, which can be directly evaluated from a simple transmission experiment. Due to various extinction effects, the accuracy is limited to about $\sim 1\%$. Better accuracy can be achieved using single crystals, but in this case, a different experimental arrangement is required, and the diffracted signal must be described by the dynamical theory of diffraction [5].

As in the case of transmission, scattering methods can also be employed to determine the incoherent scattering by using polarized neutrons [23].

A.3.3 Dynamical Diffraction

When a collimated beam is brought into a perfect crystal under near-Bragg orientation conditions, the dynamical theory of diffraction predicts a coherent splitting of the incident wave into four components, with two travelling-wave components in the Bragg direction and two components in the forward, or incident, direction (Figure A.5). The intensity of the neutron beam emanating at the exit surface depends on the incident wavelength, the density, thickness, and the scattering length of the atoms composing the crystal, in such a way that it oscillates as a function of the incident wavelength between two values, causing a phenomenon known as Pendellösung (pendulum solution). Measurement of the period of oscillation allows a very precise definition of the scattering length.

Monochromatic neutrons of reasonably high collimation are delivered to a perfect single crystal through a fine entrance slit (slit A). Dynamical theory predicts that the radiation energy is transmitted in two symmetrical directions described by the angle ε (Figure A.5) with this angle determined by the direction of the incident ray θ and its deviation from the exact Bragg angle θ_B. The angle ε ranges from zero when $\theta = \theta_B$ and energy flow is along the Bragg planes, up to a limit value of $\varepsilon = \theta_B$. The Bragg-reflected intensity emanates

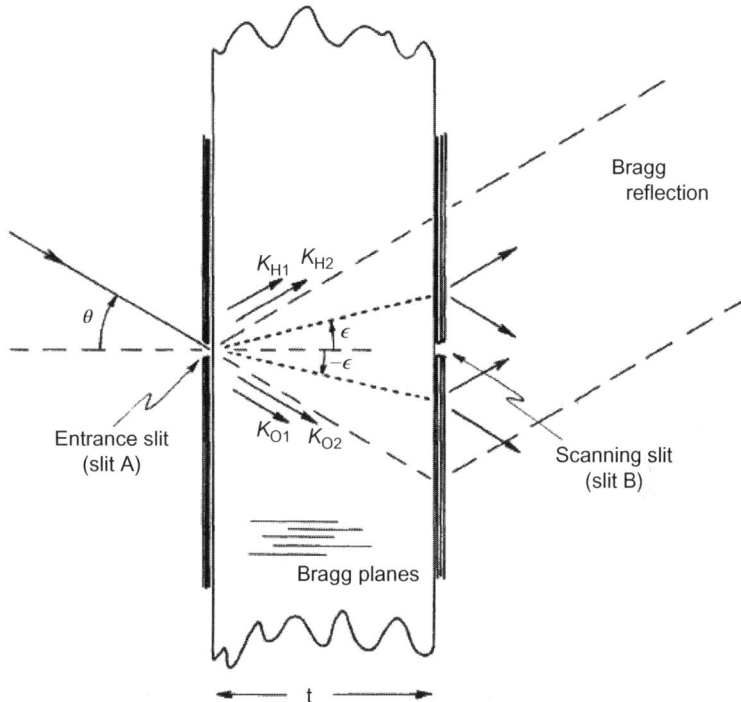

FIGURE A.5 Schematic diagram of a Bragg-reflection experiment to detect Pendellösung interference fringes. *Reprinted with permission from Ref. [24].*

from the exit surface over a band whose linear width is dictated by ε. The intensity distribution within this band is a function of the parameter $\gamma = \tan(\varepsilon)/\tan(\theta_B)$, given by

$$I(\gamma) = C(1-\gamma^2)^{-1/2} \sin^2\left((\pi t/\Delta_0)(1-\gamma^2)^{-1/2}\right) \quad (A.61)$$

with t the crystal thickness and C a normalization constant. The Pendellösung period Δ_0 is given by

$$\Delta_0 = \frac{\pi \cos\theta}{N\lambda F_{hkl}}, \quad (A.62)$$

where N is the number of unit cells per unit volume, and F_{hkl} is the crystal structure factor per unit cell as given by Equation (A.60). This intensity profile can be measured by using a scanning slit at the exit surface. For $\varepsilon = 0$ ($\gamma = 0$), intensity oscillations occur as a function of t or of λ. Figure A.6 shows the oscillations measured for Si crystals of three different thicknesses by keeping the scanning slit aligned to the incident slit, and varying the wavelength of the incident neutrons. As in the Bragg diffraction method (A.3.2), the coherent scattering length is determined through F_{hkl}.

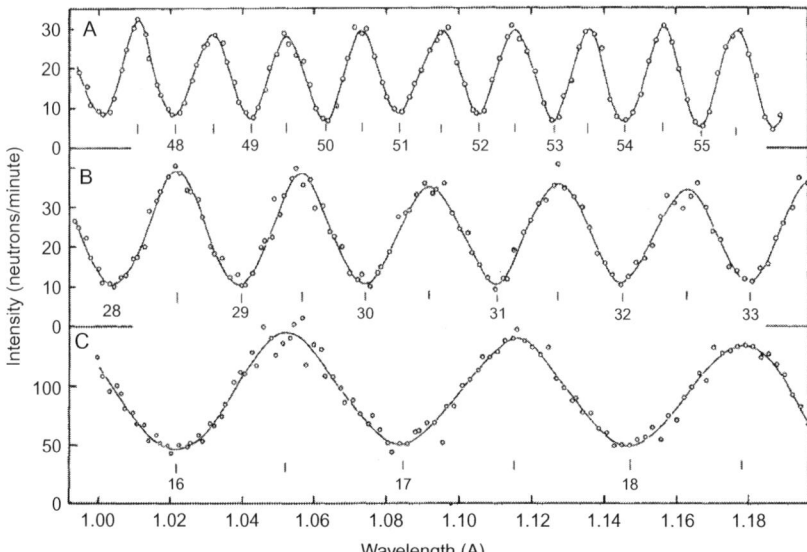

FIGURE A.6 Intensity oscillations as a function on neutron wavelength, registered at center of the Bragg reflection. The three fringe patterns correspond to different crystal thickness: (a) 1.0000 cm, (b) 0.5939 cm, and (c) 0.3315 cm. Fringe order numbers are shown at the minima positions. *Reprinted with permission from Ref. [24].*

Accurate determinations of $\bar{b}$ can be achieved with this method (of about 0.03%), but it is restricted to materials where large perfect crystals are available.

A.3.4 Prism Refraction

The index of refraction n for a thermal neutron beam passing through a material with low absorption is given by

$$n = 1 - \frac{N\bar{b}\lambda^2}{2\pi}. \qquad (A.63)$$

A prism with an apex angle α deflects the neutron by an angle

$$\beta = 2(1-n)\tan\frac{\alpha}{2} = \frac{N\bar{b}\lambda^2}{\pi}\tan\frac{\alpha}{2}, \qquad (A.64)$$

which is usually in the order of seconds of arc. Such a small deflection can be measured with a double-crystal arrangement of perfect crystals, as shown in Figure A.7. The technique was introduced by C. Shull while assessing an experimental limit for the neutron charge [26] and subsequently developed during the PhD Thesis of Schneider [27]. The method has been applied to

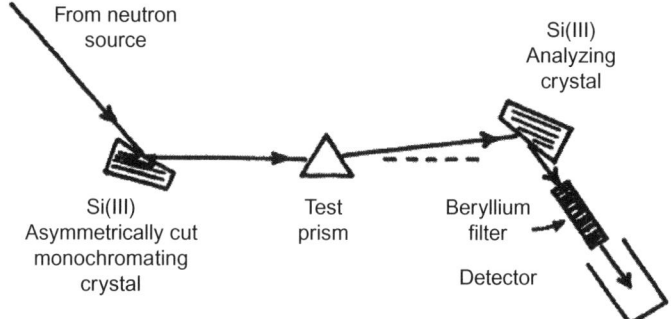

FIGURE A.7 Schematic diagram a of double-crystal spectrometer used in determining refractive bending by a prism. *Reprinted with permission from Ref. [25].*

Fe [25], and to Ge, Cu, and O [28], obtaining results in agreement with those from other techniques. Accuracies of the order of 0.03% are attained with this technique.

A.3.5 Christiansen Filter

The Christiansen filter technique is used in neutron optics in analogy with classical optics. It is based on the idea of varying the relative index between a substance (normally a finely divided powder) and the liquid in which it is immersed.

The beam experiences a broadening due to small-angle scattering effects that disappear if both refractive indices are equal. It was first presented by Christiansen [29] in 1884 with experiments on the scattering of light by mixtures of powders with liquids. In neutron optics, this technique was first employed in 1969 by Koester and Ungerer [30] to measure the scattering lengths of tungsten powder and heavy water in terms of the known scattering lengths of reference liquids. Koester *et al.* [31] measured the refractive indices of all stable nuclei with nucleon number between $A = 79$ and 96, using cold neutrons in the arrangement shown in Figure A.8.

A well-collimated neutron beam is passed through a Christiansen filter containing the sample. The relative index of refraction is given by $n = 1 \pm \frac{\lambda^2}{2\pi}(N_p \bar{b}_p - N_l \bar{b}_l)$, where $N_i \bar{b}_i$ is the mean coherent scattering length density, p labels the powder, and l the liquid. If $N_p \bar{b}_p \neq N_l \bar{b}_l$, the angular divergence of the incident neutron beam will be broadened by small-angle scattering caused by multiple refraction or diffraction, depending on the size of the powder particles. This small-angle scattering was first discussed in 1951 by Weiss [33]. If $N_p \bar{b}_p = N_l \bar{b}_l$, the sample is optically homogeneous and no broadening will occur. Reference liquids consist of mixtures such as CCl_4 and C_8H_{10}, or H_2O and D_2O in known adjustable concentrations, whose refractive index is calculable from that of its constituents.

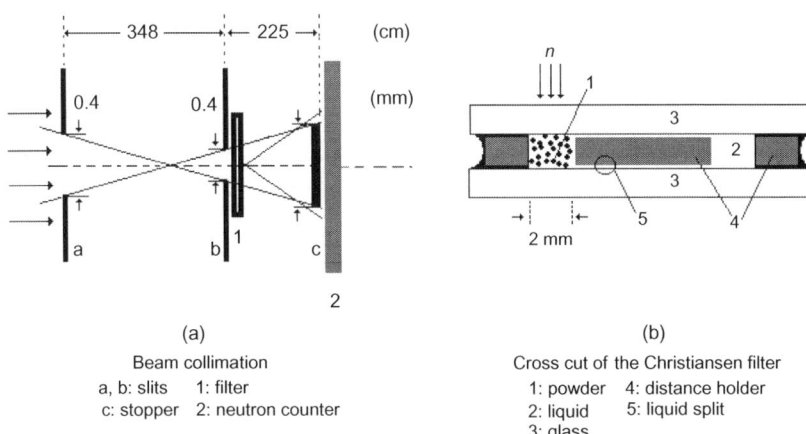

FIGURE A.8 Experimental setup for Christiansen-filter measurements. *Reprinted with permission from Ref. [32].*

In the arrangement shown in Figure A.8, a beam of cold neutrons of 13 Å wavelength enters a system of two narrow slits and hits the detector after having passed the filter [32]. The unscattered part of the beam is absorbed by a cadmium shield positioned before the detector. Neutrons scattered by the filter into small angles are able to pass the stopper and to strike the detector. The method consists in measuring the ratio of the scattered neutrons to the total number measured without stopper as a function of the scattering density of the reference liquid. In the minimum of the curves, the sought scattering density of the powder equals that of the liquid. This method is useful and reliable for determinations of scattering lengths with medium accuracy $\Delta \overline{b}/\overline{b}$ within 10^{-3} and 10^{-2}. Since it requires a small amount of sample, it is particularly applicable to measure the scattering length of separated isotopes of an element [34].

A.3.6 Neutron Gravity Refractometer

The neutron gravity refractometer was proposed by Maier-Leibnitz [35] in 1962 and constructed by Koester [36] in 1965. It is based on the fact that a freely falling slow neutron gains in the gravitational field an energy $m_n g h$, which is of the same order as the mean potential energy $\overline{V}$ of a neutron in matter. A falling neutron initially in an horizontal beam will be reflected from an horizontal plane surface (mirror) if $m_n g h$ is smaller than $\overline{V}$. Otherwise, the neutron will penetrate the mirror. A critical height h_c for total reflection will be reached when

$$m_n g h_c = \overline{V} = \frac{2\pi \hbar N \overline{b}}{m_n}, \tag{A.65}$$

where $N\overline{b}$ is the mean scattering length density.

From this expression, it can be noted that the only quantity that has to be measured to determine $N\overline{b}$ is h_c. Although the expression is independent of the neutron wavelength, it affects the experimental design because of the way the neutron trajectories are determined by the initial horizontal velocity. This basic equation allows the determination of $\overline{b}$ on an absolute scale with a high accuracy which is limited practically only by the experimental errors of the measurements of h_c and N. Accuracies of $\Delta\overline{b}/\overline{b}$ within 1×10^{-4} and 3×10^{-4} have been achieved [37]. This high accuracy can be obtained only for liquid mirrors (molten metals) with a high content of the element under investigation [4]. The centrifugal and Coriolis forces as well as Earth's curvature introduce a correction in gravity refractometry measurements which is close to 0.2% [37]. This is the most accurate method to determine $\overline{b}$ but requires samples in the form of large liquid mirrors of the order of 1-m length, hardly available.

A.3.7 Neutron Interferometry

This technique is analogous to that employed in classical optics. It is used to determine the relative phase of two coherent beams. Depending on the way the two coherent beams are derived from one source, this technique can be divided into two classes: wave front division or amplitude division.

The first neutron interferometer was of the first class, a Fresnel biprism built by Maier-Leibnitz [35] in 1962. In this experiment, the two coherent beams were obtained by refraction of a neutron beam by a biprism. Interference by division of the wavefront was also demonstrated in the experiment done by Klein *et al.* in 1981 [38]. The extremely small separation distance between the coherent beams is the main disadvantage of this method. It can be solved by means of a perfect crystal which uses the dynamical diffraction phenomena to split and to overlap two widely separated beams. This belongs to the second mentioned class. It was first applied to neutrons by Rauch *et al.* [39] in 1974.

This kind of interferometer is machined from a single crystal ingot so as to leave three blades as shown in Figure A.9. A monoenergetic beam, incident on the first blade, is split into two and recombined in the third. The test sample is introduced into one of the paths, producing a phase difference with the other beam. The intensity of the interference between two beams thus depends on the phase shift, resulting in counting rate shifts of opposite sign in the two detectors. A precise measurement of the phase shift allows the determination of $\overline{b}$. Samples can be used in solid–liquid and gaseous forms [41]. An accurate knowledge of the incident wavelength is required.

Measurements of scattering lengths with uncertainties of 0.1% $\Delta\overline{b}/\overline{b}$ have been reported using this method [42], caused mainly by the inaccuracy in the wavelength determination.

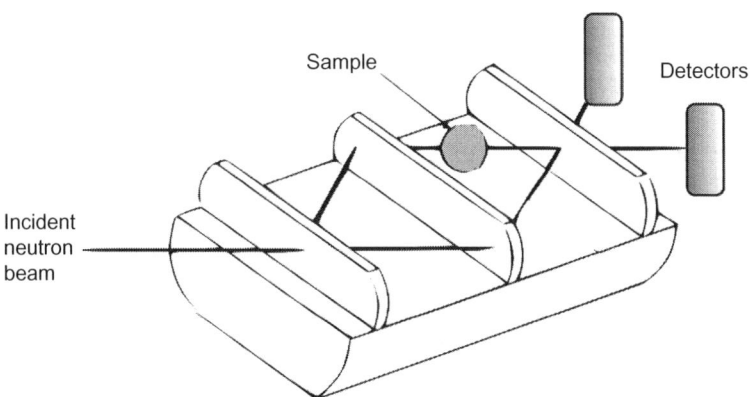

FIGURE A.9 Schematic diagram of the three-blade neutron interferometer. *Illustration based on Ref. [40].*

A.3.8 Small-Angle Scattering

Besides being a powerful experimental tool to investigate some mesoscopic properties of condensed matter (particle size, porosity, etc.), this technique is employed to explore coherent scattering lengths. In this technique, a neutron beam traverses an inhomogeneous material where diffraction, refraction, and reflection processes can occur, causing a broadening of the beam profile.

For the study of the coherent scattering length, Weiss [33] carried out an experimental study of the different regimes characterized by the phase shift of the neutron wave (φ) when it passes through a material. In the refractive regime ($\varphi = 2(1-n)kR \gg 1$, where R is the mean particle radius), the angular aperture of the neutron beam is analyzed on the basis of von Nardroff's theory [43], that gives a final distribution which is Gaussian. The broadening of the beam can be written as

$$\Delta\omega^2 - \Delta\omega_0^2 = 16N(1-n)^2 \ln(2)\left[1 + \ln\left(\frac{2}{1-n}\right)\right], \quad (A.66)$$

where $\Delta\omega$ is the broadened beam full-width-at-half-maximum, and $\Delta\omega_0$ is its initial width, N is the number of particles traversed, and n the refractive index of the material.

In the diffraction range ($\varphi \ll 1$), the broadening of the neutron beam is much smaller than in the refraction range but independent of n and can be written as

$$\Delta\omega^2 - \Delta\omega_0^2 = \frac{2.89\lambda_{\text{eff}}^2}{\pi^2 R^2}, \quad (A.67)$$

where λ_{eff}^2 is the mean square wavelength of the incident neutron beam. To measure $\overline{b}$ particles are often embedded into another substance and the relative refractive index between both substances is measured. The limitations of this method are in the unknown distribution of particle sizes and multiple scattering processes, which in many cases prevent an unambiguous interpretation of the results. Accuracies of 1% were achieved with this method.

A.3.9 Total Reflection

When total reflection of monochromatic neutrons is achieved in the plane surface of mirrors, the critical angle θ_c is directly related to the coherent scattering length

$$\theta_c = \lambda \sqrt{\frac{N\overline{b}}{\pi}}, \tag{A.68}$$

where N is the number density of atoms in the mirror. In principle, one could determine $\overline{b}$ by measuring the reflectivity as a function of the angle of incidence for a collimated monoenergetic beam. However, θ_c is so small (about 0.2°) that the beam has to be very finely collimated and polyenergetic beams are employed in order not to lose intensity. As a result, the accuracy of the method is limited by about 1%. Higher accuracies up to 0.1% can be obtained by comparing relative reflectivities with the same experimental geometry. With this method, Fermi and Marshall [2] determined that most of the nuclides have a scattering length of the same sign. Fermi included a minus sign in Equation (A.4) to define as positive the scattering lengths of the majority of the nuclides.

A.3.10 Pseudomagnetic Method

This method has been particularly useful in the determination of the spin-dependent scattering lengths. In particular, this method is particularly suitable for determining the difference $(b^+ - b^-)$ in Equation (A.25). Thus it has been employed in the determination of many of the currently accepted values of b_i [44]. The method is based on the observation that thermal neutrons in passing through a sample with polarized nuclei "see" a pseudomagnetic field

$$H^* = 4\pi N \mu^* P, \tag{A.69}$$

where N is the number density of nuclei with polarization P and μ^* is known as the pseudomagnetic moment which is proportional to $(b^+ - b^-)$

$$\mu^* = -\frac{\mu_B}{\gamma r_0} \frac{I}{\left(I + \frac{1}{2}\right)} (b^+ - b^-), \tag{A.70}$$

where μ_B is the Bohr magneton, r_0 the classical radius of the electron, and $\gamma = -1.913$ the magnetic moment of the neutron in nuclear magnetons. The magnetic field is measured by determining the precession undergone by a polarized neutron beam as a function of the nuclear polarization of the sample. The precession angle is as follows:

$$\alpha = -\gamma_n H^* \tau, \quad (A.71)$$

where γ_n is the gyromagnetic ratio of the neutron and τ is the transit time of the neutrons through the sample. Since for many nuclei μ^* is small, high nuclear polarizations are needed. Two complementary methods can be used to achieve sizable nuclear polarizations. The first consists in lowering the temperature of the sample down to the millikelvin region. The second uses microwave techniques to polarize the nuclei dynamically. The determination of the precession spin angle is done by the two-coil beam Ramsey method [45,46]. With this method, accuracies of up to 3% were achieved. In many cases, it has provided the only information about spin-dependent scattering lengths available to date.

A.3.11 High-Energy Experiments

In the field of nuclear physics, experiments at high energies (about tens of MeV) can be carried out to determine the singlet scattering length between elementary particles, namely, neutron–neutron b_{nn}^-, neutron–proton b_{np}^-, and proton–proton b_{pp}^-. Complementarily, interferometry experiments can be used to measure both singlet and triplet states. The knowledge of those magnitudes is of crucial importance for the understanding of the nuclear forces, and the depth and width of the potential interaction between nucleons. The neutron–neutron and proton–proton scattering lengths are of special interest because, in principle, they allow a very sensitive test of charge symmetry in the strong interaction [47,48]. Since 1951, numerous attempts of indirect methods have been made to determine b_{nn}^-, using mostly the reactions ^{2}H$(\pi^-, 2n)\gamma$ and ^{2}H$(n, 2n)p$ and investigating the region of the neutron–neutron final-state interactions (FSIs), where the two neutrons travel together with small relative energy. Curiously, measurements using these two reactions give significantly different values of b_{nn}^- [49–52]. The discrepancy suggests that the difference in these values has its origin in the three-nucleon force (3NF), which would act in the second reaction but not in the first [53]. Recently, experiments carried out at the Los Alamos Meson Physics Facility to determine the neutron–neutron scattering length, which takes into account the moment interaction of the two neutrons and a kinematically complete measurement of the neutron TOF spectrum of the ^{2}H$(p, 2n)p$ reaction, gives a recommended value of $b_{nn} = -37.8 \pm 0.8$ fm [54].

Since neutron–neutron scattering is between identical particles, it can be shown that the triplet state does not contribute to S-wave scattering. A schematic view of the experimental setup to study the ^{2}H$(\pi^-, 2n)\gamma$ and ^{2}H$(n, 2n)p$ reactions can be found in Refs. [52] and [54].

The neutron–proton scattering length can also be determined using the same $^2\mathrm{H}(n,2n)p$ reaction by replacing the proton detector by a second neutron detector as shown in Ref. [52]. In this case, investigating in the region of the neutron–proton FSIs, the value obtained is $b_{np} = -47.8 \pm 2.0\,\mathrm{fm}$ [52]. It is observed that b_{np} tends to the asymptotic "free" value up to about 1 keV, and from there, the neutron–proton scattering length decreases significantly up to energies about 300 MeV. Also, the proton–proton scattering length is measured directly in proton–proton scattering. Its analysis is out of the scope of this work.

A.4 TABLES OF NEUTRON SCATTERING LENGTHS AND CROSS SECTIONS

Table A.2 contains the information listed below. The scattering lengths are expressed in femtometers and the cross sections in barn ($1\mathrm{b} = 100\,\mathrm{fm}^2$).

1. Name of the element and its isotopes.
2. Atomic number Z.
3. Mass number A.
4. Nuclear spin (I) and parity (Π) for a given nucleus [55].
5. Natural abundance of the nuclear species (with a % sign) for the stable nuclei and half-life for the unstable nuclei (followed by units) [55].
6. Bound coherent scattering length $\bar{b}$.
7. Bound scattering length for the $(I + 1/2)$ state b^+.
8. Bound scattering length for the $(I - 1/2)$ state b^-.
9. Bound incoherent scattering length b_i.
10. Bound coherent scattering cross section σ_c.
11. Bound incoherent scattering cross section σ_i.
12. Total bound scattering cross section σ_s.
13. Absorption cross section for thermal ($v = 2200$ m/s) neutrons.

The trailing digits in parentheses give the standard errors.

The data were compiled from the best preexisting tables [3,4,56] updated for some specific cases. Nuclear data as spin, parity, and abundance (or half-life for unstable nuclei) were taken from Ref. [55]. Absorption cross sections were taken from Mughabghab et al. [57,58]. The imaginary parts of the scattering lengths for the strongly absorbing nuclides were calculated from the absorption cross sections. Those are the cases where a (n,γ) resonance at thermal energies exists and are indicated with an asterisk in the table. For some particular elements, it was necessary to estimate the scattering length of one or two isotopes to complete the table. Such cases were indicated by the letter E. These isotopes are of low natural abundance and these estimates have a minor effect on the final results.

The table contains the neutron–nuclei scattering lengths. It was compiled with data obtained with the experimental techniques presented in this appendix. In cases where the techniques employ low-energy neutrons, corrections

TABLE A.2 Bound Scattering Lengths and Cross Sections of the Elements and Their Isotopes

Element	Z	A	$I(\pi)$	c_a	$\bar{b}$	b^+	b^-	b_i	σ_c	σ_i	σ_s	σ_a
n	0	1	1/2(+)		−37.8(8)	0	−37.8(8)	0	44.89(4)	0	44.89(4)	0
H	1				−3.7390(11)				1.7568(10)	80.26(6)	82.02(6)	0.3326(7)
^{1}H		1	1/2(+)	99.9885	−3.7423(12)	10.817(5)	−47.420(14)	25.217(6)	1.7589(11)	79.91(4)	81.67(4)	0.3326(7)
^{2}H		2	1(+)	0.0115	6.674(6)	9.53(3)	0.975(60)	4.03(3)	5.597(10)	2.04(3)	7.64(3)	0.000519(7)
^{3}H		3	1/2(+)	12.32 y	4.792(27)	4.18(15)	6.56(37)	−1.04(17)	2.89(3)	0.14(4)	3.03(5)	<6E−06
He	2				3.26(3)				1.34(2)	0	1.34(2)	0.00747(1)
^{3}He		3	1/2(+)	0.000134	5.74(7)−1.483(2)i	4.5(3)	9.3(5)	−2.1(3)+2.568(3)i	4.42(10)	1.38(16)	5.8(2)	5333.0(7.0)
^{4}He		4	0(+)	99.999866	3.26(3)			0	1.34(2)	0	1.34(2)	0
Li	3				−1.90(3)				0.454(10)	0.92(3)	1.37(3)	70.5(3)
^{6}Li		6	1(+)	7.59	2.00(11)−0.261(1)i	0.67(14)	4.67(17)	−1.89(10)+0.26(1)i	0.51(5)	0.46(5)	0.97(7)	940.0(4.0)

^{7}Li		7	3/2(−)	92.41	−2.22(2)	−4.15(6)	1.00(8)	−2.49(5)	0.619(11)	0.78(3)	1.40(3)	0.0454(3)
Be	4	9	3/2(−)	100	7.79(1)			0.12(3)	7.63(2)	0.0018(9)	7.63(2)	0.0076(8)
B	5				5.30(4) − 0.213(2)i				3.54(5)	1.70(12)	5.24(11)	767.0(8.0)
^{10}B		10	3(+)	19.9	−0.1(3) − 1.066(3)i	−4.2(4)	5.2(4)	−4.7(3) + 1.231(3)i	0.144(6)	3.0(4)	3.1(4)	3835.0(9.0)
^{11}B		11	3/2(−)	80.1	6.65(4)	5.6(3)	8.3(3)	−1.3(2)	5.56(7)	0.21(7)	5.77(10)	0.0055(33)
C	6				6.6484(13)				5.551(2)	0.001(4)	5.551(3)	0.00350(7)
^{12}C		12	0(+)	98.93	6.6535(14)			0	5.559(3)	0	5.559(3)	0.00353(7)
^{13}C		13	1/2(−)	1.07	6.19(9)	5.6(5)	6.2(5)	−0.25(3)	4.81(14)	0.034(11)	4.84(14)	0.00137(4)
N	7				9.36(2)				11.01(5)	0.50(12)	11.51(11)	1.90(3)
^{14}N		14	1(+)	99.636	9.37(2)	10.7(2)	6.2(3)	2.1(3)	11.03(5)	0.50(12)	11.53(11)	1.91(3)
^{15}N		15	1/2(−)	0.364	6.44(3)	6.77(10)	6.21(10)	0.24(6)	5.21(5)	0.00005(10)	5.21(5)	0.000024(8)
O	8				5.805(4)				4.232(6)	0.000(8)	4.232(6)	0.00019(2)

Continued

TABLE A.2 Bound Scattering Lengths and Cross Sections of the Elements and Their Isotopes—Cont'd

Element	Z	A	$I(\pi)$	c_a	$\bar{b}$	b^+	b^-	b_i	σ_c	σ_i	σ_s	σ_a
^{16}O	8	16	0(+)	99.757	5.805(5)			0	4.232(6)	0	4.232(6)	0.00010(2)
^{17}O		17	5/2(+)	0.038	5.66(5)	5.86(7)	5.41(17)	0.17(5)	4.20(22)	0.004(3)	4.20(22)	0.236(10)
^{18}O		18	0(+)	0.205	5.84(7)			0	4.29(10)	0	4.29(10)	0.00016(1)
F	9	19	1/2(+)	100	5.654(12)	5.632(10)	5.767(10)	−0.082(9)	4.017(14)	0.0008(2)	4.018(14)	0.0096(5)
Ne	10				4.566(6)				2.620(7)	0.008(9)	2.628(6)	0.039(4)
^{20}Ne		20	0(+)	90.48	4.631(6)			0	2.695(7)	0	2.695(7)	0.036(4)
^{21}Ne		21	3/2(+)	0.27	6.66(19)			(+/−)0.6(1)	5.6(3)	0.05(2)	5.7(3)	0.67(11)
^{22}Ne		22	0(+)	9.25	3.87(1)			0	1.88(1)	0	1.88(1)	0.046(6)
Na	11	23	3/2(+)	100	3.63(2)	6.42(4)	−1.00(6)	3.59(3)	1.66(2)	1.62(3)	3.28(4)	0.530(5)
Mg	12				5.375(4)				3.631(5)	0.08(6)	3.71(4)	0.063(3)

	A	I(π)	Abundance							
²⁴Mg	24	0(+)	78.99	5.49(18)		0	4.03(4)	0	4.03(4)	0.050(5)
²⁵Mg	25	5/2(+)	10.00	3.62(14)	4.73(30)	1.48(10)	1.65(13)	0.28(4)	1.93(14)	0.19(3)
²⁶Mg	26	0(+)	11.01	4.89(15)		0	3.00(18)	0	3.00(18)	0.0382(8)
Al 13	27	5/2(+)	100	3.449(5)	3.67(2)	0.256(10)	1.495(4)	0.0082(6)	1.503(4)	0.231(3)
Si 14				4.15071(22)			2.1633(10)	0.004(8)	2.167(8)	0.171(3)
²⁸Si	28	0(+)	92.223	4.106(6)		0	2.120(6)	0	2.120(6)	0.177(3)
²⁹Si	29	1/2(+)	4.685	4.7(1)	4.50(15)	4.7(4)	2.78(12)	0.79(3)	3.57(15)	0.101(14)
³⁰Si	30	0(+)	3.092	4.58(8)		0	2.64(9)	0	2.64(9)	0.107(2)
P 15	31	1/2(+)	100	5.13(1)		0.3	3.307(13)	0.005(10)	3.312(16)	0.172(6)
S 16				2.847(1)			1.0186(7)	0.007(5)	1.026(5)	0.53(1)
³²S	32	0(+)	94.99	2.804(2)		0	0.9880(14)	0	0.9880(14)	0.54(4)
³³S	33	3/2(+)	0.75	4.74(19)		1.5(1.5)	2.8(2)	0.3(6)	3.1(6)	0.54(4)
³⁴S	34	0(+)	4.25	3.48(3)		0	1.52(3)	0	1.52(3)	0.227(5)

Continued

TABLE A.2 Bound Scattering Lengths and Cross Sections of the Elements and Their Isotopes—Cont'd

Element	Z	A	$I(\pi)$	c_a	$\bar{b}$	b^+	b^-	b_i	σ_c	σ_i	σ_s	σ_a
^{36}S	36	0(+)	0.01	3.(1.)E			0		1.1(8)	0	1.1(8)	0.15(3)
Cl	17				9.5792(8)				11.528(2)	5.3(5)	16.8(5)	33.5(3)
^{35}Cl		35	3/2(+)	75.76	11.70(9)	16.3(2)	4.0(3)	6.1(4)	17.06(6)	4.7(6)	21.8(6)	44.1(4)
^{37}Cl		37	3/2(+)	24.24	3.08(6)	3.10(7)	3.05(7)	0.02(5)	1.19(5)	0.001(3)	1.19(5)	0.433(6)
Ar	18				1.909(6)				0.458(3)	0.225(5)	0.683(4)	0.675(9)
^{36}Ar		36	0(+)	0.3336	24.9(7)			0	77.9(4)	0	77.9(4)	5.2(5)
^{38}Ar		38	0(+)	0.0629	3.5(3.5)			0	1.5(3.1)	0	1.5(3.1)	0.8(5)
^{40}Ar		40	0(+)	99.6035	1.84(3)			0	0.421(3)	0	0.421(3)	0.660(9)
K	19				3.67(2)				1.69(2)	0.27(11)	1.96(11)	2.1(1)
^{39}K		39	3/2(+)	93.2581	3.79(2)	5.15E	1.51E	1.4(3)	1.76(2)	0.25(11)	2.01(11)	2.1(1)
^{40}K		40	4(−)	0.0117	3.1(1.0)E				1.1(6)	0.5(5)E	1.6(9)	35.0(8.0)
^{41}K		41	3/2(+)	6.7302	2.69(8)			1.5(1.5)	0.91(5)	0.3(6)	1.2(6)	1.46(3)

Ca	20				4.70(2)			2.78(2)	0.05(3)	2.83(2)	0.43(2)
^{40}Ca		40	0(+)	96.94	4.78(5)		0	2.90(2)	0	2.90(2)	0.41(2)
^{42}Ca		42	0(+)	0.647	3.36(10)		0	1.42(8)	0	1.42(8)	0.68(7)
^{43}Ca		43	7/2(−)	0.135	−1.56(9)			0.31(4)	0.5(5)	0.8(5)	6.2(6)
^{44}Ca		44	0(+)	2.09	1.42(6)		0	0.25(2)	0	0.25(2)	0.88(5)
^{46}Ca		46	0(+)	0.004	3.55(21)			1.6(2)	0	1.6(2)	0.74(7)
^{48}Ca		48	0(+)	0.187	0.39(9)		0	0.019(9)	0	0.019(9)	1.09(14)
Sc	21	45	7/2(−)	100	12.1(1)	6.91(22)	18.99(28)	−6.0(2)	19.0(3)	4.5(3)	23.5(6)	27.5(2)
Ti	22				−3.370(13)			1.485(2)	2.87(3)	4.35(3)	6.09(13)
^{46}Ti		46	0(+)	8.25	4.72(5)		0	3.05(7)	0	3.05(7)	0.59(18)
^{47}Ti		47	5/2(−)	7.44	3.53(7)	0.46(23)	7.64(13)	−3.5(1)	1.66(11)	1.5(2)	3.2(2)	1.7(2)
^{48}Ti		48	0(+)	73.72	−5.86(2)		0	4.65(3)	0	4.65(3)	7.84(25)
^{49}Ti		49	7/2(−)	5.41	0.98(5)	2.6(3)	−1.2(4)	1.9(2)	0.14(1)	3.3(3)	3.4(3)	2.2(3)
^{50}Ti		50	0(+)	5.18	5.88(10)		0	4.80(12)	0	4.80(12)	0.179(3)
V	23				−0.443(14)			0.01838(12)	5.08(6)	5.10(6)	5.08(4)

Continued

TABLE A.2 Bound Scattering Lengths and Cross Sections of the Elements and Their Isotopes—Cont'd

Element	Z	A	c_a	$\bar{b}$	b^+	b^-	b_i	σ_c	σ_i	σ_s	σ_a
^{50}V		50	0.25	7.6(6)				7.3(1.1)	0.5(5)E	7.8(1.0)	60.0(40.0)
^{51}V		51	99.75	−0.402(2)	4.93(25)	−7.58(28)	6.35(4)	0.0203(2)	5.07(6)	5.09(6)	4.9(1)
Cr	24			3.635(7)				1.660(6)	1.83(2)	3.49(2)	3.05(6)
^{50}Cr		50	4.345	−4.50(5)			0	2.54(6)	0	2.54(6)	15.8(2)
^{52}Cr		52	83.789	4.914(15)			0	3.042(12)	0	3.042(12)	0.76(6)
^{53}Cr		53	9.501	−4.20(3)	1.16(10)	−13.0(2)	6.9(1)	2.22(3)	5.93(17)	8.15(17)	18.1(1.5)
^{54}Cr		54	2.365	4.55(10)			0	2.60(11)	0	2.60(11)	0.36(4)
Mn	25	55	100	−3.750(18)	−4.93(46)	−1.46(33)	−1.71(28)	1.75(2)	0.40(2)	2.15(3)	13.3(2)
Fe	26			9.45(2)				11.22(5)	0.40(11)	11.62(10)	2.56(3)
^{54}Fe		54	5.845	4.2(1)			0	2.2(1)	0	2.2(1)	2.25(18)
^{56}Fe		56	91.754	10.1(2)			0	12.42(7)	0	12.42(7)	2.59(14)

Isotope	Z	A	I(π)	% abundance								
^{57}Fe		57	1/2(−)	2.119	2.3(1)				0.66(6)	0.3(3)E	1.0(3)	2.48(30)
^{58}Fe		58	0(+)	0.282	15(7)			0	28.0(26.0)	0	28.0(26.0)	1.28(5)
Co	27	59	7/2(−)	100	2.49(2)	−9.21(10)	3.58(10)	−6.34(7)	0.779(13)	4.8(3)	5.6(3)	37.18(6)
Ni	28				10.3(1)				13.3(3)	5.2(4)	18.5(3)	4.49(16)
^{58}Ni		58	0(+)	68.077	14.4(1)			0	26.1(4)	0	26.1(4)	4.6(3)
^{60}Ni		60	0(+)	26.223	2.8(1)			0	0.99(7)	0	0.99(7)	2.9(2)
^{61}Ni		61	3/2(−)	1.1399	7.60(6)			±3.9(3)	7.26(11)	1.9(3)	9.2(3)	2.5(8)
^{62}Ni		62	0(+)	3.6346	−8.7(2)			0	9.5(4)	0	9.5(4)	14.5(3)
^{64}Ni		64	0(+)	0.9255	−0.37(7)			0	0.017(7)	0	0.017(7)	1.52(3)
Cu	29				7.718(4)				7.485(8)	0.55(3)	8.03(3)	3.78(2)
^{63}Cu		63	3/2(−)	69.15	6.477(13)			0.22(2)	5.2(2)	0.006(1)	5.2(2)	4.50(2)
^{65}Cu		65	3/2(−)	30.85	10.204(20)			1.82(10)	14.1(5)	0.40(4)	14.5(5)	2.17(3)
Zn	30				5.680(5)				4.054(7)	0.077(7)	4.131(10)	1.11(2)
^{64}Zn		64	0(+)	49.17	5.23(4)			0	3.42(5)	0	3.42(5)	0.93(9)
^{66}Zn		66	0(+)	27.73	5.98(5)			0	4.48(8)	0	4.48(8)	0.62(6)

Continued

TABLE A.2 Bound Scattering Lengths and Cross Sections of the Elements and Their Isotopes—Cont'd

Element	Z	A	I(π)	c_a	$\bar{b}$	b^+	b^-	b_i	σ_c	σ_i	σ_s	σ_a
^{67}Zn		67	5/2(−)	4.04	7.58(8)	5.8(5)	10.1(7)	−1.50(7)	7.18(15)	0.28(3)	7.46(15)	6.8(8)
^{68}Zn		68	0(+)	18.45	6.04(3)			0	4.57(5)	0	4.57(5)	1.1(1)
^{70}Zn		70	0(+)	0.61	6.0(1.0)E			0	4.5(1.5)	0	4.5(1.5)	0.092(5)
Ga	31				7.288(2)				6.675(4)	0.16(3)	6.83(3)	2.75(3)
^{69}Ga		69	3/2(−)	60.108	8.043(16)	6.3(2)	10.5(4)	−0.85(5)	7.80(4)	0.091(11)	7.89(4)	2.18(5)
^{71}Ga		71	3/2(−)	39.892	6.170(11)	5.5(6)	7.8(1)	−0.82(7)	5.15(5)	0.084(8)	5.23(5)	3.61(10)
Ge	32				8.185(20)				8.42(4)	0.18(7)	8.60(6)	2.20(4)
^{70}Ge		70	0(+)	20.57	10.0(1)			0	12.6(3)	0	12.6(3)	3.0(2)
^{72}Ge		72	0(+)	27.45	8.51(10)			0	9.1(2)	0	9.1(2)	0.8(2)
^{73}Ge		73	9/2(+)	7.75	5.02(4)	8.1(4)	1.2(4)	3.4(3)	3.17(5)	1.5(3)	4.7(3)	15.1(4)
^{74}Ge		74	0(+)	36.5	7.58(10)			0	7.2(2)	0	7.2(2)	0.4(2)
^{76}Ge		76	0(+)	7.73	8.2(1.5)			0	8.0(3.0)	0	8.0(3.0)	0.16(2)
As	33	75	3/2(−)	100	6.58(1)	6.04(5)	7.47(8)	−0.69(6)	5.44(2)	0.060(10)	5.50(2)	4.5(1)
Se	34				7.970(9)				7.98(2)	0.32(6)	8.30(6)	11.7(2)

^{74}Se	74	0(+)	0.89	0.8(3.0)	0	0.1(6)	0	0.1(6)	51.8(1.2)
^{76}Se	76	0(+)	9.37	12.2(1)	0	18.7(3)	0	18.7(3)	85.0(7.0)
^{77}Se	77	1/2(−)	7.63	8.25(8)	±0.6(1.0)	8.6(2)	0.05(25)	8.65 (16)	42.0(4.0)
^{78}Se	78	0(+)	23.77	8.24(9)	0	8.5(2)	0	8.5(2)	0.43(2)
^{80}Se	80	0(+)	49.61	7.48(3)	0	7.03(6)	0	7.03(6)	0.61(5)
^{82}Se	82	0(+)	8.73	6.34(8)	0	5.05 (13)	0	5.05 (13)	0.044(3)
Br	35			6.79(2)		5.80(3)	0.10(9)	5.90(9)	6.9(2)
^{79}Br	79	3/2(−)	50.69	6.79(7)	−1.1(2)	5.81(2)	0.15(6)	5.96 (13)	11.0(7)
^{81}Br	81	3/2(−)	49.31	6.78(7)	−0.6(1)	5.79 (12)	0.05(2)	5.84 (12)	2.7(2)
Kr	36			7.81(2)		7.67(4)	0.01(14)	7.68 (13)	25.0(1.0)
^{78}Kr	78	0(+)	0.355		0		0		6.4(9)
^{80}Kr	80	0(+)	2.286		0		0		11.8(5)
^{82}Kr	82	0(+)	11.593		0		0		29.0 (20.0)
^{83}Kr	83	9/2(+)	11.5						185.0 (30.0)

Continued

TABLE A.2 Bound Scattering Lengths and Cross Sections of the Elements and Their Isotopes—Cont'd

Element	Z	A	$I(\pi)$	c_a	$\bar{b}$	b^+	b^-	b_i	σ_c	σ_i	σ_s	σ_a
^{84}Kr		84	0(+)	56.987				0		0	6.6	0.113(15)
^{86}Kr		86	0(+)	17.279	8.07(26)			0	8.2(4)	0	8.2(4)	0.003(2)
Rb	37				7.08(2)				6.32(4)	0.5(4)	6.8(4)	0.38(1)
^{85}Rb		85	5/2(−)	72.17	7.07(10)				6.2(2)	0.5(5)E	6.7(5)	0.48(1)
^{87}Rb		87	3/2(−)	27.83	7.27(12)				6.6(2)	0.5(5)E	7.1(5)	0.12(3)
Sr	38				7.02(2)				6.19(4)	0.06(11)	6.25(10)	1.28(6)
^{84}Sr		84	0(+)	0.56	5.0(2.0)			0	6.0(2.0)	0	6.0(2.0)	0.87(7)
^{86}Sr		86	0(+)	9.86	5.68(5)			0	4.04(7)	0	4.04(7)	1.04(7)
^{87}Sr		87	9/2(+)	7	7.41(7)				6.88(13)	0.5(5)E	7.4(5)	16.0(3.0)
^{88}Sr		88	0(+)	82.58	7.16(6)			0	6.42(11)	0	6.42(11)	0.058(4)
Y	39	89	1/2(−)	100	7.75(2)	8.4(2)	5.8(5)	1.1(2)	7.55(4)	0.15(8)	7.70(9)	1.28(2)
Zr	40				7.16(3)				6.44(1)	0.02(15)	6.46(14)	0.185(3)
^{90}Zr		90	0(+)	51.45	6.5(1)			0	5.1(2)	0	5.1(2)	0.011(5)

^{91}Zr	91	5/2(+)	11.22	8.8(1)	7.9(2)	10.1(2)	−1.08(15)	9.5(2)	0.15(4)	9.7(2)	1.17(10)
^{92}Zr	92	0(+)	17.15	7.5(2)			0	6.9(4)	0	6.9(4)	0.22(6)
^{94}Zr	94	0(+)	17.38	8.3(2)			0	8.4(4)	0	8.4(4)	0.0499(24)
^{96}Zr	96	0(+)	2.8	5.5(1)			0	3.8(1)	0	3.8(1)	0.0229(10)
Nb	41	9/2(+)	100	7.054(3)	7.06(4)	7.35(4)	−0.139(10)	6.253(5)	0.0024(3)	6.255(5)	1.15(6)
Mo	42			6.715(20)				5.67(3)	0.04(5)	5.71(4)	2.48(4)
^{92}Mo	92	0(+)	14.53	6.93(8)			0	6.00(14)	0	6.00(14)	0.019(2)
^{94}Mo	94	0(+)	9.15	6.82(7)			0	5.81(12)	0	5.81(12)	0.015(2)
^{95}Mo	95	5/2(+)	15.84	6.93(6)				6.00(10)	0.5(5)E	6.5(5)	13.1(3)
^{96}Mo	96	0(+)	16.67	6.22(6)			0	4.83(9)	0	4.83(9)	0.5(2)
^{97}Mo	97	5/2(+)	9.6	7.26(8)				6.59(15)	0.5(5)E	7.1(5)	2.5(2)
^{98}Mo	98	0(+)	24.39	6.60(7)			0	5.44(12)	0	5.44(12)	0.127(6)
^{100}Mo	100	0(+)	9.82	6.75(7)			0	5.69(12)	0	5.69(12)	0.4(2)
Tc	43	9/2(+)	2.111E+5 y	6.8(3)				5.8(5)	0.5(5)E	6.3(7)	20.0(1.0)

Continued

TABLE A.2 Bound Scattering Lengths and Cross Sections of the Elements and Their Isotopes—Cont'd

Element	Z	A	$I(\pi)$	c_a	$\bar{b}$	b^+	b^-	b_i	σ_c	σ_i	σ_s	σ_a
Ru	44				7.02(2)				6.21(5)	0.4(1)	6.6(1)	2.56(13)
^{96}Ru		96	0(+)	5.54				0		0		0.28(2)
^{98}Ru		98	0(+)	1.87				0		0		<8.0
^{99}Ru		99	5/2(+)	12.76								6.9(1.0)
^{100}Ru		100	0(+)	12.6				0		0		4.8(6)
^{101}Ru		101	5/2(+)	17.06								3.3(9)
^{102}Ru		102	0(+)	31.55				0		0		1.17(7)
^{104}Ru		104	0(+)	18.62				0		0		0.31(2)
Rh	45	103	1/2(−)	100	5.90(4)	8.15(6)	6.74(6)	0.61(4)	4.34(6)	0.047(7)	4.39(7)	144.8(7)
Pd	46				5.91(6)				4.39(9)	0.093(9)	4.48(9)	6.9(4)
^{102}Pd		102	0(+)	1.02	7.7(7)E			0	7.5(1.4)	0	7.5(1.4)	3.4(3)
^{104}Pd		104	0(+)	11.14	7.7(7)E			0	7.5(1.4)	0	7.5(1.4)	0.6(3)
^{105}Pd		105	5/2(+)	22.33	5.5(3)			−2.6(1.6)	3.8(4)	0.8(1.0)	4.6(1.1)	20.0(3.0)
^{106}Pd		106	0(+)	27.33	6.4(4)			0	5.1(6)	0	5.1(6)	0.304(29)
^{108}Pd		108	0(+)	26.46	4.1(3)			0	2.1(3)	0	2.1(3)	8.5(5)

¹¹⁰Pd		110	0(+)	11.72	7.7(7)E		0	7.5(1.4)	0	7.5(1.4)	0.226 (31)
Ag	47				5.922(7)			4.407 (10)	0.58(3)	4.99(3)	63.3(4)
¹⁰⁷Ag		107	1/2(−)	51.839	7.555(11)	8.14 (9)	5.8(3)	1.00(13)	7.17(2)	0.13(3)	7.30(4)	37.6(1.2)
¹⁰⁹Ag		109	1/2(−)	48.161	4.165(11)	3.24 (8)	6.9(2)	−1.60(13)	2.18(1)	0.32(5)	2.50(5)	91.0(1.0)
Cd	48				4.83(5) − 0.70(1)i				3.04(6)	3.46(13)	6.50 (12)	2520.0 (50.0)
¹⁰⁶Cd		106	0(+)	1.25	5.0(2.0)E			0	3.1(2.5)	0	3.1(2.5)	1.0(2.0)
¹⁰⁸Cd		108	0(+)	0.89	5.31(24)			0	3.7(1)	0	3.7(1)	1.1(3)
¹¹⁰Cd		110	0(+)	12.49	5.78(8)			0	4.4(1)	0	4.4(1)	11.0(1.0)
¹¹¹Cd		111	1/2(+)	12.8	6.47(8)				5.3(2)	0.3(3)E	5.6(4)	24.0(5.0)
¹¹²Cd		112	0(+)	24.13	6.34(6)			0	5.1(2)	0	5.1(2)	2.2(5)
¹¹³Cd	*	113	1/2(+)	12.22	−8.0(1) − 5.73(11)i				12.1(4)	0.3(3)E	12.4(5)	20600.0 (400.0)
¹¹⁴Cd		114	0(+)	28.73	7.48(5)			0	7.1(2)	0	7.1(2)	0.34(2)
¹¹⁶Cd		116	0(+)	7.49	6.26(9)			0	5.0(2)	0	5.0(2)	0.075 (13)
In	49				4.065(20) − 0.0539(4)i				2.08(2)	0.54(11)	2.62 (11)	193.8 (1.5)

Continued

TABLE A.2 Bound Scattering Lengths and Cross Sections of the Elements and Their Isotopes—Cont'd

Element	Z	A	$I(\pi)$	c_a	$\bar{b}$	b^+	b^-	b_i	σ_c	σ_i	σ_s	σ_a
^{113}In		113	9/2(+)	4.29	5.39(6)				3.65(8)	0.000037(5)	3.65(8)	12.0(1.1)
^{115}In		115	9/2(+)	95.71	4.00(3) − 0.0562(6)i	2.1(1)	6.4(4)	−2.1(2)	2.02(2)	0.55(11)	2.57(11)	202.0(2.0)
Sn	50				6.225(2)				4.871(3)	0.022(5)	4.892(6)	0.626(9)
^{112}Sn		112	0(+)	0.97	6.0(1.0)E			0	4.5(1.5)	0	4.5(1.5)	1.00(11)
^{114}Sn		114	0(+)	0.66	6.0(3)			0	4.8(5)	0	4.8(5)	0.114(30)
^{115}Sn		115	1/2(+)	0.34	6.0(1.0)E				4.5(1.5)	0.3(3)E	4.8(1.5)	30.0(7.0)
^{116}Sn		116	0(+)	14.54	6.10(1)			0	4.42(7)	0	4.42(7)	0.14(3)
^{117}Sn		117	1/2(+)	7.68	6.59(8)	0.22(10)	−0.23(10)	0.19(6)	5.28(8)	0.3(3)E	5.6(3)	2.3(5)
^{118}Sn		118	0(+)	24.22	6.23(4)			0	4.63(8)	0	4.63(8)	0.22(5)
^{119}Sn		119	1/2(+)	8.59	6.28(3)	0.14(10)	0.0(1)	0.06(6)	4.71(8)	0.3(3)E	5.0(3)	2.2(5)
^{120}Sn		120	0(+)	32.58	6.67(4)			0	5.29(8)	0	5.29(8)	0.14(3)
^{122}Sn		122	0(+)	4.63	5.93(3)			0	4.14(7)	0	4.14(7)	0.18(2)
^{124}Sn		124	0(+)	5.79	6.15(3)			0	4.48(8)	0	4.48(8)	0.133(5)
Sb	51				5.57(3)				3.90(4)	0.00(7)	3.90(6)	4.91(5)

	A	I	Abundance							
^{121}Sb	121	5/2(+)	57.21	5.71(6)	5.7(2)	5.8(2)	−0.05(15)	4.10(9)	0.0003(19)	4.10(19)	5.75(12)
^{123}Sb	123	7/2(+)	42.79	5.38(7)	5.2(2)	5.4(2)	−0.10(14)	3.64(9)	0.001(4)	3.64(9)	3.8(2)
Te				5.68(2)				4.23(4)	0.09(6)	4.32(5)	4.7(1)
52											
^{120}Te	120	0(+)	0.09	5.3(5)			0	3.5(7)	0	3.5(7)	2.3(3)
^{122}Te	122	0(+)	2.55	3.8(2)			0	1.8(2)	0	1.8(2)	3.4(5)
^{123}Te	123	1/2(+)	0.89	−0.05(25) − 0.116(8)i	−1.2(2)	3.5(2)	−2.04(12)	0.002(3)	0.52(5)	0.52(5)	418.0(30.0)
^{124}Te	124	0(+)	4.74	7.95(10)			0	8.0(2)	0	8.0(2)	6.8(1.3)
^{125}Te	125	1/2(+)	7.07	5.01(8)	4.9(2)	5.5(2)	−0.26(12)	3.17(10)	0.008(8)	3.18(10)	1.55(16)
^{126}Te	126	0(+)	18.84	5.55(7)			0	3.88(10)	0	3.88(10)	1.04(15)
^{128}Te	128	0(+)	31.74	5.88(8)			0	4.36(10)	0	4.36(10)	0.215(8)
^{130}Te	130	0(+)	34.08	6.01(7)			0	4.55(11)	0	4.55(11)	0.29(6)
I											
53	127	5/2(+)	100	5.28(2)	6.6(2)	3.4(2)	1.58(15)	3.50(3)	0.31(6)	3.81(7)	6.15(6)
Xe				4.69(4)				3.04(4)	0		23.9(1.2)
54											
^{124}Xe	124	0(+)	0.0952				0		0		165.0(20.0)

Continued

TABLE A.2 Bound Scattering Lengths and Cross Sections of the Elements and Their Isotopes—Cont'd

Element	Z	A	$I(\pi)$	c_a	$\bar{b}$	b^+	b^-	b_i	σ_c	σ_i	σ_s	σ_a
^{126}Xe		126	0(+)	0.089				0		0		3.5(8)
^{128}Xe		128	0(+)	1.9102				0		0		<8.0
^{129}Xe		129	1/2(+)	26.4006								21.0(5.0)
^{130}Xe		130	0(+)	4.071				0		0		<26.0
^{131}Xe		131	3/2(+)	21.232								85.0 (10.0)
^{132}Xe		132	0(+)	26.9086				0		0		0.45(6)
^{134}Xe		134	0(+)	10.4357				0		0		0.265 (20)
^{136}Xe		136	0(+)	8.8573				0		0		0.26(2)
Cs	55	133	7/2(+)	100	5.42(2)			1.29(15)	3.69 (15)	0.21(5)	3.90(6)	29.0(1.5)
Ba	56				5.07(3)				3.23(4)	0.15(11)	3.38 (10)	1.1(1)
^{130}Ba		130	0(+)	0.106	−3.6(6)			0	1.6(5)	0	1.6(5)	30.0(5.0)
^{132}Ba		132	0(+)	0.101	7.8(3)			0	7.6(6)	0	7.6(6)	7.0(8)
^{134}Ba		134	0(+)	2.417	5.7(1)			0	4.08 (14)	0	4.08 (14)	2.0(1.6)

Nuclide	Z	A	I(π)	c (%)	b_c	b_i	b_+	b_-	σ_{coh}	σ_{inc}	σ_s	σ_a
^{135}Ba		135	3/2(+)	6.592	4.66(10)				2.74(12)	0.5(5)E	3.2(5)	5.8(9)
^{136}Ba		136	0(+)	7.854	4.90(8)	0			3.03(10)	0	3.03(10)	0.68(17)
^{137}Ba		137	3/2(+)	11.232	6.82(10)				5.86(17)	0.5(5)E	6.4(5)	3.6(2)
^{138}Ba		138	0(+)	71.698	4.83(8)	0			2.94(10)	0	2.94(10)	0.27(14)
La	57				8.24(4)				8.53(8)	1.13(19)	9.66(17)	8.97(2)
^{138}La		138	5(+)	0.08881	8.0(2.0)E				8.0(4.0)	0.5(5)E	8.5(4.0)	57.0(6.0)
^{139}La		139	7/2(+)	99.9119	8.24(4)		11.4(3)	4.5(4)	8.53(8)	1.13(15)	9.66(17)	8.93(4)
Ce	58				4.84(2)				2.94(2)	0.00(10)	2.94(10)	0.63(4)
^{136}Ce		136	0(+)	0.185	5.76(9)	0			4.23(13)	0	4.23(13)	7.3(1.5)
^{138}Ce		138	0(+)	0.251	6.65(9)	0			5.64(15)	0	5.64(15)	1.1(3)
^{140}Ce		140	0(+)	88.45	4.81(9)	0			2.94(11)	0	2.94(11)	0.57(4)
^{142}Ce		142	0(+)	11.114	4.72(9)	0			2.84(11)	0	2.84(11)	0.95(5)
Pr	59	141	5/2(+)	100	4.58(5)	−0.055(3)			2.64(6)	0.015(3)	2.66(6)	11.5(3)

Continued

TABLE A.2 Bound Scattering Lengths and Cross Sections of the Elements and Their Isotopes—Cont'd

Element	Z	A	$I(\pi)$	c_a	$\bar{b}$	b^+	b^-	b_i	σ_c	σ_i	σ_s	σ_a
Nd	60				7.69(5)				7.43(19)	9.2(8)	16.6(8)	50.5(1.2)
^{142}Nd		142	0(+)	27.152	7.7(3)			0	7.5(6)	0	7.5(6)	18.7(7)
^{143}Nd		143	7/2(+)	12.174	14.0(2.0)E			±21.(1.)	25.0(7.0)	55.0(7.0)	80.0(2.0)	337.0(10.0)
^{144}Nd		144	0(+)	23.798	2.8(3)			0	1.0(2)	0	1.0(2)	3.6(3)
^{145}Nd		145	7/2(+)	8.293	14.0(2.0)E				25.0(7.0)	5.0(5.0)E	30.0(9.0)	42.0(2.0)
^{146}Nd		146	0(+)	17.189	8.7(2)			0	9.5(4)	0	9.5(4)	1.4(1)
^{148}Nd		148	0(+)	5.756	5.7(3)			0	4.1(4)	0	4.1(4)	2.5(2)
^{150}Nd		150	0(+)	5.638	5.28(20)			0	3.5(3)	0	3.5(3)	1.2(2)
Pm	61	147	7/2(+)	2.6234 y	12.6(4)				20.0(1.3)	1.3(2.0)	21.3(1.5)	168.4(3.5)
Sm	62				0.00(5) – 1.65(2)i				0.422(9)	39.0(3.0)	39.4(3.0)	5922.0(56.0)
^{144}Sm		144	0(+)	3.07	−3.0(4.0)E			0	1.0(3.0)	0	1.0(3.0)	0.7(3)
^{147}Sm		147	7/2(−)	14.99	14.0(3.0)			±11.(7.)	25.0(11.0)	14.0(19.0)	39.0(16.0)	57.0(3.0)
^{148}Sm		148	0(+)	11.24	−3.0(4.0)E			0	1.0(3.0)	0	1.0(3.0)	2.4(6)

^{149}Sm	*	149	7/2(−)	13.82	18.7(28) − 11.7(1)i	±31.4(6) − 10.3i	63.5(6)	137.0(5.0)	200.0(5.0)	42080.0(400.0)
^{150}Sm		150	0(+)	7.38	14.0(3.0)	0	25.0(11.0)	0	25.0(11.0)	104.0(4.0)
^{152}Sm		152	0(+)	26.75	−5.0(6)	0	3.1(8)	0	3.1(8)	206.0(6.0)
^{154}Sm		154	0(+)	22.75	8.0(1.0)	0	11.0(2.0)	0	11.0(2.0)	8.4(5)
Eu		63			5.3(3) − 1.26(1)i		6.57(4)	2.5(4)	9.2(4)	4530.0(40.0)
^{151}Eu	*	151	5/2(+)	47.81	6.92(15) − 2.53(3)i	±4.5(4) − 2.14(2)i	5.5(2)	3.1(4)	8.6(4)	9100.0(100.0)
^{153}Eu		153	5/2(+)	52.19	8.22(12)	±3.2(9)	8.5(2)	1.3(7)	9.8(7)	312.0(7.0)
Gd		64			9.5(2) − 13.82(3)i		29.3(8)	151.0(2.0)	180.0(2.0)	49700.0(125.0)
^{152}Gd		152	0(+)	0.2	10.0(3.0)E	0	13.0(8.0)	0	13.0(8.0)	735.0(20.0)
^{154}Gd		154	0(+)	2.18	10.0(3.0)E	0	13.0(8.0)	0	13.0(8.0)	85.0(12.0)
^{155}Gd	*	155	3/2(−)	14.8	13.8(3) − 17.0(1)i	±5.(5.) − 13.16(9)i	40.8(4)	25.0(6.0)	66.0(6.0)	61100.0(400.0)
^{156}Gd		156	0(+)	20.47	6.3(4)	0	5.0(6)	0	5.0(6)	1.5(1.2)

Continued

TABLE A.2 Bound Scattering Lengths and Cross Sections of the Elements and Their Isotopes—Cont'd

Element	Z	A	I(π)	c_a	$\bar{b}$	b^+	b^-	b_i	σ_c	σ_i	σ_s	σ_a
^{157}Gd	*	157	3/2(−)	15.65	−1.14(2) − 72.0(2)i			±5.(5.) − 55.8(2)i	650.0 (4.0)	394.0 (7.0)	1044.0 (8.0)	259000.0 (700.0)
^{158}Gd		158	0(+)	24.84	9.0(2.0)			0	10.0 (5.0)	0	10.0 (5.0)	2.2(2)
^{160}Gd		160	0(+)	21.86	9.15(5)			0	10.52 (11)	0	10.52 (11)	0.77(2)
Tb	65	159	3/2(+)	100	7.34(2)	6.8(2)	8.1(2)	−0.17(7)	6.84(6)	0.004(3)	6.84(6)	23.4(4)
Dy	66				16.9(3) − 0.276(4)i				35.9(8)	54.4(1.2)	90.3(9)	994.0 (13.0)
^{156}Dy		156	0(+)	0.056	6.1(5)			0	4.7(8)	0	4.7(8)	33.0(3.0)
^{158}Dy		158	0(+)	0.095	6.7(4.0)E			0	5.0(6.0)	0	5.0(6.0)	43.0(6.0)
^{160}Dy		160	0(+)	2.329	6.7(4)			0	5.6(7)	0	5.6(7)	56.0(5.0)
^{161}Dy		161	5/2(+)	18.889	10.3(4)	14.5 (5)	4.2(5)	5.1(4)	13.3 (1.0)	3.0(1.0)	16.0 (1.0)	600.0 (25.0)
^{162}Dy		162	0(+)	25.475	−1.4(5)			0	0.25 (18)	0	0.25 (18)	194.0 (10.0)
^{163}Dy		163	5/2(−)	24.896	5.0(4)	6.1(5)	3.5(5)	1.3(3)	3.1(5)	0.21(19)	3.3(5)	124.0 (7.0)
^{164}Dy		164	0(+)	28.26	49.4(5) − 0.79(1)i			0	307.0 (3.0)	0	307.0 (3.0)	2840.0 (40.0)

Ho	67	165	7/2(−)	100	8.44(3)	6.9(2)	10.3(2)	−1.69(14)	8.06(8)	0.36(3)	8.42(16)	64.7(1.2)
Er	68				7.79(2)				7.63(4)	1.1(3)	8.7(3)	159.0(4.0)
^{162}Er		162	0(+)	0.139	9.01(11)		0	0	9.7(4)	0	9.7(4)	19.0(2.0)
^{164}Er		164	0(+)	1.601	7.95(14)		0	0	8.4(4)	0	8.4(4)	13.0(2.0)
^{166}Er		166	0(+)	33.503	10.51(19)		0	0	14.1(5)	0	14.1(5)	19.6(1.5)
^{167}Er		167	7/2(+)	22.869	3.06(5)	5.3(3)	0.0(3)	2.6(2)	1.1(2)	0.13(6)	1.2(2)	659.0(16.0)
^{168}Er		168	0(+)	26.978	7.43(8)		0	0	6.9(7)	0	6.9(7)	2.74(8)
^{170}Er		170	0(+)	14.91	9.61(6)		0	0	11.6(1.2)	0	11.6(1.2)	5.8(3)
Tm	69	169	1/2(+)	100	7.07(3)			0.9(3)	6.28(5)	0.10(7)	6.38(9)	100.0(2.0)
Yb	70				12.41(3)				19.42(9)	4.0(2)	23.4(2)	34.8(8)
^{168}Yb		168	0(+)	0.123	−4.07(2) − 0.62(1)i		0	0	2.13(2)	0	2.13(2)	2230.0(40.0)
^{170}Yb		170	0(+)	2.982	6.8(1)		0	0	5.8(2)	0	5.8(2)	11.4(1.0)
^{171}Yb		171	1/2(−)	14.09	9.7(1)	6.5(2)	19.4(4)	−5.59(19)	11.7(2)	3.9(2)	15.6(3)	48.6(2.5)
^{172}Yb		172	0(+)	21.68	9.5(1)		0	0	11.2(2)	0	11.2(2)	0.8(4)
^{173}Yb		173	5/2(−)	16.103	9.56(10)	2.5(2)	13.3(3)	−5.3(2)	11.5(2)	3.5(3)	15.0(4)	17.1(1.3)

Continued

TABLE A.2 Bound Scattering Lengths and Cross Sections of the Elements and Their Isotopes—Cont'd

Element	Z	A	$I(\pi)$	c_a	$\bar{b}$	b^+	b^-	b_i	σ_c	σ_i	σ_s	σ_a
^{174}Yb		174	0(+)	32.026	19.2(1)			0	46.8(5)	0	46.8(5)	69.4(5.0)
^{176}Yb		176	0(+)	12.996	8.7(1)			0	9.6(2)	0	9.6(2)	2.85(5)
Lu	71				7.21(3)				6.53(5)	0.7(4)	7.2(4)	74.0(2.0)
^{175}Lu		175	7/2(+)	97.401	7.28(9)			±2.2(7)	6.59(5)	0.6(4)	7.2(4)	21.0(3)
^{176}Lu	*	176	7(−)	2.599	6.1(2) − 0.57(1)i			±3.0(4) + 0.61(1)i	4.7(2)	1.2(3)	5.9(4)	2065.0 (35.0)
Hf	72				7.77(14)				7.6(3)	2.6(5)	10.2(4)	104.1(5)
^{174}Hf		174	0(+)	0.16	10.9(1.1)			0	15.0 (3.0)	0	15.0 (3.0)	561.0 (35.0)
^{176}Hf		176	0(+)	5.26	6.61(18)			0	5.5(3)	0	5.5(3)	23.5(3.1)
^{177}Hf		177	7/2(+)	18.6	0.8(1.0)E			±0.9(1.3)	0.1(2)	0.1(3)	0.2(2)	373.0 (10.0)
^{178}Hf		178	0(+)	27.28	5.9(2)			0	4.4(3)	0	4.4(3)	84.0(4.0)
^{179}Hf		179	9/2(+)	13.62	7.46(16)			±1.06(8)	7.0(3)	0.14(2)	7.1(3)	41.0(3.0)
^{180}Hf		180	0(+)	35.08	13.2(3)			0	21.9 (1.0)	0	21.9 (1.0)	13.04(7)
Ta	73				6.91(7)				6.00 (12)	0.01(17)	6.01 (12)	20.6(5)

^{180}Ta	*	180	9(−)	0.01201	7.0(2.0)E		6.2(3.5)	0.5(5)E	7.0(4.0)	563.0 (60.0)
^{181}Ta		181	7/2(+)	99.98799	6.91(7)	−0.29(3)	6.00 (12)	0.011(2)	6.01 (12)	20.5(5)
W	74				4.755(18)		2.97(2)	1.63(6)	4.60(6)	18.3(2)
^{180}W		180	0(+)	0.12	5.0(3.0)E	0	3.0(4.0)	0	3.0(4.0)	30.0 (20.0)
^{182}W		182	0(+)	26.5	7.04(4)	0	6.10(7)	0	6.10(7)	20.7(5)
^{183}W		183	1/2(−)	14.31	6.59(4)	6.3(4)	5.36(7)	0.3(3)E	5.7(3)	10.1(3)
^{184}W		184	0(+)	30.64	7.55(6)	0	7.03 (11)	0	7.03 (11)	1.7(1)
^{186}W		186	0(+)	28.43	−0.73(4)	0	0.065 (7)	0	0.065 (7)	37.9(6)
Re	75				9.2(2)		10.6(5)	0.9(6)	11.5(3)	89.7(1.0)
^{185}Re		185	5/2(+)	37.4	9.0(3)	±2.0(1.8)	10.2(7)	0.5(9)	10.7(6)	112.0 (2.0)
^{187}Re		187	5/2(+)	62.6	9.3(3)	±2.8(1.1)	10.9(7)	1.0(6)	11.9(4)	76.4(1.0)
Os	76				10.7(2)		14.4(5)	0.3(8)	14.7(6)	16.0(4.0)
^{184}Os		184	0(+)	0.02	10.0(2.0)E	0	13.0 (5.0)	0	13.0 (5.0)	3000.0 (150.0)
^{186}Os		186	0(+)	1.59	12.0(1.7)	0	17.0 (5.0)	0	17.0 (5.0)	80.0 (13.0)

Continued

TABLE A.2 Bound Scattering Lengths and Cross Sections of the Elements and Their Isotopes—Cont'd

Element	Z	A	I(π)	c_a	$\bar{b}$	b^+	b^-	b_i	σ_c	σ_i	σ_s	σ_a
^{187}Os		187	1/2(−)	1.96	10.0(2.0)E				13.0 (5.0)	0.3(3)E	13.0 (5.0)	320.0 (10.0)
^{188}Os		188	0(+)	13.24	7.8(3)			0	7.3(6)	0	7.3(6)	4.7(5)
^{189}Os		189	3/2(−)	16.15	11.0(3)				14.4(8)	0.5(5)E	14.9(9)	25.0(4.0)
^{190}Os		190	0(+)	26.26	11.4(3)			0	15.2(8)	0	15.2(8)	13.1(3)
^{192}Os		192	0(+)	40.78	11.9(4)			0	16.6 (1.2)	0	16.6 (1.2)	2.0(1)
Ir	77				10.6(3)				14.1(8)	0.0(3.0)	14.0 (3.0)	425.0 (2.0)
^{191}Ir		191	3/2(+)	37.3								954.0 (10.0)
^{193}Ir		193	3/2(+)	62.7								111.0 (5.0)
Pt	78				9.60(1)				11.58 (2)	0.13(11)	11.71 (11)	10.3(3)
^{190}Pt		190	0(+)	0.012	9.0(1.0)			0	10.0 (2.0)	0	10.0 (2.0)	152.0 (4.0)
^{192}Pt		192	0(+)	0.782	9.9(5)			0	12.3 (1.2)	0	12.3 (1.2)	10.0(2.5)
^{194}Pt		194	0(+)	32.86	10.55(8)			0	14.0(2)	0	14.0(2)	1.44(19)

	A	I(π)	Abundance								
¹⁹⁵Pt	195	1/2(−)	33.78	8.91(9)	9.5(3)	7.2(3)	1.00(17)	9.8(2)	0.13(4)	9.9(2)	27.5(1.2)
¹⁹⁶Pt	196	0(+)	25.21	9.89(8)			0	12.3(2)	0	12.3(2)	0.72(4)
¹⁹⁸Pt	198	0(+)	7.36	7.8(1)			0	7.6(2)	0	7.6(2)	3.66(19)
Au ¹⁹⁷	197	3/2(+)	100	7.90(7)	6.26(10)	9.90(14)	−1.76(8)	7.32(12)	0.43(5)	7.75(13)	98.65(9)
Hg				12.595(45)				20.24(5)	6.6(1)	26.8(1)	372.3(4.0)
¹⁹⁶Hg	196	0(+)	0.15	30.3(1.0)−0.85(5)i			0	115.0(8.0)	0	115.0(8.0)	3080.0(180.0)
¹⁹⁸Hg	198	0(+)	9.97				0		0		2.0(3)
¹⁹⁹Hg	199	1/2(−)	16.87	16.9(4)−0.60(1)i			±15.5(8)	36.0(2.0)	30.0(3.0)	66.0(2.0)	2150.0(48.0)
²⁰⁰Hg	200	0(+)	23.1				0		0		<60.0
²⁰¹Hg	201	3/2(−)	13.18								7.8(2.0)
²⁰²Hg	202	0(+)	29.86	11.002(43)			0	15.2108(2)	0	15.2108(2)	4.89(5)
²⁰⁴Hg	204	0(+)	6.87				0		0		0.43(10)
Tl				8.776(5)				9.678(11)	0.21(15)	9.89(15)	3.43(6)
²⁰³Tl	203	1/2(+)	29.524	8.51(8)	9.08(10)	6.62(10)	1.061(14)	6.14(28)	0.14(4)	6.28(28)	11.4(2)
²⁰⁵Tl	205	1/2(+)	70.48	8.87(7)	5.15(10)	9.43(10)	−0.242(17)	11.39(17)	0.007(1)	11.40(17)	0.104(17)

Continued

TABLE A.2 Bound Scattering Lengths and Cross Sections of the Elements and Their Isotopes—Cont'd

Element	Z	A	$I(\pi)$	c_a	$\bar{b}$	b^+	b^-	b_i	σ_c	σ_i	σ_s	σ_a
Pb	82				9.401(2)				11.115 (7)	0.0030 (7)	11.118 (7)	0.171(2)
^{204}Pb		204	0(+)	1.4	10.893(78)			0	12.3(2)	0	12.3(2)	0.65(7)
^{206}Pb		206	0(+)	24.1	9.221(78)			0	10.68 (12)	0	10.68 (12)	0.0300 (8)
^{207}Pb		207	1/2(−)	22.1	9.286(16)			0.14(6)	10.82 (9)	0.002(2)	10.82 (9)	0.699 (10)
^{208}Pb		208	0(+)	52.4	9.494(30)			0	11.34 (5)	0	11.34 (5)	0.00048 (3)
Bi	83	209	9/2(−)	100	8.532(2)	8.26 (1)	8.74(1)	0.22(4)	9.148 (4)	0.0084 (19)	9.156 (4)	0.0338 (7)
Po	84											
At	85											
Rn	86											
Fr	87											
Ra	88	226	0(+)	1600 y	10.0(1.0)			0	13.0 (3.0)	0	13.0 (3.0)	12.8(1.5)
Ac	89											
Th	90	232	0(+)	100	10.31(3)			0	13.36 (8)	0	13.36 (8)	7.37(6)

Pa	91	231	3/2(−)	32,760 y	9.1(3)		10.4(7)	0.1(3.3)	10.5 (3.2)	200.6 (2.3)
U	92				8.417(5)		8.903 (11)	0.005 (16)	8.908 (11)	7.57(2)
^{233}U		233	5/2(+)	159,200 y	10.1(2)	±1.0(3.0)	12.8(5)	0.1(6)	12.9(3)	574.7 (1.0)
^{234}U		234	0(+)	0.0054	12.4(3)	0	19.3(9)	0	19.3(9)	100.1 (1.3)
^{235}U		235	7/2(−)	0.7204	10.50(3)	±1.3(6)	13.78 (11)	0.2(2)	14.0(2)	680.9 (1.1)
^{238}U		238	0(+)	99.2742	8.407(7)	0	8.871 (11)	0	8.871 (11)	2.68(2)
Np	93	237	5/2(+)	2,144,000 y	10.55(10)		14.0(3)	0.5(5)E	14.5(6)	175.9 (2.9)
Pu	94									
^{238}Pu		238	0(+)	87.74 y	14.1(5)	0	25.0 (1.8)	0	25.0 (1.8)	558.0 (7.0)
^{239}Pu		239	1/2(+)	24,110 y	7.7(1)	±1.3(1.9)	7.5(2)	0.2(6)	7.7(6)	1017.3 (2.1)
^{240}Pu		240	0(+)	6561 y	3.5(1)	0	1.54(9)	0	1.54(9)	289.6 (1.4)
^{242}Pu		242	0(+)	375,000 y	8.1(1)	0	8.2(2)	0	8.2(2)	18.5(5)
Am	95	243	5/2(−)	7370 y	8.3(2)	±2.0(7.0)	8.74(4)	0.3(2.6)	9.0(2.6)	75.3(1.8)

Continued

TABLE A.2 Bound Scattering Lengths and Cross Sections of the Elements and Their Isotopes—Cont'd

Element	Z	A	$I(\pi)$	c_a	$\bar{b}$	b^+	b^-	b_i	σ_c	σ_i	σ_s	σ_a
Cm	96											
^{244}Cm		244	0(+)	18.1 y	9.5(3)			0	11.3(7)	0	11.3(7)	16.2(1.2)
^{246}Cm		246	0(+)	4706 y	9.3(2)			0	10.9(5)	0	10.9(5)	1.36(17)
^{248}Cm		248	0(+)	348,000 y	7.7(2)			0	7.5(4)	0	7.5(4)	3.00(26)

arising from neutron–electron interactions mentioned in Section A.2.3 were applied. A word of caution about the use of this table must be made here. When using this table, it must be taken into account the energy of the neutrons if we need to know the nuclear or the atomic scattering length. Generally speaking, the expressions shown in Section A.2.3 must be taken into account. It must be kept in mind that in scattering experiments involving thermal neutrons, the coherent atomic scattering length is described by Equation (A.47), while in transmission experiments, it does according to Equation (A.54) or Equation (A.55). In the limit where energy tends to zero, we will measure $b_c(0) = b_c$, whereas in the epithermal energy region and beyond the coherent cross section is determined by b_v.

Table A.3 shows the contributions from isotope incoherence and spin incoherence to the total incoherent cross section for a group of selected elements, in order of increasing isotope incoherence. Na and Co consist entirely of a single isotope of $I \neq 0$, so incoherence is completely due to spin. On the other hand, the three stable isotopes of Ar have $I = 0$, so incoherence is entirely isotopic.

The absorption cross section encompasses all absorption processes that occur with thermal neutrons. They are (n,γ) reactions (radiative capture), (n,α) reactions, (n,p) reactions, and (n,f) reactions (fission). Thus, the absorption cross section is $\sigma_a = \sigma_\gamma + \sigma_\alpha + \sigma_p + \sigma_f$. The only natural nuclide that has a zero absorption cross section is ^{4}He. Also, there are several radionuclides for which $\sigma_a = 0$, such as ^{3}H. For most of the nuclides, the main contribution to the absorption cross section is due to radiative processes. Table A.4 lists the

TABLE A.3 Contributions to Spin and Isotope Incoherence to the Total Bound Incoherent Scattering Cross Sections of Selected Elements, Listed in Order of Increasing Ratio of Isotope Incoherence to Total Incoherence (Cross Sections in Barns)

Element	Z	σ_i	σ_i (spin)	σ_i (isotope)	σ_i (isotope)/σ_i
H	1	80.26	80.26	0.006	<0.0001
V	23	5.08	5.06	0.02	0.00
Li	3	0.92	0.76	0.16	0.18
Cl	17	5.3	3.56	1.74	0.33
Gd	64	151	65.36	85.64	0.57
Cr	24	1.83	0.56	1.27	0.69
Ti	22	2.87	0.29	2.58	0.90
W	74	1.63	0.23	1.40	0.86
Ni	28	5.2	0.02	5.18	1.00
Ar	18	0.23	0.00	0.23	1.00

TABLE A.4 Radiative-Capture Cross Sections and Resonance Energies of Naturally Occurring Nuclides Having (n,γ) Resonances at Thermal Neutron Energies

Nuclide	σ_γ (b)	E_0 (meV)
^{113}Cd	20,600	178
^{149}Sm	42,080	97.3
^{151}Eu	9100	321
^{155}Gd	61,100	26.8
^{157}Gd	259000	31.4
^{176}Lu	2065	141.3
^{180}Ta	563	200

TABLE A.5 Absorption Cross Sections for Nuclides That Have Significant Charged-Particle Reactions with Thermal Neutrons (Cross Sections in Barns)

Nuclide	σ_a	σ_γ	σ_p	σ_α	σ_f
^{3}He	5333	0.000031	5333	0	0
^{6}Li	940	0.0385	0	940	0
^{10}B	3835	0.5	0	3834	0
^{14}N	1.91	0.075	1.83	0	0
^{17}O	0.236	0.00054	0	0.235	0
^{33}S	0.54	0.35	0.002	0.19	0
^{35}Cl	44.1	43.6	0.489	0	0
^{40}K	35	30	4.4	0.39	0
^{233}U	575	46	0	0	529
^{235}U	681	98	0	0	583
^{238}Pu	558	540	0	0	17.9
^{239}Pu	1017	269	0	0	748

radiative-capture cross sections and resonance energies, for naturally occurring nuclides with resonances at thermal energies. In Table A.2, those nuclides were marked with an asterisk. The contribution of charged-particle reactions to the absorption cross section is normally very small. Table A.5

lists some of the few nuclides that have an appreciable cross section for charged-particle reactions with thermal neutrons. Nuclides with large cross sections for charged-particle reactions such as ^{3}He, ^{6}Li, and ^{10}B are used as materials for thermal neutron detectors.

REFERENCES

[1] Sears VF. Phys Rep 1986;141:281.
[2] Fermi E, Marshall L. Phys Rev 1947;71:666.
[3] Sears VF. In: Sköld K, Price DL, editors. Neutron scattering, part A. Experimental methods in the physical sciences, vol. 23. London: Academic Press; 1987 [Appendix].
[4] Koester L, Rauch H, Seymann E. Atom Data Nucl Data Tables 1991;49:65.
[5] Sears VF. Neutron optics. New York: Oxford University Press; 1989.
[6] Parks DE, Nelkin MS, Beyster JR, Wikner NF. Slow neutron scattering and thermalization. New York: W. A. Benjamin Inc; 1970.
[7] Breit G, Wigner E. Phys Rev 1936;48:519.
[8] Marshall W, Lovesey SW. Theory of thermal neutron scattering. Oxford: Clarendon Press; 1971.
[9] Granada JR, Santisteban JR, Dawidowski J, Mayer RE. Phys B 1993;190:259.
[10] Koester L, Wachkowski W, Klüver A. Phys B 1993;137:282.
[11] Koester L, Waschkowski W, Mitsyna LV, Samosvat GS, Prokofjevs P, Tambergs J. Phys Rev C 1995;51:3363.
[12] Foldy LL. Rev Mod Phys 1958;30:471.
[13] Kopecky S, Riehs P, Harvey JA, Hill NW. Phys Rev Lett 1995;74:2427.
[14] Kopecky S, Harvey JA, Hill NW, Krenn M, Pernicka M, Riehs P, Steiner S. Phys Rev C 1997;56:2229.
[15] Huber MG. PhD Thesis, Tulane University; 2009.
[16] Ignatovich VK, Utsuro M, Ignatovich PhV. Phys Rev C 1999;59:1136.
[17] Granada JR, Santisteban JR, Dawidowski J, Mayer RE. Phys Proc 2012;26:108.
[18] Granada JR. Z Naturforsch 1984;39A:1160.
[19] Santisteban JR, Edwards L, Steuwer A, Withers PJ. J Appl Cryst 2001;34:289.
[20] Kropff F, Granada JR. CRIPO. Program for total cross section calculation of polycrystalline materials. Argentina: Centro Atómico Bariloche, CNEA; 1975.
[21] Schermer RI. Phys Rev 1963;130:1907, Phys Rev 1964;136:B1285.
[22] Shull CG, Wollan EO. Phys Rev 1951;81:527.
[23] Shull CG, Ferrier RP. Phys Rev Lett 1963;10:295.
[24] Shull CG. Phys Rev Lett 1968;21:1585.
[25] Schneider CS, Shull CG. Phys Rev B 1971;3:830.
[26] Shull CG, Billman KW, Wedgwood FA. Phys Rev 1967;153:1415.
[27] Schneider CS. PhD Thesis. Massachusetts Institute of Technology; 1969.
[28] Schneider CS. Acta Crystallogr A 1976;32:375.
[29] Christiansen C. Wied Annln 1884;23:298.
[30] Koester L, Ungerer H. Z Phys 1969;220:300.
[31] Koester L, Knopf K, Waschkowski W. Z Phys A 1981;301:215.
[32] Koester L, Knopf K, Waschkowski W. Z Phys 1977;A282:371.
[33] Weiss RJ. Phys Rev 1951;83:379.
[34] Koester L. Neutron physics. In: Springer tracts in modern physics, vol. 80. p. 1, Berlin, Heidelberg.

[35] Maier-Leibnitz H. Z Angew Phys 1962;14:738.
[36] Koester L. Z Phys 1965;182:328.
[37] Koester L, Nistler W. Z Phys A 1975;272:189.
[38] Klein AG, Kearney PD, Oplat GI, Cimmino A, Gähler R. Phys Rev Lett 1981;46:959.
[39] Rauch H, Treimer W, Bonse U. Phys Lett 1974;47A:369.
[40] Rauch H, Werner SA. Neutron interferometry. New York: Oxford University Press; 2000.
[41] Werner SA, Klein AG. In: Skold K, Price DL, editors. Neutron scattering, part A. Experimental methods in the physical sciences, vol. 23. London: Academic Press; 1987 [chapter 4].
[42] Bauspiess W, Bonse U, Rauch H. Nucl Instrum Methods 1978;157:495.
[43] von Nardroff R. Phys Rev 1926;28:240.
[44] Glättli H, Coustham J, Malinovski A, Pinot M. Z Phys A—Atomic Nuclei 1987;327:149–55.
[45] Ramsey N. Molecular beams. Oxford: Oxford University Press; 1956.
[46] Piegsa FM, van den Brandt B, Hautle P, Konter JA. Nucl Instrum Methods A 2009;605:5–8.
[47] Wong DY, Noyes HP. Phys Rev 1962;126:1866.
[48] Miller GA, Nefkens BMK, Slaus I. Phys Rep 1990;194:1.
[49] Howell CR, Chen Q, Carman TS, Hussein A, Gibbs WR, Gibson BF, et al. Phys Lett B 1998;444:252.
[50] Schori O, Gabioud B, Joseph C, Perroud JP, Ruegger D, Tram MT, Truol P, Winkelmann E, Dahme W. Phys Rev C 1987;35:2252.
[51] Gabioud B, Alder J-C, Joseph C, Loude J-F, Morel N, Perrenoud A, et al. Nucl Phys A 1984;420:496.
[52] Huhn V, Wätzold L, Weber Ch, Siepe A, von Witsch W, Witała H, Glöckle W. Phys Rev C 2001;63:014003.
[53] Šlaus I, Akaishi Y, Tanaka H. Phys Rep 1989;173:259.
[54] Chen Q, Howell CR, Carman TS, Gibbs WR, Gibson BF, Hussein A, et al. Phys Rev C 2008;77:054002.
[55] Tuli JK. Nuclear Wallet Cards 8th Edition. Brookhaven National Laboratory, http://www.nndc.bnl.gov; 2011 [last accessed 20.08.12].
[56] Sears VF. Neutron News 1992;3(3):26.
[57] Mughabghab SF, Divadeenam M, Holden NE. Neutron cross sections, vol. 1. New York: Academic Press; 1981, Part A, $Z=1$-60.
[58] Mughabghab SF, Divadeenam M, Holden NE. Neutron cross sections, vol. 1. New York: Academic Press; 1984, Part B, $Z=61$-100.

Index

Note: Page numbers followed by "*f*" indicate figures and "*t*" indicate tables.

A

ab initio, 326–328
Absorption, 8–9, 10, 14, 15, 36
Accelerator-driven neutron sources
  "buckling", 160–162
  ING project, 198
  LANL, 198
  neutron generation, 160–161
  optimal moderators, 159–160
  para-H_2 and normal hydrogen, 163, 164*f*
  and peak shapes, reactor, 163–165, 164*f*
  prototype production cyclotron, 198
  prototype working, 197–198
  semiempirical grounds, spectrum, 161
  time and associated RMS deviation, 162–163
Acoustic magnon, 109
Adsorption
  air-solution interface, 365–374
  biological membranes, structure, 378–382
  colloidal particles, 392–394
  lamellar phases and vesicles, 385–392
  liquid-solid interface, 374–378
  micelles, 382–385
  polymers, solution, melt, and thin films, 394–403
  proteins and biomacromolecules, 404–409
Air-solution interface, adsorption
  $AlCl_3$ concentration, 370–371, 371*f*
  biosurfactants, 369
  deuterium-labeled surfactant, 366
  Gibbs equation, 365–366
  glycolipids, 366–367
  isotherms, 368*f*
  labeling schemes, 369–370
  Langmuir isotherm, 368
  modified PEI/SDS mixtures, 372–373, 373*f*, 374*f*
  multivalent counterions, 370
  null-reflecting water, 366
  polyelectrolytes, 370
  polymer-surfactant mixtures, 372
  regular solution theory (RST), 368–369
  rhamnolipid and rhamnolipid/anionic surfactant mixtures, 367
  scattering length, 366
  solution composition, 370*f*
  specular reflectivity, 367*f*
  surface composition *vs.* solution composition, 369*f*
  surface phase diagram, 372*f*
  surface tension (ST), 365–366
  surfactant adsorption, 366
  surfactant concentration, 367
β-Alanine, 342–343, 343*f*
Amorphous solid, 31, 35
Analyzer
  angular ambiguity, 257
  energy focusing, 288–289
  inverted-geometry instrument, 289*f*
  line focus mode, 277, 277*f*
  setup, RITA, 276
  and single-crystal monochromator, 275
Angle-resolved photoelectron spectroscopy (ARPES), 21
Angle-scanning diffraction
  Bragg condition, 250
  integrated intensity, 250
  Lorentz factor, 250
  structure factor $F(Q)$, 250
  unit-cell volumes, 250
  of volume in Q-space, 249, 249*f*, 250
Anharmonicity, 41
Anomalous scattering, 15, 19–21, 117, 118
Antiferromagnet, 95–97, 111, 119, 121–122, 125, 126*f*
ARPES, *see* Angle-resolved photoelectron spectroscopy (ARPES)
Atomic-mass number, 7
Atomic mass units, 2, 11–12, 82–83
Atomic number, 7–8
Atomistic modelling, 418, 419, 422
Atom type, 27–28, 29, 30, 34, 38–39, 59

B

Backscattering, 425–426
Backscattering spectrometer, 20*f*, 70–71

529

Ballistic neutron guide, *see* Focusing and ballistic guides
Bandwidth choppers
  illustration, 313–315, 314f
  at J-PARC, 313–315, 315f
Barn, 7
Beam injectors
  accelerator structures, 177
  appropriate target, 178
  Buncher/RFQ, 177
  classification, accelerator structures, 167, 168f
  cyclotron machines, 173–174
  duty factor, 170–173
  electrons, relativistic kinematics, 168–169
  Gentle Buncher, 170
  HEBT, 178
  high-power accelerators, 170
  high-voltage platform, 176
  intermediate structure, 178
  ion/electron source, 175, 176f
  LEBT, 176
  linear accelerator, MEBT, 177
  optical focusing/defocusing elements, 178
  proton/ion energies, 169
  RF power, 169
  RFQ, 169–170, 172f
  RF transmitters, 177
  sequence, vane modulations, 169–170, 171f
  wakefield accelerators, 167–168
Beam transport and tailoring
  choppers, 313–318
  neutron optics, *see* Neutron optics
Benzene, molecular vibrations, 342–343
Benzoic acid, 430–431, 431f, 432f, 453
Beryllium filter, 196, 424–425
Bessel function, 69, 80, 85–86
Biological membranes, adsorption
  bilayer membranes, 378–381
  dipalmitoylphosphatidylcholine (DPPC) bilayers, 381, 382
  1,2-dipalmitoyl-sn-glycero-3-phosphoserine (DPPS), 381
  distearoylphosphatidylcholine (DSPC), 378, 379f
  "floating bilayer" method, 378–381
  generic proteinaceous material, 378
  hydrogenated phosphocholine lipids, 380f
  lipids, 378
  water layer thickness, 381f
Bohr magneton, 82–83, 85–86, 120–121
Boltzmann's constant, 3
Born approximation, 2–3, 54–56

Bose-Einstein
  condensate, 53f
  population factor, 109–110
  population number, 32–33
Boson peak, 428–429
Bragg
  condition, 12, 34, 58–59, 60, 75–76
  peaks, 34, 60, 75, 96, 125
Bragg's law, 261–262
Bravais lattice, 31, 71–72, 94, 108–109, 117–118
Bremsstrahlung
  dipole interaction, 144–145
  electromagnetic cascade, 145–146
  electron linacs, neutron production activities, 146, 147t
  energy loss, incoming electrons, 144
  fission spectrum, 146
  higher energies, g-photon, 145
  plateau and yield, 146
Brillouin
  spectroscopy, 21
  zone, 18–19, 37, 54–56, 100–101, 111f
Brownian motion, 64, 65
Brownmillerite, 439–441
Buried interface, 81–82
"Burp" phenomena, 190–192

C

Canonical solid, 31–41, 42–82
Catalysis, 73–74, 127
CH_3, 418, 446, 447
Chadwick, 3
Characteristic reflection angle, 310–311
Chemical diffusion, 72–73
China Spallation Neutron Source (CSNS)
  layout, 207–208, 208f
  monolith hosting, solid target, 208, 209f
Chopper instruments
  energy resolution, 278–281
  LRMECS, 277–278
  observable Q-E space, 282–288
  Q resolution, 281–282
  4SEASONS spectrometer at J-PARC, 277–278, 278f
  spallation neutron sources, 278
Choppers
  bandwidth choppers, 313–315
  Fermi choppers, 315–318
  T_0 choppers, 313
Chudley, C.T, 71
Circular dichroism, 15

Classical electron radius, 7–8, 9, 83–84
Classic phonons and soft modes, 435
Clathrate, 447–448, 451–453, 452f
CM, see Coupled moderator (CM)
CO_2, 436, 459–460
Coherent
  dynamic structure factor, 30, 45–47
  QENS, rotation, 456–457
  scatterer CO_2, diffusion, 459–460
  scattering, 14, 28–30, 34, 39, 59, 70, 71–73, 75, 80–81, 87, 89–90, 91f, 93–94, 107, 434–446
  scattering function, 30, 31, 33, 35, 36–37, 46f, 51, 70, 72–73, 90, 107
Cold neutrons, 3–5, 5t, 65–66, 75
Colloidal particles, adsorption
  adsorbed layer, oscillations, 392–393, 393f
  oil-water interface, 393–394
  SANS, 392, 394f
  surfactant concentrations, 392, 394
Colossal magnetoresistance, 115
Combination band, 39–41
Compact pulsed hadron source (CPHS), 210–211, 211f
Compositional disorder, 60
Compton scattering, 416–417, 423–424
Computational experiments, 22, 23t
Computer simulation, 26–27, 33, 47, 57, 70
Concrete, 462–463
Confined media, 70
Constant-energy, 423
Correlation functions, 33, 42–43, 47, 51, 57, 64, 66–67, 70, 71, 86, 87, 100–101, 106, 109–110, 115, 117–118
Corrosion, 58
Coulomb, 420, 433–434
Coulomb integral, 100–101, 114
Count rate, 6, 14, 17–18, 39–41
Coupled moderator (CM)
  advanced neutronics technology, 258
  AMATERAS, 282
  calculated pulse peak intensities, 255, 256f
  description, 258–259
  pulse-shaping chopper, 291
  pulse structures, neutrons, 260f
CPHS, see Compact pulsed hadron source (CPHS)
Critical
  angle, 81, 103
  magnetic scattering, 117, 118, 120f
  opalescence, 118–119
  scattering, 117–119

Cross section, 2–3, 8–9, 10, 29–30, 38, 39, 47, 52, 60, 82–86, 89–90, 94, 99–100, 109–110, 123–125
Cryogenic device, 344–346
Cryostat, 344–345
Crystal
  field, 87, 111–114
  field excitations, 113f
  lattice, 16–17, 34, 37, 96–97, 111
  rocking curve, 96–97, 98f
Crystallography, 10f, 12, 34, 54–56, 85, 110–111, 111f, 342–343
CSNS, see China Spallation Neutron Source (CSNS)
Cs_4Pb_4, 456–457
CuBTC, 436–438, 437f
Cyclotron, 223

D

Darwin width, 291
de Broglie wavelength, 3
Debye ring, 250–251, 253
Debye-Scherrer method, see Powder diffraction
Debye–Waller factor, 32–34, 38–39, 61–63, 71, 426, 427–428
Decoupled moderator (DM)
  calculated pulse peak intensities, 256f
  neutron decoupling layer, 258–259
  pulse structures, neutrons, 260f
  TOF measurement, 272–273
Decoupler, 258–259
Delta function, 9, 24–25, 26–27, 33, 35–36, 37, 41, 42, 49–50, 69–70
Density functional theory (DFT)
  calculation, 429–430
  codes, 465–466
  computational experiments, 22
  gas-phase benzene molecule, 429
  harmonic vibrations, 41
  matrix/MD methods, 421–422
  MD simulations, 438
  molecular spectra, 425
  normal-mode calculation, 431f
  phonon calculations, 434–435, 439–442, 443–444
Density of states, 423
Deoxyribonucleic acid (DNA), 445–446
Detailed balance, 47, 52, 57, 107
Detectors
  CYCLOPS, 342, 342f
  fast readout neutron CCD cameras, 342

Detectors (*Continued*)
  scintillator screens, 341–342
  single-crystal diffractometers, 341
Deuteration
  experimental procedure, 432–433
  ice cage, 451–453
  isotopic substitution, 418
  protons, hydrogen bond, 430–431
  QENS signal, 463
  selective, 458, 459*f*
  spectral shifts, 430
DFT, *see* Density functional theory (DFT)
Dielectric spectroscopy, 22
Differential cross section, 249, 275
Differential scattering cross section, 6, 87, 88, 110, 123
Diffraction, 3, 12–13, 13*f*, 14, 15, 18–21, 41, 44, 45–47, 50, 75, 81, 82, 95, 96–97, 100–101, 101*f*, 125
Diffraction techniques
  powder diffraction, *see* Powder diffraction
  Q-scan methods, 261–262, 262*f*
  single-crystal diffractometers, *see* Single-crystal diffractometers
Diffuse scattering, 37–38, 58–61, 97–101, 104–106
  benzyl molecule, 336–337, 337*f*
  diffraction pattern, 336–337, 338*f*
  forms, 335–336
  manganese Prussian blue analogue, 336, 336*f*
  sensitivity, 335–336
  static/dynamic, 335–336
  x-ray data, 335–337
Diffusion
  equation, 64, 65
  of water, 418
Diglyme, 434
Dihydrogen hydrides
  hydrogenation processes, 332
  inelastic neutron scattering (INS), 333
  neutron diffraction structure, 332–333, 333*f*
  nuclear magnetic resonance measurements, 332–333
  Ru pincer complexes, 332–333
  ruthenium, 332
Direct and stripping reactions
  deuterons, 144
  function, incoming proton energy, 143–144
  nuclear reactions, 142–143
  yield, neutron, 142–143, 143*f*
Direct-geometry instrument, 7, 246, 277–278
Dislocation, 58, 102
Disorder, 58–61, 63, 82, 100, 120–125

Disordered solid, 22, 44, 58, 70–74
Distinct scattering, 44, 84–85, 100–101
DM, *see* Decoupled moderator (DM)
Double differential cross section, 17–18, 25, 26–27, 28, 30, 36–37, 49–50, 52, 61–63, 83–84, 86, 90–92, 106–107, 109–110, 112, 115, 116–117
Dynamical
  matrix, 31–32, 419–420, 421*f*
  scaling theory, 118–119
  transition, 457–458
Dynamics
  light scattering, 21
  nuclear polarization, 125
Dynamic structure factor, 30, 43–44, 45–58, 63–64, 66–67, 69, 70, 71

E

EISF, *see* Elastic incoherent structure factor (EISF)
Elastic
  diffuse scattering, 35
  scan, 457–458
  scattering, 3, 12–13, 33, 35, 45–47, 50, 61–63, 75, 87–88, 93, 106–107, 109–110, 114, 121–122, 247, 265
  structure factor, 33, 34
Elastic incoherent structure factor (EISF), 453–454, 454*f*, 456, 457*f*, 463–464, 465*f*
Electromagnetic cascade, 145–146
Electron g factor, 82–83
Element, 7–8, 8*f*, 14, 15–16, 23*t*, 26–28, 47, 53–54, 56, 83–84, 89–90, 92, 93–94, 125
Elliott, S.R., 38–39, 71
Energy resolution, chopper instruments
  distances, 280
  individual component, 280, 280*f*
  installation, 281
  moderation time, 280
  Ray diagrams, 278, 279*f*
  root-mean-square sum, 280
  time uncertainty at detector, 279–280
Epithermal neutrons
  chopper instruments, 281
  liquid-amorphous diffractometer, 263, 266
  total scattering diffractometer, 313
Equilibrium site, 31, 33, 34, 42, 61–64, 67, 71
ESS Project, *see* European Spallation Source (ESS) Project
Ethylene, 331–332

European Spallation Source (ESS) Project
  liquid hydrogen moderators, 204
  parameters, 203, 204t
  proposed layout, 202–203, 203f
  rotating tungsten wheel, 203
Evaporation, 148–149
Ewald sphere, 249, 249f, 271
EXAFS, see Extended x-ray absorption fine-structure spectroscopy (EXAFS)
Exchange integral, 87
Exponential decay, 418, 462–463
Extended x-ray absorption fine-structure spectroscopy (EXAFS), 15
External reflection, 80–81

F

Fast-ion conductor, 63, 70–71, 72–73
Fermi choppers
  cross section, 315, 316f
  description, 315
  neutron transmission, 316
  sloppy choppers, 316–318
  at SNS facility, 316, 317f
  speed, rotor blade, 315
  transmission through, 316, 317f
Fermion, 2
Fermi pseudopotential, 22–24, 25
Fermi's Golden Rule, 22–25
Ferroelectric, 54, 55f, 435
Ferromagnet, 94, 95, 96–97, 104–106, 105f, 110, 115–116, 117–118, 125
FFAG synchrotrons, see Fixed-field alternating gradient (FFAG) synchrotrons
Fickian limit, 67, 71–72
First moment, 49, 50–51
First-principles methods, 22, 41, 54–56, 110–111
Fission reaction
  Maxwellian, 141
  neutron economy, 141–142
  reactors, experimentation, 141
Fission reactor, 3–5
Fission spectrum, 146
Fixed-field alternating gradient (FFAG) synchrotrons
  development, resonant circular accelerator, 221–222
  drawbacks, research, 222
  scaling and nonscaling, 222–223
Flipping ratio, 94, 95, 103–104
Fluctuation-dissipation theorem, 47, 66–67
Focusing and ballistic guides
  estimated performance, 311–312, 312f
  linearly tapered and parabolic sections, 311–312, 312f
Force constants, 31–32, 419–420, 426, 431f, 434–435, 445–446, 457
Form factor, 9, 10f, 14, 15, 69, 75, 76–77, 78, 79–80, 79f, 85–86
Four-circle diffractometer
  FONDER diffractometer at JRR3, 271, 272f
  single-crystal diffractometer, 271
  structure factor, 271
Fourier transform, 419, 426–428
Free-atom cross section, 9, 52
Free particle, 41, 50–57
Frequency spectrum, 427
Fundamental transition, 39–41

G

Garland reflection
  characteristic reflection angle, 310–311
  curved guide section, 311f
  defined, 310–311
Gas storage, 73–74
Gaussian approximation, 64
GDOS, see Generalised density of states (GDOS)
Generalised density of states (GDOS), 423
Generalized susceptibililty, 106–108, 115–116
GeTe, 438
GISANS, see Grazing/glancing incidence small-angle neutron scattering (GISANS)
Glass, 12–13, 18–19, 22, 33, 35, 45–47, 50, 56–57, 69–70, 100
Glassy system
  de-Broglie wave, 284–285
  description, 283–284
  dynamic structure factor $S(Q,E)$, 284, 284f
  IXS and BRISP, 284–285
  scanning trajectories, 284, 284f
Gluons, 148
Grazing/glancing incidence small-angle neutron scattering (GISANS), 81
Guinier approximation, 78
Gyromagnetic ratio, 296–297

H

Hadron, 2
Hadron physics, 229–230
Hall and Ross, 455, 461
Hamiltonian, 419–420, 446, 447, 451–453

Harmonic
   approximation, 32–33, 37–38, 39–41,
      54–56, 108–109, 427–428, 433f, 457
   oscillator, 32
H/D substitution, 345
HEBT, see High-energy beam transport
   (HEBT)
^{3}He-gas detectors
   array of, chopper instrument, 301, 302f
   high voltage, 301
   position-sensitive detector and electric
      circuit, 301, 301f
Heisenberg
   exchange integrals, 108–109
   ferromagnets, 108–109
^{3}He spin-echo spectroscopy
High-energy beam transport (HEBT), 178, 203
High-energy neutron background
   gamma spectrum, JSNS moderators,
      261, 261f
   intensity function, beam port of JSNS,
      259–260, 260f
   methods, 260–261
   nuclear evaporation, 259–260
Host-guest interaction, 334
Hot neutrons, 20f
Hybrid systems, 225–226
Hydrides, 431–433
Hydrogen
   adsorbates and, 459–460
   carbon, 461f, 462f
   dynamics, 457f
   hydrogen-storage materials interaction,
      447–448
   incoherent scattering, 456–457
   light atoms and, 423
   nuclei, 448
   quantum rotation, 448
   readsorption, 440f
Hydrogen-bonding
   amines, in aqueous solution, 326
   hydrogen nucleus, 326
   Laue diffraction pattern, 326
   oxygen-platinum distance, 327–328, 328f
   porous materials, see Porous materials
   Pt-H interaction, 327–328
   Pt-lone pair interaction, 327–328, 328f
   role, 326–327
   single crystal neutron diffraction,
      327, 327f
Hydrogen-storage, 438–439, 447–448, 451
Hyperfine transitions, 121–122
HYSPEC, 246

I

iBIX diffractometer, protein structural
   analysis, 272–273, 274f
Ikeda-Carpenter function, 255–256, 257f
ILL, 424–425, 426, 433–434, 439–441, 445f,
   456–457, 458, see Institute Laue
   Langevin (ILL)
iMATERIA, detector arrangement, 267–269,
   268f
Impulse approximation, 41, 50–57
IN1, 424–425, 433–434
IN5, 445f, 456
IN6, 439–442, 456–457
IN13, 426, 458
Incoherent
   approximation, 39, 50
   dynamic structure factor, 30, 71
   scattering, 8–9, 23t, 28–30, 33, 38–41, 44,
      45, 66, 67, 69–70, 71–73, 86–87, 89, 90,
      91f, 120–122, 125
   scattering function, 30, 31, 33–34, 36, 46f,
      49, 51, 69, 70, 72–73
Incommensurate, 96
Indirect Fourier transform (IFT) method, 364–365
Indirect-geometry instrument, 7
Induction machines, 220–221
Inelastic
   form factor, 39–41, 61
   neutron scattering, 16, 18, 19–21, 19f, 37f,
      38–41, 40f, 43, 54–56, 111–112, 114f,
      116f, 121–122
   one-phonon scattering, 35–39
   spin-echo, 41
   x-ray scattering (IXS), 19–21
Inelastic scattering techniques
   chopper instruments, 277–288
   inverted-geometry instrument, 288–293
   triple-axis spectrometer, 275–277
Infrared spectroscopy, 39–41
ING project, see Intense neutron generation
   (ING) project
INS
   benzoic acid, 430, 431f
   computational investigations, 430–431
   D_2O uses, 451–453
   five-dimensional Hamiltonian, 451
   measurements, 445–446
   molecular system, 430
   phonon expansion, 428
   solid benzene, spectrum, 429f
   spectrum, 438–439
   variable temperature diffraction, 430–431

Index

Institute Laue Langevin (ILL)
  D2B diffractometer, 262, 263f
  SANS instrument, 293, 293f
Intense neutron generation (ING) project, 149–150, 198
Interband transitions, 115–117
Interface, 75, 80–82, 104–106
Intermediate
  scattering function, 30, 42–43, 51, 57, 64, 71, 73f
  self scattering function, 30, 67–69
Interparticle structure factor, 80
Inverted-geometry instrument
  description, 288–289, 289f
  DNA instrument at J-PARC, 291–293
  energy resolution, 289–291
Inverted TOF, 423–426
In-vessel irradiation, 228
Ionic liquids, 463–464
IPNS facility, 277–278, 316–318
ISIS, 423–424, 425, 429, 431f
ISIS facility, 277–278, 301, 316–318
Isothermal
  compressibility, 48
  susceptibility, 107–108, 118–119
Isotope, 7–8, 8f, 27–28, 29–30, 89–90, 228
Isotopic substitution, 7–8, 15, 23t, 27–28, 29–30
Isotropic scattering, 7, 9
Itinerant electrons, 88, 102–103, 115–116

J

Japan-Proton Accelerator Research Complex (J-PARC)
  decoupling, 202
  DNA instrument at, 291–293
  iBIX diffractometer, 272–273
  layout, MLSF, 201–202, 202f
  low frequency source, 267–269
  neutron-scattering KENS facility, 200–201
  pulse time width, 259f
  4SEASONS spectrometer, 278f
J-PARC, see Japan-Proton Accelerator Research Complex (J-PARC)
JRR3, FONDER four-circle diffractometer, 271, 272f
Jump
  distance, 71–72
  rate, 71–72, 73–74

K

Kagome bilayer, 443–444
Kinematic approximation, 81–82
Kinematic constraint, 284–285

Kohlrausch–Williams–Watts, 455–456
Kramers-Kronig causality relations, 107–108

L

Lamellar phases and vesicles, adsorption
  aggregate packing parameter, 385
  alkanol addition, 388
  alkyl chain volume, 385
  bi-lamellar and multi-lamellar, 386–388
  bilayer spacing, 386–388
  Caille parameter, 386–388
  cold/thermal neutrons, 389
  core-shell micelle model, 388–389
  Couette flow cell scattering geometry, 389, 389f
  dialkyl chain surfactants, 385
  elongational flow cell, 390–392, 391f
  Nallet model, 386–388, 387f
  SANS data, 385–386, 387f
  SANS intensity maps, 390f
  scattered intensity, 386–388
Langevin theory, 65
LANL, see Los Alamos National Laboratory (LANL)
Large-scale
  magnetic structures, 101–106
  structures, 75–82
Large-scale structures
  applications, 354
  isotopic substitution, 355
  neutron wavelength, 355
  NR, see Neutron reflectivity (NR)
  SANS, 354, 358–361
  scattering lengths, 355t
  structural characteristics, 354
  thin films, interfaces, and solutions, 365–409
  X-ray and neutron scattering techniques, 354–355
Laser-driven plasmas, neutron production, 224
Lattice dynamics
  calculation, 419–420, 434–435
  dispersion, 429–430
  simulations, 427
Laue function, 250, 251–252
Laue method, see Time-of-flight (TOF) diffraction
LEBT, see Low-energy beam transport (LEBT)
LENS, see Low-energy neutron source (LENS)
Levitation, 72–73, 127
Lifetime, 2, 41

Ligand water rotation, 456
Liouville theorem, 175
Liquid-solid interface, adsorption
  Bragg scattering, 374–375
  cellulose surfaces, 376
  interference fringe, 376–378
  layer thickness, 376–378
  NR geometry, 375f
  NR measurements, 376–378
  scattering length densities, 375–376
  specular reflectivity vs. wave-vector transfer, 377f
  surfactants and mixed surfactants, 376
  thermal/cold neutrons, transmission, 374–375
Lithium-ion battery, 63
Long-range order, 31, 34, 35, 36–38, 59, 72–73, 100, 107, 108–109, 112–114, 119, 125
Lorentz factor and integrated intensity
  angle-scanning diffraction, 249–250
  powder diffraction, 250–253
  TOF diffraction, see Time-of-flight (TOF) diffraction
Lorentzian, 65, 67, 71–72, 115, 118–119
Los Alamos National Laboratory (LANL)
  experiments, 190–192
  flux-trap moderator configuration, 194
Low-energy beam transport (LEBT), 176, 177, 199–200, 203
Low-energy neutron source (LENS), 212

M

Macroscopic differential cross section, 79–80
Magnetic
  Bragg scattering, 59, 61, 95–97, 99, 118
  clusters, 111–114
  defects, 99
  diffraction, 88, 96
  diffuse scattering, 58–59, 100–101, 100f
  disorder, 97–101
  domain, 104–106
  form factor, 9, 10f, 87, 93
  interaction operator, 83, 84, 85f
  interfaces, 104–106
  large-scale structures, 75–82
  moment, 3, 9, 10, 10f, 15, 82–83, 85–86, 87, 88, 95, 96–99, 100, 115, 120–121
  neutron mirror, 103
  order, 92–94, 96, 108–109, 125
  reflectivity, 104–106

scattering, 10, 22–24, 83–84, 87, 88, 90–93, 94, 95, 97–99, 100–102, 100f, 103–104, 107
small-angle scattering, 76–80, 81f, 102–103, 127
structure, 82–106
structure factor, 88, 93–94
superlattices, 104–106
unit-cell structure factor, 34, 87–88, 93–94, 96
Magnetism, 3, 10, 26, 82, 94, 103–106, 105f
Magnetization, 84–86, 93–94, 102, 103–104, 112–114, 118
  operator, 84
Magneto-vibrational scattering, 106–107
Magnon, 108–110, 119
  annihilation, 109–110
  creation, 108–109
Mapping, 419–420
Massachusetts Institute of Technology (MIT) reactor, 187, 213
Master formula, 22–25, 82–84, 90
Materials and Life Sciences Facility (MLSF)
  Japanese, 140
  J-PARC, 138–140
  layout, J-PARC, 201–202, 202f
  moderators, 191f, 192, 193f
Maxwell distribution, 254
MD, see Molecular dynamics (MD)
Mean kinetic energy, 53–54, 56–57
Mean-square displacement, 33, 64–65
Measured neutron-scattering signal, 427
MEBT, see Medium energy beam transport (MEBT)
Medium energy beam transport (MEBT), 177, 203, 211–212
Megawatt-range sources
  ESS, 202–204
  JPARC, see Japan-Proton Accelerator Research Complex (J-PARC)
  Oak Ridge spallation neutron, 199–200
  SINQ neutron, PSI, 204–205
Metal hydride, 39–41, 40f, 63, 70–71
Metal organic frameworks (MOFs), 436
MgH_2, 438–439, 440f
Micelles, adsorption
  $AlCl_3$ concentration, 384, 384f
  deuterium labeling, 384–385
  globular micellar model, 382, 382f
  ionic distribution, 383–384
  SANS data, 382, 383f
  $SDS/C_{12}E_6$ solution compositions, 385
Microemulsion, 79f

Index

MIT reactor, *see* Massachusetts Institute of Technology (MIT) reactor
Mixed-valence compounds, 115, 116f
MLSF, *see* Materials and Life Sciences Facility (MLSF)
Mode
  coupling theory, 70
  polarization vector, 31–32
Moderators, *see also* Slowing down processes and moderators, neutron
  geometries, 192–195, 193f
  optimal thickness, 192
  temperature, 190–192
MOF5
  adsorption interaction strength, 448–449
  crystal structure, 437f
  INS spectra, 449–451
  spectra, 450f
  $Zn_4O(1,4\text{-benzenedicarboxylate})_3$, 436
MOFs, *see* Metal organic frameworks (MOFs)
Molecular dynamics (MD)
  backscattering spectrometers, 425–426
  empirical and *ab initio* energy calculation, 420–422
  Fourier transforms, 419
  incoherent scatterers, 419
  mapping potential energy surfaces, 419–420
  measured neutron-scattering signal, 427
  molecular spectrometers, 424–425
  neutron Compton scattering spectrometers, 423–424
  neutron scattering theory, 418
  neutron spin-echo instruments, 426–427
  oscillatory motion, incoherent scattering, *see* Oscillatory motion, incoherent scattering
  simulation, 420
  three-axis spectrometers, 422
  time of flight, 422–423
  types, 417, 417f
  van Hove correlation function, 418
Molecular modelling, 417
Molecular spectrometers, 424–425
Molecular spectroscopy, 422, 424–425, 429
Molecular vibration, 429–430
Momentum transfer
  elastic scattering, 247
  and initial/final wave vectors, 247, 248f
  for Ni, 308
  TOF diffraction, 264
Monochromated diffraction with beam, 262
Monte Carlo, 22
Multiphonon scattering, 39–41

Multiscale modelling, 22
Muon, 21
Muon beams, 231

N

Nafion®, 460–462
Nanotubes, 448, 449f
N_2-cryostream, 346f
NEAT, 436–438
Negative thermal expansion (NTE), 435–438
Neutrino-related phenomena, 229–230
Neutron
  Cambridge Structure Database, 323f, 339–340
  Compton profile, 52–53, 54–56
  detectors, 341–342
  diffraction, 326–338
  diffuse scattering, 335–338
  energy loss, 11, 38, 119
  energy transfer, 11, 115
  flux, 3–5, 4f, 14
  guide, 80–81, 94
  hydrogen bonding, 326–328
  magnetic moment, 3, 82–83
  optics, 340–341
  "push-button" instruments, 322–323
  radium-beryllium radioactive-decaying source, 322
  reflectivity, 80–81, 82, 105f
  refractive index, 80–81
  samples and sample environment, 342–347
  scattering length, 324
  software, 347–348
  sources, 339–340
  spin echo, 19–21, 20f, 30, 72
  transition metal hydrides, 332–333
  wave vector, 3, 11
Neutron activation analysis, 228
Neutron detectors
  description, 300
  $^{157}Gd(n,\gamma)$, 300–301
  ^{3}He-gas detectors, 301
  ^{3}He(n,p), 300
  ^{10}He(n,α), 300
  ^{6}Li(n,α), 300
  scintillation detectors, 301–304
Neutron diffusion theory, 160–161
Neutron flux from moderators
  calculated pulse peak intensities, 255, 256f
  description, energy, 255
  neutron wavelength λ, 254–255
  production, 254
  short-pulse spallation sources, 255

Neutron guides
  critical angle, 308
  critical momentum transfer, 308
  description, 307
  reflection, super-mirror, 308, 309f
  reflectivity, 308, 309f
  super-mirror, illustration, 308, 308f
Neutronics, 258, 259, 261
Neutron optics
  critical-angle coefficient α, 306–307, 307t
  critical angle, 306
  description, 305
  Fermi pseudo-potential, V, 306
  focusing and ballistic guides, 311–312
  kinetic energy, 306, 340–341
  neutron guides, 307–310
  refraction and total reflection, 306, 307f
  single-crystal monochromator, 341
  straight and curved guides, 310–311
  supermirrors, 340–341
Neutron production reactions
  Bremsstrahlung, 144–146
  fission, see Fission reaction
  separation, 140–141
  spallation, see Spallation reactions
"Neutron quantum optics", 230
Neutron reflectivity (NR)
  air-substrate interface, 356, 357f
  Born approximation, 356
  data analysis, 363
  Fresnel coefficients, 357–358
  Gaussian roughness, 357–358
  horizontal geometry, 361–362
  one-dimensional Fourier transforms, 358
  scattering length density, 356
  self partial structure factors, 358
  SURF reflectometer, 361–362, 362f
  wave-vector transfer, 356, 361–362
Neutron scattering measurements
  cross section, 247–249
  description, 247
  detector and scattering probability, 247–248
  directions, 249
  double differential cross section, 248
  elastic scattering, $k_i = k_f$, 247
  incident neutron, 248, 248f
  integrated intensity and Lorentz factor, 249–254
  momentum transfer and initial/final wave vectors, 247, 248f
  single differential cross section, 249

Neutron sources
  accelerator-driven sources, see Accelerator-driven neutron sources
  beam injectors, 167–178
  Beryllium filter, 196
  The Bilbao neutron facility (ESS-B), 212
  CPHS, 210–211, 211f
  CSNS, see China Spallation Neutron Source (CSNS)
  cyclotron, 223
  direct and stripping reactions, see Direct and stripping reactions
  electron driven sources, 178–179
  FFAG, 219–220, 221–223
  high-flux reactors, 339–340
  high-intensity pulsed neutron generator, 340
  high-power accelerator-driven sources, 179, 180t
  hybrid systems, 225–226
  induction machines, 220–221
  ISIS second target station, 205–207, 206f
  last-generation megawatt-range, see Megawatt-range sources
  LENS, 212
  neutron beam, intensity, 340
  nonneutron-scattering, 227–231
  nuclear reactors, 339–340
  operation, IPNS, 138–140
  PEFP, 211–212
  PKUNIFTY, 211
  poisoning, 195
  premoderator, 195–196
  production, laser-driven plasmas, 224
  production reactions, see Neutron production reactions
  publications, referred journals, 138–140, 139t
  reactors, 226–227
  reflector, 188–190
  research reactors, see Research reactors
  shielding strategy, 196–197
  slowing down processes and moderators, 153–166
  spallation neutron source, 140
  tabletop production, 224–225
  target cooling, 183–187
  target materials, proton drivers, 179–183
  TMR complex, 187–188, 188f
  wakefield accelerators, 221
Neutron spin-echo (NSE), 426–427
Neutron spin-echo spectrometers
  description, 296

Fourier time, 298
  illustration, 299, 300f
  IN11, schematic layout, 299, 299f
  Larmor precession, 296, 298
  magnetic moment, neutron, 296–297
  neutron energy, 298
  polarization component, P_x, 297–298
  precessions, 300
  principle of, 296, 297f
  resonance flipper, 300
Neutron transport equation, 234
Newtonian dynamics, 420
Nitrogen cryostream, 346f
Nonneutron-scattering
  hadron physics, 229–230
  HFIR, 228
  muon beams, 231
  neutron interferometry techniques, 230
  nuclear physics and engineering, 228–229
  standard model, particle physics, 230–231
Normal mode, 31–33, 35–38, 63, 427–428, 431f, 432f
NRA, see Nuclear reaction analysis (NRA)
NSE, see Neutron spin-echo (NSE)
NTE, see Negative thermal expansion (NTE)
Nuclear
  absorption, 10, 94
  magnetic resonance (NMR), 21, 29, 68f, 120–121, 125
  magneton, 2, 82–83, 85–86
  physics, 27–28
  scattering, 9, 25–27, 58–59, 82–84, 86–87, 88, 89, 90, 92, 93–94, 95, 102, 106–107, 120–121
  spin, 3, 8–10, 15–16, 27–28, 29, 120–125
  spin correlations, 3, 8–9, 27–28, 120–121
  spin disorder, 120–125
  spin order, 120–125
  spin wavefunction, 121–122
Nuclear physics and engineering, 228–229
Nuclear reaction analysis (NRA), 398, 399f

O

Observable Q-E space, chopper instruments
  glassy system, 283–285
  one-dimensional system, 285
  scattering locus and projection, 282, 283f
  three-dimensional system, 286–288
  two-dimensional system, 285
Occupation number, 59–60, 115–116
Off-specular reflectivity, 104–106, 105f
One-phonon structure factor, 36–37

Optical techniques, 21
Optic magnon, 109
Orientational disorder, 63
Oscillator mass, 425f, 428
Oscillatory motion, coherent scattering
  classic phonons and soft modes, 435
  deoxyribonucleic acid, 445–446
  nanostructured materials, 438–439
  negative thermal expansion (NTE), 338
  oxygen-ion conductors-Brownmillerites, 439–441
  pnictides, 442–443
  strontium gallium oxides, 443–444
  thermoelectrics-Skutterudites, 441–442
Oscillatory motion, incoherent scattering
  benzene, molecular vibrations, 342–343
  complex hydrides, 345–347
  Debye-Waller factor, 331
  ethylene, 331–332
  harmonic approximation, 331
  hydrogen-bonded systems, 344–345
  polymers, 347–348
  quasilocalized dynamics, 332
  VDOS, anomalous peak, 332
Overdamped, 63
Overtone, 39–41

P

Pair density function, 43, 44
Pair distribution functions (PDF), 43–47
Paramagnet, 9, 21, 86, 90–92, 117
Path-integral molecular dynamics (PIMD), 56
Pauli Exclusion Principle, 121–122
Paul Scherrer Institut (PSI), 204–205
PDF, see Pair distribution functions (PDF)
PEFP, see Proton Engineering Frontier Project (PEFP)
Peking University Neutron Imaging Facility (PKUNIFTY), 211
Penetration depth, 14, 103–104
Perovskites, 435
Phonon
  annihilation, 37, 38, 108–109
  creation, 108–109
  dispersion relations, 37, 37f, 39, 54–56
Photocorrelation laser spectroscopy, 21
Photoelectron spectroscopy, 21
Photon, 19–21, 81–82
PKUNIFTY, see Peking University Neutron Imaging Facility (PKUNIFTY)
Planck's constant, 3

Plastic crystal, 42, 63
PM, see Poisoned decoupled moderator (PM)
Pnictides, 442–443
Poisoned decoupled moderator (PM)
  advanced neutronics technology, 258
  calculated pulse peak intensities, 256f
  pulse structures, neutrons, 260f
Polarized
  muons, 21
  neutrons, 26, 82–106, 121, 127
  nuclei, 94
Polymer electrolytes, 434
Polymers, adsorption
  anionic surfactant SDS interaction, 402, 403f
  carboxyl-terminated chains, 400f
  crystalline bulk heterojunction structure, 400–401
  hydrogen and deuterium-labeled polystyrene, 396
  intermixing degree, 396
  microstructure richness, 402
  mixing parameter, 396
  morphology, 394–396
  NRA, 398, 399f
  NR application, 394–396
  optical flow birefringence, 402, 403f
  polyelectrolytes, 401–402
  pre- and postannealed NR profiles, 400–401, 401f
  reflectivity vs. momentum transfer, 397f
  SCFT, 398–400
  segment length, 396
  in solution and melts, 402
  thin films, 396
  volume fraction profiles, 398f
Porod approximation, 78
Porous materials
  Dianine's compound, 334
  extrinsically, 334
  gas storage systems, 333–334
  hydrogen-bonded channels, 334
  intrinsically, 334
  metal-organic frameworks, 333–334
  neutron diffraction determination, 334–335
  4-phenoxyphenol, 334–335, 335f
Position sensitive detector (PSD)
  with ^{3}He gas, 301
  scattering diagram, 272, 273f
Potential energy, 419–420
Powder diffraction
  crystallites, 251

Debye-Scherrer diffraction ring, powder sample, 250–251, 251f
  integrated intensity, 251
  Lorentz factor, 253
  with monochromated beam, 262, 262f, 263f
  observed intensity δI, 251
  reciprocal space δV_q, 252
  resolution-oriented instrument, see Powder diffractometer
  scattering diagram, Debye-Scherrer method, 251, 252f
  structure factor F(Q), 251–252
  TOF powder diffraction, 262–266
  total scattering diffractometer, 269–270
Powder diffractometer
  detector arrangement, 267, 267f
  iMATERIA, detector arrangement, 267–269, 268f
  layout, POWGEN at SNS, 267–269, 268f
  resolution, POWGEN, 269, 269f
  source frequency, 267–269
Protein
  back-bone, 418
  complex molecular systems, 445–446
  dehydrated, 463
  dynamical transitions, 457
  dynamics, 463
  hydrated, 418
  polymers, 417–418
Proteins and biomacromolecules, adsorption
  amphiphilic protein hydrophobin HFBII, 404, 405f
  cationic, anionic, and nonionic surfactants, 404–405, 405f
  human blood clotting enzyme, 407
  lysozyme, 404
  mucosal protein Immunoglobulin A (IgA), 407
  processed foods, and biofouling, 404
  SANS and SAXS data, 407–409
Proton conductor, 72–73
Proton drivers, target materials
  future candidates, 183
  liquid Hg target, 182
  neutron-producing reactions, 179–182
  PbBi eutectic, 182–183
  SINQ, 183
Proton Engineering Frontier Project (PEFP), 211–212
Proton migration
  biological systems, 329
  crystallographer, 329
  electron density maps, 329, 330f
  Fourier difference maps, 329, 330f

Grotthus mechanism, 331–332, 332f
  hopping-reorientation mechanism, 331
  hydrogen atoms, displacement, 329
  isotropic displacement parameters, 331–332
  manganese citrate cubane coordination polymer, 331
  molecular crystals, 329
  neutron diffraction, 329
  single-crystal transformation, 331–332
  water diffusion, 329–331
PSD, see Position sensitive detector (PSD)
PSI, see Paul Scherrer Institut (PSI)
Pulse peak structure, neutrons
  Ikeda-Carpenter function, 255–256, 257f
  neutron-time distribution, 255–256
  SINQ and PSI, 255
Pulse peak width, neutrons
  description, 257
  high-resolution instruments, 257
  for typical moderators, 257, 258f
  wavelength, 257
Pulse shape function, 255–256
3,5-Pyridine dicarboxylic acid (PDCA), 334

Q

QENS, see Quasielastic neutron scattering (QENS)
Q-resolution, 281–282, 424–425
Q-scan methods, 261–262, 262f
Quantum
  delocalization, 451
  dynamics, 448, 451–453
  hydrogen molecules, rotations, 448
  motion, 446
  nuclear dynamics, 420
  proton transfer, 430
  single-particle rotations, 446
  transitions, 451–453
Quantum solid, 31, 41, 52–53
Quasielastic neutron scattering (QENS), 416, 418, 456–457
Quasielastic scattering, 63–64, 72–73, 74f
Quasiharmonic, 54–56
Q,ω space, 424–425, 434–435

R

g-Radiation, 228
Radio frequency (RF)
  acceleration techniques, 204
  accelerators, 220
  power, 169, 170–173
  transmitters, 168–169, 177
Radio frequency quadrupole (RFQ)
  accelerator structures, 170, 172f
  modern accelerators, 169–170
Raman spectroscopy, 21, 37
Rate constant, 418
RCM, see Relaxing cage model (RCM)
Reciprocal lattice vector, 12, 34, 95
Reciprocal space reconstruction, 347f
Reflectivity, 80–82, 103–106, 105f, 127
Reflector
  decoupled H_2 cryogenic moderator, 190, 191f
  nutshell, pros and cons, 188–189
  optimal materials, 188–189
  TMR arrangement installation, MLSF, 189–190, 190f
Relaxing cage model (RCM), 462–463
Research reactors
  absorption elements, 216
  cladding materials, 217
  cooling system, 217
  dangers, proliferation, 216–217
  experiment-driven, 213
  FRM-II, 213–216, 215f
  MIT, 213
  OPAL, 216
  productive, 213, 214t
  relevant fuels, 217
  vessel, 217–219, 218f
Residence time, 67, 71
Resonance, 8–9, 10, 15–16, 21, 22–24, 94, 121, 125
Resonant x-ray inelastic scattering (RIXS), 19–21
RF, see Radio frequency (RF)
RFQ, see Radio frequency quadrupole (RFQ)
Rigid unit modes, 435–436
RITA, advanced triple-axis spectrometer
  collimators, 276
  line focus mode, 277, 277f
  NIST reactor and MACS, 277
  point focus mode, 277, 277f
  spectrometer, schematic drawing, 276, 276f
Rotational diffusion, 69
Rotational tunneling
  one-dimensional threefold rotors, 447–448
  two-dimensional rotor H_2, 448–453
Rowland circle, 288–289

S

Sample environment, 14, 127
SANS, *see* Small-angle neutron scattering (SANS) technique
scaling FFAGs, 222–223
Scattering
  cross section, 6, 7, 18, 24–25, 26–27, 29, 32–33, 34, 44, 45, 52, 61–63, 83–84, 86–87, 88, 92, 93, 94, 106–107, 109–110, 115, 120–122, 123
  function, 30, 31, 33–34, 35–37, 41, 42–43, 45–47, 49, 50–51, 52, 57, 64–65, 67–69, 70, 71–73, 73f, 78, 90, 100–101, 107
  length, 7–9, 8f, 10, 14, 23t, 25, 26–27, 29, 45, 59, 75–76, 79–81, 87, 89–90, 103, 120–121, 122
  length density, 75–77, 78, 80, 81–82
  vector, 11, 15, 75–76, 99–100
Scattering length
  Bragg peaks, 324
  nuclear interaction, 324
  properties, 324
  variation, 324, 325f
SCFT, *see* Self consistent field theory (SCFT)
Scintillation detectors
  higher spatial resolution and detection speed, 301
  specifications, venetian-type scintillation detector, 302–303, 303t
  TAKUMI, 302–303, 303f
  venetian-type geometry, 302–303, 302f
  WLSF, *see* Wavelength-shifting fiber (WLSF) detector
  ZnS/^{6}LiF, 302
Second moment, 49, 50–51, 54–56, 66–67
Selective deuteration, 458, 459f
Self-assembly
  block copolymers, 394–396
  SANS, 365
  surfactant, 385, 389
Self consistent field theory (SCFT), 398–400
Self diffusion coefficient, 45–47, 65, 67, 68f
Self intermediate scattering function, 30, 67–69
Semi-macroscopic structures
  neutron spin-echo spectrometers, 296–300
  SANS, *see* Small-angle neutron scattering (SANS) technique
SENJU diffractometer, 272–273, 274f
Short, strong hydrogen bonds (SSHBs), 430–431
Single crystal, 422–423, 434–435
Single-crystal diffractometers

four-circle diffractometer, steady-state source, 271
tofLD technique, *see* Time-of-flight Laue diffraction (tofLD) technique
Single-particle form factor, 76–77
Sjölander approximation, 39
Skutterudites, 441–442
Skyrmion, 102–103, 102f
Sloppy choppers, 316–318, 318f
Slowing down processes and moderators, neutron
  accelerator-driven sources, 159–163
  "age equations", 157–158
  classification, energy ranges, 153, 154t
  collision law, nucleus, 155–156, 156f
  collision parameters, 155f
  energies, neutron beams, 153
  energy and momentum, equations, 155
  Fermi distribution, 158
  moderation process, 157
  nuclei scattering, 157–158
  production, very cold and ultracold neutrons, 165–166
  properties, moderator materials, 156–157, 156t
  reactors, 158–159
Small-angle neutron scattering (SANS) technique
  amplitude-weighted phase shifts, 358–359
  cold neutrons, 363
  collimation and resolution, 293–295
  converging collimation, 295–296
  data analysis, 364–365
  decoupling approximation, 359–360, 360f
  description, 293
  detector, 293
  focusing, 296
  Guinier expansion, 361
  illustration, D11, 293, 293f
  indirect Fourier transform (IFT) method, 364–365
  intensity, 363–364
  length density, 359
  Porod's law, 361
  pulsed source, instrument, 364f
  sample-detector distances, 363
  scattered waves, coherent superposition, 358–359
  scattering geometry, 359f
  volume and number density, 361
Snell's law, 306
SNS, *see* Spallation Neutron Source (SNS)
Spallation Neutron Source (SNS)

Index

Hg target, 200, 201f
  layout, 199–200, 199f
  power upgrade, beam energy, 200
Spallation reactions
  ab initio approaches, 148
  energy distribution, 149, 150f
  evaporation, 148–149
  free particle-particle collisions, nucleus, 149
  incident particles, 148
  intranuclear cascade, 148–149
  measured distribution, products, 152, 152f
  neutron multiplicities, target materials, 150, 150t
  neutron yield, 149–150
  optimal incident energies, protons, 151
  proton beam kinetic energies, 151–152
  research, 153
Spallation sources
  CM, DM and PM, 258–259
  concepts, advanced neutronics technology, 258, 259f
  conceptual design, moderators, 257
  hydrogen moderators at 20 K, 259
  pulse structures, neutrons, 259, 260f
  scientific objectives and users' demands, 257
  temperature, neutronic structure and materials, 258
Spectral density, 66–67
Spectroscopy, 3, 6t, 11–12, 15, 16–22, 37, 39–41, 67, 72, 73–74
Spin
  dependent scattering length, 89
  dynamics, 3, 21, 106–119
  filter, 94
  fluctuations, 114–115
  order, 87–88, 120–125
  waves, 106, 108–112, 111f, 115–117, 121–122
Spin-echo, 417f, 426–427, 443–444, see also Neutron spin-echo spectrometers
$SrTiO_3$, 435
SSHBs, see Short, strong hydrogen bonds (SSHBs)
Static approximation, 49–50, 118, 266
Stern-Gerlach effect, 296
Stochastic
  diffusion, 61–74
  rotational diffusion, 69
  translational diffusion, 118–119
Stochastic relaxation/dynamics
  coherent QENS, rotation, 456–457
  coherent scatterer CO_2, diffusion, 459–460

complex diffusion, 455–456
dynamical transitions, elastic scans, 457–458
elastic incoherent structure factor (EISF), 346–347
ionic liquids, 463–464
ligand water rotation, 456
water and complex diffusion, 460–463
Stoner modes, 117
Straight and curved guides
  curved guide section, 310–311, 311f
  effective solid angle subtended, 310, 310f
  high-energy neutrons and gammas, 310
  line-of-sight length, 311
Stretched exponential, 455–456, 462–463
Stripping reactions, see Direct and stripping reactions
Strontium gallium oxides, 443–444
Structure factor $F(Q)$
  advantages, accurate structure analysis, 271
  in Al powder sample, 284, 284f
  smooth function, Q, 250
Subdiffusion, 455–456
Superconductivity (SC), 19–21, 103–106, 105f, 110–111, 442–443
Supercooled water, 460
Supersolidity, 54
Surface, 21, 43, 56, 73–74, 78, 80–82, 103–104, 117
  magnetism, 103–104
Susceptibility, 22, 106–108, 115–119, 116f
S-wave scattering, 22–24, 121

T

Target cooling, neutron sources
  coefficient, heat-transfer, 184
  convective/heat-transfer coefficient, 184
  gallium-cooled system, 187
  heat-transfer mechanisms, 183–184
  megawatt-range applications, 185–186
  molten metals, 185
  NBR, 184–185
  single-phase coolant, 184–185
  windowless liquid metal, 187
Target-moderator-reflector (TMR) complex, 187–188, 188f
TASs, see Three-axis spectrometers (TASs)
T_0 choppers
  with an in-wheel motor, 313, 313f
  background suppression, 313, 314f
  inelastic signals, 313
  lower rotation speeds, 313

Terahertz window, 21
Thermal neutrons, 3, 5–10, 12, 15–16, 23t, 25, 39, 50, 54, 56–57, 115, 120–121, 125–127
Thin films, adsorption, see Adsorption
Three-axis spectrometers (TASs), 422
Time-dependent diffusion theory, 161–162
Time focusing, 264–265
Time-of-flight (TOF) diffraction
 derivatives, 254
 implications, 254
 inelastic effects with, 265–266
 Lorentz factor, 254
 neutron energies, 253
 principle of, 262–264
 scattering diagram, 253–254, 253f
 time focusing, 264–265
Time-of-flight Laue diffraction (tofLD) technique
 Bragg reflections, 272
 description, 272
 iBIX diffractometer, protein structural analysis, 272–273, 274f
 Lorentz factor, 272
 at pulsed source, 272–273
 scattering diagram and reciprocal space region, 272, 273f
 SENJU diffractometer, small crystal, 272–273, 274f
Time reversal, 47
TMR complex, see Target-moderator-reflector (TMR) complex
tofLD, see Time-of-flight Laue diffraction (tofLD) technique
TOSCA, 425, 429, 430, 431f
Total scattering, 12–13, 14, 18, 28–29, 42, 44, 47, 49–50, 52, 61, 82, 94
Total scattering diffractometer
 detector arrangement, NOVA, 270, 271f
 energy integration, 269–270
 free gas model and scattering locus, TOF measurement, 270, 270f
 liquid and amorphous systems, 269–270
Trajectories, 420, 426–427, 436–438
Transition rate, 22–24
Translational diffusion constant, 69, 118–119
Translational tunneling, 453
Transmission, 10, 18
Triple-axis spectrometer
 Brock house instrument and principle, 275, 275f
 incident and scattered neutron wave vectors, 275
 RITA, advanced, 276–277
 scan, 276
 scattering vector Q, 275
Tunnelling
 rotational, 446–453
 translational, 453

U

Umklapp scattering, 441–442
Unit-cell structure factor, 34, 87–88, 93–94, 96
Unpolarized
 beam, 26, 95
 neutron, 83–84, 89–90, 93–94
Urea phosphoric acid (UPA), 430–431
Useful neutron flux, 3–5

V

van der Waals, 420, 433–434
Van Hove, 42–43, 64, 71, 118–119, 418
VDOS, see Vibrational density-of-states (VDOS)
Velocity autocorrelation function, 66–67
Venetian-type scintillation detector, 302–303, 303t
VESUVIO, 423–424
Vibration, 39–41, 45–47, 63–64, 67–69, 71, 106–107, 109–110, 114
Vibrational density-of-states (VDOS), 38–41, 423
Vibrational dynamics, 431–432
Vineyard convolution approximation, 70
VIVALDI, diffraction image, 345f

W

Wakefield accelerators, 221
Water and complex diffusion, 460–463
Wave function, 9, 24, 25, 50, 85–86, 112, 121–122
Wavelength scanning method, 261–262
Wavelength-shifting fiber (WLSF) detector
 cross section, 303, 304f
 iBIX protein diffractometer, 304, 305f
 and scintillator plate, 303, 304f
 specification, 304, 305t
Wave-vectors, 424–425

Index

White beam
  Q-scan methods, 262f
  tofLD, 272
  and TOF method, 253
WLSF, see Wavelength-shifting fiber (WLSF) detector

X
X-ray
  absorption near-edge spectroscopy (XANES), 15
  form factor, 14, 15
  free-electron laser (XFEL), 19–21
  photon correlation spectroscopy (XPCS), 19–21
  scattering, 7–8, 15, 58–59

Z
Zeeman splitting, 87
Zeroth moment, 48, 50–51, 107–108

Edwards Brothers Malloy
Ann Arbor MI. USA
November 22, 2013